Reactive and Flexible Molecules in Liquids

NATO ASI Series

Advanced Science Institutes Series

A Series presenting the results of activities sponsored by the NATO Science Committee, which aims at the dissemination of advanced scientific and technological knowledge, with a view to strengthening links between scientific communities.

The Series is published by an international board of publishers in conjunction with the NATO Scientific Affairs Division

A Life Sciences	Plenum Publishing Corporation
B Physics	London and New York
C Mathematical	Kluwer Academic Publishers
and Physical Sciences	Dordrecht, Boston and London
D Behavioural and Social Sciences	
E Applied Sciences	
F Computer and Systems Sciences	Springer-Verlag
G Ecological Sciences	Berlin, Heidelberg, New York, London,
H Cell Biology	Paris and Tokyo

Series C: Mathematical and Physical Sciences - Vol. 291

Reactive and Flexible Molecules in Liquids

edited by

Th. Dorfmüller

Fakultät für Chemie,
Universität Bielefeld,
Bielefeld, F.R.G.

Kluwer Academic Publishers

Dordrecht / Boston / London

Published in cooperation with NATO Scientific Affairs Division

Proceedings of the NATO Advanced Study Institute on
Reactive and Flexible Molecules in Liquids
Nafplion, Greece
September 23 – October 2, 1988

Library of Congress Cataloging in Publication Data

```
NATO Advanced Study Institute on Reactive and Flexible Molecules in
  Liquids (1988 : Nauplion, Greece)
   Reactive and flexible molecules in liquids : proceedings of the
NATO Advanced Study Institute on Reactive and Flexible Molecules in
Liquids, held in Nafplion, Greece, September 23-October 2, 1988 /
edited by Th. Dorfmüller.
      p.   cm. -- (NATO ASI series. Series C, Mathematical and
physical sciences ; vol. 291)
   Includes bibliographical references.
   ISBN 978-94-010-6961-8      ISBN 978-94-009-1043-0 (eBook)
   DOI 10.1007/978-94-009-1043-0
   1. Solution (Chemistry)--Congresses.  2. Molecular dynamics-
-Congresses.  3. Chemical reactions, Rate of--Congresses.
I. Dorfmuller, Th. (Thomas), 1928-    . II. Title.  III. Series:
NATO ASI series.  Series C, Mathematical and physical sciences ; no.
291.
QD540.N37  1988
541'.0422--dc20                                          89-19969
```

ISBN 978–94–010–6961–8

Published by Kluwer Academic Publishers,
P.O. Box 17, 3300 AA Dordrecht, The Netherlands.

Kluwer Academic Publishers incorporates the publishing programmes of
D. Reidel, Martinus Nijhoff, Dr W. Junk and MTP Press.

Sold and distributed in the U.S.A. and Canada
by Kluwer Academic Publishers,
101 Philip Drive, Norwell, MA 02061, U.S.A.

In all other countries, sold and distributed
by Kluwer Academic Publishers Group,
P.O. Box 322, 3300 AH Dordrecht, The Netherlands.

Printed on acid free paper

Table of Contents

Preface

The observable phenomena in liquids which distinguishes this state of matter from other types of condensed matter can be mainly assigned to a) the configurational disorder and b) the random motion of molecules. Both, the static and the dynamic aspect of randomness are typical for the liquid state and serve as a useful guideline in the attempts to theoretically understand this state. These two basic features, however, introduce in liquid state theory a number of apparently unsurmountable technical and conceptual problems so that progress in the last decades has only been made by small steps. In order not to complicate the situation even more, the tacit assumption was made that we could neglect internal motions of the molecules and that the molecular interactions which had to be taken into account are as simple as possible. We thus became accustomed to visualize molecules in the liquid as a dense assembly of classical rigid particles interacting with a potential which basically is represented by a Lennard–Jones type relation.

In the last decade, it has become obvious that with these restrictions we disregard many interesting effects in those liquids which are the most important ones. We thus see a serious gap developing between the refinements of liquid state theory and the exciting experiments being carried out in many laboratories. There is certainly no ideal remedy which would help us in bridging this gap, except perhaps, to intensify as much as possible the discussion which is going on between theorists and experimentalists from various fields. A conference such as the present NATO ASI whose initiative goes back to the "European Molecular Liquids Group" can be considered to have been successful if it has contributed in stimulating this discussion.

Elementary chemical reactions in liquids can be considered from two different but interrelated aspects, the electronic and the configurational aspect. In the former we focus our attention on the evolution of orbitals and in the latter on the dynamics of the relative configuration of atoms within molecules. Both effects are a consequence of intermolecular interactions. The two points of view are linked by the central role played by the time–dependent random intermolecular interactions typical for the liquid state. Although this is also the case for other states of aggregation, the situation is especially complex in the liquid state where the dynamics are determined by the continuous and simultaneous interaction of each molecule to several neighbouring molecules and by the high temperature as compared to the depth of potential energy wells.

The electronic aspect is the main topic of the article of B. Guillot and G. Birnbaum. In fact, the polarization which results from intermolecular interactions is the basis of the elementary act of chemical reactions and is reflected in the collision–induced spectra as well. These considerations have been extended to a charge transfer reaction in which the evolution in time of the resulting dipole is manifested in the spectra.

A different point of view is adopted in the study of S. Bratos and P. Viot, in which the elementary chemical act is analyzed in terms of a pre–reactive and a post–reactive step. It is shown that fast pre–reactive and post–reactive effects are manifested in isotropic Raman spectra and the correlation functions calculated from the spectral bands display two chemical modes.

The Monte Carlo simulation is used by B. Guillot, Y. Guissani, D. Borgis and S. Bratos to study clustering reactions of ionic species with protic and aprotic solvent molecules.

With the contribution of J. Yarwood, A. Whitley, D.G. Gardiner and M.P. Dare–Edwards the complex effects observed in the shapes of infrared, far–infrared and Raman spectral bands as a consequence of internal rotation are analyzed and compared to the situation in rigid molecules.

W. A. Steele studies a fundamental aspect of molecular dynamics in liquids by calculating expressions for time–correlation functions of pairs of molecules with translational and orientational degrees of freedom.

J. McConnel describes in his paper a rotational diffusion theory for molecules which have no special symmetry and which can be applied to molecules of various shapes.

In the theoretical paper of G.J. Moro, A Ferrarini, A. Polimeno and P.L. Nordio, the authors study the intramolecular motion of a linear chain by using a model of a chain of rotors with one torsional degree of freedom. Using the Smoluchowski equation they extend the validity of the Kramers model to the case of high potential barriers and to multi–dimensional diffusion equations. The role of the frictional coupling between reactive and non–reactive modes is stressed. It is shown that the cooperative transitions during saddle point crossings arise as a consequence of frictional coupling.

W. Schroer, D. Labrenz and C. Rybarsch deal in their contribution with the dielectric permittivity of molecules with internal rotational degrees of freedom and the effect of the reaction field on the equilibrium conformer distribution.

The short time internal dynamics of styrene oligomers in bulk phase below the glass point is presented in the article of M. Buchner and Th. Dorfmüller. They are using a combination of molecular dynamics simulation and harmonic analysis. The properties of two different mechanical models are illustrated and the results are compared to Raman and far infrared spectral analysis reflecting the rotational motion of the phenyl group.

The dynamics of polymer molecules in polymer melts was studied by a Monte–Carlo technique by J. Skolnick. It is shown that the dynamics display two distinct regimes as a function of the degree of polymerization. At a high degree of polymerization the motion is highly constrained without, however, displaying the features of reptation.

The diffusion of polymer molecules through porous media relies on molecular flexibility. The study of A. Baumgärtner, carried out by Monte Carlo simulation, shows that this system displays anomalous features which are discussed in terms of entanglement and the presence of entropic barriers separating cavities in the disordered medium.

G. Floudas and G. Fytas report experimental results on the polymer chain conformation as monitored by optical anisotropy.

Polymer blends are industrially important systems and the knowledge of their molecular dynamics is an important clue to their macroscopic properties. The contribution of J. Kanetakis, A. Rizos and G. Fytas deals with the diffusion in polymer blends as a function of the degree of polymerization by photon correlation measurements.

Biological macromolecules generally display a complex and quite specific behaviour related to their structure and to various interactions such as hydrogen bonding, amphiphilic effects etc.. Thus, the dynamics of DNA, presented by D. Pörschke, reflects the complex structure of this molecule and its different intra– and intermolecular interactions. It also becomes clear that the observed dynamics correlate with several biological functions of this molecule. It also becomes apparent how conformational dynamics and quasi chemical intramolecular interactions both correlate with the function. The study of the dynamics of such molecules which have passed the filter of evolution is only at its beginning and promises to be extremely rewarding.

Another important biological macromolecule is the muscle protein f–actin which is studied by J. Seils and Th. Dorfmüller. The photon correlation spectra were analyzed by inverse numerical Laplace transform and the observed features of the frequency distribution could be assigned to the dynamics of crosslinking between the macromolecules.

Membrane and lipid systems are flexible systems in themselves and their properties depend on the flexibility of the constituents as well. In the paper of L.J. Korrstanje, E.E. van Faassen and Y.K. Levine the orientational order and the reorientational dynamics of a lipid bilayer was studied with a nitroxide spin label.

The paper of M. Drifford P.J. Derian and L. Belloni presents experimental results in which the counterions in surfactant micellar solutions can be directly studied by X–ray and neutron scattering. The micelles unter study present themselves as soft flexible aggregates as a result of the dominant hydrophobic effect. Amphiphilic chains in the micelle are considered as flexible and reactive molecules in liquids.

The kinetics of polymerization of an epoxy resin was studied by B. Chu and C. Wu using light scattering and small–angle x–ray scattering. With these two experimental techniques the authors were able to obtain interesting new data on structural changes of the branched epoxy polymer during the curing process.

A. Bick and Th. Dorfmüller report on a study of the polymerization kinetics of an epoxy resin using a catalyst. This reaction was monitored from the liquid to the fully polymerized glassy state through the glass point by Brillouin light scattering. The relaxation times as well as mechanical parameters were determined along the curing process.

G. Wylie presents a theoretical study of the kinetic properties of the hydrated electron. Random walks over short distances give rise to non–Gaussian distributions and to anomalous diffusion which could be related to fractal structures.

Both molecular flexibility and chemical reactivity are properties which evolve in a large interval of time scales. On the short–time side of the presently available techniques we have femtosecond laser spectroscopy. In the paper of K. Carpenter and G. Kenny–Wallace an outlook is given on the present possibilities of this method.

A particular experimental aspect is addressed in the paper of H. Versmold and T. Palberg with the method of electrophoretic light scattering. The paper indicates some of the pitfalls inherent in this method and the authors come to the conclusion that, if the method is properly used, it can be valuable in the study of electrokinetic parameters of macromolecular solutions.

The articles included in the present volume address a large variety of problems and of methods used to study them. Despite this variety, however, a common ground becomes clear. Quite different aspects of the common topic, i.e. reactivity and flexibility in liquids, are studied on rather simple liquids as well as on polymers, biological macromolecules and supramolecular aggregates. As far as the methods are concerned, these include spectroscopic studies, computer simulation and analytic theory.

The intention in organizing the ASI and in publishing this volume was to reveal existing connections between flexibility and reactivity in liquids insofar as both are mediated by stochastic intermolecular interactions. The effort of all participants and authors to shed some light on this aspect of liquid state dynamics are gratefully acknowledged.

Thomas Dorfmüller

INTERACTION INDUCED ABSORPTION IN SIMPLE TO COMPLEX LIQUIDS

Bertrand Guillot
Université Pierre et Marie Curie
Laboratoire de Physique Théorique des Liquides
75230 Paris, France

George Birnbaum
National Institute of Standards and Technology
Gaithersburg, Maryland 20899, USA

ABSTRACT. An attempt is made to provide a theoretical basis for dealing with the collision-induced absorption spectra of complex liquids from a consistent viewpoint. One of the most important approximations is the decoupling of translational and rotational motions. This makes it possible to compute separately the spectrum resulting from each of these motions by a three-variable Mori-Zwanzig theory. The parameters in the equation for the translational spectrum are estimated by a lattice gas model that is extended to include two component mixtures. This theory, in which there are no adjustable parameters, is first tested on the far infrared (FIR) spectrum of liquid N_2 by assuming that the spectrum arises from the point quadrupole and point polarizability induced-dipole (QID) mechanism. Satisfactory agreement is obtained with a computer simulation, although both theory and simulation have somewhat less high frequency intensity than experiment. The FIR spectra of liquid C_6H_6 and I_2 is also computed on the basis of the point QID mechanism. In both liquids, the theory grossly underestimates the experimental high frequency absorption. Finally, the FIR spectrum of a dilute solution of I_2 in C_6H_6 is computed and satisfactory agreement with experiment is obtained only in the low frequency part of the spectrum. To improve agreement at high frequencies, an attempt is made to include the effect of a contact (transient) charge transfer induced-dipole, but further work is required before any definite conclusions can be drawn regarding its influence. A general conclusion of this work is that the point QID mechanism in the strongly quadrupolar molecules investigated here, although providing a useful and convenient starting point for describing FIR absorption in such liquids, does not provide enough intensity in the high frequency part of the spectrum. Possible reasons for this are discussed.

1

Th. Dorfmüller (ed.), Reactive and Flexible Molecules in Liquids, 1–36.
© *1989 by Kluwer Academic Publishers.*

1. INTRODUCTION

Collision or interaction induced absorption (IIA) has been studied for
many years in simple liquids composed of small diatomic and linear
molecules, and mixtures of rare gas atoms, particles which by symmetry
have no permanent dipole moments. Much has been learned about such
absorption from experiment, molecular dynamics simulations, and approximate analytic theories (Birnbaum, 1985). However, progress in quantitatively understanding the phenomenon, even in simple liquids, has been
frequently impeded by the difficulty of treating analytically, in their
full complexity, all the ingredients of the phenomenon, i.e., the
induced dipole models and anisotropic potential functions. Nevertheless,
the growing interest in the spectroscopy of more complex liquids composed of larger, and strongly anisotropic molecules, which moreover may
react chemically, has made it necessary to examine the role of induced
absorption in such liquids. Indeed, studies have already appeared on
the far infrared (FIR) absorption of large molecules in liquids
(Dorfmüller, 1985), IIA spectra of iodine in benzene which has been
considered to form a weak charge-transfer complex (Lascombe and Besnard,
1986; Yarwood and Catlow, 1987), and chemical reactivity in weak charge-
transfer complexes (Besnard et al., 1988). Moreover, problems relating
to the roles of induced absorption and induced depolarized light scattering arose a number of times in this Advanced Studies Institute (for
example, Yarwood, 1989; Dorfmüller, 1989; and Floudas and Fytas, 1989).
   This presentation aims to see how one may apply what has been learned
in the study of IIA in simple liquids to the case of complex liquids.
To make this presentation tractable, we limit it to induced absorption
in the FIR region due to molecules that lack permanent dipole moments.
Our treatment is based on a relatively simple theoretical approach which
has been shown to provide a qualitative understanding of a variety of
interaction induced phenomena in liquids (Bratos et al., 1985; Guillot
and Birnbaum, 1985). We believe that there is heuristic value in presenting a theoretical framework to deal with the problems of concern
here in a uniform and systematic way, and that much can be learned about
the phenomenon under study even if a quantitative accuracy cannot be
claimed for the theory in all circumstances. In fact, an important aim
of this study is to determine what needs to be done to improve the
theory.
   Section 2 presents the general correlation function (CF), and
develops a simplified version based on an approximation for decoupling
translational and rotational (TR) motions. The theory of FIR induced
band shapes in neat fluids is presented in Section 3. Since the molecules investigated here (N_2, and particularly C_6H_6 and I_2) have a large
quadrupole moment, the quadrupole induced-dipole (QID) mechanism is
assumed. A three variable Mori-Zwanzig equation is used to represent
the spectrum arising from translational motions, and another such equation is used to represent the rotational motions, the total spectrum
being a convolution of both. The parameters in the translational
spectrum are evaluated with the aid of a lattice gas model, and thus
there are no adjustable parameters. This theory is compared with experimental results of liquid N_2, C_6H_6, and I_2 and with computer simulations

of just liquid N_2 for which such results are available.

The extension of the above theory to two component molecular fluids is presented in Section 4. These results are applied to a study of the FIR band shape of a dilute solution of I_2 in C_6H_6 and compared with experiment. The role of contact charge transfer is briefly considered. A summary and conclusions are presented in Section 5.

2. THE INDUCED DIPOLE CORRELATION FUNCTION

2.1 General Relations

The induced dipole CF is given by

$$G(t) = \sum_{\substack{i \neq j \\ k \neq l}} <\vec{\mu}_{ij}(t) \cdot \vec{\mu}_{kl}(0)> \tag{1}$$

where $\vec{\mu}_{ij}(t)$ is the dipole induced in a pair of molecules at time t, and it is assumed that only pairwise interactions occur. It is instructive to write Eq. (1) in the form (Steele, 1985; Samios et al., 1986)

$$G(t) = G_{2a}(t) + G_{2b}(t) + G_{3a}(t) + G_{3b}(t)$$

$$+ 2G_{3c}(t) + G_4(t) \tag{2}$$

where

$$G_{2a}(t) = \sum_{i \neq j} <\vec{\mu}_{ij}(t) \cdot \vec{\mu}_{ij}(0)> \tag{3}$$

$$G_{2b}(t) = \sum_{i \neq j} <\vec{\mu}_{ij}(t) \cdot \vec{\mu}_{ji}(0)> \tag{4}$$

$$G_{3a}(t) = \sum_{i \neq j \neq k} <\vec{\mu}_{ij}(t) \cdot \vec{\mu}_{ik}(0)> \tag{5}$$

$$G_{3b}(t) = \sum_{i \neq j \neq k} <\vec{\mu}_{ij}(t) \cdot \vec{\mu}_{kj}(0)> \tag{6}$$

$$G_{3c}(t) = \sum_{i \neq j \neq k} <\vec{\mu}_{ij}(t) \cdot \vec{\mu}_{ki}(0)> \tag{7}$$

$$G_4(t) = \sum_{i \neq j \neq k \neq l} <\vec{\mu}_{ij}(t) \cdot \vec{\mu}_{kl}(0)> \tag{8}$$

In these expressions $\vec{\mu}_{ij}$ is the dipole induced in molecule i by molecule j, and the angular brackets denote an ensemble average.

In the case of induced vibrational spectra, μ_{ij} is a function of vibrational, rotational, and translational coordinates. However, by assuming that the vibrational coordinates (v) of different molecules are uncorrelated, an expression for the CF is obtained in which $<\cos\omega_v t>$ multiplies Eqs. (3) to (7), each multiplied by an appropriate intensity factor (Steele, 1985). Equations (3) to (8) express the TR motions which must be decoupled in order to obtain analytical results. Unfortunately, a usually sufficient basis for such decoupling may not apply, namely, that translational and rotational correlations decay on very different time scales.

2.2 Translational-Rotational Decoupling Approximation

For the sake of clarity, consider just $G_{3a}(t)$, and specifically just one term of the sum, which may be written as

$$<\vec{\mu}_{ij}(t)\cdot\vec{\mu}_{ik}(0)> = <T_{ij}(t)\theta_j(t)T_{ik}(0)\theta_k(0)> \tag{9}$$

where T_{ij} contains the relative translational coordinates of molecules i and j and θ_j contains the orientational coordinates of molecule j, and similarly for T_{ik}, θ_k. The most important induction mechanism for diatomic, linear, and symmetric top molecules is the dipole induced in molecule i by the quadrupolar field of molecule j,

$$\mu_{ij}^\alpha = \alpha E_{ij}^\alpha \quad (\alpha = x, y, z) \tag{10a}$$

where

$$E_{ij}^\alpha = \frac{1}{3} T_{ij}^{\alpha\beta\gamma}\theta_j^{\beta\gamma} \tag{10b}$$

with

$$T_{ij}^{\alpha\beta\gamma} = \frac{\partial}{\partial R_\alpha} \frac{\partial}{\partial R_\beta} \frac{\partial}{\partial R_\gamma} R^{-1} \tag{10c}$$

$$\theta_j^{\beta\gamma} = \frac{1}{2} Q(3u_j^\beta u_j^\gamma - \delta_{\beta\gamma}) \tag{10d}$$

Q is the quadrupole moment, α in Eq. (10a) is the isotropic polarizability, R_α is the α component of the vector connecting molecules i and j, and u_j^β is the β component of the unitary vector along the axis of molecule j. Induction via the anisotropic polarizability is ignored since this contribution to the induced dipole is very small $(\gamma/\alpha)^2 \ll 1$, see Table 5). In Eq. (9) and a number of others to follow, we omit for convenience an intensity factor containing $(\alpha Q)^2$ and some other constant factors. The decoupling approximation consists in writing Eq. (9) as

$$<T_{ij}(t)\Theta_j(t)T_{ik}(0)\Theta_k(0)> \simeq <T_{ij}(t)T_{ik}(0)> <\Theta_j(t)\Theta_k(0)>_\ell \quad (11a)$$

where

$$<\Theta_j(t)\Theta_k(0)>_\ell \propto <P_2(\cos\theta_{jk}(t))>_\ell \quad (11b)$$

Here $\theta_{jk}(t)$ is the angle between the orientation of molecule j at time t and molecule k at a time zero. It is clear that for a term such as $G_{3b}(t)$, Eq. (6), that the decoupling approximation yields

$$<T_{ij}(t)T_{kj}(0)> <P_2(\cos\theta_j(t))>.$$

The subscript ℓ in Eqs. (11a) and (11b) expresses local cross-orientational correlations in the liquid, and implies that the orientational correlations are weighted by the range of the translational functions T_{ij} and T_{ik}. Thus, only those orientational correlations between molecules j and k separated by no more than the range probed by the induction mechanism, for example, R^{-8} for the QID mechanism, need to be considered. In other words, the average $<P_2(\cos\theta_{jk}(t))>$ means that we take into account the orientational correlation between molecules j and k only if molecule k belongs to a sphere centered on molecule j, and whose radius is equal to or smaller than the range of the QID mechanism. This has a very important consequence. While the depolarized Rayleigh spectrum gives information on cross-orientational correlations between molecules of the liquid regardless of their separation, the FIR spectrum gives information on cross-orientational correlations only at a local level.

It is instructive to examine the Kirkwood g_2 factor

$$g_2 = 1 + \sum_{j \neq 1} <P_2(\cos(\theta_{1j})> = 1 + (N - 1)<P_2(\cos\theta_{12})> \quad (12)$$

where N is the number of molecules in the liquid sample, and the sum is over all molecules of the liquid regardless of their separation from molecule 1. Thus, due to the isotropy of the liquid, $<P_2(\cos\theta_{12})>$ is very small and is of the order of N^{-1}. For example, for liquid N_2 molecular dynamics (MD) simulation gives $g_2 \simeq 1.10$ (or $(N-1)<P_2(\cos\theta_{12})> \simeq 0.1$). On the contrary, we can define a local g_2 factor such as

$$g_{2\ell} = 1 + \sum_{\substack{j \neq 1 \\ r_{ij} \leq R_c}} <P_2(\cos\theta_{1j})> \quad (13)$$

where R_c is the range of the induction mechanism. For the QID mechanism, 90% of the integrated intensity in the liquid is located roughly in the region 0.8σ to 1.3σ (σ is the molecular diameter), which corresponds roughly to the first shell of neighbors. Thus equivalent to Eq. (13) we have

$$g_{2\ell} \approx 1 + N_c <P_2(\cos\theta_{12})>_\ell \qquad (14)$$

where N_c is the mean number of molecules in the first shell. It is thus clear that $<P_2(\cos\theta_{12})>$ can be significant at the local level even if the Kirkwood g_2 factor probed by Rayleigh scattering shows only small correlations at the macroscopic level. This conclusion in confirmed by MD calculations on CH_3CN (Edwards et al., 1984) and CS_2 (Impey et al., 1981).

2.3 The Translation-Rotation Correlation Function in the TR Decoupling Approximation

With the decoupling approximation discussed previously, the partial correlation functions given by Eqs. (3) to (8) become, omitting the intensity factor $(\alpha Q)^2$ as before,

$$G_{2a}(t) = N(N - 1)<P_2(\cos\theta(t))> <T_{12}(t)T_{12}(0)> \qquad (15)$$

$$G_{2b}(t) = - N(N - 1)<P_2(\cos\theta_{12}(t))>_\ell <T_{12}(t)T_{12}(0)> \qquad (16)$$

$$G_{3a}(t) = N(N - 1)(N - 2)<P_2(\cos\theta_{12}(t))>_\ell <T_{12}(t)T_{13}(0)> \qquad (17)$$

$$G_{3b}(t) = N(N - 1)(N - 2)<P_2(\cos\theta(t))> <T_{12}(t)T_{13}(0)> \qquad (18)$$

$$G_{3c}(t) = - G_{3a}(t) \qquad (19)$$

$$G_4(t) = 0 \qquad (20)$$

In obtaining these results, we used the property of invariance by permutation of identical particles in $<T_{ij}(t)T_{ik}(0)>$ and the odd character of T_{ij}, namely

$$T_{ij} = - T_{ji}$$

and

$$<T_{ij}T_{ik}> = <T_{ik}T_{ij}> = - <T_{ik}T_{ji}>$$

$G_4(t) = 0$ is a consequence of these properties. The theory has several aspects that merit attention: $G_{2a}(t)$ and $G_{2b}(t)$ are opposite in sign; and the ratio

$$G_{2b}(0)/G_{2a}(0) = - <P_2(\cos\theta_{12}(0))>_\ell \qquad (21)$$

is a measure of angular cross-correlations in the first shell.

Moreover, the lattice gas model allows us to compute three-body translational correlations

$$(N - 2)<T_{12}(0)T_{13}(0)> = - (\rho/\rho_0)<T_{12}(0)T_{12}(0)> \qquad (22)$$

where ρ_0 is a reference density usually taken equal to the solid state density at the temperature of investigation.

Equations (15) to (20) are particularly interesting because they permit a direct comparison with simulation data to check the reliability of the decoupling approximation. For example, from Steele's (1985) MD results (see Figure 1) we find that $G_{2b}(0)/G_{2a}(0) = -0.03$. Introducing this result into Eqs. (15) to (19) and evaluating the various zero-time partial CFs by using a pair distribution function for Lennard-Jones liquids (Marteau et al., 1986), we obtain the results shown in Table 1. The comparison of the values in Table 1 is encouraging although there are discrepancies. As predicted by our theory, the MD results show that $|G_{2b}(0)| << |G_{2a}(0)|$, and $|G_{3a}(0)|$ and $|G_{3c}(0)| << |G_{3b}(0)|$; these inequalities come from the weakness of the cross-correlations ($<P_2(\cos\theta_{12})>_\varrho = 0.03$) in the case of N_2. The theory predicts that $G_{3c}(0) = - G_{3a}(0) = +0.02$, whereas Steele (1985) obtains -0.03 and $+0.15$, respectively. (Note in Fig. 1 that $G_{3c}(t)$ changes sign at

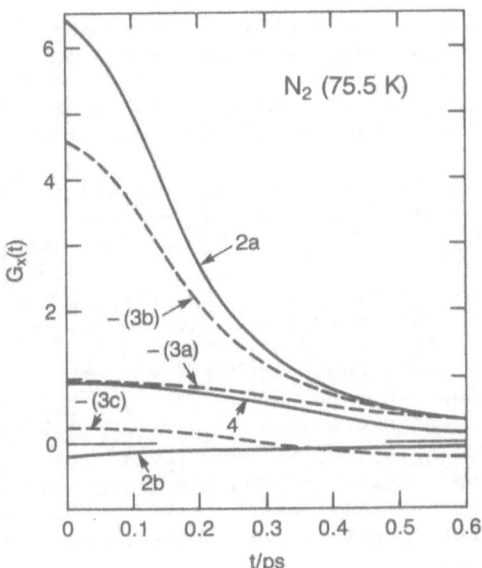

Figure 1. Simulated induced dipole time-correlation functions are shown for liquid nitrogen at 75.5 K. The correlation functions, $G_x(t)$, are defined by Eq. (3) for x = 2a, Eq. (4) for x = 2b, Eq. (5) for x = 3a, Eq. (6) for x = 3b, Eq. (7) for x = 3c, and Eq. (8) for x = 4. Note that the dashed lines are the negative of the correlation function (Steele, 1985).

TABLE 1. Comparison of correlation functions $G_x(0)$ from Eqs. (15) - (20) with those obtained from the MD simulations of Steele (1985).

liquid N_2 T=75 K ρ=650 amagat	$G_{2a}(0)$	$G_{2b}(0)$	$G_{3a}(0)$	$G_{3b}(0)$	$G_{3c}(0)$	$G_4(0)$	$\sum G_i$
Steele MD	1.0	−0.03	−0.15	−0.71	−0.03	+0.15	0.20
theory ρ_0=792 amagat	1.0	−0.03	−0.02	−0.82	+0.02	0	0.17

0.4 ps.) Furthermore, the MD results give for $G_4(0)$ the non negligible value 0.15, whereas our theory predicts the value zero. However, we note that the strong cancellations that exist in the induced dipole CF make it difficult to obtain MD results with good accuracy, in general, and that is difficult to obtain accurate results for a small partial contribution to the CF, in particular. The total CFs at t = 0, $G(0)$, are in fact close to each other in the simulation and theory, 0.20 and 0.17, respectively.

We show in Appendix A that the orientational function $\langle P_2(\cos\theta_{12}(t))\rangle$ is expected to have a much slower decay than $\langle P_2(\cos\theta(t))\rangle$. Then one would expect that $G_{2b}(t)$, $G_{3a}(t)$ and $G_{3c}(t)$ should also exhibit a slow decay compared with $G_{2a}(t)$ and $G_{3b}(t)$ (see Eqs. (15) to (19)). Steele's (1985) simulation results for N_2 (Fig. 1) are in very good agreement with these predictions. The same general behavior is observed for the more anisotropic fluids CS_2 (Fig. 2) and CO_2 (Fig. 3). As predicted by our theory $G_{3a}(t)$ and $G_{3c}(t)$ are opposite in sign for CO_2, and although of the same order of magnitude they are not equal. For CS_2, $G_{3a}(t)$ and $G_{3c}(t)$ are opposite in sign and nearly equal, and are close to zero. As for N_2, the individual CFs, $G_{3a}(t)$, $G_{3c}(t)$ and $G_{2b}(t)$, for CO_2 and CS_2 are slowly decaying functions. Cross-orientational correlations seem much more important for these systems (see the ratio between $G_{2a}(0)$ and $G_{2b}(0)$). We note for CS_2 that $G_{2b}(t)$ is relatively large, and in fact is not too different from $G_{3b}(t)$.

What can be said about the validity of the decoupling approximation? Steele (1985) addressed this problem in a computer simulation study of N_2, and he concluded that terms involving orientational correlations between two molecules are non-negligible, and that their time dependence cannot be adequately approximated by the decoupling approximation. One assumes in this approximation that the rotational and translational motions are decoupled even in the presence of static orientational

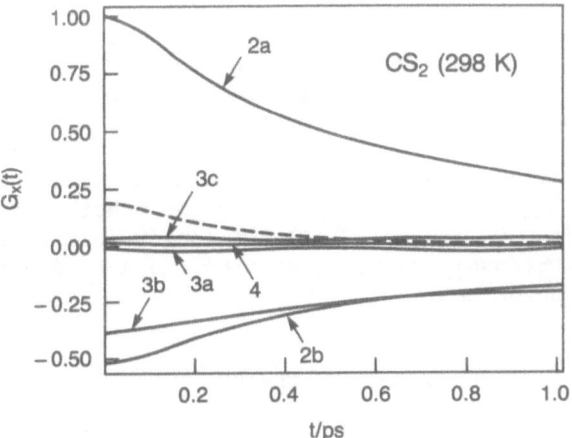

Figure 2. Simulated induced-dipole time correlation functions for liquid CS_2 at 298 K. The correlation functions, $G_x(t)$, are defined in Figure 1 (Samios et al., 1986).

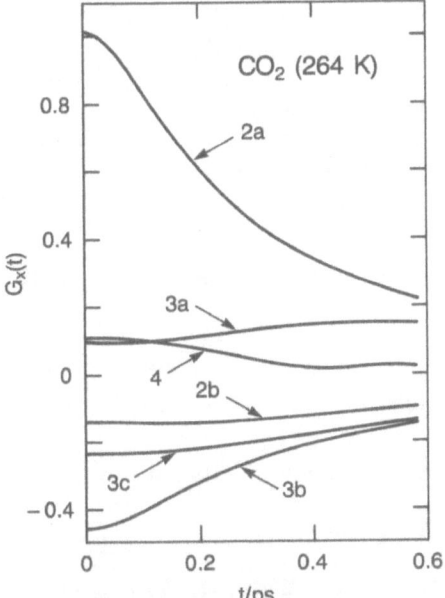

Figure 3. Simulated induced-dipole time correlation functions for liquid CO_2 at 264 K. The correlation functions, $G_x(t)$, are defined in Fig. 1. Distributed quadrupole moment and polarizability were used (Steele and Posch, 1987).

correlations and hindered reorientation. However, he found that the CFs not explicitly requiring mutual orientational correlations, i.e., $G_{2a}(t)$ and $G_{3b}(t)$, can be reproduced by the product of separate translational and rotational functions even in the presence of the static orientational correlations and hindered rotations. It appears that the decoupling approximation works well for $G_{2a}(t)$ and $G_{3b}(t)$ because the self CF $<P_2(\cos(\theta(t)))>$ decays on a time scale much smaller than $<T_{ij}(t)T_{ik}(0)>$ (see Fig. 6, Steele, 1985). On the other hand, it appears that for $G_{2b}(t)$, $G_{3c}(t)$ and $G_4(t)$ the longer relaxation time of the distinct CF, $<P_2(\cos\theta_{12}(t))>$, might preclude a good agreement with the decoupling approximation.

3. FIR BAND SHAPES OF SINGLE COMPONENT FLUIDS

3.1 Analytical Theory

The starting point in the development of an analytical theory of band shape is the expression for $G(t)$, Eqs. (2) and (15) to (19), which may be written in the useful form

$$G(t) = N(N-1)(\alpha Q)^2 [<P_2(\cos\theta(t))> - <P_2(\cos\theta_{12}(t))>_{\ell}]$$
$$x[<T_{12}(t)T_{12}(0)> + (N-2)<T_{12}(t)T_{13}(0)>] \qquad (23)$$

where we assume the induction to arise from quadrupole induced dipoles and insert the intensity factor $(\alpha Q)^2$. We ignore the contributions due to the anisotropic polarizability for the sake of simplicity, although these terms which are usually small can be included without any difficulty. If we suppose that $<P(\cos\theta_{12}(0))>$ is $\ll 1$, which is certainly true for N_2 (Levesque and Weis, 1975; Singer et al., 1979), we may make the approximation (see App. A)

$$<P_2(\cos(\theta(t)))> - <P_2(\cos\theta_{12}(t))>_{\ell} \simeq A<P_2(\cos\theta(t))> \qquad (24)$$

where

$$A = 1 - <P_2(\cos\theta_{12}(0))>_{\ell}$$

which is temperature and density dependent. Then we obtain

$$G(t) = N(N-1)(\alpha Q)^2 A<P_2(\cos\theta(t))>$$
$$x[<T_{12}(t)T_{12}(0)> + (N-2)<T_{12}(t)T_{13}(0)>]$$
$$= N(N-1)(\frac{\alpha Q}{2})^2 A G_{rot}(t) \text{ x } G_{tr}(t) \qquad (25)$$

where $G_{rot}(t) = <P_2(\cos\theta(t))>$, and $G_{tr}(t)$ represents the terms in the square brackets. The spectral density, $G(\omega)$ is given by the convolution

$$\widehat{G}(\omega) = \int_{-\infty}^{+\infty} \widehat{G}_{rot}(\omega')\widehat{G}_{tr}(\omega - \omega')d\omega' \tag{26}$$

where $G_{rot}(\omega)$ and $G_{tr}(\omega)$ are the Laplace transforms of $G_{rot}(t)$ and $G_{tr}(t)$, respectively.

By using a three-variable Mori theory for $G_{rot}(t)$ (see, for example, Bratos and Guillot, 1982), we obtain

$$\widehat{G}_{rot}(\omega) = \frac{1}{\pi} \frac{\tau_{rot}}{\left[1 - \frac{\omega^2}{\omega_0^2}\right]^2 + (\omega\tau_{rot})^2 \left[1 - \frac{1 - \omega^2/\omega_0^2}{1 - \omega_1^4/(\omega_0^2)^2}\right]^2} \tag{27}$$

where

$$\tau_{rot} = \int_0^\infty dt\, <P_2(\cos\theta(t))> \tag{28a}$$

$$\omega_0^2 = -\ddot{G}_{rot}(0) = \frac{6kT}{I_\perp} \quad \text{(for linear and symetric} \tag{28b}$$
$$\text{top molecules)}$$

$$\omega_1^4 = \overset{....}{G}(0) = \begin{cases} 96\left(\frac{kT}{I}\right)^2 + 3\frac{<N^2>}{I^2} & \text{linear} \tag{28c} \\ \\ \left(\frac{kT}{I_\perp}\right)^2 \left(102 - \frac{6\eta}{1+\eta}\right) + 3\frac{<N^2>}{I^2} & \text{symmetric top} \tag{28d} \end{cases}$$

$$\eta = \frac{I_\perp}{I_\parallel} - 1 \quad (= -1/2 \text{ for } C_6H_6)$$

and $<N^2>$ is the mean squared torque. We shall see later that the rotational spectrum exhibits a diffusive character at low frequencies and a librational character at intermediate frequencies. Whenever $<N^2>$ is important $(<N^2>/(kT)^2 = 101$ for $C_6H_6)$, $\omega^2\widehat{G}_{rot}(\omega)$ then is maximum for $\omega \simeq 2\{\omega_1^4 - (\omega_0^2)^2\}^{1/4}$. A crude estimate of the correlation time τ_{rot} is given by the well known Hubbard relation, $\tau_{rot} = (6D_{rot})^{-1}$, valid in the rotational diffusion limit. Although this expression is useful, we prefer in the following to extract τ_{rot} from MD simulations or from Raman and Rayleigh spectra.

For the translational spectrum, we have

$$\hat{G}_{tr}(\omega) = \frac{1}{\pi} \frac{\tau_{tr} \; G_{tr}(t=0)}{(1 - \frac{\omega^2}{\omega_0^2})^2 + (\omega\tau_{tr})^2 \left[1 - \frac{1 - (\omega^2/\omega_0^2)}{1 - \omega_1^4/(\omega_0^2)^2}\right]^2} \tag{29}$$

where (Guillot and Birnbaum, 1985)

$$\tau_{tr} = \int_0^\infty \frac{\sum\limits_{i\neq j,k}<T_{ij}(t)T_{ik}(0)>}{\sum\limits_{i\neq j,k}<T_{ij}(0)T_{ik}(0)>} dt = \frac{(1 - \frac{\rho}{\rho_0})\int d\vec{r}g(r)r^{-8}}{28D_r(1 - \frac{\rho}{2\rho_0})\int d\vec{r}g(r)r^{-10}} \tag{30a}$$

($D_r \simeq 2D_s$, where D_r is the relative diffusion coefficient and D_s is the single particle diffusion coefficient)

$$\omega_0^2 = \frac{M_2^{tr}}{M_0^{tr}} \; ; \; \omega_1^4 = \frac{M_4^{tr}}{M_0^{tr}} \tag{30b}$$

$$M_0^{tr} = 12V\rho^2(1 - \rho/\rho_0) \int d\vec{r}g(r)r^{-8} \tag{30c}$$

$$M_2^{tr} = 336(kT/m_{12})V\rho^2(1 - \rho/2\rho_0) \int d\vec{r}g(r)r^{-10} \tag{30d}$$

$$M_4^{tr} = \frac{V\rho^2}{m_{12}^2} \left\{ \begin{array}{l} 192(1 - \rho/\rho_0) \int d\vec{r}g(r)r^{-10}(\partial u(r)/\partial r)^2 \\ + 97920(kT)^2(1 - \frac{71}{272}(\rho/\rho_0)) \int d\vec{r}g(r)r^{-12} \end{array} \right. \tag{30e}$$

Note that τ_{tr} is a collective correlation time that differs markedly from the 2-body correlation time τ_{12} ($\tau_{tr} \ll \tau_{12}$). Moreover, the spectral moments are evaluated in the present study with the help of a pair distribution function for Lennard-Jones liquids (Marteau et al., 1986). In future work, an approach which takes into account anisotropic interactions should be developed.

3.2 Comparison with Experiment

3.2.1 Liquid N_2. The kinetic and thermodynamic constants used to calculate the FIR spectrum of N_2 and the various computed quantities that appear in the theory are gathered together in Tables 2 and 5. Similar data are shown in Table 3 for C_6H_6 and Table 4 for I_2. Once the reference lattice gas density, ρ_0, is chosen, approximately the solid density at the indicated temperature, there are no adjustable parameters in the theory. Values for τ_{rot}, $<N^2>/(kT)^2$, and D_s were obtained from computer simulation. The agreement between theory and experiment in

TABLE 2. Quantities used in calculating the FIR spectrum of liquid N_2. T = 87.3 K, ρ = 743 amagat, ρ_0 = 792 amagat.

$\langle N^2\rangle/(kT)^2 = 17^{[a]}$,	$D_s = 1.8 \times 10^{-5}cm^2/s^{[a]}$,	$\tau_{rot} = 0.24ps^{[a]}$ $(22cm^{-1})^{[a]}$	$\tau_{tr} = 0.18ps$ $(29\ cm^{-1})$

$M_0(cm^{-1})$	$M_2(cm^{-3})$	$M_2/M_0(cm^{-2})$	$6kT/I(cm^{-2})$	$M_2^{tr}/M_0^{tr}(cm^{-2})$
2.04	7088	3475	1464	2113

			$M_4^{rot}/(M_2^{rot})^2$	$M_4^{tr}/(M_2^{tr})^2$
			4.08	2.00

[a] Cheung and Powles (1985).

TABLE 3. Quantities used in calculating the FIR spectrum of liquid C_6H_6. T = 300 K, ρ = 251 amagat, ρ_0 = 300 amagat.

$\langle N^2\rangle/(kT)^2 = 101^{[a]}$,	$D_s = 2.8 \times 10^{-5}cm^2/s$,	$\tau_{rot} = 2.8ps,$ $(1.9cm^{-1})$	$\tau_{tr} = 0.59ps$ $(9\ cm^{-1})$

$M_0(cm^{-1})$	$M_2(cm^{-3})$	$M_2/M_0(cm^{-2})$	$6kT/I_\perp(cm^{-2})$	$M_2^{tr}/M_0^{tr}(cm^{-2})$
142.7	145.9×10^3	1022	476	546

			$M_4^{rot}/(M_2^{rot})^2$	$M_4^{tr}/(M_2^{tr})^2$
			11.4 (libration)	5.4

[a] Linse et al. (1985).

TABLE 4. Quantities used in calculating the FIR spectrum of liquid I_2. T = 393 K, ρ = 353 amagat, ρ_0 = 400 amagat.

$\langle N^2\rangle/(kT)^2 = 40^{[a]}$,	$D_s = 1.0 \times 10^{-5}cm^2/s$,	$\tau_{rot} = 5ps,$ $(1\ cm^{-1})^{[a]}$	$\tau_{tr} = 1.0ps$ $(5.3\ cm^{-1})$

$M_0(cm^{-1})$	$M_2(cm^{-3})$	$M_2/M_0(cm^{-2})$	$6kT/I(cm^{-2})$	$M_2^{tr}/M_0^{tr}(cm^{-2})$
205.8	97.6×10^3	474	122	352

			$M_4^{rot}/(M_2^{rot})^2$	$M_4^{tr}/(M_2^{tr})^2$
			6.0 (libration)	4.2

[a] Lynden-Bell and Steele (1984).

TABLE 5. Molecular constants of N_2, C_6H_6, and I_2. ϵ and σ are the Lennard-Jones parameters (N_2, Egelstaff et al., 1978; C_6H_6, I_2, Hirschfelder et al., 1954). $\alpha = 1/3(\alpha_\perp + 2\alpha_\|)$, $\gamma = \alpha/\beta$ ($\beta = \alpha_\perp - \alpha_\|$), Q and Φ are from Gray and Gubbins (1984).

	ϵ(K)	σ(Å)	Q(10^{-26}esu)	Φ(10^{-42}esu)	α(10^{-24}cm^3)	γ
N_2	98	3.6	-1.47 ± 0.09	-2.93	1.740	0.400
C_6H_6	440	5.27	-8.69 ± 0.50	$+76.1$	10.6	0.530
I_2	550	4.98	$+5.61$		12.3	0.544

Fig. 4 is considered to be satisfactory, even if at high frequencies the theoretical spectrum decreases too rapidly. The same behavior is observed for the MD result of Levesque et al. (1985) in comparison with experiment. The MD result of Steele (1985) may be in better agreement with experiment, but it is difficult to make such a judgement because his results are given only for the CFs and were not Fourier transformed. In any case, excess high frequency absorption in computer simulations is not always so marked. Samios et al. (1986), for example, obtained good agreement with experiment for CS_2.

Several approximations in our calculation could account for at least some of the high frequency discrepancy. These include neglect of the induced dipole due to the anisotropy of polarizability, neglect of induced dipole contributions coming from the hexadecapole moment and

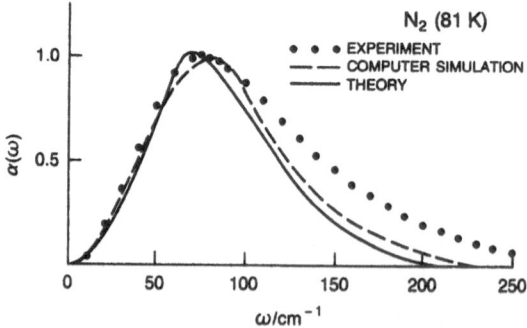

Figure 4. Absorption coefficient (normalized to unity at maximum) of liquid N_2 at 81 K. Experimental and computer simulation results are from Levesque et al. (1985); the theoretical spectrum is computed from Eqs. (26), (27), and (29) with the parameters given in Tables 2 and 5. Note that $\alpha(\omega) \equiv \omega \tanh(\beta\hbar\omega/2) \hat{G}(\omega)$. In the theory $\hat{G}(\omega)$ is given its classical value, $G_{cl}(\omega)$, while the computer simulation result is corrected for quantum effects. However, if the theoretical spectrum is corrected by the same approximate formula then, the MD spectrum and the theoretical spectrum are almost undistinguishable from each other beyond 100cm^{-1}.

anisotropic overlap induction, and neglect of anisotropic forces in the calculation of the translational spectral moments (Eqs 30c-30e). All these contributions to the induced dipole moment, although generally much weaker than the QID, contribute relatively much more intensity at the higher frequencies. However, we note that in the case of gaseous N_2, excellent agreement was obtained between experiment and a quantum mechanical computation of the band shape on the basis of binary collisions, a QID induction mechanism (with the addition of a very small anisotropic overlap dipole), and a refined isotropic potential (Borysow and Frommhold, 1986). We next consider the reliability of the decoupling approximation, and the Mori and lattice gas calculations. Regarding the first, it was shown that this approximation appears to work well for liquid N_2. The three-variable Mori theory has proven to be reasonably adequate for describing the spectral shape of a variety of phenomena, even if the spectra have the tendency to fall off too quickly at high frequencies. Concerning the ability of the lattice gas model for estimating the moments (M_0^{tr}, M_2^{tr}, M_4^{tr}) in the Mori theory, previous work (Marteau et al., 1986) has shown that this model describes the general trend with density of the experimental spectral moments. Although the lattice gas model does not appear to give accurate quantitative results (Ladanyi et al., 1986; Briganti et al., 1986), and further refinements would be desirable, this model nevertheless makes it possible to compute liquid spectra in a rather simple way. However, we believe that even with very accurate values of the Mori parameters, the high frequency discrepancy might not be solved.

Finally, the close similarity between the theory and MD simulation (which includes the full RT coupling, anisotropy of polarizability, anisotropic forces, and only the QID mechanism) suggests that the neglect of hexadecapolar induction could be the most serious assumption. As emphasized by Moon and Oxtoby (1986), if 89 percent of the integrated absorption in N_2 gas is due to the quadrupolar induction, 8 percent is due to the hexadecapolar induction which takes place mostly in the high frequency wing. This estimate neglects, however, the contribution of induction due to anisotropic overlap forces (Borysow and Frommhold, 1986).

3.2.2 Liquid C_6H_6. The experimental and computed spectrum of liquid C_6H_6 are shown in Fig. 5, where it is seen that the computed spectrum underestimates the experimental spectrum much more than for N_2. Moreover, the former peaks at 50 cm^{-1} whereas the latter peaks at 80 cm^{-1}. The parameters used in the computed spectrum are given in Tables 3 and 5. Although the potential for N_2 was tailored for the liquid, that for C_6H_6 is the simple isotropic gas potential, clearly a very crude approximation. The mean squared torque was obtained from a computer simulation. The same procedure was used in the calculation of I_2 and I_2-C_6H_6 liquids. We note that the C_6H_6 FIR spectra measured earlier by Garg et al. (1968), Kettle and Price (1972), and Davies (1974) are in general agreement with that of Besnard (1988), except that the latter shows small features at about 118 and 170 cm^{-1}. As emphasized by Besnard (1988), these features can be attributed to vibrational difference bands, following the early work of Barnes et al. (1935). We find

16

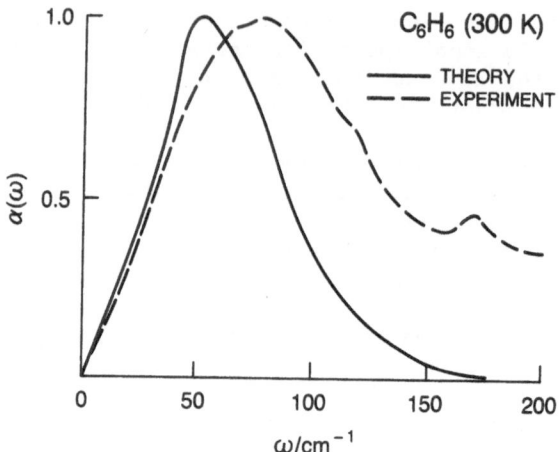

Figure 5. Absorption coefficient (normalized to unity at maximum) of liquid C_6H_6 at 300 K. The experimental spectrum is from Besnard (1988). The theoretical spectrum is computed from Eqs. (26), (27) and (29) with the parameters given in Tables 3 and 5. $\alpha(\omega)$ is defined in Fig. 4.

$M_2/M_0 = 1022$ cm^{-2} whereas $M_2^{rot} = 6kT/I_\perp = 476$ cm^{-2}: thus the translational component is very important. However, this result could be due in part to the decoupling approximation. It is interesting to note that in induced depolarized Rayleigh scattering Dardy et al. (1973) and Perrot et al. (1981) found that $M_2^{exp}/M_0^{exp} \simeq 1100$ cm^{-2}, which is very close to what we obtained from the calculation of the FIR absorption.

Perrot et al. (1981) and Lund et al. (1978) found that the product $\omega^2 I^{ray}(\omega)$ peaks at 75 cm^{-1}, indicating a librational motion for C_6H_6 molecules. In our case, the three-variable Mori theory for the rotational spectrum peaks at nearly the same wave number, 72 cm^{-1}, confirming that our treatment of the rotation of C_6H_6 is essentially correct. Notice, as mentioned earlier, that $\omega^2 G_{rot}(\omega)$ is a maximum when

$$\omega_{max} \simeq 2[72(kT/I)^2(1 + <N^2>/24(kT)^2)]^{1/4}$$

Although the depolarized induced light spectrum does not reflect only molecular rotation (the collision-induced contribution is important as seen in the second spectral moment), the underlying dynamics is governed essentially by the rotational correlation function (Landanyi and Levinger, 1984).

One may suppose that the discrepancy between theory and experiment resides in the inadequacy of the decoupling approximation which upsets some delicate cancellations between several correlation functions, or that the point quadrupole mechanism is too crude to describe the real induction mechanism for such a molecule. A possible method of assessing the validity of the decoupling approximation is to examine the time scales of $<P_2(\cos\theta(t))>$ and $<T_{12}(t)T_{12}(0)>$ (see Eq. 25). Although one may obtain the former, τ_{rot}, from a Rayleigh scattering experiment (Dardy et al., 1973), or a computer simulation (Steinhauser, 1982; Linse et al., 1985), the time constant of $<T_{12}(t)T_{12}(0)>$, τ_{12}, is not so readily available. If the MD results of depolarized light scattering from liquid argon (Ladd et al., 1979) are at all representative of what may be expected in other collision-induced spectra, then one would expect that $\tau_{tr} \ll \tau_{12}$, a result of the dynamical cancellation effect (Guillot, 1987). The present theory gives τ_{tr} (Eq. 30a) but not τ_{12}, although it is possible to estimate this quantity in the framework of the Mori theory

$$\tau_{tr} = \tau_{12}(1 - \frac{\rho}{\rho_0})/(1 - \frac{\rho}{2\rho_0})$$

For liquid C_6H_6 at room temperature the above relation gives (see Table 3), $\tau_{12} \simeq 4\tau_{tr} = 2.4$ ps, a time of the same order as $\tau_{rot} = 2.8$ ps. Thus contrary to liquid N_2 where $\tau_{12}/\tau_{rot} \simeq 6.5$, the decoupling approximation may fail for liquid C_6H_6 where $\tau_{12}/\tau_{rot} = 0.75$. But, Steele and Vallauri (1987) raised the point that in liquid Br_2, where rotation is strongly hindered, MD simulations show that in <u>the first shell</u> the reorientation time of the molecular axis is much shorter than that of pair translational dynamics. This could explain why the decoupling approximation may be reliable in describing induced spectra in a large variety of liquids, since only the dynamical behavior of the first shell is probed.

A part of the discrepancy between experiment and theory may be due to the inadequacy of the induced dipole model in C_6H_6, i.e. the assumption that the FIR absorption is due entirely to the point QID mechanism. There are in fact higher order multipole induced dipoles which are smaller than the quadrupole induced dipole, but whose absorption spectra are displaced to higher frequencies (Pringle et al., 1983).

Another source of inadequacy is the point multipole expansion itself which is known to converge slowly or even not at all, and is particularly poor when the intermolecular separation is roughly equal to the molecular diameter or less (see Stone and Alderton, 1985). This problem has been discussed for FIR absorption in the gas phase by Pringle et al. (1983, 1987). Distributed multipoles and polarizabilities for computing FIR absorption in the liquid phase by MD was employed by Steele and Posch (1987) in their investigation of CO_2. Thus many higher order multipoles and polarizabilities are implicitly included, although with coefficients that may not correspond well with the actual molecular values. In particular, they used distributed dipoles as a representation of the non-ideal quadrupole moment and applied the Applequist model (see, for example, Applequist, 1985) for distributed polarizabilities. They found significant differences between the point quadrupole and polarizability and distributed quadrupole and polarizability models with each other, and significant disagreement of each of these models with experiment. In the case of N_2, such deviations although significant are much smaller (Moon and Oxtoby, 1986). Since Steele and Posch (1987) did not Fourier transform their CFs, we were unable to ascertain the effect of using distributed quadrupoles and polarizabilities on the FIR spectrum. In any event, it is not at all clear how to incorporate distributed multipoles and polarizabilities in our theoretical approach. Rather, it appears that to obtain analytical results we may have to use a mutipole expansion with suitably modified multipole moments and supplemented by a model anisotropic overlap dipole which terminates the expansion.

In an attempt to obtain better agreement between MD computations and experiment, Steele and Posch (1987) decreased the value of the molecular diameter σ by 5%. They found that this produced large changes in the CFs and produced changes by over 30% in $G(0)$ for the case of induction due to distributed-Q distributed-α and point-Q point-α. Such a change in $G(0)$ with σ is not surprising since the moments given by Eqs. (30c) to (30e) are very sensitive to the value of σ. There have been relatively few studies on the role of the potential on collision-induced band shapes in liquids, and further studies could provide some valuable insights. Furthermore, some new developments on models of the potential have been advanced (Stone and Price, 1988; Price, 1988).

We conclude this section on C_6H_6 by showing its FIR spectral density in a logarithmic plot (Fig. 6). The experimental result can be described as roughly consisting of a low-frequency Lorentzian and a high-frequency exponential decay. However, the main point of Fig. 6 is to emphasize that whereas the low frequency absorption can be accounted for by the QID mechanism, there appears to be other mechanisms producing high frequency absorption, which as yet have not been identified. As seen in Fig. 5, the contribution of vibrational difference bands should be included in the calculated spectrum. Furthermore, C_6H_6 is a symmetric top and reorientation about both an axis perpendicular to the plane of the C_6H_6 ring and an axis lying in the plane (Steinhauser, 1982) may become involved in the induced absorption (Pringle et al., 1983, 1987).

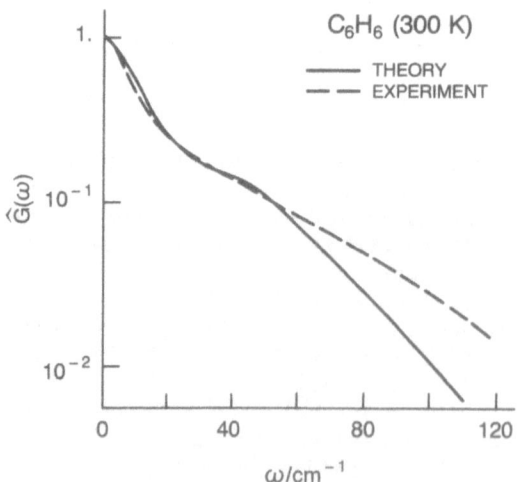

Figure 6. Spectral density (normalized to unity at $\omega = 0$) of liquid C_6H_6 at 300 K. The experimental spectrum is from Besnard (1988). The theoretical spectrum is computed from Eqs. (26), (27), and (29) with the parameters in Tables 3 and 5.

3.2.3. Liquid I_2. By using values of the parameters of liquid I_2 given in Tables 4 and 5, we obtain the computed FIR absorption spectrum shown in Fig. 7, and for comparison an experimental result is shown. The disagreement between theory and experiment is great; there is not even agreement in the low frequency part of the spectrum as for C_6H_6. Furthermore, the theoretical spectrum is not very sensitive to a large variation of the aforementioned parameters.

Before speculating on the reasons for this large discrepancy between theory and experiment, let us consider the comparison between MD results and experiment for the IR spectrum of liquid Cl_2 shown in Fig. 8 (Murthy et al., 1982). In their MD simulation, they used a two center L-J potential plus a point quadrupole, assumed a point QID mechanism, and included the anisotropy of the polarizability. To compare Figs. 7 and 8, the ordinates in Fig. 8 should be multiplied by $\omega[1-\exp(\hbar\omega/kT)]$. We see in Fig. 8 that the experimental results for Cl_2 are greater at high frequency and lower at low frequencies compared with the MD simulation, as in the case of I_2. We also note that the measured FIR spectrum of liquid Br_2 at room temperature resembles that of the liquid FIR I_2 spectrum at 393 K. The maximum of the absorption of the former is at 57 cm^{-1}, whereas that of the latter is at 50 cm^{-1} (Wagner, 1969). Since the discrepancy between experiment and theory for I_2 is qualitatively the same as the discrepancy between MD results and experiment for Cl_2, we are led to believe that the discrepancy in the I_2 computation is probably not due (at least entirely) to the inadequacy of the TR

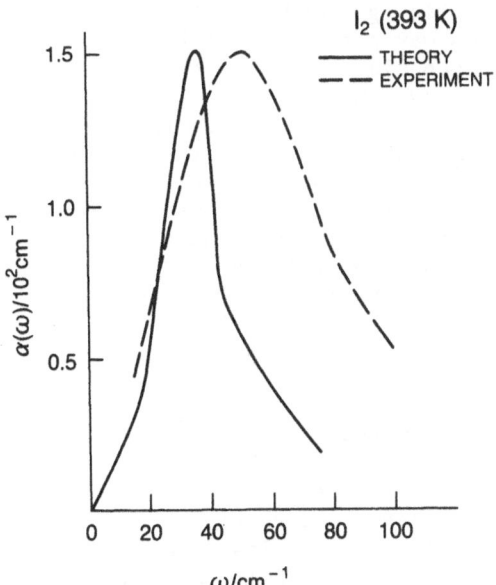

Figure 7. Far infrared absorption spectrum of liquid I_2 at 393 K. Experimental spectrum is from Wagner (1969). Theoretical spectrum (normalized to experiment at maximum) is computed from Eqs. (26), (27), and (29) with the parameters in Tables 4 and 5.

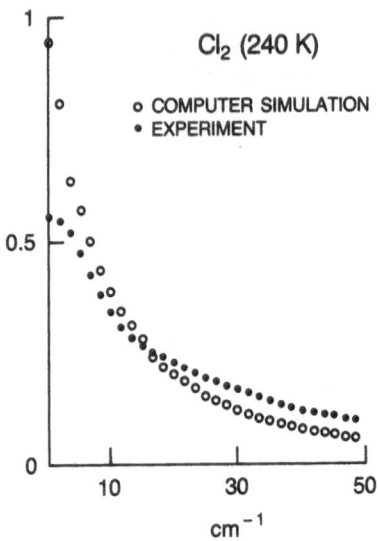

Figure 8. Absorption of the fundamental vibrational band of liquid Cl_2 at 240 K, normalized to unit area. (Murthy et al., 1982).

decoupling approximation or the approximations involved in the three-variable Mori theory and lattice gas model. The problem with the MD and theoretical computations may possibly reside in the use of a point multipole (QID) induction mechanism and an inadequate potential function. Finally, we note that there are two phonon modes in solid I_2 at 39 cm^{-1} and 60 cm^{-1} at T = 300 K that apparently merge into one broad band at 50 cm^{-1} after melting, which suggests the possibility of external vibrational modes in the liquid (Wagner, 1972). Furthermore, a Raman study (Magaña and Lannin, 1985) provides some evidence for the presence of clustered $(I_2)_n$ species in the liquid. However, one would think that such clusters would also give rise to low frequency motions and enhance the very low frequency part of the spectrum.

4. MOLECULAR SOLUTIONS

4.1 Correlation Function of a Mixture

In most cases, large molecules will be dissolved in a suitable solvent and, consequently, it is important to investigate the case of dilute mixtures. For a dilute mixture of molecules I in a solvent consisting of molecules B, the expression for the CF in the framework of the decoupling approximation is

$$G_{IB}(t) = N_I N_B \, (\alpha_I Q_B)^2 [<P_2(\cos\theta_B(t))>$$

$$+ \, (\frac{\alpha_B Q_I}{\alpha_I Q_B})^2 <P_2(\cos\theta_I(t))> + 2 \, \frac{\alpha_B}{\alpha_I} \frac{Q_I}{Q_B} <P_2(\cos\theta_{IB}(t))>_\ell]$$

$$\times <T_{I_1 B_1}(t) T_{I_1 B_1}(0)> + 2 \, \frac{\alpha_B}{\alpha_I} \, (N_B - 1) \, [<P_2(\cos\theta_B(t))>$$

$$- <P_2(\cos\theta_{B_1 B_2}(t))>_\ell] <T_{I_1 B_2}(t) T_{B_1 B_2}(0)>$$

$$+ \, (N_B - 1)[<P_2(\cos\theta_{B_1 B_2}(t))> \, + \, (\frac{\alpha_B Q_I}{\alpha_I Q_B})^2 <P_2(\cos\theta_I(t))>$$

$$- 2 \, \frac{\alpha_B}{\alpha_I} \frac{Q_I}{Q_B} \, <P_2(\cos\theta_{IB}(t))>_\ell] \, <T_{I_1 B_1}(t) T_{I_1 B_2}(0)> \quad (32)$$

The contribution due to the solvent molecules alone is omitted since it is assumed that it is subtracted from the absorption coefficient of the mixture. The solute contribution is neglected since it is negligible at great dilutions. Then, if one neglects the cross-angular correlations and three-body terms, one recovers the expressions obtained by Lascombe and Besnard (1986)

$$G_{IB}^{2\text{-body}}(t) = N_I N_B (\alpha_I Q_B)^2 [<P_2 \cos\theta_B(t)>$$

$$+ \, (\frac{\alpha_B Q_I}{\alpha_I Q_B})^2 <P_2(\cos\theta_I(t))>] <T_{I_1 B_1}(t) T_{I_1 B_1}(0)> \quad (33)$$

It is clear at first glance that Eq. (33) is a gross oversimplification of Eq. (32) and conclusions drawn from its use must thus be viewed with some caution. However, to obtain a usuable analytical expression such as Eq. (33), some approximations must be made.

A first approximation takes

$$<T_{I_1 B_2}(t) T_{B_1 B_2}(0)> \text{ and } <T_{I_1 B_1}(t) T_{I_1 B_2}(0)>$$

to have the same time dependence. Both CFs are purely translational, and we believe that the relative diffusion of two C_6H_6 molecules is not very different from that of an I_2-C_6H_6 pair. Thus we may take

$$<T_{I_1 B_2}(t) T_{B_1 B_2}(0)> \simeq \gamma <T_{I_1 B_1}(t) T_{I_1 B_2}(0)>$$

where γ is a factor equal to the ratio of the two bracketed expressions at zero time. Then one obtains

$$G_{IB}(t) = N_I N_B (\alpha_I Q_B)^2 \Big\{ [<P_2(\cos\theta_B(t))> + (\frac{\alpha_B Q_I}{\alpha_I Q_B})^2 \times <P_2(\cos\theta_I(t))>$$

$$+ 2\frac{\alpha_B}{\alpha_I}\frac{Q_I}{Q_B}<P_2(\cos\theta_{IB}(t))>_\ell]<T_{I_1 B_1}(t)T_{I_1 B_1}(0)>$$

$$+ (N_B - 1)<T_{I_1 B_1}(t)T_{I_1 B_2}(0)>[2\gamma\frac{\alpha_B}{\alpha_I}<P_2(\cos\theta_B(t))>$$

$$+ (1 - 2\gamma\frac{\alpha_B}{\alpha_I})<P_2(\cos\theta_{B_1 B_2}(t))>_\ell + (\frac{\alpha_B Q_I}{\alpha_I Q_B})^2<P_2(\cos\theta_I(t))>$$

$$- 2\frac{\alpha_B}{\alpha_I}\frac{Q_I}{Q_B}<P_2(\cos\theta_{IB}(t))>_\ell]\Big\} \qquad (34)$$

Now let us specialize Eq. (34) for an I_2-C_6H_6 mixture. We note that $\alpha_B/\alpha_I \simeq 1$ but $Q_I/Q_B \simeq 0.6$, where I now represents I_2 and B represents C_6H_6. Supposing that the terms

$$(1 - 2\gamma\frac{\alpha_B}{\alpha_I})<P_2(\cos\theta_{B_1 B_2}(t))>_\ell \quad \text{and} \quad \frac{\alpha_B}{\alpha_I}\frac{Q_I}{Q_B}<P_2(\cos\theta_{IB}(t))>_\ell$$

are negligible with respect to $<P_2(\cos\theta_I(t))>$ and $<P_2(\cos\theta_B(t))>$, we obtain the trial correlation function,

$$G_{IB}(t) = N_I N_B (\alpha_I Q_B)^2 \Big\{ [<P_2(\cos\theta_B(t))>[<T_{I_1 B_1}(t)T_{I_1 B_1}(0)>$$

$$+ 2(N_B - 1)\gamma\frac{\alpha_B}{\alpha_I}<T_{I_1 B_1}(t)T_{I_1 B_2}(0)>] + (\frac{\alpha_B Q_I}{\alpha_I Q_B})^2<P_2(\cos\theta_I(t))>$$

$$\times [<T_{I_1 B_1}(t)T_{I_1 B_1}(0)> + (N_B-1)<T_{I_1 B_1}(t)T_{I_1 B_2}(0)>]\Big\} \qquad (35)$$

One notices that the 3-body term of the benzene contribution is characterized by the coupling parameter

$$\gamma = \frac{<T_{I_1 B_2}(0)T_{B_1 B_2}(0)>}{<T_{I_1 B_1}(0)T_{I_1 B_2}(0)>}$$

Since we have no obvious information regarding this function, we will treat it in what follows as a free parameter.

The calculation of the spectrum associated with Eq. (35) proceeds in the same way as that for the neat liquid. The correlation functions $<P_2(\cos\theta_I(t)>$ and $<P_2(\cos\theta_B(t))>$ are described by the 3-variable Mori theory. However, since we deal with a dilute ($\simeq 2\%$) mixture of I_2 in C_6H_6, $<P_2(\cos\theta_I(t))>$ refers to the self angular correlation of an I_2 molecule surrounded by a shell of a C_6H_6 molecules, and $<P_2(\cos(\theta_B(t))>$ refers to the self angular correlation of a C_6H_6 molecule belonging to the first shell of an I_2 molecule. Thus the associated correlation times $\tau_{rot}(I_2)$ and $\tau_{rot}(C_6H_6)$, respectively, do not necessarily correspond to the values expected in the respective neat liquids. Raman spectroscopy indicates that $\tau_{rot}(I_2-C_6H_6) \simeq 5$ ps (Besnard et al., 1988), which is greater than $\tau_{rot}(C_6H_6) \simeq 2.8$ ps. However, we expect that the reorientation of C_6H_6 molecules that surrounds an I_2 molecule are slower than in the neat liquid. Consequently, we assign to $\tau_{rot}(C_6H_6)$ the same value of 5 ps as for $\tau_{rot}(I_2)$.

The translational CF is also given by the 3-variable Mori theory, but with a modification in the expression of the translational moments (and correlation time) due to the presence of a mixture of molecules of different mass. These results are

$$M_0^{tr} = 12 V \rho_I \rho_B (1 - \frac{\rho_B}{\rho_0}) \int d\vec{r} g_{IB}(r) r^{-8} \tag{36a}$$

$$M_2^{tr} = 336 (\frac{kT}{m_{IB}}) V \rho_I \rho_B (1 - \frac{m_{IB}}{m_I} \frac{\rho_B}{\rho_0}) \int d\vec{r} g_{IB}(r) r^{-10} \tag{36b}$$

$$M_4^{tr} = (\frac{kT}{m_{IB}})^2 V \rho_I \rho_B (192(1 - \frac{\rho_B}{\rho_0}) \int d\vec{r} g_{IB}(r) r^{-10} \frac{1}{(kT)^2} (\frac{\partial u_{IB}}{\partial r})^2$$
$$+ 97920 (1 - \frac{1 + 67 (m_{IB}/m_I)^2}{68} \frac{\rho_B}{\rho_0}) \int d\vec{r} g_{IB}(r) r^{-12} \tag{36c}$$

Note that Eqs. (36a) to 36c) apply to mixtures in general (see Guillot et al., 1988). The presence of the ratio m_{IB}/m_I in the M_2^{tr} and M_4^{tr} moments can affect in a strong way their density dependence. For the CF which includes the term containing $2\gamma(\alpha_B/\alpha_I)$, its moments are given by the above equations provided ρ_0^{-1} is replaced by $2\gamma(\alpha_B/\alpha_I)\rho_0^{-1}$. We also obtain

$$\tau_{tr} = \frac{(1 - \rho_B/\rho_0)}{28(D_I + D_B)(1 - \frac{D_I}{D_I + D_B} \frac{\rho_B}{\rho_0})} \frac{\int d\vec{r} g_{IB}(r) r^{-8}}{\int d\vec{r} g_{IB}(r) r^{-10}} \tag{37}$$

where D_I is the self diffusion coefficient of molecules I in the mixture, and D_B is the same for molecules B.

With the values of the constants for the $I_2-C_6H_6$ mixture given in Table 6, we obtain the results shown in Table 7. Although there is a large discrepancy in the experimental results, there is also a large discrepancy between the theory with either of the experimental results. The disagreement of theory with experiment for the $I_2-C_6H_6$ mixture was not at all unexpected in view of the discrepancies noted in comparisons of theory and experiment for the neat liquids I_2 and C_6H_6. In both cases, the theory based on a point QID mechanism failed to account for the large amount of high frequency absorption observed experimentally. Table 7 shows the same type of discrepancy, i.e., our values of M_2 are much less than the experimental values. We note that Eq. (33) gives

TABLE 6. Values of the quantities used in calculating the spectral moments and translational correlation time of an I_2-C_6H_6 mixture. T = 300 K, ρ_I = 4.5 amagat, ρ_B = 251 amagat, ρ_0 = 300 amagat. I signifies I_2 and B signifies C_6H_6.

$<N^2_I>/(kT)^2 = <N^2_B>/(kT)^2 = 101;\ \tau_{rot}(B) = \tau_{rot}(I) = 5\ ps$

$D_S(I) = 1.7 \times 10^{-5} cm^2/s^a;\ D_S(B) = 2.3 \times 10^{-5} cm^2/s^b$

---

[a] Chang and Wilke (1955).
[b] Falcone et al. (1967).

TABLE 7. Values of the moments M_0 and M_2 of the I_2-C_6H_6 mixture at T = 300 K, $\rho(I_2)$ = 4.5 amagat and $\rho(C_6H_6)$ = 251 amagat.

	$M_0(cm^{-1})$	$M_2(10^5 cm^{-3})$	$M_2/M_0(cm^{-2})$
experiment			
LB[a]	32	0.7	2188
YC[b]	21	0.26	1238
theory, Eq. (33)			
2-body	27	0.13	475
theory, Eq. (35)			
$2\gamma(\alpha_B/\alpha_I) = 0$	21	0.12	583
theory, Eq. (35)			
$2\gamma(\alpha_B/\alpha_I) = 0.5$	13	0.08	637

[a] Lascombe and Besnard (1986), and Besnard (1988).
[b] Yarwood and Catlow (1987).

values of M_0 and M_2 that are not too different from those obtained with Eq. (35) and $\gamma = 0$. In this case, the three-body term $(N_B - 1)<T_{I_1B_1}(t)T_{I_1B_2}(0)>$ is small compared with the two-body terms. However, the other three-body term containing $<T_{I_1B_1}(t)T_{I_1B_2}0)>$ is significant since the values of M_0 and M_2 change significantly with $2\gamma(\alpha_B/\alpha_I)$.

Table 8 gives two-and three-body correlation times for various partial CFs. We note the strong dynamical cancellation effect (Guillot, 1987), a decrease in the total correlation time compared with that for just two-body collisions. However, the weighing factor of 0.36 of the I_2 contribution (see Eq. (35)) means that this will have a rather small effect on the spectral shape.

TABLE 8. Translational correlation times for 2- and 3-body interactions in a dilute I_2-C_6H_6 mixture at 300 K, (see Eqs. (33) and (35). I signifies I_2 and B signifies C_6H_6 in the table.

partial cor. function	correlation time τ ps
$\langle T_{I_1B_1}(t)T_{I_1B_1}(0)\rangle$	$\tau(\text{2-body}) = 2.7$
$\langle T_{I_1B_1}(t)T_{I_1B_1}(0)\rangle +$ $(N_B-1)\langle T_{I_1B_1}(t)T_{I_1B_2}(0)\rangle$	$\tau(\text{2 and 3-body}) = 0.7$
$\langle T_{I_1B_1}(t)T_{I_1B_1}(0)\rangle +$ $2\gamma\dfrac{\alpha_I}{\alpha_B}(N_B-1)\langle T_{I_1B_1}(t)T_{I_1B_2}(0)\rangle$	$\tau(\text{2 and 3-body}) = 2.0$ if $2\gamma = 0.5$ $(\alpha_I/\alpha_B=1)$

4.2 The FIR Band Shape of a Dilute Solution of I_2 in C_6H_6

The FIR spectrum of I_2 dissolved in C_6H_6 was measured by several investigators (Fig. 9) (Yarwood and Catlow, 1987; Besnard et al., 1988). The high frequency band is due to the induced fundamental vibration of I_2. The low frequency band resembles the FIR C_6H_6 band (see Fig. 5) but has its maximum value at a higher frequency than the C_6H_6 band. For comparison, the FIR spectrum of I_2 and 2-methyl pyridine (MP) in cyclohexane is shown in Fig. 10. The band marked b is much sharper than the low frequency (LF) band in I_2-C_6H_6, and it is attributed to an I_2-MP complex, i.e., a tightly bound dimer in which charge transfer constitutes a significant portion of the binding energy. The band at approximately 170 cm^{-1} is the fundamental band of I_2, highly perturbed by the complex.

In view of the resemblance of the LF I_2-C_6H_6 band to the LF C_6H_6 band, and because it has been shown that the electrostatic interactions play an important role in the I_2-C_6H_6 system (Hanna and Lippert, 1973), we first calculate the LF FIR I_2-C_6H_6 spectrum on the basis of the QID mechanism. The initial 20 cm^{-1} of the I_2-C_6H_6 spectrum is shown in Fig. 11. One notes that the spectral densities computed on the basis of Eq. (33) (two-body approximation) and Eq. (35) with $\gamma = 0$ (simplified three-body approximation) are the same. Different results are obtained when we let $2\gamma = 0.5$ in Eq. (35). In either case, the overall agreement with experiment is satisfactory in the light of the many approximations that are involved in the theory. It is interesting to observe that

$$\Delta\omega_{1/2} \simeq \tau_{rot}^{-1}(B) + \tau_{tr}^{-1}(I\text{-}B) = \begin{cases} \simeq 1.0cm^{-1} + 2.0cm^{-1}(\text{Eq. (33)}) \\ \simeq 1.0cm^{-1} + 2.6cm^{-1}(\text{Eq. (35)}, \\ \qquad\qquad\qquad\qquad 2\gamma = 0.5) \end{cases}$$

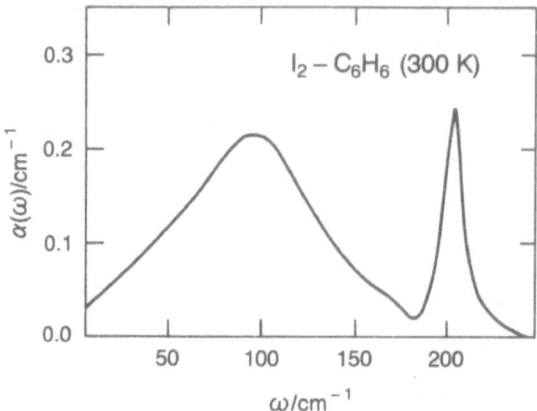

Figure 9. Absorption coefficient of I_2 (0.2 mol l^{-1}) dissolved in C_6H_6 at room temperature. The absorption due to C_6H_6 alone was subtracted. The broad band is the induced rotational spectrum and the sharp band is the induced vibrational spectrum (Besnard et al., 1988).

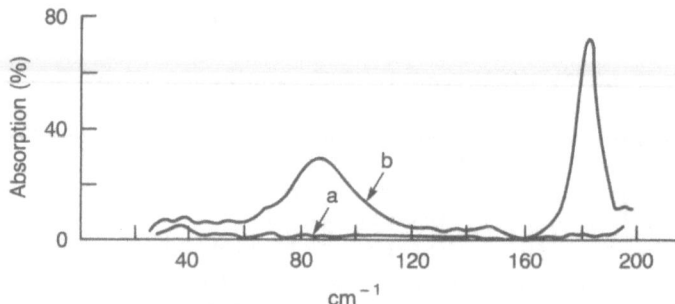

Figure 10. Spectrum at room temperature due to 2-methyl pyridine-iodine in cyclohexane. Path length, 9.35 mm. (b) concentration of iodine 0.01 mol/l, 2-methyl-pyridine 0.00954 mol/l, (a) 2-methyl pyridene, 0.00954 mol/l (Lake and Thompson, 1967).

namely, $\tau_{rot}(B)$ and $\tau_{tr}(I-B)$ are not too different, thus throwing the TR decoupling approximation into question. Furthermore, the theory predicts two librational modes, one at 40 cm^{-1} due to I_2 and the other at 78 cm^{-1} due to C_6H_6 shown in Fig. 12. These modes, due to torques produced by the anisotropy of the potential, exhibit pronounced peaks that are not seen in the experimental spectrum. However, this spectrum exhibits a broad shoulder in the same frequency domain as the peaks in the computed spectrum.

We now briefly consider the role of charge transfer in I_2-solvent systems. For strong complexes (large binding energies) like the I_2-MP system, the donor-acceptor separations are close to those of covalent bonds. A well-defined structure results, and presumably the same

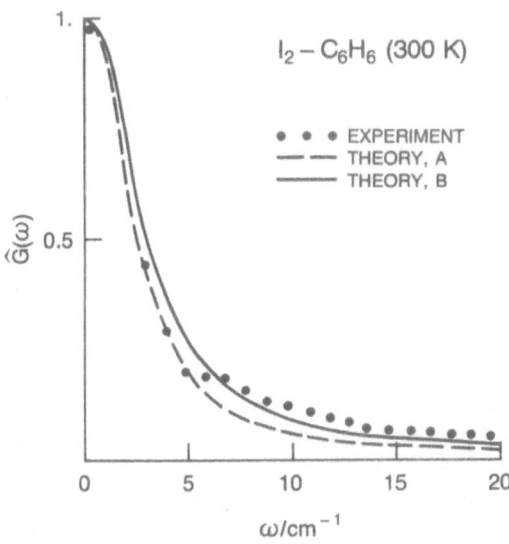

Figure 11. Spectral density of 0.2 mol l^{-1} of I$_2$ dissolved in C$_6$H$_6$ at 300 K. Experiment (Besnard et al., 1988); theory A, Eq. (35) with $\gamma = 0$; theory B, Eq. (35) with $2\gamma = 0.5$.

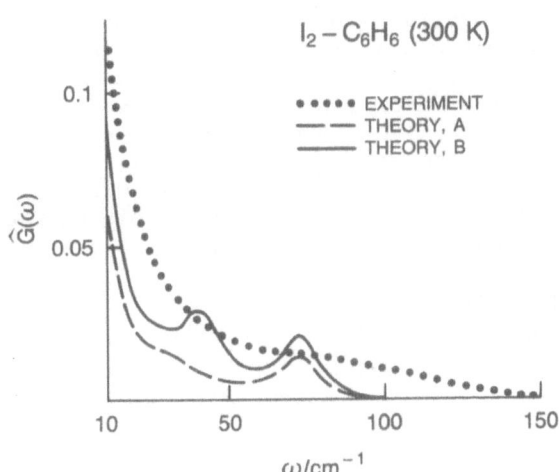

Figure 12. Spectral density of 0.2 mol l^{-1} of I$_2$ dissolved in C$_6$H$_6$ at 300 K. Experiment (Besnard et al., (1988); theory A, Eq. (35) with $\gamma = 0$; theory B, Eq. (35) with $2\gamma = 0.5$. All spectra are normalized to unity at $\omega = 0$.

configuration exists in solution and in a crystalline form made up of 1:1 complexes. With decreasing magnitude of the interaction energy, the complex becomes less stable and its geometry becomes less defined so that its properties become the statistical average over all configurations of the two molecules. As the binding energy approaches kT, there can be only a brief contact between donor and acceptor. The charge transfer in this case, which is thought to be the situation in I_2-C_6H_6 (Hanna and Lippert, 1973), is called contact charge transfer. A calculation of the dipole moment of Br_2-C_6H_6 (that for I_2-C_6H_6 is not available), gives the following results (Hanna and Lippert, 1973): μ_E = 0.42 Debye, where μ_E is due to electrostatic interactions, here principally due to distributed-quadrupole distributed-polarizability; and μ_{CT} = 0.28 to 0.82 Debye, where μ_{CT} is due to charge transfer (CT). This result, which is not expected to be too different for I_2-C_6H_6, shows that the two contributions are of the same order of magnitude, and that therefore both must be considered.

In order to obtain a rough idea regarding the possible role of contact CT on the FIR spectrum of I_2-C_6H_6, we assume an isotropic overlap mechanism given by $\mu_{IB} = \mu(\sigma)\exp[-(r - \sigma)/\lambda]$ and take $\lambda/\sigma = 0.1$ (appropriate for rare gas mixtures), and two trial values of $\mu(\sigma)$, 0.4D and 0.04D. The spectral moments (see Guillot, 1987) that are obtained, M_0 and M_2, are shown in Table 9. In the case of weak overlap, the contribution to the spectral moments is insignificant. Since the isotropic overlap and QID mechanisms have different symmetries, their effect on the spectrum is simply additive. The contribution of the former creates a small shoulder around 17 cm^{-1}, which for our purpose is of small consequence. In the case of strong overlap, essentially a model for the effect of contact CT, its spectrum becomes so intense that it dominates the total absorption below 40 cm^{-1}, although it is not sufficient to give the experimental value of the second moment when added to M_2(QID). However, it is reasonable to assume a contact CT induction with quadrupolar symmetry, just as one assumes an anisotropic overlap induction with quadrupolar symmetry; these then mix with the QID induction because they have the same symmetry. It is expected that this anisotropic overlap induction will give improved results.

Even if good agreement with experiment can be achieved for the LF part of the spectrum, we have failed to account for the HF part. The

TABLE 9. Spectral moments due to an isotropic overlap induced dipole, Eq. (39), with $\lambda/\sigma = 0.1$. ρ_I = 4.5 amagat and ρ_B = 251 amagat.

$\mu(\sigma)$ (D)	M_0 (cm^{-1})	M_2 (cm^{-3})
0.04	5 x 10^{-2}	105
0.4	5	1.05 x 10^4

following questions are raised by such disagreement here and elsewhere in this work. (1) Are the assumptions used in arriving at Eq. (35), starting from Eq. (32) justified? (2) Is the decoupling approximation reliable for this system? (3) Is it appropriate to model the electrostatic induction by a point-quadrupole point-polarizability mechanism. The neglect of the anisotropic polarizability, although not thought to be as serious as the above assumptions and approximations, should nevertheless be investigated. In the same category is the neglect of additional induced dipole terms arising from the couplings provided by the anisotropy of the potential (Guillot and Birnbaum, 1985). Also worthy of study is the reliability of the potential and the pair distribution functions used in the computations. To study these questions, it is important to compare the theory with MD simulations where the input data are known.

5. SUMMARY AND CONCLUSIONS

Rather satisfactory representations of the FIR I_2-C_6H_6 spectrum were obtained (Yarwood and Catlow, 1987; Besnard et al., 1988) by fitting various relations to experiment. However, this procedure is of value in providing an understanding of the mechanisms responsible for such spectra and gives information about the experiment only when the relations are justified and the assumptions on which they are based are clear. To this end, we developed a theoretical approach for obtaining analytical relations for the study of IIA in liquids, and compared these relations primarily with experiments for the liquids of interest, since MD results were not generally available.

We used a three-variable Mori theory, whose parameters were estimated by a lattice gas model. There were no adjustable parameters in the final result, with one exception regarding the magnitude of a partial CF. To obtain tractable results, we introduced an approximation to decouple translation from rotation and found that it worked reasonably well for liquid N_2, where such coupling is expected to be small. This theory was extended to mixtures, which play a dominant role in the study of "large" molecules. A three-body term was found that was previously overlooked, but calculations for the mixture studied here indicated that this term (whose value had to be parameterized) and the usual three-body terms are of secondary importance. Nevertheless, the effect of the former on the spectrum, while not large, is not negligible. We found that a point-QID point-polarizability mechanism in all the liquids studied here, except N_2, cannot account for a large amount of high frequency absorption, which, as it happens, also appears on comparing computer simulations of induced absorption in liquids with experiment. However, we indicated how an overlap induction mechanism, which can enhance the high frequency absorption, can be included in the theory in a straightforward way. Given that the theory presented here provides a computationally convenient way of computing and describing IIA in liquids, improvements in the theory are worthy of further investigation.

APPENDIX. CROSS ORIENTATIONAL CORRELATIONS

We calculate here the two correlation functions $<P_2(\cos\theta(t))>$ and $<P_2(\cos\theta_{12}(t))>$ on the basis of a Markov-Gauss process (Bratos and Tarjus, 1981), which is equivalent to the Zwanzig-Mori theory with a δ-truncation at the order n (Keyes and Kivelson, 1972; Kivelson and Madden, 1975; also see Steele, 1989). In this formalism, the N variables a_i and their first n derivatives follow a Markov-Gauss process when the associated correlation matrix is exponential, namely,

$$R^{(n)}(t) = R^{(n)}(0)\exp(-\Gamma^{(n)}(t))$$

where $\Gamma^{(n)}$ is the transport matrix of order (n). For the sake of simplicity, we will restrict ourselves in what follows to the zero order (no derivatives), the generalization being straightforward but lengthy. Thus considering the variables $a_i = D_{00}^{(2)}(\theta_i)$, $a_i = i = 1, \ldots, N$, the diagonalization of the transport matrix $\Gamma^{(0)}$ yields the following expression:

$$<P_2(\cos\theta_{11}(t))> = <D_{00}^2(\theta_1(0))D_{00}^2(\theta_1(t))>$$

$$= \frac{1}{N}[1 + (N-1)<D_{00}^2(\theta_1(0))D_{00}^2(\theta_2(0))>]e^{-t/\tau_M}$$

$$+ \frac{N-1}{N}[1 - <D_{00}^2(\theta_1(0))D_{00}^2(\theta_2(0))>]e^{-t/\tau_m} \qquad (A1)$$

the self part, and

$$<P_2(\cos\theta_{12}(t))> = <D_{00}^2(\theta_1(0))D_{00}^2(\theta_2(t))>$$

$$= \frac{1}{N}[1 + (N-1)<D_{00}^2(\theta_1(0)D_{00}^2(\theta_2(0))>]e^{-t/\tau_M}$$

$$- \frac{1}{N}[1 - <D_{00}^2(\theta_1(0))D_{00}^2(\theta_2(0))>]e^{-t/\tau_m} \qquad (A2)$$

the distinct part, where τ_M is the macroscopic relaxation time given by

$$\tau_M = \int_0^\infty \frac{<P_2(\cos\theta_{11}(t))> + (N-1)<P_2(\cos\theta_{12}(t))>}{1 + (N-1)<P_2(\cos\theta_{12}(0))>} dt \qquad (A3)$$

and τ_m is the microscopic relaxation time given by

$$\tau_m = \int_0^\infty \frac{<P_2(\cos\theta_{11}(t))> - P_2(\cos\theta_{12}(t))>}{1 - <P_2(\cos\theta_{12}(0))>} dt \qquad (A4)$$

Introducing in Eqs. (A3) and (A4) the definition of the Kirkwood g_2 factor and also that of the self, τ_1, and distinct, τ_2, correlation times yields, respectively,

$$\tau_M = \frac{\tau_1 + (g_2 - 1)\,\tau_2}{g_2} \tag{A5}$$

$$\tau_m = (\tau_1 - \frac{\tau_2 g_2}{N})/(1 - (g_2/N)) \tag{A6}$$

Expressions (A1, A2) have several consequences. If one neglects all terms of the order 1/N, the self part is mostly governed by τ_m while the distinct part by τ_M. Furthermore, due to their definitions, Eq. (A5) and Eq. (A6) can exhibit very different time scales, particularly when cross correlations in the first shell are important (then $\tau_2 \gg \tau_1$); this is confirmed by MD simulations (Singer et al., 1979), Fig. 13. But we need in our theory (see Eq. 23), the difference correlation function CF(self) − CF(distinct). By subtracting (A1) and (A2) one obtains

$$<P_2(\cos\theta_{11}(t))> - <P_2(\cos\theta_{12}(t))> \simeq$$
$$(1 - <P_2(\cos\theta_{12}(0))>)e^{-t/\tau_m} \tag{A7}$$

an approximation where the difference CF has the same time dependence as the self part (governed by τ_m) irrespective of the magnitude of the correlations. However, as may be noted from Fig. 13, subtracting the constant $<P_2(\cos\theta_{12}(0))>$ requires that the resulting CF have a relaxation time smaller than τ_m at long times. In practice we equate τ_m with τ_{rot} extracted from MD simulation or from Raman and Rayleigh spectra.

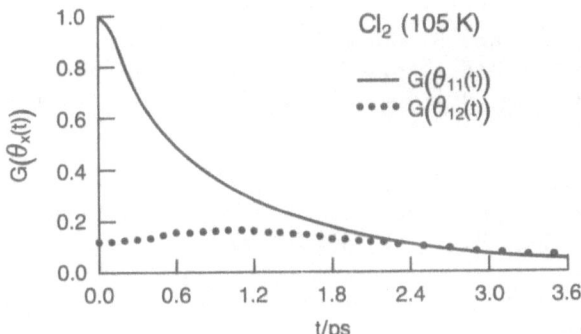

Figure 13. Self correlation function, $G(\theta_{11}(t)) = <P_2(\cos\theta_{11}(t))>$, and cross-correlation function, $G(\theta_{12}(t)) = <P_2(\cos\theta_{12}(t))>$, of liquid Cl_2 at 105 K (Singer et al., 1979).

REFERENCES

Applequist, J. (1985) 'A multipole interaction theory of electric polarization of atomic and molecular assemblies', J. Chem. Phys. $\underline{83}$, 809-826.

Besnard, M., DelCampo, N, and Lascombe, J. (1988) 'Chemical reactivity in weak charge transfer complexes: Analysis of induced far infrared profiles and Raman scattering profiles', in M. Moreau and P. Turq (eds.), Chemical Reactivity in Liquids: Fundamental Aspects, Plenum Press, New York, pp. 33-54.

Besnard, M. (1988) private communication.

Birnbaum, G. (1980) 'Determination of molecular constants from collision-induced far-infrared spectra and related methods,' in J. Van Kranendonk (ed.), Intermolecular Spectroscopy and Dynamical Properties of Dense Systems, North-Holland Publishing Co., Amsterdam, pp. 111-145.

Birnbaum, G. (ed.) (1985) Phenomena Induced by Intermolecular Interactions, NATO ASI Series B: Physics Vol. 127, Plenum Press, New York.

Borysow, A., and Frommhold, L. (1986) 'Collision-induced rototranslational absorption spectra of N_2-N_2 pairs for temperatures from 50 to 300 K', Ap. J. 311, 1043-1057.

Bowling Barnes, R., Benedict, W.S., and Lewis, C.M. (1935) The far infrared absorption of benzene', Phys. Rev., 47, 129-130.

Bratos, S., and Guillot, B. (1982) 'Theoretical investigation and experimental detection of rattling motions in atomic and molecular fluids', J. of Mol. Struct., 84, 195-203.

Bratos, S. and Tarjus, G. (1981) 'Raman scattering from pure liquids. Theory of band profiles', Phys. Rev. A24, 1591-1600.

Bratos, S., Guillot, B., and Birnbaum, G. (1985) 'Theory of collision-induced light scattering and absorption in dense rare gas fluids', in G. Birnbaum (ed.), Phenomena Induced by Intermolecular Interactions, pp. 363-381.

Briganti, G., Rocca, D., and Nardone, M. (1986) 'Interaction induced light scattering. First and second spectral moments in the superposition approximation', Mol. Phys. 59, 1259-1272.

Chang, P. and Wilkes, C.R. (1955) 'Some measurements of diffusion in liquids', J. Phys. Chem. 59, 592-596.

Cheung, P.S.Y., and Powles, J.G. (1975) 'The properties of liquid nitrogen. IV. A computer simulation,' Mol. Phys. 30, 921-949.

Cheung, P.S.Y., and Powles, J.G. (1975) 'The properties of liquid nitrogen. V. Computer simulation with quadrupole interaction', Mol. Phys. 32, 1383-1405.

Dardy, H.D., Volterra, V., and Litovitz, T.A. (1973) 'Rayleigh scattering: Orientational motion in highly anisotropic liquids', J. Chem. Phys, 59, 4491-4500.

Davies, M. (1974) 'Far infrared absorptions in non-dipolar liquids', in J. Lascombe (ed.), Molecular Motions in Liquids, Reidel Publishing Company, Dordrecht, pp. 615-635.

Dorfmüller, Th. (1985) 'Interaction induced spectra of "large" molecules in liquids', in G. Birnbaum (ed.), Phenomena Induced by Intermolecular Interactions, pp. 661-676.

Dorfmüller, Th. (1989) 'Looking at internal motions of molecules with dynamic light scattering', this volume.

Edwards, M.F.E., Madden, P.A., and McDonald, I.R. (1984) 'A computer simulation study of the dielectric properties of a model of methyl cyanide', Mol. Phys, 51, 1141-1161.

Egelstaff, P.A., Litchinsky, D., McPherson, R., and Hahn, L. (1978) 'Correlations in nitrogen gas at room temperature', Mol. Phys., 36, 445-451.

Falcone, D.R., Douglass, D.C., and McCall, D.W. (1967) 'Self diffusion in benzene', J. Phys, Chem. 71, 2754-2755.

Floudas, G., and Fytas, G. (1989) 'Optical anisotropy of macromolecular systems by depolarized Rayleigh sattering', this volume.

Garg, S.K., Bertie, J.E., Kilp, H., and Smyth, C.P. (1968) 'Dielectric relaxation, far infrared absorption and intermolecular forces in non-polar liquids', J. Chem. Phys, 49, 2551-2562.

Gray, C.G. (1971) 'Theory of collision-induced absorption for spherical top molecules', J. Phys . B: Atom Molec. Phys. 4, 1661-1669.

Gray, C.G. and Gubbins, K.E. (1984) Theory of Molecular Fluids, Vol. 1: Fundamentals, Clarendon Press, Oxford.

Guillot, B. and Birnbaum, G. (1985) 'Theoretical interpretation of the far infrared absorption spectrum in molecular liquids, nitrogen', in G. Birnbaum (ed.), Phenomena Induced by Intermolecular Interactions, pp. 437-455.

Guillot, B. (1987) 'Theoretical investigation of the dip in the far infrared absorption spectrum of dense rare gas mixtures', J. Chem. Phys. 87, 1952-1961.

Guillot, B., Ph. Marteau, and Obriot, J. (1988) 'Far-infrared absorption in nitrogen-rare gas compressed mixtures: an experimental and theoretical survey', Mol. Phys. 65, 765-784.

Hanna, M.W., and Lippert, J.L. (1973) 'Theory of the ground state structure of molecular complexes', in R. Foster (ed.), Molecular Complexes, Vol. 1, Crane, Russak and Company, Inc., New York, pp. 1-48.

Hirschfelder, J.O., Curtiss, C.F., and Bird, R.B. (1954) Molecular Theory of Gases and Liquids, John Wiley and Sons, Inc., New York.

Impey, R.W., Madden, P.A., and Tildesley, D.J. (1981) 'On the calculation of the orientational correlation parameter g_2', Mol. Phys. 44, 1319-1334.

Kettle, J.P., and Price, A.H. (1972) 'Far infra-red measurements on solutions of iodine and bromine', J. Chem. Soc., Faraday Trans. II, 68, 1306-1311.

Keyes, T. and Kivelson, D. (1972) 'Depolarized light scattering: Theory of the sharp and broad Rayleigh lines,' J. Chem. Phys. 56, 1057-1065.

Kivelson, D. and Madden, P. (1975) 'Theory of dielectric relaxation,' Mol. Phys. 30, 1749-1780.

Ladd, A.J.C., Litovitz, T.A., and Montrose, C.J. (1979) 'Molecular dynamics studies of depolarized light scattering from argon at various fluid densities', J. Chem. Phys., 71, 4242-4248.

Ladanyi, B.M. and Levinger, N.E. (1984) 'Computer simulation of Raman scattering from molecular fluids', J. Chem. Phys. 81, 2620-2633.

Ladanyi, B.M., Barreau, A., Chave, A., Dumon, B., and Thibeau, M. (1986) 'Collision-induced light scattering by fluids of optically isotropic molecules: Comparison of results of two model studies', Phys. Rev. A34, 4120-4130.

Lake, R.F., and Thompson, H.W. (1967) 'Far infrared spectra of charge-transfer complexes between iodine and substituted pyridines', Proc. Roy. Soc., A297, 440-448.

Levesque, D. and Weis, J.J. (1975) 'Collective motion and depolarized light scattering from diatomic fluids,' Phys. Rev. A12, 2584-2586.

Lascombe, J. and Besnard, M. (1986) 'Far infrared study of the benzene-iodine complex', Mol. Phys. 58, 573-592.

Levesque, D., Weis, J.J., Marteau, Ph., Obriot, J., and Fondere, F. (1985) 'Collision induced far infrared spectrum of liquid N_2. Comparison between computer simulations and experiment', Mol. Phys. 54, 1161-1172.

Linse, P., Engström, S., and Jonssön, B. (1985) 'Molecular dynamics simulation of liquid and solid benzene', Chem. Phys. Letters 115, 95-110.

Lund, P.-A., Faurskov Nielsen, O., and Praestgaard, E. (1978) 'Comparison of depolarized Rayleigh wing scattering and far-infrared absorption in molecular liquids', Chem. Phys. 28, 167-173.

Lynden-Bell, R.M. and Steele, W.A. (1984) 'A model for strongly hindered molecular reorientation in liquids', J. Phys. Chem. 88, pp. 6514-6518.

Magaña, R.J. and Lannin, J.S. (1985) 'Observation of clustered molecules and ions in liquid iodine', Phys. Rev. B32, 3819-3823.

Marteau, Ph., Obriot, J., Fondere, F., and Guillot, B. (1986) 'Density effects and relative diffusion in the far infrared absorption spectrum of compressed liquid nitrogen', Mol. Phys., 59, 1305-1328.

Moon, M. and Oxtoby, D.W. (1986) 'Collision-induced absorption in gaseous N_2', J. Chem. Phys. 84, 3830-3842.

Murthy, C.S., Singer, K., Steele, D., Tindle, J.J., and Vallauri, R. (1982) 'Interaction-induced infrared absorption in liquid Cl_2: an experimental and molecular-dynamics investigation', Chem. Phys. Letters 90, 95-98.

Perrot, M., Brooker, M.H., and Lascombe, J. (1981) 'Raman light scattering studies of the depolarized Rayleigh wing of liquids and solutions', J. Chem. Phys, 74, 2787-2794.

Price, S.L., 1988, 'Is the isotropic atom-atom model potential adequate?', Mol. Simulation 1, 135-156.

Pringle, W.C., Jacobs, S.M., and Rosenblatt, D.H. (1983) 'Collision-induced rotational spectrum of allene', Mol. Phys, 50, 205-215.

Pringle, W.C., Gronlund, W.R., and Cohen, R.C. (1987) 'Collision induced far infrared spectrum of cyclopropane', Mol. Phys. 62, 669-678.

Samios, J., Mittag, V., and Dorfmüller, Th. (1986) 'A molecular dynamics simulation of interaction-induced fir absorption spectra of liquid CS_2', Mol. Phys, 59, 65-79.

Singer, K, Singer, J.V.L., and Taylor, A.J. (1979) 'Molecular dynamics of liquids modelled by '2-Lennard-Jones centres' pair potentials. II. Translational and rotational autocorrelation functions', Mol. Phys. 37, 1239-1262.

Steele, W.A. (1985) 'Computer simulation study of the forbidden absorption spectra of liquid nitrogen', Mol. Phys. 56, 415-430.

Steele, W.A. (1989) 'Relative motion of pairs of molecules,' this volume.

Steele, W.A. and Posch, H.A. (1985) 'Workshop report: Liquids and liquid state interactions', in G. Birnbaum (ed.), Phenomena Induced by Inter- molecular Interactions, pp. 549-556.

Steele, W.A. and Posch, H.A. (1987) 'A simulation study of the induced infrared absorption in liquid CO_2', J. Chem. Soc, Faraday Trans, II, 83, 1843-1858.

Steele, W.A. and Vallauri, R. (1987) 'Computer simulations of pair dynamics in molecular fluids', Mol. Phys. 61, 1019-1030.

Steinhauser, O. (1982) 'On the structure and dynamics of liquid benzene', Chem. Phys. 73, 155-167.

Stone, A.J. and Alderton, M. (1985) 'Distributed multipole analysis: methods and applications', Mol. Phys, 56, 1047-1064.

Stone, A.J. and Price, S.L. (1988) 'Some new ideas in the theory of intermolecular forces: anisotropic atom-atom potentials', J. Phys. Chem. 92, 3325-3335.

Wagner, V. (1969) 'Infrared absorption of liquid Br_2, I_2, and $Br_2:I_2$', Z. Physik 224, 353-363.

Wagner, V. (1972) 'Lattice absorption of solid and liquid iodine', Phys. Stat. Solid. B., 50, 585-592.

Yarwood, J. and Catlow, B. (1987) 'Far-infrared interaction-induced spectra of the halogens', J. Chem. Soc. Faraday Trans. II, 83, 1801- 1814.

Yarwood, J. (1989) 'Investigation of intermolecular interactions using raman and infrared spectroscopy', this volume.

Isotropic Raman Study of Pre-reactive, Reactive and Post-reactive Processes in Liquids

S. BRATOS and P. VIOT
Laboratoire de Physique Théorique des Liquides
Université Pierre et Marie Curie
4, place Jussieu
75252 Paris Cedex 05
France

ABSTRACT. Spectral manifestations in isotropic Raman Spectra of fast pre-reactive, reactive and post-reactive motions in the liquid phase are examined theoretically. The kinetics of the reaction process is analysed in the frame of a one-dimensional Smoluchowski equation. The resulting correlation function exibits two chemical modes and two modes generated by the pre-reactive and post-reactive processes. The possibilities of Raman spectroscopy to detect these latter motions are critically evaluated.

1 Introduction

A number of spectroscopic techniques can be employed to study chemical reactions in gases and liquids, the techniques such as dielectric relaxation, ultrasonic absorption, nuclear magnetic resonance, infra-red and Raman spectroscopies. More recently, an increasing interest is being focused on the study of the elementary chemical act, involving its pre-reactive, reactive and post-reactive steps. For exemple, by using femtosecond laser pulses it was possible to watch molecules in the act of formation or fragmentation, which yielding insight into the most crucial steps of a chemical reaction[1]. It is of fundamental importance for chemists to know how atoms and molecules get together to form new molecules.

The purpose of the present paper is to describe the possibilities of the Raman spectroscopy in studying the pre-reactive, reactive and post-reactive processes in liquids. The time domain covered by this spectroscopy extends from 10^{-10} to 10^{-14} sec, which corresponds to the requirements. Three sorts of reactions have been studied from this point of view : conformational changes[2-7], proton transfer reactions[8-11], and charge transfer complex formation[12]. Three main theories are currently employed to interpreted the experimental data. The first is due to Lascombe and Cavagnat[2] who proposed a Smoluchowski-type description of the reaction process. In another theory, elaborated by Strauss and MacPhail [13], the standard NMR theories of chemically reacting liquids were modified to study infrared and Raman band shapes of chemically reacting liquids. Finally, Bratos, Tarjus and Viot[14,15] transposed the general theory of Raman band shapes of chemically inert liquids to the chemically reactive liquids. Unfortunately, no of these theories deals with pre-reactive and post-reactive processes; they will be examined in the present paper.

Th. Dorfmüller (ed.), Reactive and Flexible Molecules in Liquids, 37–45.
© *1989 by Kluwer Academic Publishers.*

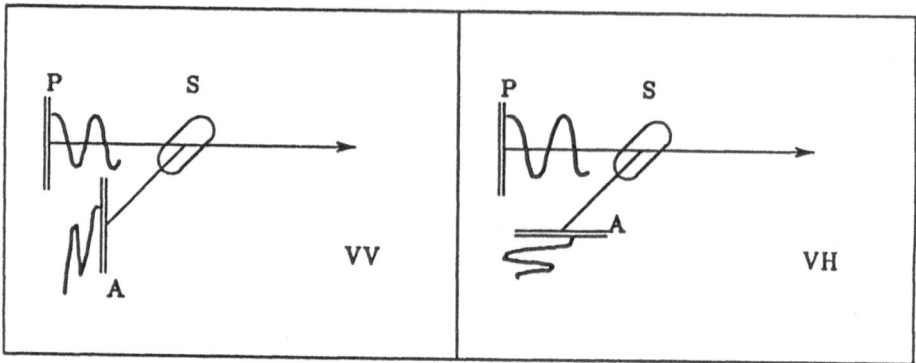

Figure 1: Schematic representation of the VV and VH experiments. P is the polarizer, A the analyser and S the liquid sample.

2 Experimental procedure

The basic experiment consists in recording, under appropriate conditions, the isotropic Raman spectra of chemically reactive liquid mixtures. The experimental procedure is the standard VV-VH separation procedure[16] (Fig.1). The spectrum is first measured by keeping the polarizer and analyser perpendicular to the scattering plane. It is then measured again by rotating the analyser into the plane. The differential cross-section of the isotropic scattering is defined as follows :

$$\left(\frac{\partial^2\sigma}{\partial\omega\partial\Omega}\right)_{iso} = \left(\frac{\partial^2\sigma}{\partial\omega\partial\Omega}\right)_{VV} - \frac{4}{3}\left(\frac{\partial^2\sigma}{\partial\omega\partial\Omega}\right)_{VH} \tag{2.1}$$

It can easily be shown that only vibrational and chemical, but not rotationnal relaxation processes are active in the isotropic Raman spectra. The competing rotationnal relaxation effects may thus be eliminated by simply using this sort of spectroscopy[14]. This is an important advantage if the fast chemical kinetics in liquids is to be studied. Unfortunately, vibrational and chemical relaxation processes can never be disentangled completely.

The most frequently observed effect of a chemical reaction is the splitting of spectral bands into as many components as there are reacting species. Alternatively, new bands may be appear and some others disappear. The interesting information may be extracted from the study of a pair of bands, or a triplet of bands, etc, associated with a normal mode localized in an unmodified fragment of the molecule, but perturbed by the reaction; the bands assigned to the reaction coordinate itself are generally not useful. However, observing spectral splittings, or noticing the appearance of new bands, does not provide, by itself, an evidence of the kinetics of chemical transformations. In fact, there are three principal signatures of chemical kinetics in Raman .(i) The first of them is the collapse of a doublet, or a triplet, when the temperature is increased (Fig.2A). This effect was fisrt observed, in 1982, in low temperature solids by Cavagnat and Lascombe[2] and by MacPhail, Snyder and Strauss[3]. One year later, in 1983, it was also reported for liquids by Lascombe, Cavagnat et al[4]. (ii) The second signature is the the concentration induced frequency shift of a Raman band between positions characteristic of the reacting species(Fig.2B).

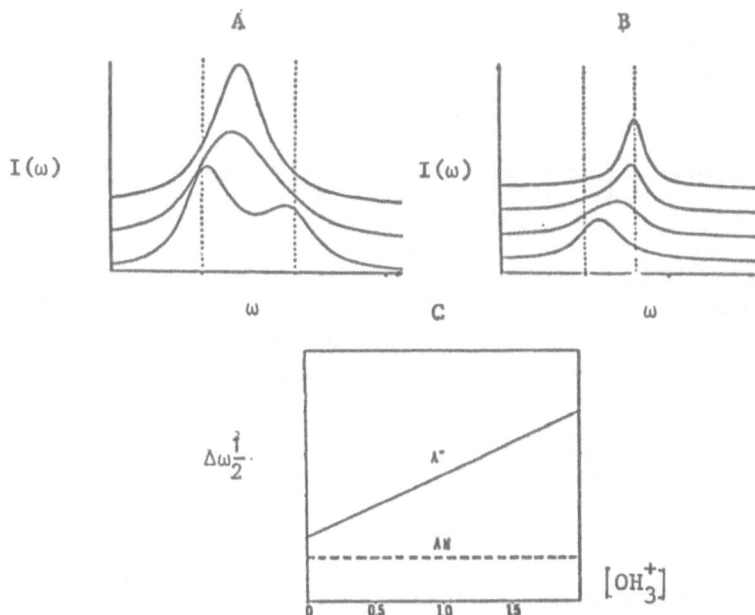

Figure 2: The three signatures of the chemical kinetics in isotropic Raman spectra of liquids.

This signature was reported in 1978, by Lassègues ans Devaure[11]; however, in abscence of an adequate theory, it was not recognized as such by these authors. More recently, in 1988, Besnard, Del Campo and Lascombe presented a more complete discussion of this effect[12]. (iii) The third signature is the lifetime broadening of spectral bands(Fig.2C). It was described long time ago, in 1962, by Kreewoy and Mead[8]; later on, it was reported by several other authors; the careful study by Chen and Irish in 1971 merits a special attention[10]. Unfortunately, no experimental analysis of pre-reactive and post-reactive motions is available at the present time.

It is important to realize that considering systems in statistical equilibrium does not prevent the study of chemical kinetics. In fact, the existence of statistical equilibrium does not imply the abscence of chemical reactions, but entails the equality of reactive fluxes in the forward and backward directions. The lifetime of reacting molecules is shortened independently of whether the system is in equilibrium, or not. However, it is probable that the chemical kinetics effects are better studied in non-equilibrium conditions. This question will not be discussed in the present paper.

3   Theory

a)GENERALITIES. The main purpose of the present work is to study the spectral manifestations of the pre-reactive, reactive and post-reactive steps of a chemical reaction in the frame of a single theory. In order to avoid all non-essential complications, the theory will be limited to the conformationnal changes, the simplest class of chemical reactions. Many results reached in this way should be transposable to bimolecular reactions.

b)DESCRIPTION OF THE MODEL. The system submitted to the study is a diluted solution of two conformers in an inert solvent. The model employed to study its spectral behaviour involves the following assumptions. (i) Molecular vibrations of a reference conformer are described by a normal coordinate n and a semi-classical mean polarizability $\alpha(n, t)$

$$H(n,t) = \left[\frac{1}{2}p^2 + \frac{1}{2}\lambda n^2 + \frac{1}{6}\mu n^3 + \cdots\right] + U(n,t) \tag{3.1}$$

$$\alpha(n,t) = \alpha'(t)n \tag{3.2}$$

The potential energy $U(n, x(t), X(t))$ takes account of the coupling between the normal coordinate n and the reaction coordinate x and of that between n and the bath coordinates X. (ii) The dynamics of the reaction process is pictured as a diffusion along x. This latter motion is considered as a Markovian process controlled by the following one-dimensionnal Smoluchowski equation[17].

$$\frac{\partial W(x_0 \mid x,t)}{\partial t} = L_{FP}W(x_0 \mid x,t) \tag{3.3}$$

$$L_{FP} = \frac{\partial}{\partial x}\left[D\left(\beta\frac{\partial U}{\partial x} + \frac{\partial}{\partial x}\right)\right] \tag{3.4}$$

where $W(x_0 \mid x,t)$ is the probabilty for the system to be in the point x at time t, if it was in the point x_0 at time 0, the L_{FP} the well-known Fokker-Planck operator, D the diffusion constant and $U(x)$ the potential of the mean force along the reaction coordinate. (iii) The time scale of the bath fluctuations is very short with respect to time scale of the vibrational relaxation and of the diffusion along x.

The assumptions of the above model may be commented as follows. The main novelty, as compared with the Bratos-Tarjus-Viot model[14], is that the reaction is pictured in a continous way. The present model is thus much less rudimentary than the jump model. The time scale considerations are similar in both theories. It is assumed that bath fluctuations are fast as compared to the vibrational relaxation, i.e. that vibrations are rapidly modulated. Though this asssumption does not necessarily hold true for all Raman bands, it does for narrow bands appropriate to study chemical kinetics. On the contrary, postulating the bath fluctuations to be fast as compared with the reactive motions along the reaction coordinate, is more restrictive than the corresponding asssumption of Ref[14]. However, such an asssumption is inherent to the Smoluchowski equation; this latter equation is only valid if the bath fluctuations are fast as compared with diffusion.

c)CALCULATION OF THE CORRELATION FUNCTION. Building the Fokker-Planck formalism into the theory of Raman band shapes of liquids requires some effort. This technique was employed only occasionally in spectroscopic problems, e.g. by Johnson and Oxtoby in studying the infrared band shapes of hydrogen bonded liquids[18]. The details of the calculation are given elsewhere[19]. The result is

$$\left(\frac{\partial^2 \sigma}{\partial\omega\partial\Omega}\right) = \frac{K_s^4}{2\pi}\int_{-\infty}^{+\infty} dt e^{-i\omega t} G(t) \tag{3.5}$$

$$G(t) = \mid n_{01}\mid^2 e^{i\omega_0 t}\alpha^+ exp\left[(i\omega + L_{FP})\,t\right]\alpha \tag{3.6}$$

$$L_{FP}\phi_n = -\lambda_n\phi_n \tag{3.7}$$

$$\alpha_n = \int_{-\infty}^{+\infty} dx\,\overline{\phi}_0(x)\,\alpha(x)\,\phi_n(x) \quad , \quad \omega_{mn} = \int_{-\infty}^{+\infty} dx\,\overline{\phi}_m(x)\,\omega(x)\,\phi_n(x) \tag{3.8}$$

where $\phi_n(x)$ and λ_n are the n'th eigenfunction and eigenvalue, respectively, of the Fokker-Planck operator, $\overline{\phi}_n(x) = \phi_n(x)\phi_0(x)^{-1}$, ω_0 the non-pertubed frequency of the normal mode under investigation and $\omega(x) = \overline{\omega}(x) + iT_2^{-1}(x)$ the complex solvent induced frequency at the point x of the reaction coordinate. Once these quantities have been calculated, the matrices α, ω, $\mathbf{L_{FP}}$ as well as the correlation function $G(t)$ can readily be determined. This correlation function is formally similar to that of the jump model, but is entirely non-empirical. Its construction requires the solution of the Fokker-Planck equation. In fact, $\phi_n(x)$ and λ_n contain all information about chemical kinetics in the liquid phase.

A particulary convenient way of treating a one-dimensionnal Fokker-Planck equation consists in transforming it into a Schrödinger-type equation by the help of the substitution $\Psi_n(x) = \phi_0(x)^{-\frac{1}{2}}\phi_n(x)$ [17]. Then, if $V(x) = \frac{1}{4}D(\beta U'(x))^2 - \frac{1}{2}D\beta U''(x)$, on finds:

$$\frac{\partial}{\partial x}\left[D\left(\beta\frac{\partial U}{\partial x} + \frac{\partial}{\partial x}\right)\right]\phi_n(x) = -\lambda_n\phi_n(x) \iff$$
$$\iff \left[D\frac{d^2}{dx^2} - V(x)\right]\Psi_n = -\lambda_n\Psi_n \tag{3.9}$$

Usual methods of quantum mechanics may easily be employed. In numerical work, the "Schrödinger" equation(3.9) is discretized by choosing an appropriate length step h. A value of $\Psi_n(x)$ is fixed at large negative x's and the equation is solved, step by step, by going from the left to right. The procedure is then repeated again by choosing another value of $\Psi_n(x)$ at large positive x's and going from right to left. The two solutions are matched at a suitably chosen encounter point by comparing, for a given λ_n, their logarithmic derivatives.

In analytical work, the "Schrödinger" equation was treated, approximately, by developping $\Psi_n(x)$'s into following basic set:

$$\chi_{10} = C_{10}e^{-\frac{1}{4}\beta f_1(x-x_1)^2} \quad ; \quad \chi_{11} = C_{11}e^{-\frac{1}{4}\beta f_1(x-x_1)^2}H_1\left[\sqrt{\frac{\beta f_1}{2}}(x-x_1)\right]$$

$$\chi_{20} = C_{20}e^{-\frac{1}{4}\beta f_2(x-x_2)^2} \quad ; \quad \chi_{21} = C_{21}e^{-\frac{1}{4}\beta f_2(x-x_2)^2}H_2\left[\sqrt{\frac{\beta f_1}{2}}(x-x_2)\right] \tag{3.10}$$

These functions are eigenfunctions of the operator $L = \phi_0(x)^{-\frac{1}{2}}L_{FP}\phi_0(x)^{\frac{1}{2}}$ for a harmonic oscillator centered at x_1 or x_2, respectively; $f_1 = U''(x_1)$ and $f_2 = U''(x_2)$; $H_n(x)$ is the Hermite polynomials and C_{nm} are the normalization factors. The motivation for this particular choice of the basic set is the same as in treating spectroscopic double minimum problems. Its quality is comparable to that of the Kramers theory of reaction rates[20]. Combining numerical and analytical theories provides accurate results and the possibility of their interpretation.

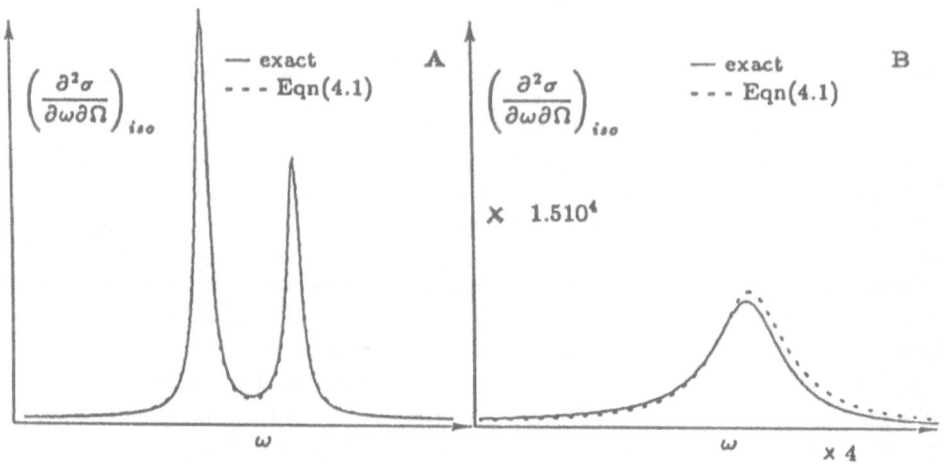

Figure 3: Isotropic Raman spectra of chemically reactive liquids in the case of slow conformation changes . Note the difference of the frequency and intensity scale in Fig.3A and Fig.3B.

4   Results and discussion

a) GENERAL CONSIDERATIONS. The problem is exactly soluble, independently of whether the conformational changes are slow or fast. Nevertheless, the discussion may conveniently be organized by treating the limiting cases of the slow and fast conformational changes first. The former of these limits is characterized, as in NMR, by the inequality $(\omega_1 - \omega_2)/\lambda_1 \gg 1$ and the latter by the inequality $(\omega_1 - \omega_2)/\lambda_1 \ll 1$, where $\omega_1 = \overline{\omega}(x_1) + iT_2^{-1}(x_1)$ and $\omega_2 = \overline{\omega}(x_2) + iT_2^{-1}(x_2)$. The subsequent discussion is limited to the case of an asymmetric double minimum potential. The symmetric systems are less representative in this context and will not be treated here; see however Ref.[19].

b) ISOTROPIC RAMAN SPECTRA. If the conformational dynamics is slow, $(\omega_1 - \omega_2)/\lambda_1 \gg 1$, the correlation function $G(t)$ of Eqn(3.6) contains four modes. Two of them show up vibrational and chemical relaxation effects, and the two others the effects of pre-reactive and post-reactive oscillations of the two conformers around the equilibrium configurations. One finds :

$$
\begin{aligned}
G(t) &= \frac{\hbar}{2\omega_0} e^{i\omega_0 t} \left\{ \xi_1 \alpha_1^2 exp\left[\left(i\overline{\omega}_1 - \frac{1}{T_{21}} - \xi_2\lambda_1 \right) t \right] + \xi_2 \alpha_2^2 exp\left[\left(i\overline{\omega}_2 - \frac{1}{T_{22}} - \xi_1\lambda_1 \right) t \right] \right. \\
&+ \left. \frac{\xi_1 \alpha_1'^2}{\beta f_1} exp\left[\left(i\overline{\omega}_1 - \frac{1}{T_{21}} - \lambda_2 \right) t \right] + \frac{\xi_2 \alpha_2'^2}{\beta f_2} exp\left[\left(i\overline{\omega}_1 - \frac{1}{T_{22}} - \lambda_3 \right) t \right] \right\}
\end{aligned}
$$

$$(4.1)$$

where ξ_1 and ξ_2 indicate the fraction of conformers 1 and 2 in the mixture, $\alpha_1 = \alpha(x_1)$, $\alpha_2 = \alpha(x_2)$, $\alpha_1' = \left(\frac{d\alpha}{dx}\right)_1$, $\alpha_2' = \left(\frac{d\alpha}{dx}\right)_2$, $\lambda_2 \sim D\beta f_1$ and $\lambda_3 \sim D\beta f_2$. The corresponding Raman spectrum is illustrated on (Fig.3). It consists of a narrow doublet and a broad

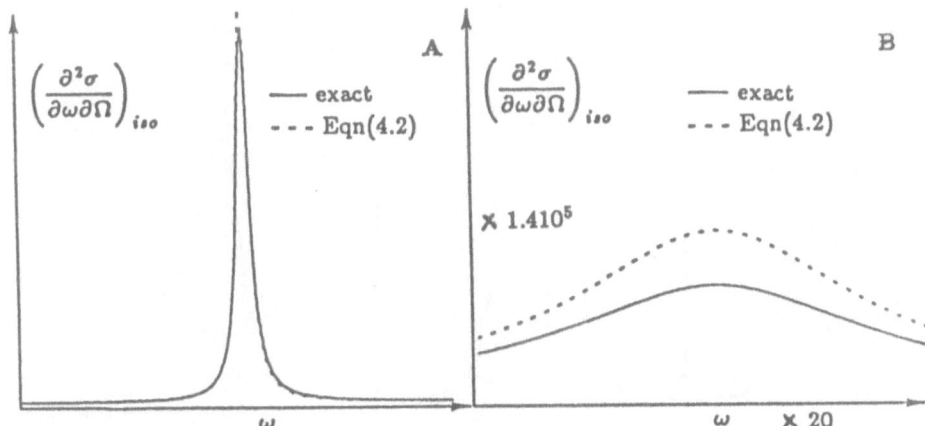

Figure 4: Isotropic Raman spectra of chemically reactive liquids in the case of fast conformation changes . Note the difference of the frequency and intensity scale in Fig.4A and Fig.4B.

wing. The two components of the former exibit, as expected, the lifetime broadening due to the chemical reaction. The reaction rates in the forward and the backward directions are $k_{12} = \xi_2\lambda_1$ and $k_{21} = \xi_1\lambda_1$. The wing is very similar to those generated by the collision induced processes. This similarity is not accidental; the site oscillations are involved in both these phenomena.

If the conformational dynamics is fast, $(\omega_1 - \omega_2)/\lambda_1 \ll 1$, the correlation function takes a very different form. The first mode exibits only vibrational relaxation effects, the second mode shows up both chemical and vibrational relaxation effects, whereas the third and the fourth manifest the presence of pre-reactive and post-reactive oscillations. One has

$$
\begin{aligned}
G(t) = \; & \frac{\hbar}{2\omega_0} e^{i\omega_0 t} \left\{ (\xi_1\alpha_1 + \xi_2\alpha_2)^2 exp\left[\left(i(\xi_1\overline{\omega}_1 + \xi_2\overline{\omega}_2) - \left(\frac{\xi_1}{T_{21}} + \frac{\xi_2}{T_{22}} \right) \right) t \right] \right. \\
& + \; \xi_1\xi_2(\alpha_1 - \alpha_2)^2 exp\left[\left(i(\xi_2\overline{\omega}_1 + \xi_1\overline{\omega}_2) - \left(\frac{\xi_2}{T_{21}} + \frac{\xi_1}{T_{22}} \right) - \lambda_1 \right) t \right] \\
& + \; \frac{\xi_1\alpha_1'^2}{\beta f_1} exp\left[\left(i\overline{\omega}_1 - \frac{1}{T_{21}} - \lambda_2 \right) t \right] + \frac{\xi_2\alpha_2'^2}{\beta f_2} exp\left[\left(i\overline{\omega}_1 - \frac{1}{T_{22}} - \lambda_3 \right) t \right] \left. \right\}
\end{aligned}
\tag{4.2}
$$

The corresponding Raman sectrum is illustrated on Fig. 4. It consists of a simple narrow band, rather than of a doublet, and of a broad wing. This difference of the spectral behaviour is entirely expected and is due, as in NMR, to the motional narrowing. Nothing special can be said about the wing, the pre-reactive and post-reactive motions producing similar spectral densities here as in the slow reaction case.

If the rate of conformational changes is neither large nor small, $(\omega_1 - \omega_2/)\lambda_1 \sim 1$, the spectral behaviour is intermediate between those of the two limits. The theory shows how the doublet transforms into a single band when the temperature is gradually increased

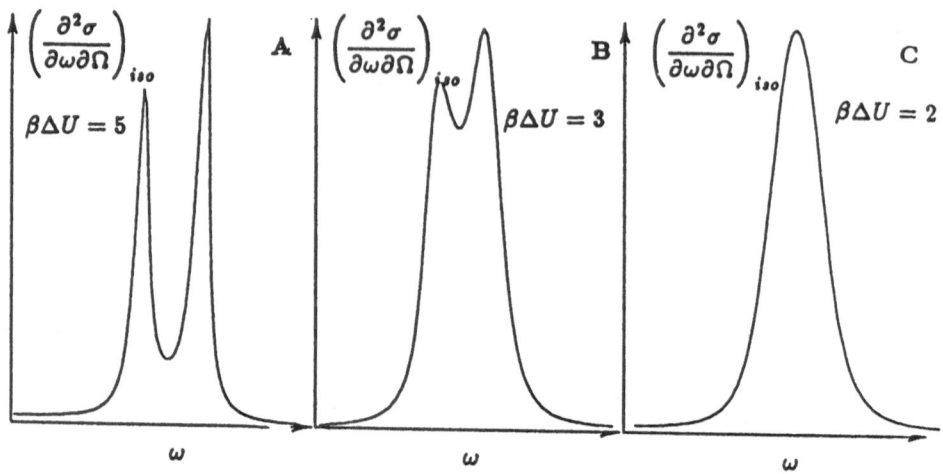

Figure 5: Collapse of the Raman doublet with temperature. The profiles were determined by solving the Fokker-Planck equation exactly.

(Fig.5). Although expected, this result proves that collapsing of a multiplet in the motional narrowing conditions is not a spurious result of the jump theories, a statement which was occasionally contested.

c) DISCUSSION. The present paper may be closed by briefly discussing the possibilities of the experimental observation of pre-reactive and post-reactive motions. As the number of methods which permit this sort of study is more than limited, the evaluation of the Raman spectroscopy in this case is of a considerable interest.

It should first be emphasized that, contrary to a superfici l impression, some effects of pre-reactive and post-reactive motions are quite large, although they are necessarily recognized as such. The most important, and often easily observable, effect is the reduction of the vibrational relaxation times T_{21}, T_{22} through the n, x coupling. In fact, the potential energy $U(n, x, X)$ of Eqn(3.1), responsible for the vibrational relaxation, depends not only on n, X as in the case of chemically inert liquids, but also on the reaction coordinate x; the vibrational coupling is enhanced correspondingly. In practice, this effect can be studied by measuring the half-width of the two components of the Raman doublet for various molecules of a homologous series. The calculations values of T_{21}, T_{22} may then be interpreted in terms of the available chemical information. The anomalous width of the ν OH bands of hydrogen bonded liquids is a particularly spectacular exemple of this kind. However, as mentioned earlier, it is generally not analysed in this way.

The detection of the pre-reactive and post-reactive modes and of the corresponding spectral densities in the wings is a much more difficult task. Considering the overall similarity between this process and those involved in the collision-induced scattering, the spectroscopic techniques in this latter field may be envisaged. However, the accessible spectral region in which the wings can be explored is limited if the number of the degrees of freedom of the reacting molecule is moderately large, or large. This is expected to be the case in many systems of practical importance. Nevertheless, whatever the solution of these problems may be, facing them is the price to pay to study the pre-reactive and post-reactive motions by the help of the Raman spectroscopy.

References

[1] M. Dantus, M. Rosker, A.H. Zewail, J. Chem. Phys. (1987), **87**, 2395.

[2] D. Cavagnat and J. Lascombe J. Chem. Phys. (1982), **76**, 4336.

[3] R.A. MacPhail, R.G. Snyder, and H.L. Strauss, J. Chem. Phys. (1982), **77**, 1118.

[4] J. Lascombe, D. Cavagnat, J.C. Lassègues, C. Rafilipomanana, in *Symmetries and Properties of Non-Rigid Molecules: A Comprehensive Study*, edited by J. Maruani and J. Serre (Elsevier, Amsterdam, 1983), p. 237.

[5] B. Cohen and S. Weiss, J. Phys. Chem. (1984), **88**, 3974.

[6] R. J. Sension and H. L. Strauss, J. Chem. Phys. (1987), **86**, 6665.

[7] W. Richter, D. Schiel, and W. Wöger, Mol. Phys. (1987), **60**, 691.

[8] M.M. Kreevoy and C.A. Mead, J. Am. Chem. Soc. (1962), **84**, 4596.

[9] A.K. Covington, J.G. Freeman, and L.Wynne-Jones, Discuss. Faraday Soc. (1965), **39**, 172.

[10] H. Chen and D.E. Irish, J. Phys. Chem. (1971), **75**, 2672.

[11] J-Cl. Lassègues and J. Devaure, in *Protons and Ions Involved in Fast Dynamic Phenomena*, edited by P. Laszlo (Elsevier, Amsterdam, 1978), p.157.

[12] M. Besnard, N. del Campo and J. Lascombe, in *Chemical Reactivity in Liquids. Fundamental Aspects*, edited by M. Moreau and P. Turq (Plenum Press, NewYork, 1988), p.33.

[13] R.A. MacPhail and H.L. Strauss, J. Chem. Phys. (1985), **82**, 1156.

[14] S. Bratos, G. Tarjus, and P. Viot, J. Chem. Phys. (1986), **85**, 803.

[15] P. Viot, G. Tarjus and S. Bratos, J. Mol. Liquids (1987), **36**, 185.

[16] S. Bratos and E. Maréchal Phys. Rev. A (1971), **4**, 1078.

[17] Risken, *The Fokker-Planck Equation*, (Springer-Verlag, Berlin, 1984).

[18] W. G. Johnson, D. W. Oxtoby, J. Chem. Phys. (1987), **87**, 781.

[19] P. Viot, G. Tarjus, D. Borgis, and S. Bratos, to be published.

[20] H.A. Kramers Physica (1940), **7**, 284.

STRUCTURES AND ENERGETICS OF ION-SOLVENT MICROCLUSTERS (n = 1,..,8) ;
Cl⁻, Br⁻ AND I⁻ WITH H₂O, CH₃OH, CH₃CN AND (CH₃)₂CO

B. Guillot, Y. Guissani, D. Borgis and S. Bratos
Laboratoire de Physique Théorique des Liquides ,
Université Pierre et Marie Curie, 4, Place Jussieu,
75252 Paris Cedex 05, France

ABSTRACT

The gas phase clustering reactions of chloride, bromide and iodide ion with water, methanol, acetonitrile and acetone molecules (n = 1,..,8) are investigated by Monte Carlo simulation. Using semi-empirical potential functions which include polarization effects, we obtain good agreement with experimental gas phase ion solvation enthalpies. Moreover, the coordination number, the geometry of the first shell and their relation with ionic radius are discussed in detail.

1. INTRODUCTION

The gas phase clustering reactions of ionic species with protic and aprotic solvent molecules play a key role in many fields such as chemistry of planetary atmospheres, chemistry of flames and of related combustion processes, radiolysis and nucleation phenomena, or ionic solvation in the liquid phase. The advent of new experimental techniques, e.g. the ion cyclotron resonance method, has prompted experimentalists to investigate in a systematic way the thermodynamics of the clustering reactions : $I.(S)_{n-1} + S \rightleftharpoons I.(S)_n$ (for reviews see [1,2]). From a theoretical point of view, the use of ab initio techniques permits to investigate the stability of ionic clusters at zero temperature by specifying their structure and energetics (for a review see [3]). Thermal motions can be taken into account by performing a Monte Carlo or a molecular dynamics simulation of these clusters provided realistic interaction potentials are available. Moreover, these computer simulations may yield the cluster structure by calculating the radial distribution functions, the coordination number of the central ion etc.. The main goal of all these calculations is to elucidate the role of the solvent in the ionic solvation process and to understand how the solvent modifies chemical reactivity in solution.

In the following, we present a study by Monte Carlo simulation of the gas phase solvation of halide ions (Br⁻, Cl⁻, I⁻) by protic (H₂O,

* Unité Associée au CNRS.

Th. Dorfmüller (ed.), Reactive and Flexible Molecules in Liquids, 47–59.
© 1989 by Kluwer Academic Publishers.

CH_3OH) and aprotic (CH_3CN, $(CH_3)_2CO$) solvent molecules. This choice was suggested by the great number of experimental data [4-13] on the gas phase clustering reactions,

$X^-.(S)_{n-1} + S \rightleftharpoons X^-.(S)_n$, where $X^- = Cl^-$, Br^-, I^- and $S = H_2O$, CH_3OH, CH_3CN, $(CH_3)_2CO$, and by the importance the solvated halide ions play in thermochemistry (see [14]). In section 2 we present the ion-solvent and solvent-solvent potential functions. The Monte Carlo algorithm is briefly described as is also the sampling method used to investigate the lowest energy configurations. The results are presented in section 3. The enthalpies of formation of the complexes are given first. Next, the structure of ionic complexes, the geometry and the coordination number of the ion are discussed in detail.

2. COMPUTATIONAL DETAILS

2.1. Potential functions

A prerequisite for a successful Monte Carlo simulation of ionic clusters is the knowledge of the intermolecular forces acting between molecular entities. As far as hydrated halide ion complexes are concerned, a number of potential functions are available in the literature [15-25]. They are either empirical [16,20-24] or result from a fit of Hartree Fock SCF calculations [15,17-19,25]. Most of these potentials are pairwise additive, but recently some attempts were made to take into account non additive contributions such as polarization effects [21-24] and three-body repulsion [21,22]. For halide ions interacting with CH_3OH, CH_3CN and $(CH_3)_2CO$ the situation is rather different since, to our knowledge, no systematic investigation of the intermolecular potential has been published, although ab initio calculations have been performed on some of these complexes [26-28].

For the solvent-solvent interactions we have used the empirical potential functions already available in the literature, namely, the SPC model of Berendsen et al. [29] for water, the model of Haughney et al. [30] for methanol, the three site potential of Edwards et al. [31] for acetonitrile and a four site model potential for acetone derived from the one of Evans et al.[32]. The molecules are considered as rigid bodies, and the methyl group is assumed spherical, the force field emanating from the carbon atom. The interactions are pairwise additive and obey the following analytic form,

$$U_{S/S} = \sum_{i<j} u_{ij} = \sum_{i<j} \sum_{\alpha,\beta} (4\,\epsilon_{\alpha\beta} \left[(\frac{\sigma_{\alpha\beta}}{r_{i\alpha j\beta}})^{12} - (\frac{\sigma_{\alpha\beta}}{r_{i\alpha j\beta}})^6 \right] + \frac{q_\alpha q_\beta}{r_{i\alpha j\beta}}) \tag{1}$$

where u_{ij} expresses the intermolecular potential between the pair (ij), α, and β running over all sites of the molecules (3 for H_2O, CH_3OH and CH_3CN, and 4 for $(CH_3)_2CO$). The Lennard-Jones parameters $\sigma_{\alpha\beta}$ and $\epsilon_{\alpha\beta}$ follow the usual Lorentz-Berthelot rules : $\sigma_{\alpha\beta} = (\sigma_{\alpha\alpha}+\sigma_{\beta\beta})/2$, $\epsilon_{\alpha\beta} = \sqrt{\epsilon_{\alpha\alpha}\,\epsilon_{\beta\beta}}$. The Lennard-Jones parameters $\sigma_{\alpha\alpha}, \epsilon_{\alpha\alpha}$ and the fractionary charges q_α are adjusted to yield, via MC ou MD simulations, the best agreement with experimental thermodynamical and structural properties of the corresponding liquids (Table 1). However, the polarization effects are not taken explicitly into account and the effective dipole moments assigned to water and methanol molecules $\mu(H_2O) = 2.27$ D and $\mu(CH_3OH) = 2.37$ D, are larger than in the gas phase $(\mu(H_2O) = 1.83$ D, $\mu(CH_3OH) = 1.7$ D) ; this compensates to some extent the absence of polarizability. However, although the effect of these enhanced dipole moments on the energetics of small ionic clusters is rather small, it may be quite large when the solvation in the liquid phase is concerned.

Table 1. Intermolecular potential parameters. The Lennard-Jones parameters associated with cross interactions are deduced from the Lorentz-Berthelot rules. The indices 1,2,3 have the following meaning :

1 $\diagdown$ 2 for H $\diagdown$ H and H $\diagdown$ CH$_3$; 1-2-3 for CH$_3$-C≡N ; 1=2$<^3$ for O=C$<^{CH_3}_{CH_3}$.
    3   O   O   3

	H₂O (a)	CH₃OH (b)	CH₃CN (c)	(CH₃)₂CO (d)
q_1 (e)	0.410	0.431	0.269	-0.512
q_2 (e)	0.410	0.297	0.169	0.512
q_3 (e)	-0.820	-0.728	-0.398	0.000
σ_1 (A)	—	—	3.500	3.08
σ_2 (A)	—	3.861	3.400	3.20
σ_3 (A)	3.165	3.083	3.300	3.86
ϵ_1 (K)	0.0	0.0	137.5	87.9
ϵ_2 (K)	0.0	91.2	50.0	50.0
ϵ_3 (K)	78.22	87.9	50.0	91.2
α(A³)	1.444	3.30	4.46	6.42

a)Ref.[29]; b)Ref.[30]; c)Ref.[31]; d)Ref.[32]

Comparatively, the presence of a full charge on halide ions requires an explicit description of the polarization effects in ion-solvent potential interactions. In a recent study of the pH of water [33], we proposed a model potential to describe the strong-hydrogen bonds between H_3O^+, OH^- and water molecules. A similar expression is used in the present study to represent the interaction potential between a spherical ion I and a cluster of n solvent molecules, each of them consisting of m centers of forces,

$$u_{I/S} = \sum_{j=1}^{n} \sum_{\beta=1}^{m} (4 \epsilon_{I,j\beta} \left[(\frac{\sigma_{I,j\beta}}{r_{I,j\beta}})^{12} - (\frac{\sigma_{I,j\beta}}{r_{I,j\beta}})^{6} \right] - \frac{q_{j\beta}e}{r_{I,j\beta}})$$

$$- \frac{1}{2} \alpha_S \sum_{j=1}^{n} E_j^2 - \frac{1}{2} \alpha_I F^2 \qquad (2)$$

where $\sigma_{I,j\beta}$ and $\epsilon_{I,j\beta}$ are the Lennard-Jones parameters between the ion and the site β of the molecule j deduced from the Lorentz-Berthelot rules, and $q_{j\beta}$ is the fractionary charge at site β of the molecule j. α_I and α_S are the mean polarizabilities of the ion and the solvent molecules, respectively, E_j is the electric field emanating from the ion and measured at the center of mass of the molecule j, whereas F is the total electric field generated by the solvent molecules and polarizing the ion. This latter contribution to the potential energy (eq.2) is not pairwise additive. It is worthwhile to notice that in the case of the ions H_3O^+ and OH^-, the polarization effects are so important [33] that they are responsible, to a large extent, of the p_H value of pure water !

In order to determine the Lennard-Jones parameters relative to ions we use the following procedure. The parameters σ_{ion} and ϵ_{ion} are adjusted in such a way that the energies calculated by Monte Carlo simulation for the complexes $Cl^-.H_2O$, $Br^-.H_2O$ and $I^-.H_2O$ coincide with the corresponding experimental hydration enthalpies [4,8,13]. The ionic parameters are left unchanged in further calculations whatever the size of the cluster (n = 1,..,8) and the solvent molecules may be (see table 2). It is generally justified to compare experimental enthalpy and theoretical energy changes. This is due to a cancellation between quantum contributions to the vibrational energy and the PV term (see [17,22]).

Table 2. Ion-ion interaction potential parameters

	Cl^-	Br^-	I^-
$\sigma(A)$	4.6	4.8	5.4
$\epsilon(K)$	53.9	52.0	49.1
$\alpha(A^3)$	3.94	5.22	7.81

2.2. Monte Carlo simulations

The Monte Carlo method being a standard theoretical tool of the liquid state physics, it is not necessary to describe it in detail (for a review see [34]). However, the following points merit attention. The usual Metropolis algorithm is performed by moving all solvent molecules and by keeping the ion fixed, when creating a trial configuration. The use of multiparticle trial moves [35] is prefered because, in small systems like clusters, it has been shown that the convergence is accelerated while the cost in computer time diminishes when the acceptance ratio is kept around 0.5.

A peculiarity of ionic clusters is their low stochasticity due to a binding energy of the order of 10 kcal/mole per particle. Thus, the small translational and rotational displacements generated by the Metropolis algorithm do not produce a strong rearrangment of the molecules around the ion as it is the case in a large system composed of weakly bound particles (e.g. Van der Waals liquids). Hence it is crucial to choose the starting configuration as close as possible to the configuration of lowest energy to be sure to investigate thermal excitations around this minimum. If not, the system can be blocked in a metastable state of higher energy, or worse, can be unstable (see for example [24]). We will illustrate this point in the following. In practice, the random walk is restricted during the first thousands of steps to a small region which corresponds to energies very close to the lowest energy of the cluster. Next, this contraint is released and a phase of equiblibration is established during $n.10^4$ steps ($n = 1,..,8$), the average values being evaluated over the $n.10^4$ following steps. If a great number of different starting configurations is tested, it is possible to find the stable configuration of lowest energy for each cluster under investigation. In the present calculations, we estimate the error in the determination of the lowest energy to be less than 5%, the statistical uncertainties being much smaller (~ 1%).

3. RESULTS

3.1. Energetics

The energies of different clusters calculated by Monte Carlo at $T = 300K$ are collected in table 3 and are compared to the experimental enthalpies given in the literature [4-13]. It should be noticed that the dispersion of experimental enthalpies measured by different authors is sometimes larger than that corresponding to the standard deviation claimed by them ; this is particularly true for the first step solvation. For example, the usually quoted value of 13.1 kcal/mol for the (0,1) Cl hydration by Kebarle et al. [4] is in disagreement with the more recent value of 14.9 ± 0.2 kcal/mol given by Keesee and Castleman [8], a value reproduced by several recent measurements (see [13] and therein). Moreover, several Monte Carlo simulations performed at zero temperature indicate that the thermal motions at room temperature increase the cluster energy by roughly 1 kcal/mole per molecule. The vibrational quantum contributions to the cluster energy are thus expected to be of this order of magnitude. Taking into account all sources of inaccuracy, experimental and theoretical, the results of table 3 may be qualified as satisfactory. Three points merit attention. (i) For a given solvent (protic or aprotic) and for a fixed size of the ionic cluster ($n = 1,..,8$), the binding energies follow the hierarchy of ionic radii, i.e. $E_n(Cl) > E_n(Br) > E_n(I)$. (ii) The water and methanol molecules are more tightly bound to halide ions than acetonitrile and acetone molecules. (iii) The cumulative effect of the size of the ion and of intermolecular forces can be considerable ; for instance, the complexes $Cl .(H_2O)_n$ have a binding energy which is two times larger than that of $I .((CH_3)_2CO)_n$ for a given value of n.

Table 3. Comparison between calculated energies and experimental enthalpies (in kcal/mol) of $X^-.(S)_n$ microclusters at 300K. The experimental values are quoted a,b,c,d,e,f or g.

n	H_2O			CH_3OH			CH_3CN			$(CH_3)_2CO$		
	Cl^-	Br^-	I^-	Cl^-	Br^-	I^-	Cl^-	Br^-	I^-	Cl^-	Br^-	I^-
1	13.6 13.1a 14.7b	12.7 12.6a	10.2 10.2a 11.1c	13.4 17.4b 14.2d	12.3	9.4 11.3e	11.3 13.4f	10.9 12.9f	9.5 11.9f	9.2 12.4g	8.6	7.3
2	26.8 25.8a 27.7b	25.4 24.9a	20.4 20.0a 21.0c	26.7 31.5b 27.2d	24.2	18.8	21.8 25.6f	20.8 24.7f	18.3 22.4f	18.5 23.9g	17.2	14.2
3	40.7 37.5a 39.5b	38.4 36.4a	30.0 29.4a 30.3c	39.0 43.3b 39.5b	35.6	27.2	31.4 36.2f	30.0 34.7f	26.3 31.7f	27.8 34.3g	26.1	21.0
4	53.8 48.6a 50.1b	50.5 47.3a	41.9	51.1 53.8b 50.7d	45.9	37.1	40.4 42.4f	38.3 40.2f	33.6	37.0	34.0	27.1
5	66.4 59.6b	61.8	50.7	61.1 63.0b 61.2d	57.7	45.0	49.5	46.6	40.2	46.5	42.2	33.5
6	76.0 68.4b	74.1	62.4	72.1 71.9b	66.3	53.7	58.5	55.2	47.9	55.6	50.0	41.0
7	85.8 76.5b	84.9	73.8	82.2 79.9b	77.1	62.1	68.6	65.3	53.8	59.8	56.0	46.5
8	96.7	94.3	84.0	91.8 87.5b	84.4	70.2	79.5	74.2	61.6	65.0	60.7	52.0

a)Ref.[4]; b)Ref.[13]; c)Ref.[8]; d)Ref.[6]; e)Ref.[7]; f)Ref.[5]; g)Ref.[12].

Table 4. Energy components (kcal/mol) for $X^-.(S)_8$ microclusters at 300K. E_{CB+LJ} is the portion of total complex formation energy due to Lennard-Jones and coulombic terms, $E_{POL}(Sol.)$ is the polarization energy of the solvent molecules, $E_{POL}(ion)$ is the polarization energy of the ion and E_{TOT} is the total energy of the complex.

Solvent	ion	E_{TOT}	E_{CB+LJ}	E_{POL}(Solvent)	E_{POL}(ion)
H_2O	Cl^-	− 96.7	− 79.4	− 13.4	− 3.9
	Br^-	− 94.3	− 77.0	− 11.8	− 5.5
	I^-	− 84.0	− 71.6	− 7.5	− 4.8
CH_3OH	Cl^-	− 91.8	− 69.7	− 20.7	− 1.4
	Br^-	− 84.4	− 64.7	− 18.9	− 0.9
	I^-	− 70.2	− 56.5	− 12.4	− 1.3
CH_3CN	Cl^-	− 79.5	− 62.0	− 17.3	− 0.15
	Br^-	− 74.2	− 58.8	− 15.2	− 0.19
	I^-	− 61.6	− 51.5	− 9.8	− 0.33
$(CH_3)_2CO$	Cl^-	− 64.9	− 45.1	− 19.8	− 0.05
	Br^-	− 60.7	− 43.3	− 17.3	− 0.07
	I^-	− 52.0	− 39.0	− 12.9	− 0.08

Table 5. Mean occupation number in the first solvation shell for clusters $X^-.(S)_8$. The symbol ~ indicates that the first and the second solvation shell overlap each other.

Solvent	Cl^-	Br^-	I^-
H_2O	6.0	5.5	~ 4.5
CH_3OH	5.0	6.0	~ 5.0
CH_3CN	8.0	8.0	7.0
$(CH_3)_2CO$	6.0	6.0	6.5

The analysis of energy components (repulsion, coulombic, polarization) to the stability of ionic clusters show the following. (i) The stabilization of a cluster results essentially from the competition between repulsive forces and coulombic forces, irrespective of the nature of the solvent (protic or aprotic). (ii) The ion-hydrogen interaction exacerbates the coulombic energy in the case of protic solvents (H_2O, CH_3OH). (iii) The energy due to the polarization of the solvent molecules by the ion is far from being negligible, particularly for the highly polarizable aprotic solvents (CH_3CN, $(CH_3)_2CO$). (iv) The back-polarization of the ion by the solvent molecules gives a significant contribution only in the case of the solvation by protic solvents. These effects are illustrated in table 4. The points (i) and (iv) permit to understand why it is possible to describe with a good accuracy the energetics of the ionic clusters by using only pairwise additive potential functions. However, although the polarization of the solvent molecules by the ion can be mimicked by pairwise additive potentials, this is not possible for the back-polarization. In fact, the role of polarization effects is much more important in solution than in the gas phase. In particular, the entropic contributions which take place in bulk solvation can only be described quantitatively by introducing the polarizability of the molecules. This must be kept in mind when using purely electrostatic potentials.

3.2. Structure

The geometry of the complex formed by a chloride ion and one solvent molecule at zero temperature is shown in fig.1. In the case of a water molecule, the Cl^-...H-O bond is not quite linear ($\Phi = 16^o 3$). This deviation is smaller with methanol ($\Phi = 8^o 5$) because the methyl group has a smaller positive charge than hydrogen (see table 1). In the case of acetonitrile, the molecular axis is also the symmetry axis of the complex, the methyl group facing the ion. In turn, in the case of acetone the symmetry is broken ($\Phi = 143^o$), the ion lying out of the plane of the molecule ; in our model potential there is a competition between the coulombic attraction Cl^-...C^+-O$^-$ and the repulsion exerted by the electronic clouds of the methyl groups on the ion. Recent ab initio calculations [28] yield geometries very similar to what we find for the complexes $Cl^-.H_2O$, $Cl^-.CH_3OH$ and $Cl^-.CH_3CN$ while for $Cl^-.(CH_3)_2CO$ other ab initio calculations [9] suggest that the Cl^- is hydrogen bonded to one of the hydrogens of the methyl group. However, these calculations also emphasize that the latter structure is energetically equivalent to the one where Cl^- is in line with the dipole of CO, a structure compatible with our prediction. The complex is highly flexible at room temperature, and this property is well described by our model. As far as the complexes involving the bromide and iodide anions are concerned, the features described above are still present, but the ion-solvent distances are somewhat dilated.

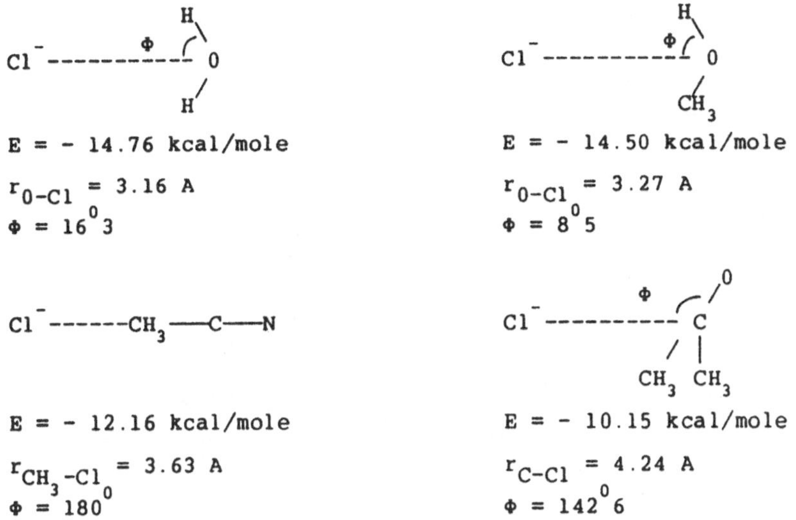

$$Cl^- \text{-----------} \Phi \quad C \begin{array}{c} H \\ \diagdown \\ O \\ \diagup \\ H \end{array}$$

E = - 14.76 kcal/mole

$r_{O-Cl} = 3.16$ A

$\Phi = 16^0 3$

$$Cl^- \text{-----------} \Phi \quad C \begin{array}{c} H \\ \diagdown \\ O \\ \diagup \\ CH_3 \end{array}$$

E = - 14.50 kcal/mole

$r_{O-Cl} = 3.27$ A

$\Phi = 8^0 5$

$$Cl^- \text{------} CH_3 \text{----} C \text{----} N$$

E = - 12.16 kcal/mole

$r_{CH_3-Cl} = 3.63$ A

$\Phi = 180^0$

$$Cl^- \text{-----------} C \begin{array}{c} \Phi \quad O \\ \diagup \\ \diagup \quad | \\ CH_3 \quad CH_3 \end{array}$$

E = - 10.15 kcal/mole

$r_{C-Cl} = 4.24$ A

$\Phi = 142^0 6$

Fig. 1   Calculated energy and geometry of Cl^-.S complexes at
zero temperature.

An important question about the structure of ionic clusters
concerns the ordering of the first solvation shell. The latter can be
described by the radial distribution functions $g_{ion-sol}$ and $g_{sol-sol}$ of
the cluster, this infering the coordination number of the central ion.
Some of these radial distribution functions and coordination numbers
are presented in figs. 2-4 for the clusters $Cl^-.(S)_8$, whereas the
average number of neighbours in the first shell around each halide ions
is listed in table 5. For hydrates, the first shell contains 6 water
molecules around Cl^-, 5.5 around Br^- and 4.5 around I^- respectively,
the distinction between the first and the second shell being unclear
for I^- (see fig.6). Moreover, the water molecules form a small hydrogen
bond network which tends to stabilize the cluster, and most of the
molecules reside on one side of the ion (see also [24]). However, the
occupation number in the first shell tends to be somewhat higher in the
present study than in other investigations [22,24]. Several causes can
explain this variation : (i) the potential interactions which are
different, (ii) the size of the simulated clusters which may be too
small (only n = 4 in [22] and n = 6 in [24]), (iii) the choice of the
starting configuration which may not permit to reach the configuration
of lowest energy for the cluster under investigation. We believe that
point (iii) is the most severe source of error. To illustrate this
point, we present in fig.5 the results of two simulation runs of the
cluster $Cl^-.(H_2O)_4$ starting from two different initial configurations.
In the final state of highest energy (E = -50.3 kcal/mol) the first
shell contains 3 molecules and the fourth molecule oscillates between
the first and the second solvation shell (<n> ~ 3.5 in the first
shell). On the contrary, in the state of lowest energy (-53.8 kcal/mol)

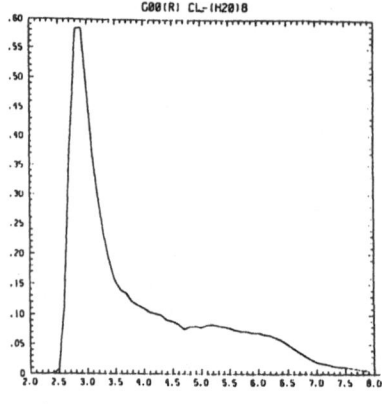

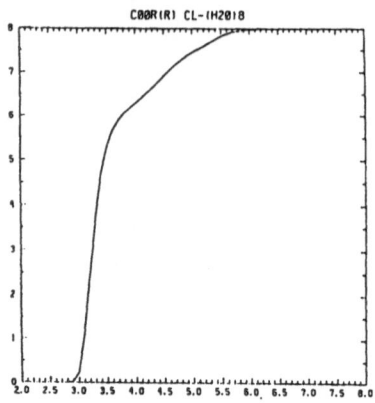

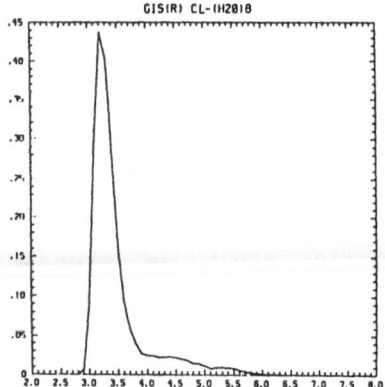

Fig. 2 Computed radial distribution functions $G_{is}(R)$ and $G_{oo}(R)$
between Cl⁻ and the oxygens of water molecules, and between
oxygen-oxygen respectively, for the cluster Cl⁻.$(H_2O)_8$. The
coordination number COOR(R) is deduced from $G_{is}(R)$ by
integration.

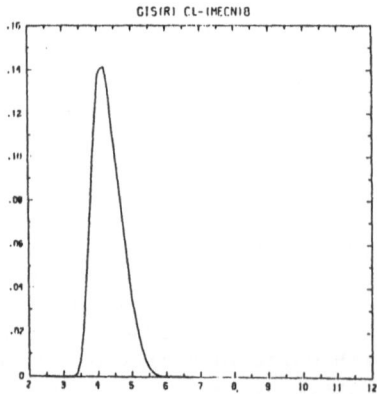

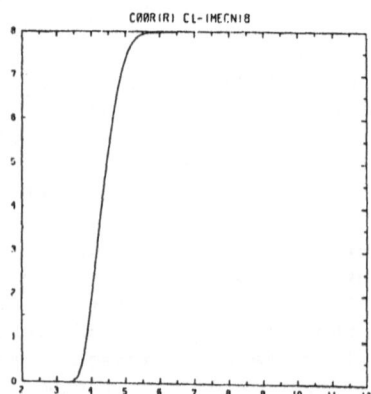

Fig. 3 ion-center of mass radial distribution function and
coordination number of Cl⁻.$(CH_3CN)_8$.

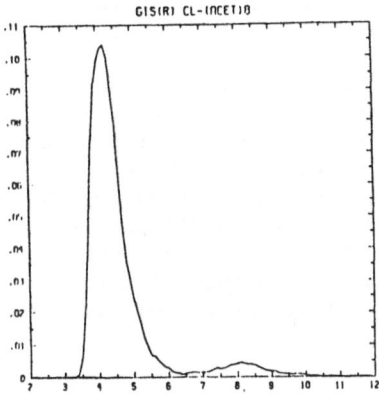
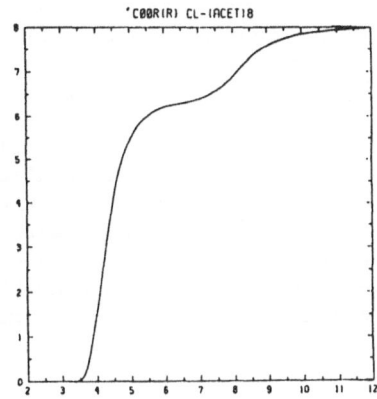

Fig. 4 ion-center of mass radial distribution function and
coordination number of $Cl^-.((CH_3)_2CO)_8$.

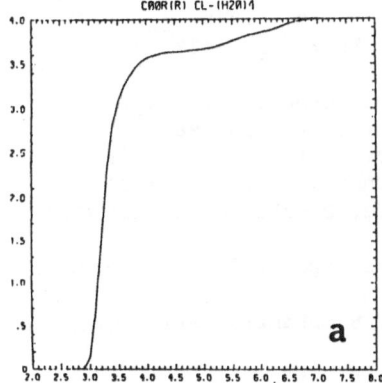
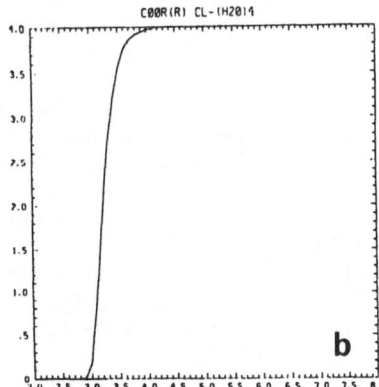

Fig. 5 (a) Coordination number for $Cl^-.(H_2O)_4$ in the metastable
state E = − 50.3 kcal/mol (b) coordination number for
$Cl^-.(H_2O)_4$ in the state of lowest energy E = − 53.8 kcal/mol.

58

the four molecules all belong to the first shell. It is interesting to notice in this context that Kollman et al. [22] find for the same cluster a 3 + 1 structure.

The structure of the clusters composed of methanol molecules is similar to a large extent to that of hydrates. In particular, the hydrogen bonds between the methanol molecules tend to stabilize the cluster, but this effect is smaller than in the case of water. For acetonitrile, the interactions between solvent molecules being mostly repulsive, all the eight molecules stay closely around Cl⁻ and Br⁻, whereas with I⁻ one of these molecules is loosely bound to the central ion. For acetone, the size of the methyl groups and the thermal motions restrict the number of molecules in the first shell to approximately 6.

ACKNOWLEDGEMENTS

The computer simulation have been performed with the support of the scientific committee of the CCVR (Palaiseau, France).

REFERENCES
[1] P.Kebarle, Ann. Rev. Phys. Chem, 1977, 28, 445
[2] A.W.Castleman Jr., R.G.Keesee, Chem. Rev., 1986, 86, 589
[3] A.Karpfen, P.Schuster, in The Chemical Physics of Solvation, Part A : Theory of Solvation, eds. R.R.Dogonadze, E.Kalman, A.A.Kornyshev, J.Ulstrup (Elsevier, Amsterdam), 1985, p.298 and references therein
[4] M.Arshadi, R.Yamdagni, P.Kebarle, J. Phys. Chem., 1970, 74, 1475
[5] R.Yamdagni, P.Kebarle, J. Am. Chem. Soc., 1971, 93, 7139 ; ibid, 1972, 94, 2940
[6] R.Yamdagni, J.D.Payzant, P.Kebarle, Can. J. Chem., 1973, 51, 2507
[7] G.Caldwell, P.Kebarle, J. Ann. Chem. Soc., 1984, 106, 967
[8] R.G.Keesee, A.W.Castleman Jr., Chem. Phys. Lett., 1980, 74, 139
[9] M.A.French, S.Ikuta, P.Kebarle, Can. J. Chem., 1982, 60, 1907
[10] T.F.Magnera, G.Caldwell, J.Sunner, S.Ikuta, P.Kebarle, J. Am. Chem. Soc., 1984, 106, 6140
[11] K.Hiraoka, K.Morise, T.Shoda, Int. J. Mass. Spectr. and Ion Process, 1985, 67, 11 ; ibid, 1986, 68, 99
[12] K.Hiraoka, H.Takimoto, K.Morise, T.Shoda, S.Nakamura, Bull. Chem. Soc. Japan, 1986, 59, 2247
[13] K.Hiraoka, S.Mizuse, Chem. Phys., 1987, 118, 457
[14] C.E. Klots, J. Phys. Chem., 1981, 85, 3585
[15] H.Kistenmacher, H.Popkie, E.Clementi, J. Chem. Phys., 1973, 59, 5842 ; ibid, 1974, 61, 799
[16] C.L.Briant, J.J.Burton, J. Chem. Phys., 1974, 60, 2849 ; 1976, 64, 2888
[17] M.R.Mruzik, F.F.Abraham, D.E.Schreiber, J. Chem. Phys., 1976, 64, 481
[18] J.Chandrasekhar, D.C.Spellmeyer, W.L.Jorgensen, J. Am. Chem. Soc., 1984, 106, 903
[19] R.W.Impey, P.A.Madden, I.R.Mc Donald, J. Phys. Chem., 1983, 87, 5071
[20] K.Heinzinger, Physica, 1985, 131B, 196
[21] T.P.Lybrand, P.A.Kollman, J. Chem. Phys., 1985, 83, 2923
[22] P.Cieplak, T.P.Lybrand, P.A.Kollman, J. Chem. Phys., 1987, 86, 6398
[23] S.S. Sung and P.C. Jordan, J. Chem. Phys., 1986, 85, 4045
[24] S. Lin and P.C. Jordan, J. chem. Phys., 1988, 89, 7492

[25] W.F. Van Gunsteren, Gromos Reference Mannual, Biomos B.V. (Nijenborgh 16, Groningen, the Netherlands, 1987)

[26] S.Yamabe, K.Hirao, Chem. Phys. Lett., 1981, 84, 598

[27] S.Yamabe, N.Ihira, Chem. Phys. Lett., 1982, 92, 172

[28] S.Yamabe, Y.Furumiya, K.Hiraoka, K.Morise, Chem. Phys. Lett., 1986, 131, 261

[29] H.J.C.Berendsen, J.P.M.Postma, W.G.Von Gunsteren, J.Hermans, in "Intermolecular Forces", B.Pullman Editor, Reidel, Dordrecht, 1981, p.331

[30] M.Haughney, M.Ferrario, I.R.Mc Donald, Mol. Phys., 1986, 58, 849

[31] D.M.F.Edwards, P.A.Madden, I.R.Mc Donald, Mol. Phys., 1984, 51, 1141 We have slightly modified the Lennard-Jones parameters $\sigma(CH_3)$ and $\epsilon(CH_3)$ (see table I)

[32] G.J.Evans, M.W.Evans, J. Chem. Soc. Faraday Trans. II, 1983, 79, 153 In order to obtain a better agreement with the thermodynamic properties of the liquid, we have modified the parameters of this potential.

[33] Y.Guissani, B.Guillot, S.Bratos, J. Chem. Phys., 1988, 88, 5850

[34] Monte Carlo Methods in Statistical Physics, K.Binder Ed., Springer Verlag, Berlin, 1979

[35] W.Chapman, N.Quirke, Physica, 1985, 131B, 34

INFRARED, FAR-INFRARED AND RAMAN INVESTIGATIONS OF THE MOLECULAR DYNAMICS AND INTERACTIONS OF NON-RIGID MOLECULES IN LIQUIDS

J. Yarwood*, A. Whitley*, D.G. Gardiner† and
M.P. Dare-Edwardsφ
* Department of Chemistry, University of Durham, South Road,
  Durham, DH1 3LE, UK.
† Department of Chemistry, Newcastle upon Tyne Polytechnic,
  Ellison Building, Ellison Place, Newcastle upon Tyne, UK.
φ Thornton Research Centre, Shell Research Ltd., P.O. Box 1,
  Chester, CH1 3SH, UK.

ABSTRACT. Non-rigidity which arises from internal rotational or librational flexibility adds complications in the band shape analysis normally carried out for rigid molecules. In particular, additional band broadening may occur and it is possible that new features may arise in the spectrum if the different processes are separated in time by more than one order of magnitude. Such are the difficulties envisaged in a full analysis that only a few studies have been attempted or published so far. However, such is the importance of molecules, for example with long alkyl chains in biological and industrially important fluids, that future studies will necessarily include attempts to disentangle the intermolecular and intramolecular dynamics. Some very encouraging reports have recently been made. We present a review of the work carried out on alkanes, alkynes, ketones, esters and small rings in an attempt to point to directions of approach which may be viable. We then outline our recent work on the model lubricant molecule 2-ethylhexyl benzoate (EHB) which has been tackled using infrared, far-infrared and Raman methods as well as by ^{13}C NMR. We show that direct evidence for internal rotation/libration is available from properly chosen experiments. We further show how the intermolecular forces are modulated by such motions and how they depend in a unique way on the surrounding environment

1. INTRODUCTION

Although spectroscopic techniques may now used almost routinely (but carefully!) to determine molecular relaxation times for small rigid molecules [1-4], very little work of this type has been reported for non-rigid systems. This is not particularly surprising since most of the methods of analysis and interpretation are designed to deal with the (simpler) rigid molecule case. In particular, most models for molecular reorientational motion in dense fluids [1,3-6] assume (either

Th. Dorfmüller (ed.), Reactive and Flexible Molecules in Liquids, 61–82.
© *1989 by Kluwer Academic Publishers.*

explicitly or implicitly) a rigid molecular framework. Nevertheless, non-rigid molecules, especially those containing medium to long alkyl chains (say C_5 to C_{20}) are very important to a wide body of practically and technically-minded scientists. Such molecules form the bulk of materials found, for example in biological systems (membranes, lipids, fats, etc.) [7] and they find important uses in industry as fuels and lubricants, etc [8]. In short, they are the 'functional' fluids on which we are all dependent. Their 'functional' behaviour must depend ultimately on their microscopic interactions and dynamics, and so it is important that we use all available techniques to elucidate their fundamental relaxation processes. Some work has been already started on these more complicated systems (section 2) and it is the purpose of this paper to outline the methods of approach taken and to assess their success so far. In section 2 we summarise the work already published and in section 3 we describe some of our recent work on a model lubricant, EHB.

2. METHODS OF APPROACH

Non-rigid molecules are usually more complicated than non-rigid ones and so a more complex spectral response is expected. The simplifications based upon molecular symmetry [9] may be lost and, of course, internal motions lead to an increase in the number of dynamic processes which can occur.

Different relaxation processes may be obviously separable if they occur on different time scales. Even in simple liquids this has been achieved for CS_2 [10], while for polymers band broadening may be eliminated on a psec timescale [11]. However, in most flexible molecules, when rapid (psec) internal rotation is expected [12,13], it is likely that the different effects will be, to a greater or lesser extent, 'overlapping'. Judicious choice of technique combination will be of crucial importance if useful information is to be obtained. Vibrational and rotational spectroscopies, ranging from a few GHz (0.1 cm^{-1}) to 5000 cm^{-1}, have already been used to tackle problems associated with non-rigid systems. For the purposes of this short review the work has been divided into effects on (i) vibrational band shapes and frequencies, (ii) reorientation processes in flexible systems.

(i) Vibrational band shapes and frequencies

Several studies have been made on the influence of external environment on the conformational populations [14-19] and conformational dynamics [20-26] of molecules containing long alkyl chains have been made (see Zerbi, Chapter   and Moro, Chapter   , Bellemans, Chapter   ). Furthermore, the rates of vibrational relaxation of different conformers have been determined in a few cases [22-26]. Wunder et al [22] have compared the band-width behaviour of suitably deuterated alkanes (propane, butane, pentane and nonadecane) as a function of chain length. They have concluded that the reorientational contribution to the Raman anisotropic band-widths for short chains are dominated by end over end rotation. However, for nonadecane (fig. 1)

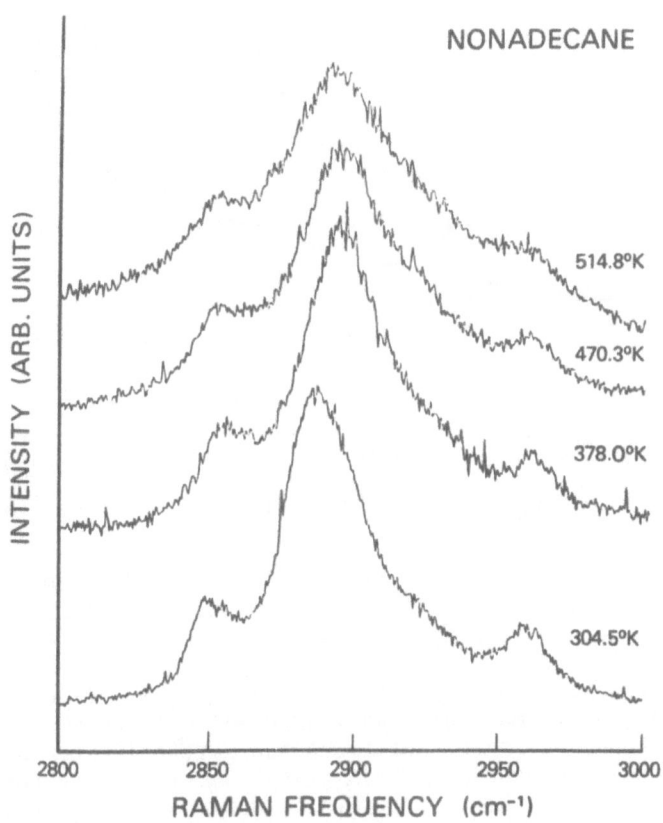

Fig. 1. I_{VH} Raman bands caused by the $\nu_{as}(CH_2)$, d⁻ modes of nonadecane at four temperatures in the melt. (Reproduced, by permission, from J. Chem. Phys., 85, 3837 (1986).)

broadening of the d⁻, antisymmetric CH_2 stretching bands (which is independent of chain length and therefore of the number of possible conformations, fig. 2) is thought to be due to coupling of backbone torsional modes. These internal hindered rotational modes become more highly populated as the temperature is increased. Because the d⁻ modes give rise to depolarised bands and because of the lack of symmetry (for long chains) and band overlap, a number of simplifying assumptions have to be made. These tend to cast uncertainty on the detailed interpretation. However, there is no doubt that the vibrational contributions to the d⁻ bands increase (at the expense of rotational broadening) as the chain length increases, and that -CH_2- group

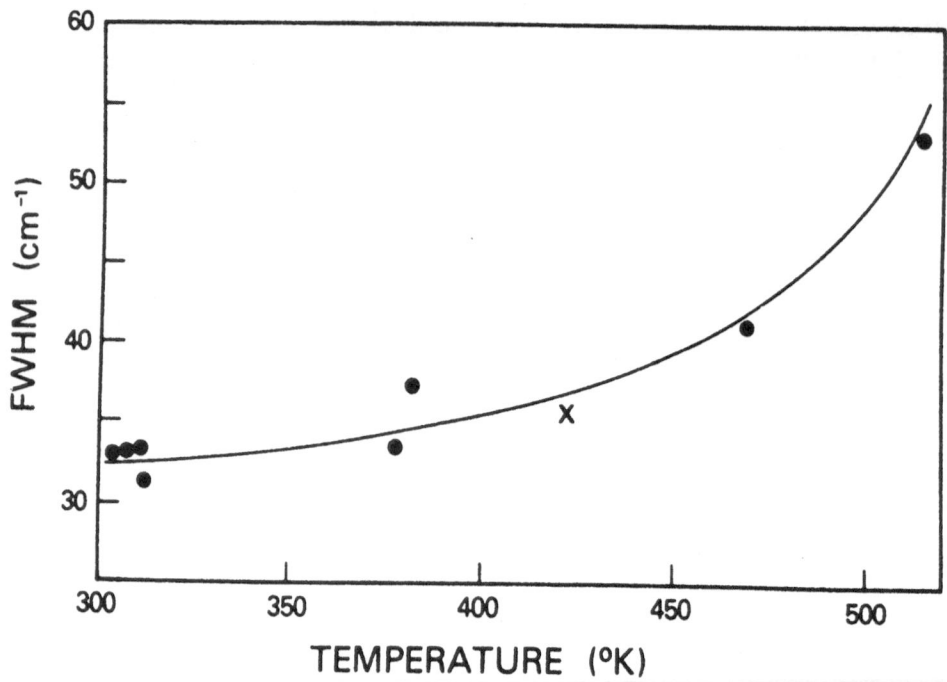

Fig. 2. Comparison of Raman band half-widths (FWHM) of the d⁻ bands of nonadecane (•) and polyethylene (x). (Reproduced, by permission, from J. Chem. Phys., **85**, 3838 (1986).)

'flexing' or torsional motion is important for the longer chains.

For rather simple molecules with internal degrees of freedom it is rather easy to demonstrate that trans and gauche isomers have different degrees of vibrational relaxation. Schwartz and co-workers [23,24] have shown that this is so for 1,2 dibromoethane. Rates of relaxation are interpreted first of all via the isolated binary collision (IBC) model [27], which turns out to be too crude a measure of the interactions of these molecules in the liquid (fig. 3). However, the environmental modulation times τ_m, obtained using the Kubo lineshape formalism [28] (eqn. 1) do show temperature trends compatible

$$C_{iso}(t) = \exp\{-M_2 v[\tau_m t + \tau_m^2 [\exp(-t/\tau_m)-1]]\} \tag{1}$$

with those calculated using the Enskog model [28] based on density (ρ) and hard sphere diameter (σ) - see fig. 3. The model predicts that,

$$\tau_E \propto 1/(T)^{\frac{1}{2}}\rho\sigma^2 \tag{2}$$

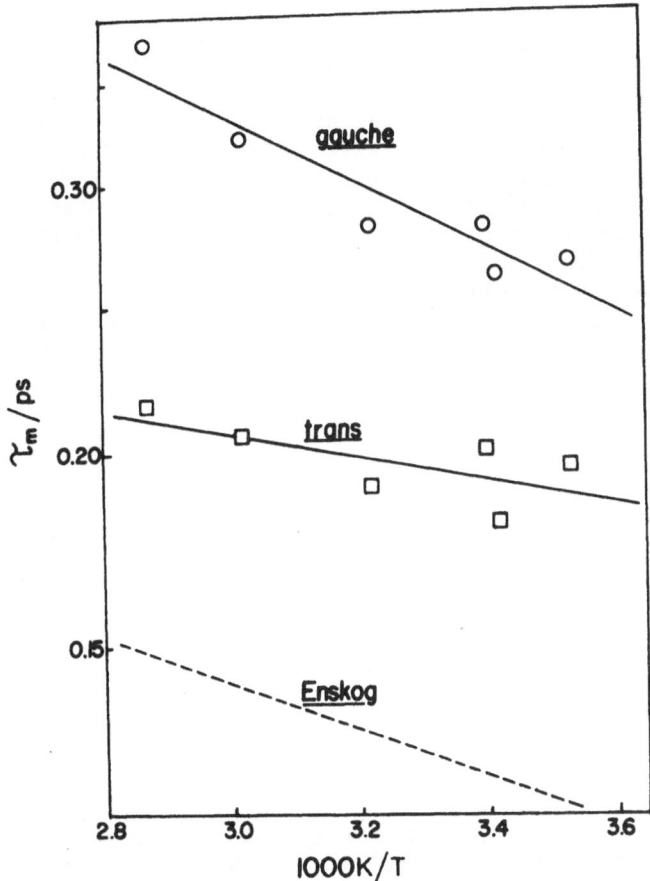

Fig. 3. Environmental modulation times calculated for the ν(C-Br) modes of 1,2-dibromoethane (trans and gauche conformers). The Enskog IBC model values (eqn. 2) are shown as a dashed line. (Reproduced, by permission, from Mol. Phys., 43, 586 (1981).)

since $\sigma_t > \sigma_g$, the modulation times τ_m are smaller for the trans isomer (fig. 3) (the rotational times are also slower for the trans isomer). The measured moments M_{2V}^{iso} are also much higher for the trans isomer and this explains why vibrational relaxation times, τ_{iso}, are lower

$$\tau_{iso} = \frac{1}{M_{2V}\tau_m} \qquad (3)$$

for the trans isomer even though τ_m is also smaller. Why the
vibrational second moments are so different is not clear (nor indeed is
the failure of these bands to exhibit a Lorentzian wing), but clearly
the intermolecular forces are different for the two isomers. This
approach is a promising one but it is applicable only to molecules with
a few (two?) isomers present under given conditions and has not been
pursued. The situation is much more complicated, for example, for
molecules containing small flexible rings. A detailed study of liquid
cyclopentene [29,30] using different techniques has shown that most of
the vibrational bands have a simple (- single band -) structure even
though most of the corresponding gas phase spectra show clear evidence
for vibration-ring puckering interactions and (small) deviations of the
molecule from its planar C_{2V} configuration (fig. 4) by puckering

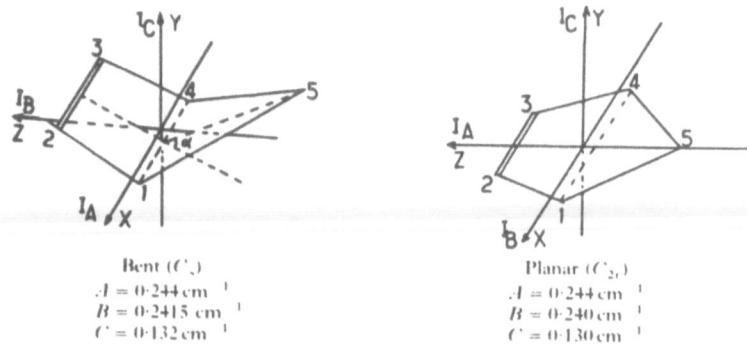

Fig. 4. The two configurations of the cyclopentene molecules.
(Reproduced, by permission, from Mol. Phys., <u>53</u>, 1145 (1984).)

motion. The ring puckering interaction is assumed to be included in
$C_{vib}(t)$ and the reorientational motion analysed in the usual way with
aid of a rigid molecule, (Debye) diffusion model. The degree of
rotational anisotropy is small and the diffusion model gives a
reasonable description provided that one recognises that rapid
modulation is not complete. The agreement between the diffusion
coefficients of cyclopentene and those of the (rigid) furan (fig. 5)
would appear to be good evidence for the 'rigid' behaviour of
cyclopentene. However, the width behaviour of infrared and Raman
profiles of the A_2 'twisting' mode of the molecule (figs. 6 and 7)
confirm that ring puckering motion, $x(t)$, contributes to the total
(infrared) correlation function,

$$C_T^{IR}(t) = \left[\frac{\partial^2 \mu}{\partial q \partial x}\right]^2 \langle q(o).q(t)\rangle\langle u_x(o).u_x(t)\rangle\langle x(o).x(t)\rangle \qquad (4)$$

in the form of a 22 cm^{-1} contribution to the infrared bandwidth (i.e.
$\langle x(o).x(t)\rangle$ decays with a time constant of < 0.5 psec.) Note that it is

necessary to assume that $\tau_{vib}^{iso} = \tau_{vib}^{IR}$ in order to deduce this

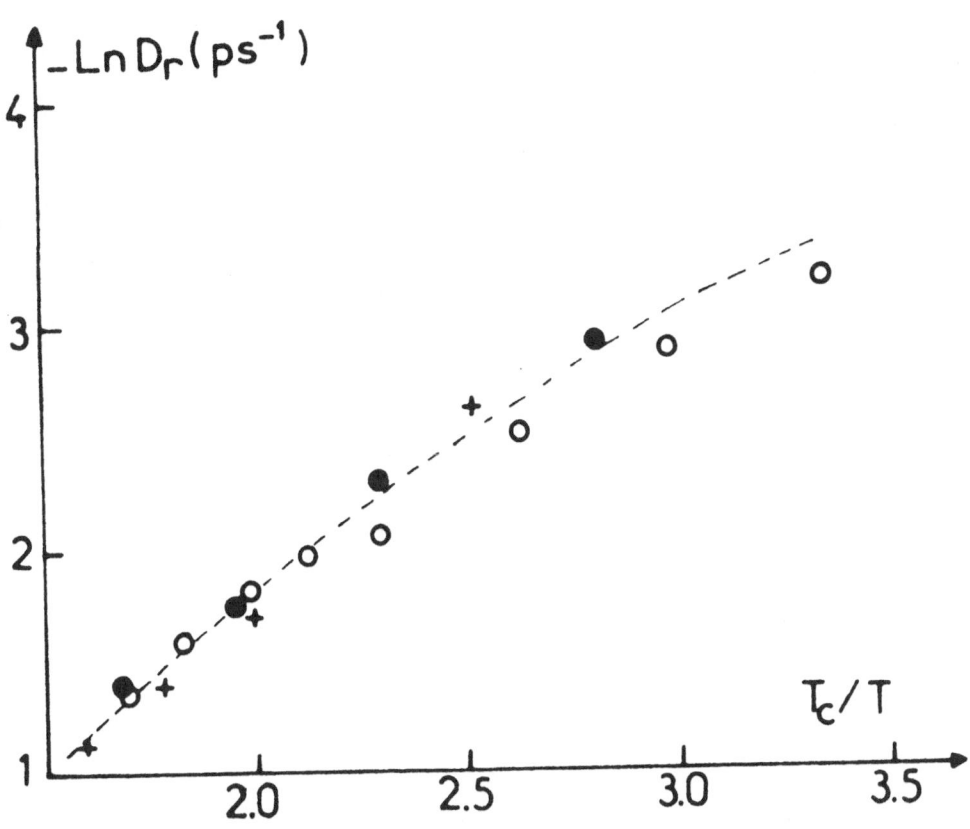

Fig. 5. Comparison of the rotational diffusion constants of cyclopentene (0, NMR; ● QENS data) and furan (crosses) as a function of reduced temperature T_c/T. (Reproduced, by permission, from Mol. Phys.. 53, 1145 (1984).)

contribution (an assumption sometimes found to be unjustified [31]). However, it does seem likely that the ring puckering dynamics - with $\tau_p^{-1} > 10^{12}$ sec^{-1} deduced from N.M.R. and neutron scattering measurements [29] - is supported by the infrared measurements.

Finally in this section, there are several interesting features of the work on 1-n-alkynes [32] and other long flexible molecules [33]. Goulay et al [32] have found that their rotational correlation times, τ_{1R} - obtained using the rigid molecule approximation in n-heptane solutions showed a very different dependence on temperature and viscosity (fig. 8) from the data for the (rigid) acetylene molecule. The experimental slopes are quite different for the long chain alkynes from that for the small rigid acetylene molecule. For C_2H_2 free volume

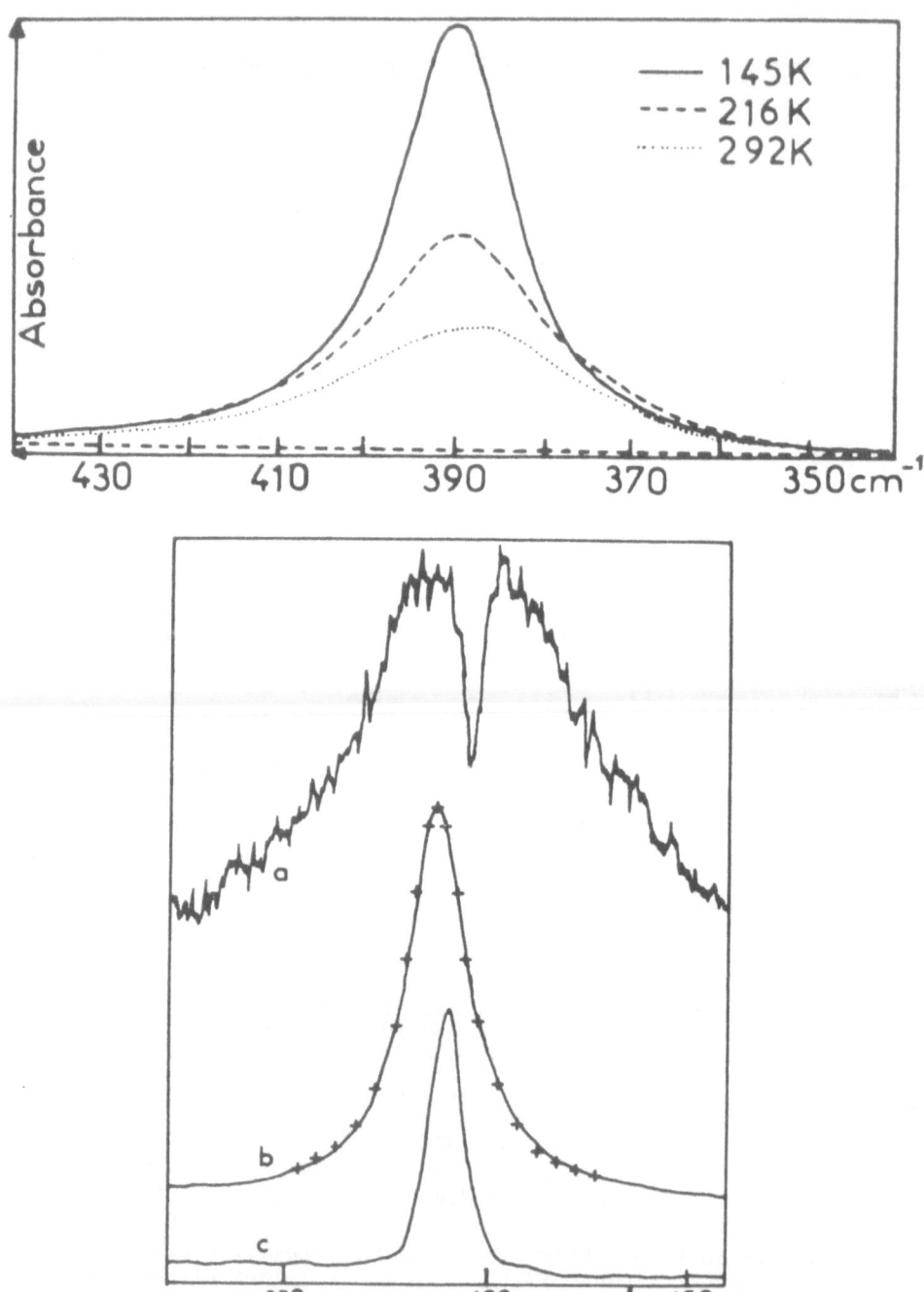

Figs. 6 and 7. Infrared (top) and Raman (bottom) spectra of the A_2 twisting mode of cyclopentene. In fig. 7 (bottom) the data in all three phases are shown. In fig. 6 (top) the liquid phase data are shown at different temperatures. (Reproduced, by permission, from Mol. Phys., 53, 1145 (1984).)

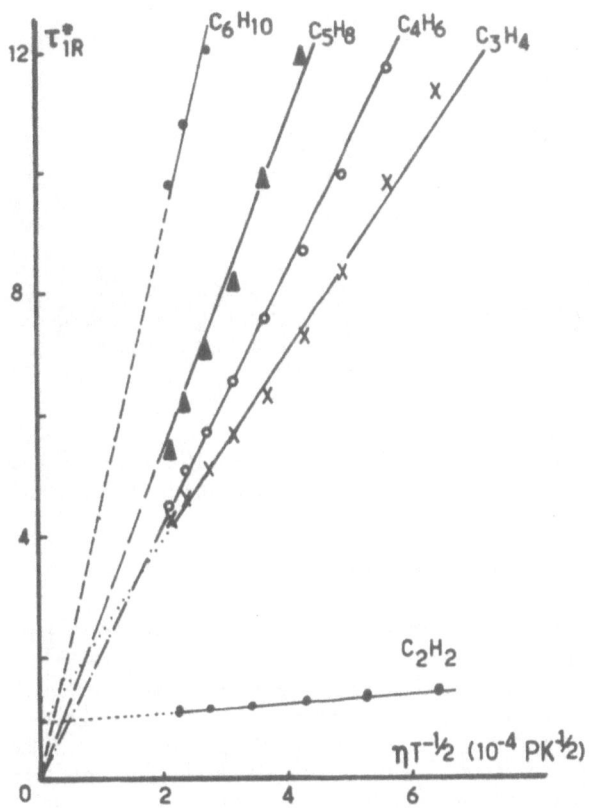

Fig. 8. Comparison of the viscosity and temperature dependence of the
reorientational correlation time of acetylene with those of longer
chain n-alkynes. (Reproduced, by permission, from Chem. Phys., 96, 333
(1985).

effects are dominant and the 'collisional' model of Hynes et al [34]
seems to fit well. For the larger molecules the slopes are close to
the hydrodynamic 'slip' limit, and are indicative of strong collective
effects. The origins of such 'collective' and 'collisional' effects
are unclear but it seems not unreasonable to speculate that they may
somehow be related to both solute and/or solvent flexibility since, as
pointed out by Tatam and Champion [35], flexing of alkyl chains is
intimately related to surrounding 'collisions' in a liquid. The
results of Moses and Baglin [33] on $(CH_3)_3SiC\equiv CSi(CH_3)_3$ are even more
puzzling. Although reorientation is, as expected, very slow and the
r.m.s. torque is very large, the $C_{2R}(t)$ functions (fig. 9) show
distinct oscillations which are removed in dilute solution. Clearly

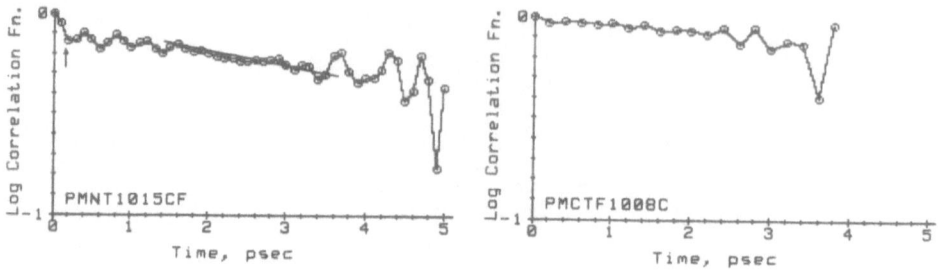

Fig. 9. Reorientational correlation function for MSA liquid (on the left) and dilute solution in CCl₄ (on the right) at 296 K. The liquid phase data are compared with the Gaussian Cage model of Lynden-Bell and Steele. (Reproduced, by permission, from J. Phys. Chem., 91, 1944 (1987).)

these arise from rapidly fluctuating collective effects, maybe connected with internal flexibility - a phenomenon at least following up!

(ii) Pure Reorientational Spectra

There have been a number of studies on flexible molecules in liquids using far infrared and/or microwave spectroscopy [36-45]. Because of the wide frequency range (and therefore broad time regime) this seems the region where one has the best chance of separating the effects of internal and overall rotation. The spectral response between 1 cm⁻¹ and 200 cm⁻¹ is particularly sensitive to both the intramolecular and intermolecular dynamics and there have been several reports [38,40,41] of multiple bands which appear to have multi-process origins. Richter and Schiel [38] noticed that the dipole correlation function for n-hexane decays with two relaxtion times (fig. 10) (3 and 8 psec) indicating almost an order of magnitude time separation of the two processes. The short relaxation is said to describe rotations about the long axis - including interchange between different gauche isomers while rotational diffusion about the short axis is said to occur on the longer time scale. However, for more polar molecules especially ketones [41-43], aldehydes [39,40] and esters [43] there have been firm reports of additional bands claimed to be due to internal rotational (librational) processes. For example, Vij and Hufnagel [41] report that all alkyl ketones (except for acetone) show a band at ~250-300 cm⁻¹ (fig. 11) in addition to the (Poley) dipolar librational band near 50 cm⁻¹ (which is not an internal mode) and which is assigned to an intra-librational motion of the -CO and CH₃CO- about the backbone to which they are attached. The assignment is supported by a calculation based upon Coffey's recent work [46] on the solution of a generalised itinerant oscillator model in terms of the far-infrared

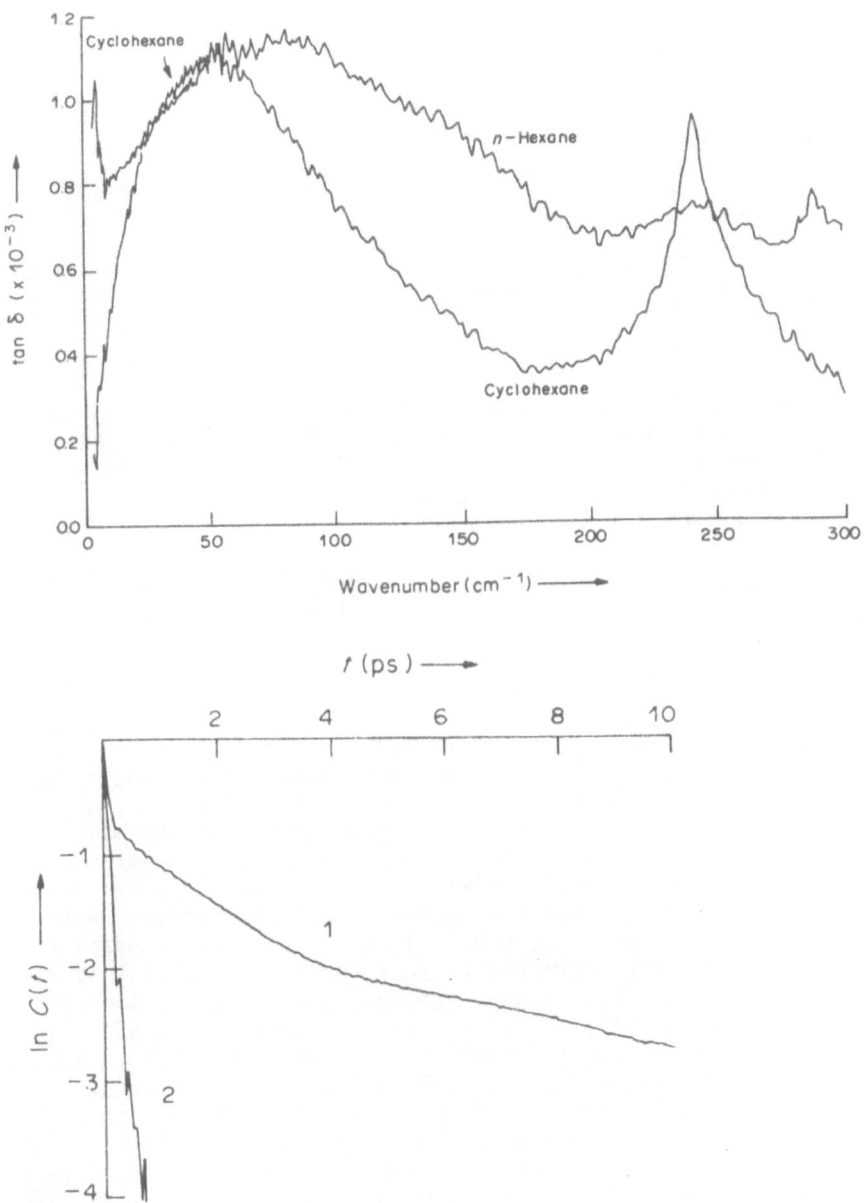

Fig. 10. Comparison of the far-infrared spectra (top) and associated correlation functions (bottom) for liquid hexane and cyclohexane. (Reproduced, by permission, from Infrared Phys., 24, 227 (1984).)

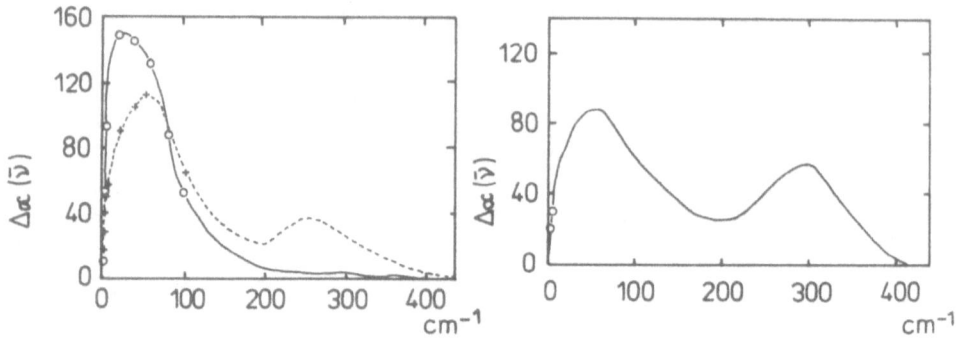

Fig. 11. Far-infrared spectra of liquid acetone (o) and hexanone
(----) at ambient temperature. The right hand spectrum demonstrates
(for undecanone-6) that the two bands change in relative intensity as
the chain lengthens. (Reproduced, by permission, from Chem. Phys.
Lett., $\underline{139}$, 77 (1987).)

torsional frequency between two dipoles (Ω_{FIR}^{ij}). It is shown that

$$\Omega_{FIR}^{2,3} \cong 3\Omega_{FIR}^{1,2} \tag{5}$$

where dipoles 1 and 2 are the <u>intermolecular</u> pair (giving rise to the
band at 50 cm^{-1}) and dipole 3 is the <u>intramolecular</u> dipole. The
additional band is roughly 4 times the frequency of the Poley band.
Even when two bands are not obvious in the spectrum it is often
considered possible to infer the presence of intramolecular motions
from band shapes or relaxation times. For example, internal rotational
motion has been claimed to be responsible for the very low Debye
relaxation time (4.1 psec) of diphenyl ether and the far-infrared
spectrum (fig. 12) can certainly be reproduced more readily if the
model parameters (K_0, K_1 and γ in the Mori continued fraction
approximation to solution of the generalised Langevin equation [41])
are derived for <u>coupled</u> motions of internal and overall rotation (fig.
13). Many approximations are involved and the interpretation may notbe
unique but the far-infrared spectrum is clearly not explainable on the
basis of overall inertia and volume of rotation (fig. 12).
        Microwave data for non-rigid molecules have also been interpreted
in terms of two processes [39-41] and the broad asymmetric dispersion
curve has been fitted to a sum of Debye diffusion processes [40]. For
differently substituted toluylaldehydes these two processes have
relaxation times in dilute solution between 10-20 psec (τ_1) and between
2-5 psec (τ_2) corresponding to overall rotation and internal -CHO
rotation respectively. The τ_2 relaxation time is relatively
independent of -CH$_3$ substitution pattern (table 1) and environment

whereas τ_1 is dependent, as expected, upon η/T. The amplitude S_1

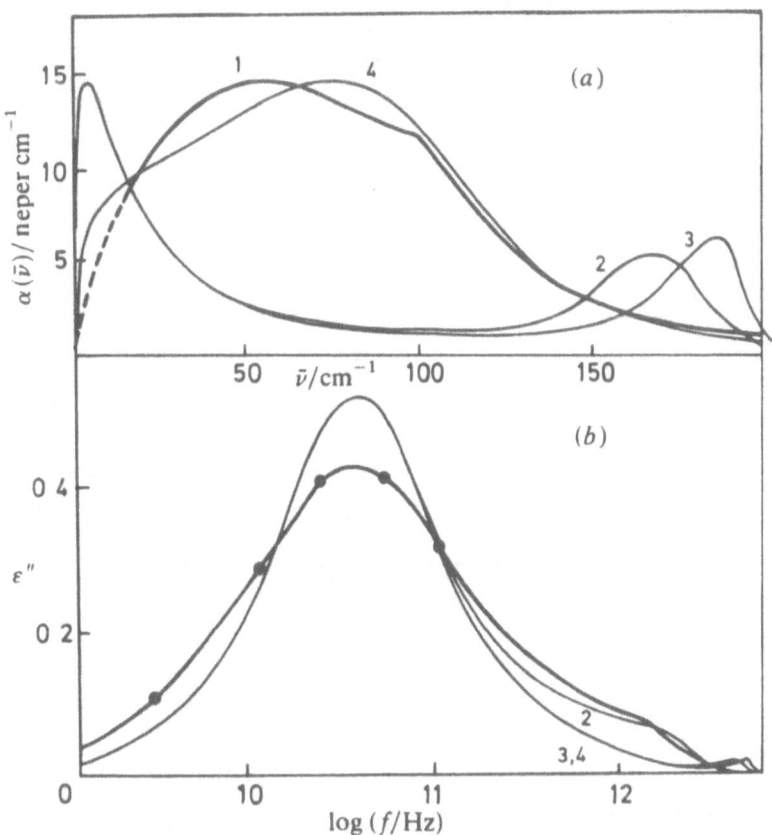

Fig. 12. Far infrared spectra of diphenyl ether (308 K) represented as (a) power absorption, (b) dielectric loss (• microwave data of Vaughan). (1) expt., (2) analytical model assuming overall rigid rotation, (3) assuming internal rotation, (4) assuming coupled internal and overall rotation - by mechanism shown in fig. 13. (Reproduced, by permission, from J. Chem. Soc., Faraday Trans. II, _78_. 1652 (1982).)

(a measure of the integrated spectral intensity) varies, also as expected, when the molecular dipole moment changes. However, the amplitude S_2 <u>also</u> varies on substitution, whereas one might expect the dipole of the internally rotating -CHO group to be roughly the same for each molecule. Furthermore, the values of S_2 do not correlate with the far-infrared intensities (I) in the region (100-150 cm^{-1})

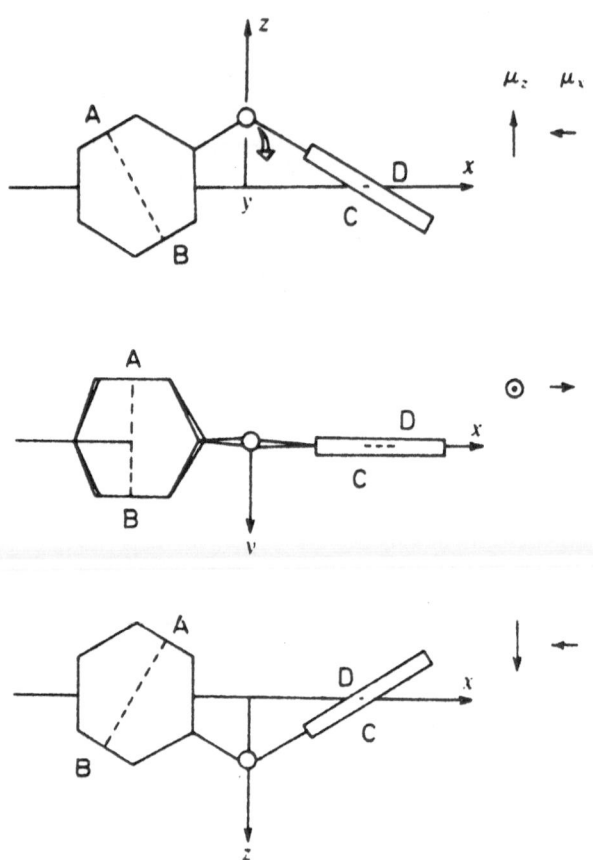

Fig. 13. Coupled inter and intra-molecular rotation in diphenyl ether. Line AB always remains in the plane of the paper and line CD always remains in the plane normal to the paper. It is assumed that $\mu_z \gg \mu_x$. (Reproduced, by permission, from J. Chem. Soc., Faraday Trans. II, 78, 1654 (1982).)

where dipolar libration spectral density is thought to arise. It is therefore speculated that the S_2 process is one of high amplitude 'diffusive' motions between the two potential minima with aldehyde and aromatic ring co-planar. The far-infrared band is then envisaged as a torsional oscillation of the aldehyde group in one of these potential wells by molecules not involved in diffusive motion, across a flattened barrier. Although the measured relationships between S_2, τ_2 and I can be explained qualitatively there is clearly still much work to be done in fully understanding such systems.

TABLE 1. Summary of the dispersion amplitudes (S) and relaxation time (τ) of the two reorientation processes observed for a series of substituted benzaldehydes. (Reproduced by J. Mol. Liquids, <u>33</u>, 213 (1987).)

Substance	S_1	S_2	τ_1	τ_2
Benzaldehyde	8.45 ± .08		7.42 ± .12	
p-Toluylaldehyde	9.66 ± .17	1.54 ± 1.7	17.0 ± .25	3.05 ± .65
m-Toluylaldehyde	7.72 ± .70	1.79 ± .70	15.9 ± 1.0	5.00 ± 1.9
o-Toluylaldehyde	6.62 ± .79	1.20 ± .79	9.19 ± .55	3.47 ± 1.8
4-Fluorbenzaldehyde	3.04 ± .18	.968 ± .18	9.51 ± .50	2.35 ± .57
3-Fluorobenzaldehyde	3.01 ± .71	4.86 ± .71	14.5 ± 2.5	5.40 ± .55
2-Fluorobenzaldehyde	7.09 ± 1.2	2.24 ± 1.2	9.93 ± 1.1	3.39 ± 1.6
Pyridin-4-aldehyde	1.41 ± .10	1.62 ± .10	9.52 ± .79	2.92 ± .19
Pyridin-3-aldehyde	2.28 ± .21	3.33 ± .21	9.93 ± .58	4.09 ± .16
Furfurol	7.05 ± .56	4.25 ± .56	9.83 ± .58	3.71 ± .36
Thiophen-2-aldehyde	7.62 ± 1.4	4.65 ± 1.4	9.24 ± 1.1	3.78 ± .73

3. APPLICATION TO FUNCTIONAL FLUIDS

As indicated in the introduction, a major class of 'flexible' molecules is that which includes lubricants and fuels - usually comprising long alkyl chains with or without other chemical groups [8]. 2-ethylhexylbenzoate (EHB) - shown in fig. 14 is a model lubricant with a wide liquid temperature range (-65°C to 250°C, at ambient pressures) and hence a wide viscosity range [61]. The molecular behaviour of this material over a range of operating temperatures and pressures is the subject of a Shell sponsored project in our laboratory. Three results are of importance in the current context.

(a) <u>Ring rotation</u>

We have measured the aromatic ring rotational correlation time over a range of temperature using the Raman 'ring stretching' mode at 1004 cm^{-1} [47]. Figures 15 and 16 show that the correlation times decrease with increasing temperature (as expected) and that the data agree well with those determined from ^{13}C n.m.r. measurements [48]. It would appear that the two techniques measure the same motional process. However, whether this motion is overall rotation, internal rotation of the ring or simply a librational motion of the ring is still to be established. Certainly the (T$_1$) spin-lattice relaxation process - dominated by intramolecular dipole-dipole coupling with directly bonded protons [48-50] - does not <u>require</u> complete rotation of the C-H bond vector and such measurements <u>may</u> reflect [29,47] both overall rotational <u>and</u> local motions (<u>i.e.</u> flexibility or mobility of the local environment of the ^{13}C nucleus via librations or conformational changes). The relaxation times of 10-100 psec (see fig. 16) seem too

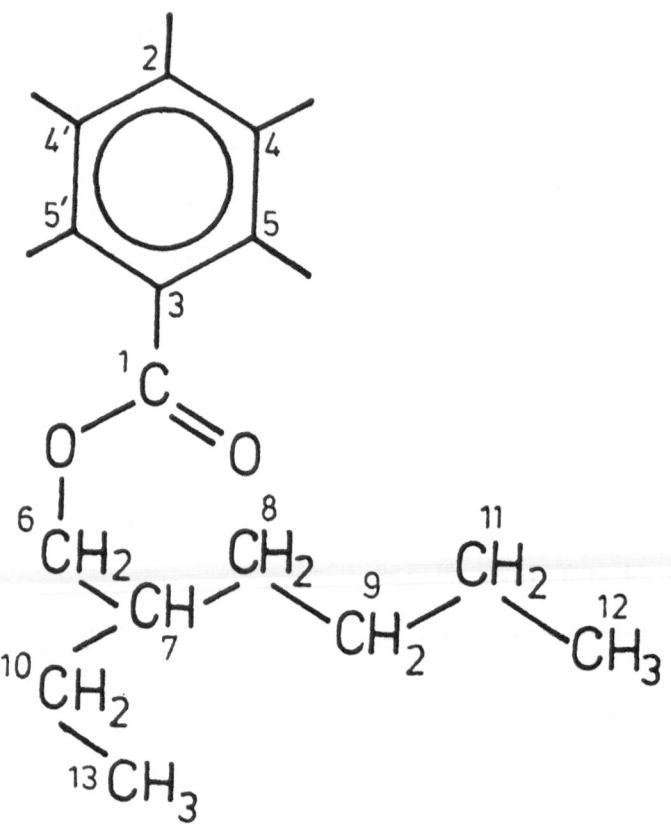

Fig. 14. The 2-ethylhexylbenzoate molecule (EHB).

short to measure overall molecular rotation and the activation energy for the motion [48] is only 20 kJ mol^{-1} so internal motion is strongly inferred. On the other hand, the correlation times appear to follow the macroscopic viscosity [51] quite well (fig. 16) and the activation energies derived for ring and chain carbon atoms are the same. Both these properties would be expected for overall rotational motion so the situation is not clear and it maybe [49,50] that both types of motion contribute to the relaxation times measured.

(b) C=O librational motion

The interplay of internal and overall molecular rotation arises also for the dipole 'librational' band observed for all polar molecules in the far-infrared [3,37,52] (fig. 17). This profile which, as may be

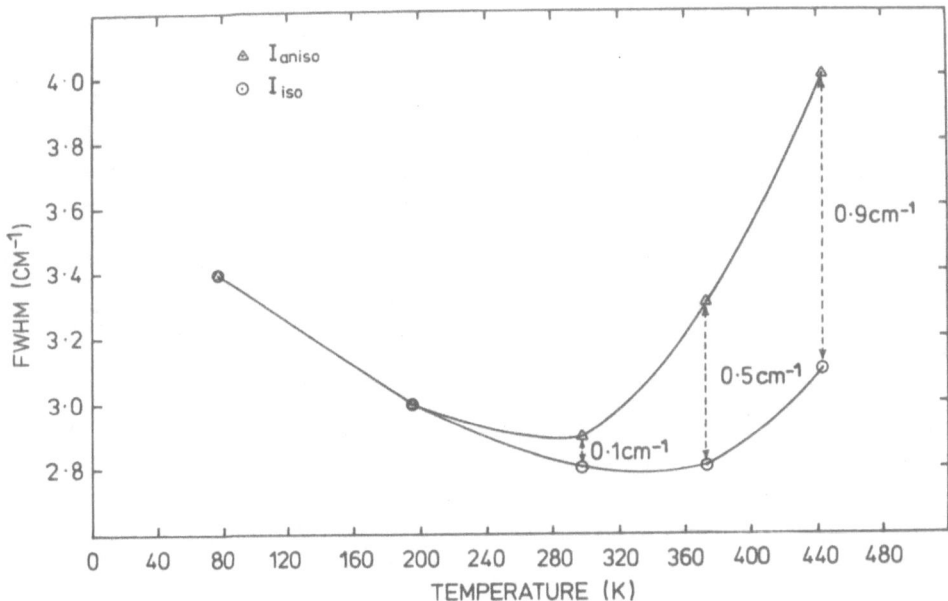

Fig. 15. Raman band halfwidths for the 1004 cm^{-1} band of EHB as a function of temperature.

seen, is insensitive to dilution in non-polar solvents appears to be composed of two overlapping spectral densities, one centred near 80 cm^{-1} and the other near 150 cm^{-1}. The low frequency band is almost certainly connected with the rotation/libration of the ester/carbonyl group of EHB but again it is not clear whether this high frequency (-short time) spectral density arises from a rapid 'librational' motion (similar to a torsional motion) or whether it represents the short time part of the overall rotational motion about the long axis. The high frequency component <u>could</u> be associated with an out-of-plane mode [53] of the substituted benzene ring. So again it is conceivable that we may be observing both internal and external librational motions. The frequencies do not, however, conform to the prediction of equation 5.

(c) <u>Interchain coupling?</u>

Finally, it is tempting to speculate on the origins of the extremely non-linear variations in infrared ν(C=O) band frequencies and widths on dilution, as shown in fig. 18. Addition of non-polar solvent molecules will cause the strongly interacting C=O dipoles to be separated. These dilution effects have been found before, for example, for N,N-dimethyl formamide [54] and acetonitrile [55] but the extent of non-linearity (and maybe nonideality) seems to be greater here with the possibility of flexible alkyl chain involvement in the molecular interactions.

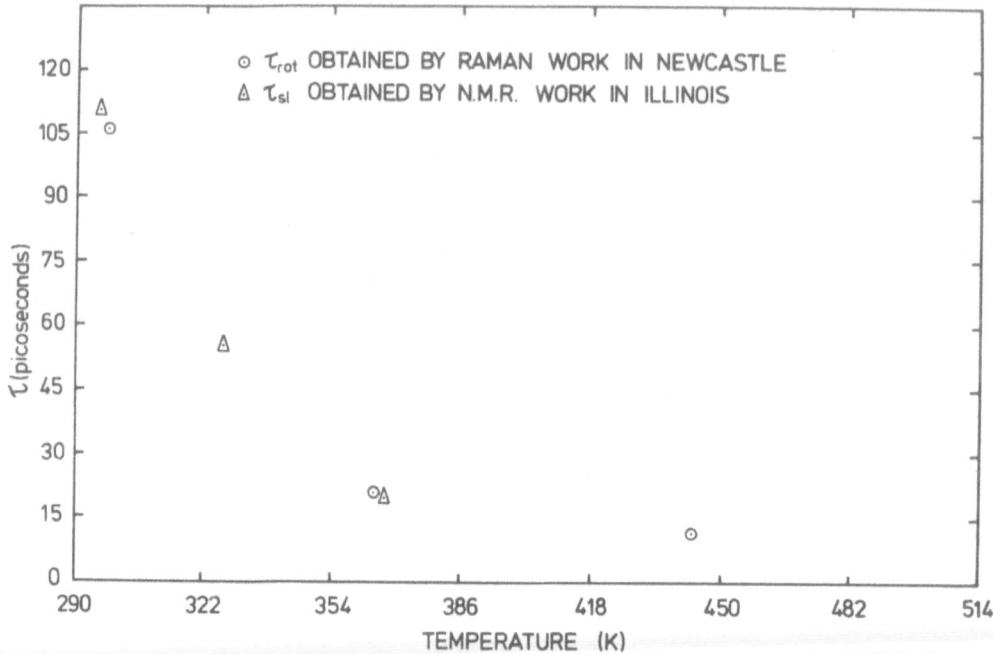

Fig. 16. Comparison of Raman (τ_{2R}) and NMR ^{13}C spin-lattice relaxation times as a function of temperature.

SUMMARY AND CONCLUSIONS

The analysis of molecular interactions and dynamics of flexible molecules in liquids is clearly still in its infancy. In some special cases, when internal and overall rotational motions are on different timescales, it is possible to observe separated or partially separated spectra features. In most cases, however, the effects of internal flexibility - although sometimes detectable - are superimposed on those due to overall rotation. Nevertheless, by a suitable choice of molecular size, shape and environment, it ought to be possible to observe and analyse the spectral consequences of internal motions occurring on a psec timescale.

ACKNOWLEDGEMENTS

The work on EHB is supported by Shell U.K. (TRC) to whom we express our sincere thanks. We also gratefully acknowledge the SERC support via equipment grants.

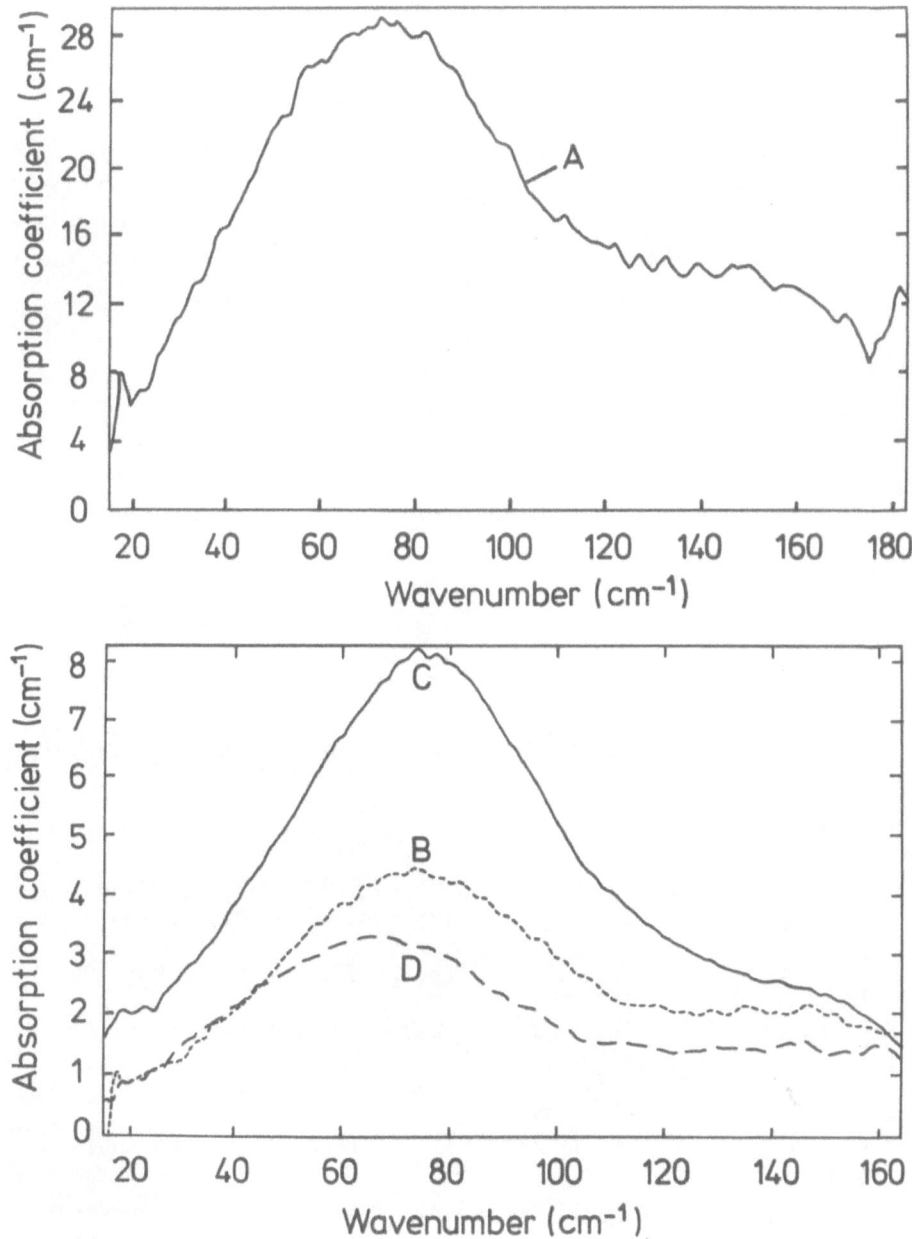

Fig. 17. Far-infrared spectra of EHB at 298 K (A) Liquid, (B) 10% mf in benzene, (C) 10% mf in CS_2, (D) 10% mf in hexane.

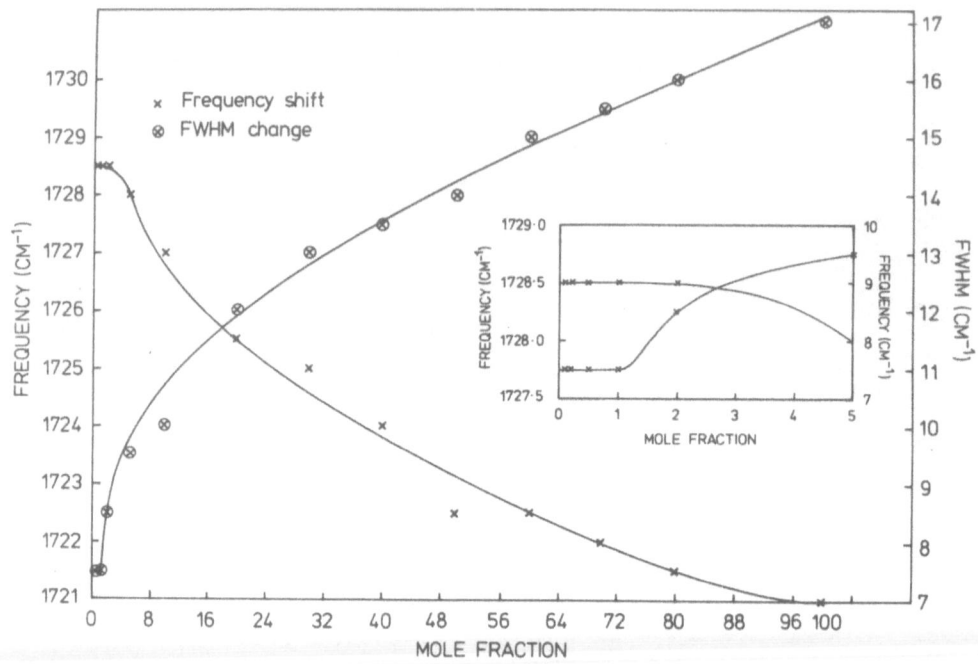

Fig. 18. Variation in frequency and half-widths of the ν(C=O) infrared band of EHB diluted in hexane.

REFERENCES

[1] W.G. Rothschild, *Dynamics of Molecular Liquids*, J. Wiley, 1984.
[2] J. Yarwood, *NATO ASI Ser. C.*, **135** (Mol. Liquids), Reidel, 1984,
     pp.357-82.
[3] J. Yarwood, *Ann. Rep. Prog. Chem. C*, **79**, The Royal Soc. Chem.,
     London, 1983, pp.157-197.
[4] D. Steele, *Stud. Phys. Theor. Chem.*, **20** (Vib. Intensities
     Infrared Raman Spectrosc.) 1982, pp.398-416.
[5] W.A. Steele, *NATO ASI Ser. C.*, **135** (Mol. Liquids) Reidel, 1984,
     pp.357-84; *Adv. Chem. Phys.*,
[6] W.A. Steele and R.M. Lynden-Bell, *J. Phys. Chem.*, **88**, 6514
     (1984).
[7] D.F.H. Wallach, S.P. Verma and J. Fookson, *Biochem. Biophys.
     Acta*, **559**, 153 (1979).
[8] A.G.M. Michell, *Lubrication*, Blackie and Son, London, 1950,
     pp.49-52.

[9]   L.F. Nafie and W.L. Peticolas, *J. Chem. Phys.*, **57**, 3145 (1972).
[10] J. Jonas in *Phenomena Induced by Intermolecular Interactions* (Ed. G. Birnbaum) *NATO ASI Ser. B*, **127**, Plenum, 1985, p.541.
[11] G.D. Patterson, *Macromolecules*, **15**, 204 (1982).
[12] M. Davies in *Dielectric Properties and Molecular Behaviour*, N.F. Hill, W.E. Vaughan, A.H. Price and M. Davies, Van Nostrand, 1969, p.357 et seq.
[13] J. Crossley and N. Koigumi, *J. Chem. Phys.*, **60**, 4800 (1974).
[14] D.G. Gardiner, M.P. Dare-Edwards and N.A. Walker, *Nature*, **313**, 614 (1985).
[15] H. Namura, Y. Udagawa and K. Murasawa, *J. Mol. Structure*, **126**, 229 (1985).
[16] P.T. Wong, T. Chagwedera and H.H. Mautsch, *J. Chem. Phys.*, **87**, 4487 (1987); H.L. Casal, P.W. Yang and H.H. Mantsch, *Can. J. Chem.*, **64**, 1544 (1986).
[17] G. Zerbi, *J. Mol. Struct.*, **126**, 209 (1985).
[18] H. Schwickert, G.R. Strobl and R. Eckel, *Coll. Polym. Sci.*, **260**, 588 (1982).
[19] E.W. Fischer, G.R. Strobl, M. Dettenmaier, M. Stramon and H. Steidle, *Faraday Diss.*, **68**, 26 (1980).
[20] D.L. Hasha, T. Eguchi and J. Jonas, *J. Amer. Chem. Soc.*, **104**, 2290 (1982).
[21] F. Coletto, G. Moro and P.L. Nordio, *Mol. Phys.*, **61**, 1259 (1987).
[22] S.L. Wunder, M.I. Bell and G. Zerbi, *J. Chem. Phys.*, **85**, 3827 (1986).
[23] P.A. McGregor and M. Schwartz, *J. Mol. Struct.*, **95**, 273 (1982).
[24] M. Schwartz, A. Moradi-Araghi and W.H. Köhler, *Mol. Phys.*, **43**, 581 (1981).
[25] A. Reklat, T. Steiger and P. Reich, *Z. Phys. Chem.*, **266**, 602 (1985); *Acta Phys. Polon.*, **A58**, 665 (1980).
[26] R.G. Snyder and H.L. Strauss, *J. Chem. Phys.*, **77**, 1118 (1982).
[27] S.F. Fischer and A. Laubereau, *Chem. Phys. Lett.*, **35**, 6 (1975).
[28] J. Schroeder, V.H. Scherinau and J. Jonas, *Mol. Phys.*, **34**, 1501 (1977) and references therein.
[29] M. Besnard, J.C. Lassegues, M. Favassier, H. Jobic, A.J. Dianoux, A. Lichanot and H. Nery, *J. Physique (Paris)*, **45**, 487, 497 (1984).
[30] M. Besnard, J.C. Lassegues, Y. Guissani and J.C. Leicknam, *Mol. Phys.*, **53**, 1145 (1984).
[31] J. Yarwood, P.L. James, G. Döge and R. Arndt, *Faraday Diss.*, **66**, 252 (1978).
[32] A.M. Goulay, A. Migdal-Mikuli, E. Mikuli, J. Saussen-Jacob and J. Vincent-Geisse, *Chem. Phys.*, **96**, 333 (1985).
[33] D.G. Moses and F.G. Baglin, *J. Phys. Chem.*, **91**, 1942 (1987).
[34] J.T. Hynes, R. Kapral and M. Weinberg, *J. Chem. Phys.*, **67**, 3256 (1977).
[35] R.V. Tatam and J.V. Champion, *Mol. Phys.*, **60**, 291 (1987).
[36] C.P. Smyth, *Molecular Relaxation Processes*, Special Publication No. 20 (The Chem. Soc., London 1966) pp.1-13.
[37] C.J. Reid and J.K. Vij, *J. Chem. Soc., Faraday Trans. II*, **78**, 1649 (1982).

[38] W. Richter and D. Schiel, *Infrared Phys.*, **24**, 227 (1984).
[39] C. Grosse, *J. Mol. Liquids*, **33**, 71, 79 (1986).
[40] C. Grosse and F. Hufnagel, *J. Mol. Liquids*, **33**, 213 (1987).
[41] J.K. Vij and F. Hufnagel, *Chem. Phys. Lett.*, **139**, 77 (1987) and references therein.
[42] O. Adamedine, *Mol. Phys.*, **47**, 1203 (1982).
[43] M.A. Kashem, H.A. Khwaja and S. Walker, *J. Mol. Liquids*, **25**, 129 (1983); *Adv. Mol. Relax. Int. Processes*, **22**, 27 (1982).
[44] J.M. Gaudhi and G.L. Sharma, *J. Mol. Liq.*, **38**, 23 (1988).
[45] J. Crossley and N. Koizumi, *J. Chem. Phys.*, **60**, 4800 (1974).
[46] W.T. Coffey and P.M. Corcoran, *Chem. Phys. Lett.*, **123**, 416 (1986); ibid **139**, 290 (1987); ibid **144**, 172 (1988).
[47] F.A. Miller, *J. Raman Spectroscopy*, **19**, 219 (1988).
[48] N. Walker and J. Jonas (in preparation).
[49] M. Baldo, K.J. Irgolic and G.C. Pappalardo, *Mol. Phys.*, **38**, 1467 (1979).
[50] S.R. Kempsall, J.R. Barnes, C.J. Craven and P.D and F.D. Wayne, Proc. 6th Int. Colloq. on Tribology, Esslingen, Germany, 12-14 January 1988.
[51] N.A. Walker, D.M. Lamb, S.T. Adamy, J. Jonas and M.P. Dare-Edwards, *J. Phys. Chem.*, **92**, 3675 (1988).
[52] R. Buchner, V.M. Shelley, A. Talintyre and J. Yarwood, *J. Chem. Soc.*, *Faraday Discussion*, **85**, 0000 (1988).
[53] P.R. Griffiths and H.W. Thompson, *Proc. Roy. Soc.*, A298, 51 (1967).
[54] J. Yarwood, V. Shelley and J. Yarwood (unpublished data).
[55] R. Arndt, G. Döge and J. Yarwood, *Chem. Phys.*, **25**, 387 (1977).

RELATIVE MOTION OF PAIRS OF MOLECULES IN DENSE FLUIDS

WILLIAM A. STEELE
Department of Chemistry
The Pennsylvania State University
University Park, PA 16802, U.S.A.

ABSTRACT

The application of cumulant theory to formulate expressions for time-correlation functions of pairs of molecules with translational and orientational degrees of freedom is discussed. As an illustration of the approach, new expressions are derived for the generalized time and distance dependent Kirkwood g-factors. It is shown that mutual angular velocity correlations play an important role in determining these g-factors. These results are compared with the expressions obtained previously for the g-factors by Kivelson, et al.

1.　　Introduction

The dynamical behavior of pairs of molecules in dense phases is known to play an important role in determining a variety of experimental quantities. Perhaps the earliest example occurs in the theory of the magnetic dipole-dipole contributions to spin relaxation times[1], where it is relative motion of pairs of spins (nuclear or electronic) that produces a fluctuating magnetic field at each spin. The inverse of the rate of change in the spin quantum state induced by this field is called the intermolecular part of the spin relaxation time. Of course, if each spin is attached to a rigid molecular frame, the relative spin motion is produced by the translations[2] and rotations of both molecules of the pair.

Another observable quantity which is dependent upon the relative translational velocities of pairs of molecules is the Onsager coefficient associated with the relative diffusion constant D_{nm} for components n and m in a mixture[3].

Of particular interest in this paper are two other quantities that may have a significant dependence upon the relative reorientation of pairs of molecules. The first of these is the generalized Kirkwood g-factor $g_{\ell}(r_0, t)$ which we define as

$$g_{\ell}(r_0, t) = \langle\, P_{\ell}(\cos\theta_{ij}(r_0, t)\,\rangle \qquad (1.1)$$

where $\theta_{ij}(t)$ the orientation angle between some relevant axes of two molecules i and j at time t which were separated by a distance r_0 at time zero, and P_{ℓ} is a Legendre polynomial. Observable quantities are given by the summation of eq. (1.1) over all i, j. For axes parallel to the molecular dipole moments, theory shows that the integral of $g_1(r_0, t)$ over r is related to the dielectric absorption of a fluid[4]; if the axes are those for the polarizability anisotropy, the integral of $g_2(r_0, t)$ is related to the depolarized light scattering[5]. In each case, a Fourier transform with respect to time produces the frequency-dependent spectral intensity due to correlated pairs of molecules.

Th. Dorfmüller (ed.), Reactive and Flexible Molecules in Liquids, 83–95.
© 1989 by Kluwer Academic Publishers.

The second property to be discussed is the "distinct" part of the inelastic neutron scattering from a liquid, which is generally denoted as $S_D(k, \omega)$[6]. Here, we discuss the Fourier-transform of this quantity $F_D(k, t)$ which is

$$F_D(k, t) = \left\langle e^{i\mathbf{k}\cdot(\mathbf{r}_{i\alpha}(0) - \mathbf{r}_{j\beta}(t))} \right\rangle \qquad (1.2)$$

Here, $\mathbf{k}$ is the wave vector change for the scattered neutron beam and $\mathbf{r}_{i\alpha}(0)$, $\mathbf{r}_{j\beta}(t)$ are the position vectors for a pair of scattering nuclei, one on molecule i and the second on molecule j. Of course, the total intermediate scattering factor is given by a sum of $F_D(k, t)$ over all nuclei in each molecule, weighted by the scattering cross-sections of the nuclei. From the theoretical viewpoint, we consider only a particular pair of nuclei α, β or at best, a sum over all equivalent nuclei of types α, β in molecules i and j. (For example, if the molecules are CS_2, one would have a one-term $F_D(k, t)$ for pairs of C atoms; an $F_D(k, t)$ which is a sum over the four C-S pairs; and a S-S scattering factor that is a sum over the four S-S pairs.) The dependence of these factors upon relative translational and rotational motion can be explicitly shown by writing

$$\mathbf{r}_{i\alpha} = \mathbf{r}_{icm} + \mathbf{L}_\alpha \qquad (1.3)$$

where $\mathbf{L}_\alpha$ is the vector between the molecular center of mass (cm) and the α nucleus. Then

$$F_D(k, t) = \left\langle e^{i\mathbf{k}\cdot(\mathbf{r}_{icm}(0)-\mathbf{r}_{jcm}(t)} \quad \times \quad e^{i\mathbf{k}\cdot(\mathbf{L}_\alpha(0)-\mathbf{L}_\beta(t)} \right\rangle \qquad (1.4)$$

The first factor within the brackets is purely translational and, since the time-dependence of $\mathbf{L}$ is due solely to rotation of the molecular frame, the second factor reflects only orientational dynamics. However, the brackets denote an average over all initial configurations of the pair and over all displacements from the initial configuration. A decoupling approximation would be extremely helpful in making further progress in performing these averages, as will be discussed in detail below.

The appearance of both translational and orientational variables in the time-correlation function of eq. (1.4) is by no means limited to the examples mentioned so far. The Kubo time-correlation-function (t.c.f.) expressions for transport properties such as viscosity and thermal conductivity also involve both sets of variables, if one has a fluid in which intermolecular forces and energies depend upon orientation as well as separation distance. For example, the largest terms in the viscosity t.c.f. for a liquid are the correlations of so-called potential contributions that depend upon off-diagonal terms in the tensor $\mathbf{r}_{ij}\,\mathbf{F}_{ij}$ where $\mathbf{r}_{ij}$, $\mathbf{F}_{ij}$ are the center-of-mass separation and force for the ij pair.[3] However, if one uses the popular site-site representation of the intermolecular potential and force, one has

$$F_{ij} = -\sum_{\substack{\alpha \text{ in } i \\ \beta \text{ in } j}} \frac{r_{\alpha\beta} \cdot r_{ij}}{r_{\alpha\beta} \ r_{ij}} \left(\frac{\partial u (r_{\alpha\beta})}{\partial r_{\alpha\beta}} \right) \qquad (1.5)$$

where $r_{\alpha\beta}$, $u(r_{\alpha\beta})$ are the site-site separation and potential, respectively. Since $r_{\alpha\beta} = r_{icm} - r_{jcm} + L_\alpha - L_\beta$, we see that orientational variables $L_\alpha - L_\beta$ are present in addition to the center-of-mass separations that are the only quantities appearing in the analogous correlation functions for spherical or monoatomic particles.

Finally, one should note that the theory of interaction-induced absorption and light-scattering [7] shows that the associated spectra for these phenomena are related to the relative motions of pairs of molecules in the fluid; however, the calculation of the spectral intensities is greatly complicated by the presence of three and four body terms in addition to the pair terms which are the subject of discussion here.

2. Atomic Fluids

In studies of relative translational motion of pairs of atoms, it has proved useful to consider the velocity cross-correlation functions[8,9].

$$C_v^{cross} (t) = \langle \sum_{i \neq j} v_i(0) \cdot v_j(t) \rangle \qquad (2.1)$$

One can relate this function to the mean square relative displacement $\delta r_{ij}(t)$ by writing

$$\langle \sum_{i \neq j} (\delta r_{ij}(t))^2 \rangle = \langle \sum_{i \neq j} [r_i(t) - r_j(0)] \cdot [r_i(t) - r_j(0)] \rangle$$

$$= 4 \int_0^t [C_v^{self}(\tau) - C_v^{cross}(\tau)] \cdot (t-\tau) \ d\tau \qquad (2.2)$$

where $C_v^{self}(t)$ is the (unnormalized) velocity self-correlation function

$$C_v^{self}(t) = \langle \sum_i v_i(0) \cdot v_i(t) \rangle \qquad (2.3)$$

Cross velocity correlations have been simulated [8,10] and generalized Langevin equations written for the equations of motion of these functions[9]. The results of these calculations help to show that eqs. (2.1) and (2.2) are too simple to describe this phenomenon. In the first place, the relative displacement depends upon the initial relative separation. For example, if this is large the relative motion can be expressed in terms of the independent diffusive displacements of both atoms in the pair. (In other words, the cross correlation function in eq. (2.1) is zero for separations large compared to the range of the intermolecular force, and only the self-correlations contribute in eq. (2.2).) Only for small separations does one anticipate seeing the effect of the mutual interactions upon the dynamics. A second complication is produced by the fact that the relative displacement (or the relative velocity) depends upon direction relative to r_{ij}.

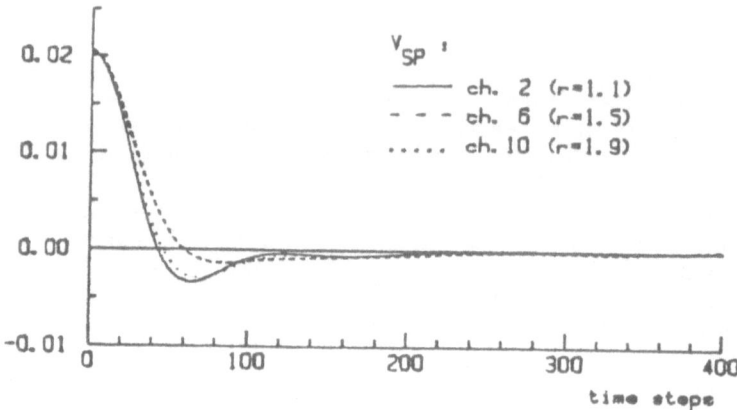

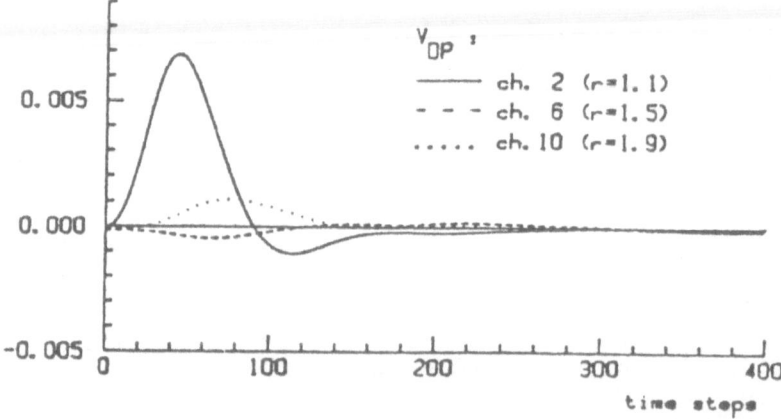

1.   Velocity cross-correlation functions are shown for a pair of atoms in liquid Ar. In the simulations, pair separations were selected from channels of width 0.1σ, centered at $1.0 + $ (ch. no.)/10. Time step = 0.034 picosec. Both the self-correlation (S) and the cross-correlation (D) are shown for components of velocity initially parallel to r_{ij}. Note that the self-correlation for atom i is not equal to the usual velocity auto-correlation function because a second atom j is known to be located at $r_{ij}(0)$ in the self-case, whereas one averages over all positions of j in the auto-calculation. At time zero, the self-correlation function is equal to $\langle v_z^2 \rangle = kT/m$, whereas the cross-correlation is precisely zero.

To see this explicitly, we can write

$$v_{ij}(t) = v_i(t) - v_j(t) = v_i(0) - v_j(0) + \int_0^t F_{ij}(\tau)\, d\tau \qquad (2.4)$$

But $F_{ij}(\tau)$ is parallel to the center-of-mass vector $r_{ij}(\tau)$ and in addition, depends strongly upon the magnitude of this separation. (Incidentally, the time dependence of F_{ij} is not easily calculated. The generalized Langevin equation or analogous equations of motion can be viewed as an attempt to write the time-dependence of the average of $v_{ij}(t)$ over all atoms except i, j in terms of a more tractable expression than that for the average of F_{ij}.)

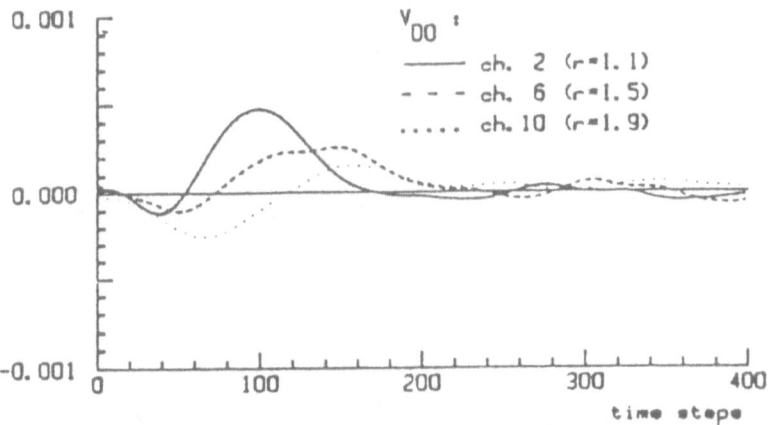

2. Same as Figure 1, except that self correlation for a component of the velocity perpendicular to $r_{12}(0)$ is shown.

To illustrate these points, Figures 1 and 2 show a few computer simulations of the cross velocity correlation $\langle v_i(0)\, v_j(t) \rangle_{ch}$ for liquid argon[11]. Here, the channel ch denotes the range of initial separation distance. Figure 1 shows the correlation function for the components of velocity initially parallel to r_{ij} and Figure 2 shows the behavior for velocities perpendicular to r_{ij}. We can gain some insight concerning the initial time dependence of these functions by considering a generalized version of the relation between displacement and velocity. In particular, it is possible to show rigorously that a correlation function $C_{fg}(t) = \langle f(0)\, g(t) \rangle$ can be written as

$$C_{fg}(t) = \langle f(0)\, g(0) \rangle - \int_0^t \langle \dot{f}(0) \cdot \dot{g}(\tau) \rangle (t-\tau)\, d\tau \qquad (2.5)$$

We now apply this expression to the cross velocity correlations $C_v^{DP}(t)$ and $C_v^{DO}(t)$.

Since the initial values are zero and since $\dot{v}_i = F_i/m$, one has

$$C_v^{DP}(t)\Big]_{ch} = \frac{1}{m^2}\int_0^t \langle F_i^P(0)\, F_j^P(\tau)\rangle_{ch}\, d\tau \tag{2.6}$$

with a similar expression for $C_v^{DO}(t)$. Here F_i^P and F_I^O denote components of the force on atom i that are parallel and perpendicular, respectively, to $r_{ij}(0)$. Now

$$\left.\begin{aligned} F_i &= \sum_{k\neq i} F_{ik} \\ F_{ij} &= -F_{ji} \end{aligned}\right\} \tag{2.7}$$

Thus,

$$\langle F_i^P(0)\, F_j^P(\tau)\rangle_{ch} = -\langle F_{ij}^P(0)\, F_{ij}^P(\tau)\rangle_{ch} + 2\langle \sum_{\substack{\ell\neq i \\ \ell\neq j}} F_{ij}^P(0)\, F_{j\ell}(\tau)\rangle_{ch} +$$

$$\langle \sum_{i\neq k,\, j\neq \ell} F_{ik}^P(0)\, F_{j\ell}^P(\tau)\rangle_{ch} \tag{2.8}$$

with a similar equation for the orthogonal components of F.

Neglecting the final three- and four-body force-force correlation functions in eq. (2.8) for the moment, substitution of (2.8) in eq. (2.6) leads to the conclusion that $C_v^{DP}(T)]_{ch}$ should initially increase from zero as a quadratic function of time with proportionality constant equal to the square of the mean force between atoms i and j at the separation distance of the given channel, in qualitative agreement with Figure 1. In the case of the perpendicular force components, the initial mean square force is zero and, aside from predicting a smaller time-dependence than for the parallel velocity correlation, one can say little about the behavior of $C_v^{DO}(t)$.

3.   Relative Molecular Motion

In the introduction, the generalized Kirkwood g-factors and the distinct part of inelastic neutron scattering were mentioned as two cases where the relative reorientations of pairs of molecules plays a significant role in determining an observable quantity. We further develop the theory for the distinct intermediate scattering factor by writing

$$e^{i \mathbf{k} \cdot \mathbf{q}} = \sum_{\ell} i^{\ell} (2\ell + 1) j_{\ell}(k q) \, D_{m,0}{}^{\ell}(\hat{k}) \, D_{m,0}{}^{*\ell}(\hat{q}) \tag{3.1}$$

where j_{ℓ} is a spherical Bessel function, $\hat{k}, \hat{q}$ are unit vectors parallel to $\mathbf{k}$ and $\mathbf{q}$ respectively and $D_{k,m}{}^{j}(\Omega)$ is the usual Wigner function of the Euler angles of a rigid body (or the polar angles of a vector, with $m = 0$). We here use Rose's notation. We write $\mathbf{r}_{1cm}(0) - \mathbf{r}_{2cm}(t)$ as $\mathbf{r}_{12}(t)$ and expand all three exponentials in eq. (1.4). We then take a coordinate system with Z axis parallel to $\mathbf{r}_{12}(0)$ and remember that the distribution of $\mathbf{r}_{12}(t)$ in this system is independent of azimuthal angle. After some reduction, one finds

$$F_D(k, t) = \sum_{m \ell \ell' \ell''} f(\ell \ell' \ell''; m) j_{\ell}(k L_{\alpha}) j_{\ell'}(k L_{\beta}) \langle j_{\ell''}(k r_{12}(t)) \tag{3.2}$$

$$d_{00}{}^{\ell}(\delta \beta(t)) D^*{}_{m,0}{}^{\ell}(\hat{L}_{\alpha}(0)) D_{m,0}{}^{\ell'}(\hat{L}_{\beta}(t)) \rangle$$

where $\delta \beta(t)$ is the angle between $\mathbf{r}_{12}(0)$ and $\mathbf{r}_{12}(t)$ and

$$f(\ell \ell' \ell''; m) = i^{\ell + \ell' + \ell'' + 2m} (2\ell + 1)(2\ell' + 1) \cdot \tag{3.3}$$

$$C(\ell \ell' \ell''; 0, 0) \, C(\ell \ell' \ell''; m, -m)$$

with C defined the usual way as a vector coupling coefficient. The brackets in eq. (3.2) surround the time-dependent quantities to be averaged, which now split into an

orientational factor dependent upon $\hat{L}_{\alpha}$ and $\hat{L}_{\beta}$ and a translational factor depending upon $\mathbf{r}_{12}(t)$ (i.e., $\mathbf{r}_{12}(t)$ and $\delta\beta(t)$). Of course, the average includes all these variables and it is not easy to tell whether any separation or simplification is feasible.

We now consider the somewhat simpler problem presented by the g-functions. Thus, eq. (1.1) is written as

$$g_{\ell}(r_0, t) = \langle \sum_m D_{m,0}{}^{\ell}(\Omega_i(0)) D^*{}_{m,0}{}^{\ell}(\Omega_j(t)) \rangle_{r_0} \tag{3.4}$$

We apply eq. (2.5) to find

$$g_\ell(r_0, t) = \langle P_\ell(\Omega_{ij}(0)) \rangle_{r_0} - \int_0^t (t-\tau) \cdot$$

$$\sum_m \langle \frac{d}{dt} D_{m,0}{}^\ell(\Omega_i(0)) \frac{d}{dt} D^*{}_{m,0}{}^\ell(\Omega_j(\tau)) \rangle_{r_0} d\tau$$

(3.5)

The time derivatives in eq. (3.5) can be written as[13]

$$\frac{d}{dt} = \omega^* \cdot \frac{\partial}{\partial \theta^*}$$

(3.6)

where ω is the angular velocity and $\partial/\partial\theta$ represents derivatives that are essentially the quantum mechanical angular momentum (or raising and lowering) operators.
From here on, we concentrate on the case $i \neq j$. Although it is usual to work with body-fixed angular velocities in one-molecule problems, this frame is unsuitable when molecules i and j are both involved. In an arbitrary space-fixed frame, one finds[13]

$$\frac{d}{dt} D_{m,0}{}^\ell(\Omega_j) = i\omega_1{}^*(j) \left(\frac{(\ell+m)(\ell-m+1)}{2} \right)^{1/2} D_{m-1,0}{}^\ell(\Omega_j)$$

$$- i\omega_0{}^*(j) \, m \, D_{m,0}{}^\ell(\Omega_j)$$

(3.7)

$$- i\omega_{-1}{}^*(j) \left(\frac{(\ell-m)(\ell+m+1)}{2} \right)^{1/2} D_{m+1,0}{}^\ell(\Omega_j)$$

where the components of $\omega(j)$, the angular velocity of molecule j, are expressed in the spherical basis:

$$\left. \begin{array}{l} \omega_{\pm 1} = \mp \frac{1}{\sqrt{2}} (\omega_x \pm i\omega_y) \\[2mm] \omega_0 = \omega_z \end{array} \right\}$$

(3.8)

We also use

$$D_{k,0}{}^\ell(\Omega_j(\tau)) = \sum_\mu D_{\mu,0}{}^\ell(\delta\Omega_j) D_{k,\mu}{}^\ell(\Omega_j(0))$$

(3.9)

An average over orientations of i and j at time zero can be performed with the aid of the irreducible expansion of the pair correlation function $g(r, \Omega_i, \Omega_j)$. For simplicity, we write this for linear molecules only:

$$g(r, \Omega_i, \Omega_j) = \sum_{\ell \ell' m} [(2\ell+1)(2\ell'+1)]^{1/2} \, g_{\ell \ell', m}(r_{ij}) \quad x$$

$$D_{m,0}^{\ell}(\Omega_i) \, D^{*}_{m,0}^{\ell'}(\Omega_j)$$

(3.10)

This expression is valid in the "ij" coordinate frame; i.e., one in which the z-axis is parallel to r_{ij}. Substitution of eqs. (3.7) and (3.9) into (3.5) followed by averages over $\Omega_i(0)$ and $\Omega_j(0)$ yields

$$\gamma_\ell(r_0, t) = \frac{1}{2\ell+1} \langle \omega_1(i, 0)\omega_1^{*}(j, t) \sum_\mu \frac{(\ell+\mu)(\ell-\mu+1)}{2} \cdot g_{\ell \ell, \mu-1}(r_{ij}(t))$$

(3.11)

$$+ \;\; \omega_0(i, 0)\, \omega_0(j, t) \sum_\mu \mu^2 \, g_{\ell \ell, \mu}(r_{ij}(t))$$

$$+ \omega_{-1}(i, 0)\, \omega_{-1}^{*}(j, t) \sum_\mu \frac{(\ell-\mu)(\ell+\mu+1)}{2} \, g_{\ell \ell, \mu+1}(r_{ij}(t)) \rangle_{r_0}$$

where we have indicated that the separation between molecules i and j is time-dependent and that the initial separation is r_0. The angular velocity cross-correlation functions are all in the ij system, so that $\omega_0 (\equiv \omega_z)$ is the velocity of rotation of the molecular axis around $r_{ij}(0)$. Nothing is known about these angular velocity cross-correlations at present. They clearly depend upon r_0, and decrease to zero at large separations. Although $\langle \omega_0(i,0)\, \omega_0^{*}(j,t) \rangle$ should be equal to $\langle \omega_{-1}(i,0) \omega_{-1}^{*}(j,t) \rangle$ it is likely that $\langle \omega_1(i,0)\, \omega_1^{*}(j,t) \rangle$ will be different, just as the parallel and orthogonal translational velocity cross correlations for argon were different from each other. Note that

$$g_\ell(r_0, t) = g_\ell(r_0, 0) - \int_0^t \gamma_\ell(r_0, \tau) \cdot \langle D_{00}^{\ell}(\delta\Omega) \rangle (t-\tau)\, d\tau$$

(3.12)

For moderately short times, it is reasonable to neglect the time-dependence of the $g_{\ell \ell', m}(r_{ij}(t))$ compared to the time-dependence of the velocity correlations. We denote this approximate quantity by $\gamma_\ell^{*}(r_0, t)$. For the commonly encountered cases of $\ell = 1$ and $\ell = 2$, it is easy to do the sums over μ in eq. (3.11) explicitly. We denote the velocity correlations by

$$C_{xy}(r_0, t) = \langle \omega_1(i,0)\, \omega_1^*(j, t)\rangle_{r_0} = \langle \omega_{-1}(i,0)\, \omega_{-1}^*(j,t)\rangle_{r_0}$$

$$C_z(r_0, t) = \langle \omega_0(i,0)\, \omega_0(j, t)\rangle_{r_0}$$

$$(3.13)$$

The (approximate) results can now be written down. For $\ell = 1$

$$\gamma_1^*(t) = \frac{1}{3}\left\{ C_{xy}(t)\,[\,2g_{11,0}(r) + g_{111}(r) + g_{11,-1}(r)\,] + \right.$$
$$\left. C_z(t)\,[\,g_{11,1}(\gamma) + g_{11,-1}(r)\,]\right\}$$

$$(3.14)$$

For $\ell = 2$,

$$\gamma_2^*(t) = \frac{1}{5}\left\{ C_{xy}(t)\,[\,2\,(g_{22,2}(r) + g_{22,-2}(r))\right.$$
$$+ 5\,(g_{22,1}(\gamma) + g_{22,-1}(r)) + 6\,g_{22,0}(r)\,]$$
$$+ C_z(t)\,[\,4\,(g_{22,2}(\gamma) + g_{22,-2}(r))$$
$$\left. + g_{22,1}(\gamma) + g_{22,-1}(r)\,]\right\}$$

$$(3.15)$$

In contrast to the previous theory for the g-coefficients due to Kivelson, et al.[14,15], we have obtained only a framework for the actual calculation, since some knowledge of the behavior of the angular velocity cross-correlations is needed to progress further. Nevertheless, some comments are in order. In the first place, it is important to realize that the rate of decay of $g_\ell(r_0, t)$ depends upon the magnitude of the angular velocity cross correlations as well as the strength of the initial angular correlations. (Of course, $g_\ell(r_0, 0)$ depends upon the same set of (insert image), but with different weighting factors.) We can write down an approximation for the entire time-dependence of the g-function by using the cumulant approximation. Some care must be exercised here because $g_\ell(r_0, 0)$ passes through zero for some r_0. This will cause great problems for this approximation, which can be schematically written as follows: Suppose

$$g(t) = g(0) - h(t) + \ldots\ldots \qquad (3.16)$$

As used here, the cumulant approximation is then

$$g(t) = g(0)\exp[-\,h(t)/g(0)] \qquad (3.17)$$

In our case, several approximations have been made in deriving eq. (3.12); the most significant of these is the neglect of the time dependence of the $g_{\ell\ell'm}(r,t)$ in eq. (3.11). If one now uses a cumulant approximation, it is hoped that the changes in $g_{\ell\ell'm}(r,t)$ will give rise to higher terms in a time-dependent series which are taken into account by the cumulant. One can skirt the problem of the zero values of $g_\ell(r_0,t)$ by the simple expedient of summing over all channels to obtain the (experimentally observable) quantities $g_\ell(t)$ before introducing the approximation. Thus,

$$g_\ell(t) = \sum_{ch} g_\ell(r_0, t) \tag{3.18}$$

$$= g_\ell(0) \, \exp[-\Gamma_\ell(t)] \tag{3.19}$$

with

$$\Gamma_\ell(t) = \frac{1}{g_\ell(0)} \int_0^t (t-\tau) \sum_{ch} \gamma_\ell^*(r_0, \tau) \cdot C_\ell(\tau) \, d\tau \tag{3.20}$$

$$C_\ell(\tau) = \langle D_{00}^\ell(\delta\Omega) \rangle \tag{3.21}$$

In principle, the displacement $\delta\Omega_i$ will depend upon the initial separation distance r_0 and orientation of molecule i, but neglect of this dependence is probably minor compared to the other approximations made.

Note that this argument can, with minor alterations, also be applied to obtain an expression for $F_D(k, t)$. The main differences are: a) there will be a significant translational time-dependence which, if one uses eq. (2.5), will be uncoupled to the rotational part as long as translational and rotational velocities are uncorrelated; b) the cumulant approximation should be applied with care, since $F_D(k, 0)$ can becomes very small for certain values of k.

4.   Discussion

We have attempted to show some of the complexities that arise when one considers pair dynamics in atomic and in molecular fluids. The fact that this is a difficult problem should not deter one from further analysis, since there are numerous observable quantities that are in theory dependent upon such dynamics.

We should compare the results obtained here for the generalized Kirkwood g-factors with the widely used alternative theory due to Kivelson and coworkers.[14,15] This theory actually exists in several forms, depending upon the level of Mori theory used, but the most common case is a first-order coupled variable form for the time dependence of both the self or autocorrelation of an appropriate Legendre function

$\langle\langle \cos\theta_1(t) \cos\theta_1(0) \rangle\rangle$ for example and the total $(\langle \sum_j \cos\theta_1(t) \cos\theta_j(0) \rangle\rangle$. As an

intermediate step; the cross-correlation $\langle\langle \cos\theta_1(t) \cos\theta_2(0) \rangle\rangle$ is also evaluated. Since Mori theory is used, the answer is expressed in terms of time-integrals over memory functions which are essentially correlations of the time-derivatives of the $\cos\theta_i$. In this regard, their theory is quite close to that given in this paper. Of course, the memory functions involved a projected time-dependence which is absent here. However, the cumulant approximation is equivalent setting the time-dependence of correlations

of $\dfrac{d}{dt}$ cos θ_i (t) equal to sin $\theta_i(0)$ $\dfrac{d}{dt}$ θ_i (t) and exponentiating the result. Perhaps the

most important difference between the treatments is the level of detail of the treatment of the cross angular velocity correlation. We feel that the distance and orientation dependence of this function should be explicitly included, in contrast to the treatment of Kivelson, et al.

The result most often quoted by other workers is that τ_M, the exponential relaxation time for the total orientational correlation, is related to τ_1, the single-molecule time, by

$\tau_M / \tau_1 = (1 + Nf) / (1 + N\dot{f})$ where f is the static Kirkwood factor and $\dot{f}$ (for $\ell = 1$) is

the ratio of the time integrals of $\left\langle \dfrac{d}{dt} \cos \theta_1 (t_p) \dfrac{d}{dt} \cos \theta_i (0) \right\rangle$ with i = 2 in the

numerator and 1 in the denominator and t_p denotes a projected time dependence. Many

workers have made the reasonable suggestion that N $\dot{f}$ is negligible compared to unity.

Since it is often found that Nf also is small compared to unity, τ_M is predicted to be close to τ_1. In the theory presented here, it is argued that the decay of the

cross-correlation part is given by cumulant involving a generalized form of N $\dot{f}$ in

which the distance and orientation dependence of this factor are included but the projection of the time-dependence is not. We expect the cross-correlation to decay

slowly because N $\dot{f}$ is small. Addition of the cross- and the auto-correlations to get the

total is then expected to yield a function containing two rather distinct terms: a large, relatively rapidly decaying contribution from the auto-correlation and a small but slowly decaying cross-correlation. Finally, we speculate that this cross-correlation may never show pure exponential decay even at long times because it is the resultant of a summation (more precisely, an integration) of many exponentials with decay constants and weight factors that depend upon the pair separation distance. It would be of considerable interest to simulate the rotational dynamics in liquids such as carbon dioxide or nitromethane where strong orientational correlations should be present. Such data could provide an interesting test of the theories presented in this paper and by Kivelson and coworkers. In a previous computer simulation study of mutual reorientation[10] attention was directed primarily to the time-dependence of coordinate-dependent functions. It is interesting to note that Singer et al.[16] have observed that the cross-correlation function decays slowly relative to the auto-correlation in simulated liquid Cl_2 - their result is reproduced in the article by Guillot and Birnbaum in this volume. Simulations of the velocity cross-correlations are now under way; it is hoped that generalizations relating such functions to intermolecular torques can be deduced in future work. In this way, one might hope to make progress toward the ultimate goal, which is of course to relate observables at a given temperature and density to functions of the intermolecular interaction law.

References

[1]   Abragam, A. (1961) The Principles of Nuclear Magnetism, Oxford University Press

[2]   Oppenheim, I. and Bloom, M. (1961) Can. J. Phys., 63, 845; Huang, L.P. and Freed, J.H., (1975) J. Chem. Phys., 63, 4017.

[3]   Mori, H., Oppenheim, I. and Ross, J. (1962) Studies in Statistical Mechanics, Vol. I, Ed. J. deBoer and G. F. Uhlenbeck, North Holland; Zwanzig, R. (1965) Ann. Rev. Phys. Chem., 16, 67.

[4]   Wertheim, M.S. (1979) Ann. Rev. Phys. Chem., 30, 471; Schroer, W. (1984) Adv. Chem. Phys. 56, 467; Adelman S. and Deutch, J. (1975) Adv. Chem. Phys. 31, 103

[5]   Berne, B. J. and Pecora, R., (1976) Dynamic Light Scattering, J. Wiley & Sons.

[6]   Lovesey, S. W. (1984) "Theory of Neutron Scattering from Condensed Matter", Clarendon Press.

[7]   See "Phenomena Induced by Inter-molecular Interactions" (1985) Ed. G. Birnbaum, NATO ASI series, Physics, Vol. 127, Plenum Press.

[8]   Posch, H. A., Balucani, U. and Vallauri R. (1984) Physica A, 123, 516.

[9]   Vesely, F. J. and Posch, H. A. (1988) Molec. Phys. 64, 97.

[10]   Vallauri, R. and Steele, W. A. (1987) Molec. Phys. 61, 1019

[11]   Vesely, F. J., private communication

[12]   Steele, W. (1987) Molec. Phys. 61, 1031.

[13]   St. Pierre, A. G. and Steele, W. A. (1981) Molec. Phys. 43, 123.

[14]   Madden, P. A. and Kivelson, D. (1984) Adv. Chem. Phys. 56, 467

[15]   Keyes, T. and Kivelson, D. (1972) J. Chem. Phys. 56, 1057; Kivelson, D. and Madden, P., (1975) Mol. Phys. 72, 1749; Keyes, T. (1972) Mol. Phys. 23, 737.

[16]   Singer, K., Singer, J.V.L. and Taylor, A.J., (1979) Mol. Phys. 37, 1239.

ROTATIONAL DIFFUSION THEORY OF NUCLEAR MAGNETIC SPIN–ROTATIONAL RELAXATION

JAMES Mc CONNELL
School of Theoretical Physics
Dublin Institute for Advanced Studies
Dublin 4
Ireland

ABSTRACT. A rotational diffusion theory of nuclear magnetic relaxation by spin—rotational interaction is described and is applied to the case when the molecule containing the relaxation nucleus has no special symmetry. The theory is based on Langevin—type equations and the use of a stochastic rotation operator. Explicit expressions for relaxation and correlation times are obtained for various molecular shapes.

1. Introduction

A theory of the rotational Brownian motion of a rigid asymmetric top molecule was proposed about ten years ago by Ford, Lewis and Mc Connell [1,2]. This theory took account of the inertial effects of the molecule and was based on the Euler—Langevin equations. These are referred to a set of cartesian axes with origin at the centre of mass of the molecule and coordinate axes coinciding with the principal axes of inertia. We call this set the inertial frame. In writing down the Euler—Langevin equations it is presupposed that the inertial tensor and the friction tensor have the same principal axes.

The rotational motion of the molecule was described in terms of the stochastic operator R(t) which specifies the rotation of the molecule about its centre of mass from its orientation at time zero to its orientation at time t. For dielectric relaxation [2] and most nuclear magnetic relaxation processes [3] the value of the ensemble average <R(t)> is sufficient to provide the physical quantities related to the relaxation. The notable exception to this general rule is nuclear magnetic relaxation by spin—rotational interaction, for which the value of R(t) itself is required.

In the investigation of dielectric relaxation the inclusion of the effect of the inertia of the molecule is crucial. This is not so for nuclear magnetic relaxation at its present stage of experimental accuracy. For simplicity of exposition we shall employ a theory in which inertial effects are neglected. Such a theory may be referred to as a rotational diffusion theory or the Debye approximation of the more general inertial theory. However we shall not, as in previous studies [4,5,6], restrict the calculation of spin—rotational relaxation times to the extreme narrowing approximation, in which the arguments of spectral densities are taken to be zero.

Th. Dorfmüller (ed.), Reactive and Flexible Molecules in Liquids, 97–105.

2.   BASIC THEORY OF SPIN–ROTATIONAL INTERACTION

Working in the inertial frame we express the spin–rotational interaction Hamiltonian $\hbar G(t)$ by [7]

$$\hbar\, G(t) = \hbar \sum_{\mu,\nu=1}^{3} S_{\mu} C_{\mu\nu} J_{\nu}, \tag{2.1}$$

where S is the spin operator of the nucleus under examination, $\hbar J$ is the angular momentum operator of the molecule to which the nucleus belongs and $C_{\mu\nu}$ is a component of the real three–by–three spin–rotation tensor. Adopting a semi–classical approach we replace $\hbar J_{\nu}$ by $I_{\nu}\omega_{\nu}(t)$, where $\omega_{\nu}(t)$ is the νth component of angular velocity of the molecule at time t and I_{ν} the corresponding moment of inertia.

It is convenient to define $b_{m\nu}$ $(m = -1,0,1;\ \nu = 1,2,3)$ by

$$b_{0\nu} = C_{3\nu};\ b_{\pm1,\nu} = \mp 2^{-1/2}\,(C_{1\nu} \mp iC_{2\nu}). \tag{2.2}$$

Then the spin–lattice relaxation time T_1 and the spin–spin relaxation time T_2 are given by

$$\frac{1}{T_1} = 2j\,(\omega_0),\ \frac{1}{T_2} = j(0) + j(\omega_0) \tag{2.3},$$

where ω_0 is the Larmor angular frequency,

$$j(\omega) = \tfrac{1}{2}\,[\,c(i\omega) + c\,(-i\omega)] \tag{2.4}$$

$$c(s) = \frac{1}{3\hbar^2} \sum_{\mu,\nu=1}^{3} \sum_{m,n=-1}^{1} (-)^{m} b_{n\mu}\, b_{m\nu}\, I_{\mu} I_{\nu}$$

$$\cdot \int_{0}^{\infty} e^{-st}\, <R(t)\,\omega_{\mu}(t)\,\omega_{\nu}(0)>_{n,-m}\ dt \tag{2.5}$$

The subscripts n, −m signify the n,−m element of the matrix representative with respect to the basis constituted by the spherical harmonics $Y_{1,-1}$, Y_{10}, Y_{11} referred to the inertial frame.

It is possible to express (2.1) in a laboratory coordinate system as [7]

$$\hbar \, G(t) = \sum_{q=-1}^{1} (-)^q \, H_{-q}(t) \, S_q.$$

Then the spin–rotational correlation time τ_{sr} is defined as the correlation time of $H_q(t)$, and it may be shown that [8]

$$\tau_{sr} = \frac{3\hbar^2}{kT} \frac{c(0)}{\displaystyle\sum_{\mu=1}^{3} \sum_{m=-1}^{1} (-)^m b_{m\mu} b_{-m\mu} I_\mu}. \tag{2.6}$$

Thus it is clear that the problem of calculating relaxation and correlation times reduces to evaluating

$$\int_0^{\infty} e^{-st} <R(t)\, \omega_\mu(t)\omega_\nu(0)> dt, \tag{2.7}$$

deducing from (2.5) the value of $c(s)$ and substituting it into (2.4) and (2.6).

3.   CALCULATION OF $c(s)$

In a basic study of spin–rotational relaxation [4] a lengthy expression was derived for (2.7) when the molecule in question has no special symmetry. It was then believed that the mathematical difficulties were such as to preclude a further analytical treatment of the problem. Recently a different approach has been made [9], namely, to study the conditions under which an analytical solution could be derived. It has been found that the necessary and sufficient condition is that the spin–rotational tensor be diagonal in the inertial frame.

Under this condition (2.2) reduces to

$$b_{03} = C_{33}, \quad b_{\pm 1,1} = \mp 2^{-\frac{1}{2}} C_{11}, \quad b_{\pm 1,2} = 2^{-\frac{1}{2}} i \, C_{22} \tag{3.1}$$

and in rotational diffusion theory (2.5) becomes

$$c(s) = \frac{kT}{3\hbar^2} \sum_{\mu=1}^{3} I_\mu \sum_{m,n=-1}^{1} (-)^m b_{n\mu} b_{m\nu} [B_\mu + s]^{-1} \delta_{n,-m}, \tag{3.2}$$

where $I_\mu B_\mu$ is the coefficient of rotational friction about the μth axis of the inertial frame. Now, by (3.1),

$$\sum_{m,n=-1}^{1} (-)^m \, b_{n\mu} \, b_{m\nu} \, \delta_{n,-m} = \sum_{m=-1}^{1} (-)^m \, b_{-m\mu} \, b_{m\mu} = \sum_{m=-1}^{1} b_{m\mu}^{*} \, b_{m\mu}$$

$$= \sum_{m=-1}^{1} |b_{m\mu}|^2. \tag{3.3}$$

Hence (3.1) – (3.3) yield

$$c(s) = \frac{kT}{3\hbar^2} \left[\frac{I_1 \, C_{11}^2}{B_1 + s} + \frac{I_2 \, C_{22}^2}{B_2 + s} + \frac{I_3 \, C_{33}^2}{B_3 + s} \right], \tag{3.4}$$

and in particular

$$c(0) = \frac{kT}{3\hbar^2} \left[\frac{I_1 \, C_{11}^2}{B_1} + \frac{I_2 \, C_{22}^2}{B_2} + \frac{I_3 \, C_{33}^2}{B_3} \right]. \tag{3.5}$$

4. RELAXATION AND CORRELATION TIMES FOR MOLECULAR MODELS

4.1. THE ASYMMETRIC ROTATOR MOLECULE

On substituting (3.4) into (2.4) we find that

$$j(\omega) = \frac{kT}{3\hbar^2} \left\{ \frac{I_1 \, C_{11}^2}{B_1 \, [1+(\omega/B_1)^2]} + \frac{I_2 \, C_{22}^2}{B_2 \, [1+(\omega/B_2)^2]} \quad \frac{I_3 \, C_{33}^2}{B_3 [1+(\omega/B_3)^2]} \right\}. \tag{4.1}$$

Hence, from (2.3),

$$\frac{I}{T_1} = \frac{2kT}{3\hbar^2} \left\{ \frac{I_1 \, C_{11}^2}{B_1 \, [1+(\omega_0/B_1)^2]} + \frac{I_2 \, C_{22}^2}{B_2 \, [1+(\omega_0/B_2)^2]} + \frac{I_3 \, C_{33}^2}{B_3 \, [1+(\omega_0/B_3)^2]} \right\} \tag{4.2}$$

$$\frac{I}{T_2} = \frac{kT}{3\hbar^2} \left\{ \frac{(2B_1^2+\omega_0^2)\ I_1\ C_{11}^2}{B_1(B_1^2 + \omega_0^2)} + \frac{(2B_2^2+\omega_0^2)\ I_2\ C_{22}^2}{B_2\ (B_2^2 + \omega_0^2)} + \right.$$

$$\left. \frac{(2B_3^3+\omega_0^2)\ I_3\ C_{33}^2}{B_3\ (B_3^2 + \omega_0^2)} \right\} \tag{4.3}$$

and in the extreme narrowing approximation where the argument of $j(\omega)$ is taken to be zero

$$\frac{I}{T_1} = \frac{I}{T_2} = \frac{2kT}{3\hbar^2} \left\{ \frac{I_1\ C_{11}^2}{B_1} + \frac{I_2\ C_{22}^2}{B_2} + \frac{I_3\ C_{33}^2}{B_3} \right\}. \tag{4.4}$$

In order to find the correlation time τ_{sr} from (2.6) we substitute from (3.3) and obtain

$$\sum_{\mu=1}^{3} \sum_{m=-1}^{1} (-)^m b_{m\mu} b_{-m\mu} I_\mu = \sum_{\mu=1}^{3} I_\mu \sum_{m=-1}^{1} |b_{m\mu}|^2$$

$$= I_1\ C_{11}^2 + I_2\ C_{22}^2 + I_3\ C_{33}^2.$$

Taking the value of $c(0)$ from (3.5) we then deduce that

$$\tau_{sr} = \frac{I_1\ C_{11}^2/B_1 + I_2\ C_{22}^2/B_2 + I_3\ C_{33}^2/B_3}{I_1\ C_{11}^2 + I_2\ C_{22}^2 + I_3\ C_{33}^2}. \tag{4.5}$$

4.2.   THE SYMMETRIC ROTATOR MOLECULE

We consider a molecule, for which the third coordinate axis of the inertial frame is an axis of rotational symmetry C_n with $n \geq 3$, and we examine the relaxation of a nucleus situated on this axis. Then clearly

$$I_2 = I_I, \quad I_2 B_2 = I_1 B_1, \tag{4.6}$$

and it may be shown [10] that $C_{\mu\nu}$ is a diagonal tensor with

$$C_{22} = C_{11}. \tag{4.7}$$

On substituting (4.6) and (4.7) into (4.2), (4.3) and (4.5) we deduce that

$$\frac{I}{T_1} = \frac{2kT}{3\hbar^2}\left\{\frac{2\ I_1\ C_{11}^2}{B_1\ [1+(\omega_0/B_1)^2]} + \frac{I\ C_{33}^2}{B_3\ [1+(\omega_0/B_3)^2]}\right\} \tag{4.8}$$

$$\frac{I}{T_2} = \frac{kT}{3\hbar^2}\left\{\frac{2(2B_1^2+\omega_0^2)\ I_1\ C_{11}^2}{B_1\ (B_1^2+\omega_0^2)} + \frac{2(2B_3^3+\omega_0^2)\ I_3\ C_{33}^2}{B_3\ (B_3^2+\omega_0^2)}\right\} \tag{4.9}$$

$$\tau_{sr} = \frac{2\ I_1\ C_{11}^2/B_1 + I_3\ C_{33}^2/B_3}{2I_1\ C_{11}^2 + I_3\ C_{33}^2}. \tag{4.10}$$

If the relaxing nucleus were not on the axis of rotational symmetry, it would be difficult to assert that $C_{\mu\nu}$ is a diagonal tensor in the inertial frame.

4.3.   THE SPHERICAL MOLECULE

When the molecule is spherical

$$I_3 = I_1 = I, \quad B_3 = B_1 = B$$

and (4.8) − (4.10) yield

$$\frac{I}{T_1} = \frac{2kTI}{3\hbar^2 B}\ \frac{2\ C_{11}^2 + C_{33}^2}{I+(\omega_0/B)^2} \tag{4.11}$$

$$\frac{I}{T_2} = \frac{kTI}{3\hbar^2 B}\ \frac{(2\ C_{11}^2+C_{33}^2)\ (2B^2+\omega_0^2)}{B^2+\omega_0^2} \tag{4.12}$$

$$\tau_{sr} = B^{-1}. \tag{4.13}$$

4.4.   THE CIRCULAR PLATE MOLECULE

Results for a circular plate molecule with the relaxing nucleus situated at the centre may be deduced from those for the symmetric top molecule by introducing the relation $I_3 = 2\ I_1$. Thus from (4.8) − (4.10) we deduce that

$$\frac{I}{T_1} = \frac{2kTI_3}{3\hbar^2}\left\{\frac{C_{11}^2}{B_1\ [1+(\omega_0/B_1)^2]} + \frac{C_{33}^2}{B_3\ [1+(\omega_0/B_3)^2]}\right\} \tag{4.14}$$

$$\frac{I}{T_2} = \frac{KTI_3}{3\hbar^2} \left\{ \frac{(2B_1^2+\omega_0^2)\ C_{11}^2}{B_1\ (B_1^2 + \omega_0^2)} + \frac{(2B_3^2+\omega_0^2)\ C_{33}^2}{B_3\ (B_3^2 + \omega_0^2)} \right\} \tag{4.15}$$

$$\tau_{sr} = \frac{C_{11}^2/B_1 + C_{33}^2/B_3}{C_{11}^2 + C_{33}^2}. \tag{4.16}$$

4.5   THE LINEAR MOLECULE

Analytical inertial theory for nuclear magnetic relaxation by spin–rotational inter-action for linear molecules has already been presented but its application was restric-ted to the extreme narrowing approximation [6]. We shall now evaluate c(s). Denot-ing by I the moment of inertia of the molecule about an axis through its centre and perpendicular to the line of the molecule and by IB the corresponding coefficient of rotational friction we take the line of the molecule as the third axis of the inertial frame. Thus the first and second axes are perpendicular to the molecule and to one another. According to eq. (22) of ref. 6

$$c(s) = -\frac{I^2}{3\hbar^2} \sum_{\substack{\mu,\nu=1,2 \\ m,n=-1,1}} b_{n\mu} b_{m\nu} \left(\int_0^\infty e^{-st}\ <R(t)\ \omega_\mu(t)\ \omega_\nu(0)>\ dt \right)_{n,-m}. \tag{4.17}$$

The value of the integral obtained from eq. (36) of ref. 6 by allowing γ to tend to zero and neglecting G in comparison with B, is

$$\frac{kT}{I}\ \frac{\delta_{\mu\nu}I}{B + s}. \tag{4.18}$$

On substituting (4.18) into (4.17) and noting that $C_{22} = C_{11}$ we find that

$$c(s) = \frac{2kTI\ C_{11}^2}{3\hbar^2\ (B + s)}. \tag{4.19}$$

We see that we can go from (3.4) to (4.19) by putting

$$I_1 = I_2 = I, \quad C_{22} = C_{11}, \quad C_{33} = 0, \quad B_2 = B_1 = B.$$

Thus we may deduce from (4.1), (4.2), (4.3) and (4.5) that the rotational diffusion results for the linear molecule are

$$j(\omega) = \frac{2kTI\ C_{11}^2}{3\hbar^2 B\ [1+(\omega/B)^2]}$$

$$\frac{I}{T_1} = \frac{4kTI\ C_{11}^2}{3\hbar^2 B\ [1+(\omega_0/B)^2]}$$

$$\frac{I}{T_2} = \frac{2kTI\ C_{11}^2\ (2B^2 + \omega_0^2)}{3\hbar^2 B\ (B^2 + \omega_0^2)}$$

$$\tau_{sr} = B^{-1}.$$

5. CONCLUSION

A rotational diffusion theory of nuclear magnetic relaxation by spin–rotational interaction based on a Langevin model has been presented for the asymmetric top molecule. To establish an analytical theory it is necessary to assume that the inertial, friction and spin–rotational tensors are simultaneously diagonalizable. This condition, which is commonly employed also in non–analytic calculations [11], restricts the range of validity of application, as we noted in subsection 4.2 when discussing the symmetric rotator molecular model. Explicit expressions have been derived for the spin–lattice and spin–spin relaxation times and for the spin–rotational correlation time, when the molecule is modelled as a sphere, circular plate, linear rotator, symmetric rotator or asymmetric rotator.

The present communication extends beyond the extreme narrowing approximation results of ref. 7. It may also complete the series of analytic papers on spin–rotational relaxation initiated in 1963 by Hubbard [12] and developed later in refs. 4,5,6,9 by employing the stochastic rotator operator technique.

REFERENCES

1. Ford, G.W., Lewis, J.T. and McConnell, J. (1979) "Rotational Brownian Motion of an asymmetric top", Physical Review A 19, 907–919.
2. McConnell, J. (1980), Rotational Brownian Motion and Dielectric Theory, Academic Press, London.
3. McConnell, J. (1987), The Theory of Nuclear Magnetic Relaxation in Liquids, Cambridge University Press.
4. McConnell, J. (1982), "Stochastic differential equation study of nuclear magnetic relaxation by spin–rotational interaction", Physica 111A, 85–113.
5. Mc Connell, J. (1982) "Nuclear magnetic spin–rotational relaxation times for symmetric molecules". Physica 112A, 479–487.
6. Mc Connell, J. (1982) "Nuclear magnetic spin–rotational relaxation times for linear molecules". Physica 112A, 488–504.

7.   Ref. 3, chap. 10.
8.   Ref. 3, eq. (10,29).
9.   McConnell, J. (1988) "Theory of nuclear magnetic spin–rotational relaxation times for asymmetric molecules". Physica 152A, 309–327.
10. Hubbard, P.S. (1974) "Nuclear relaxation in spherical–top molecules undergoing rotational Brownian motion", Physical Review A9, 481–494.
11. Lee, D.H. and McClung, R.E.D. (1987) "The Fokker–Planck–Langevin model for rotational Brownian motion IV. Asymmetric top molecules" Chem.Phys. 112, 23–41.
12. Hubbard, P.S. (1963) "Theory of nuclear magnetic relaxation by spin–rotational interaction in liquids", Physical Review 131, 1152–1165.

MODELS OF CONFORMATIONAL DYNAMICS

Giorgio J. Moro
Institute of Physical Chemistry, University of Parma,
viale delle Scienze, 43100 PARMA, Italy

Alberta Ferrarini, Antonino Polimeno and Pier Luigi Nordio
Department of Physical Chemistry, University of Padua,
via Loredan 2, 35131 PADOVA, Italy.

ABSTRACT. Discrete master equations can be obtained from diffusion-Smoluchowski equation in the presence of large barriers separating the potential minima, the treatment being equivalent to the derivation of the Kramers transition rates in the overdamped regime. The one-dimensional problem is considered as a test case to illustrate the projection procedure onto the subspace of localized functions. This method, however, generalizes the Kramers theory to intermediate potential barriers. The inertial effects are shortly discussed in relation to the numerical solutions for a bistable problem. The coupling between the overall rotation and the conformational transitions is analyzed in a molecule with one torsional degree of freedom. The generalization of the Kramers theory to multi-dimensional diffusion equations is presented with particular emphasis on the frictional coupling between reactive and non-reactive modes of the potential function. The model system of a linear chain of rotors is used to demonstrate that cooperative transitions during saddle point crossing arise as a consequence of the frictional coupling. The parametrization of the transition rates for an alkyl chain attached to a rigid core is summarized, together with the main results concerning the relaxation of conformer populations and the methylene rotational relaxation.

1. Introduction

In the past, Molecular Dynamics simulations were confined to liquids composed of rigid molecules. Improvements of the numerical algorithms and the availability of faster computers make now feasible simulations with flexible molecules [1]. Thus, fully microscopic descriptions of the conformational processes can be generated starting from elementary informations concerning inter- and intra-molecular potentials. Still, theoretical models based on the representation of few relevant degrees of freedom are required to understand the nature of conformational processes. In this way, for example, one can study the slowing down of conformational transitions with increasing potential barriers when Molecular Dynamics calculations become prohibitively difficult.

Because of their different purposes, Molecular Dynamics simulations and theoretical models should be considered as complementary methods. In opposition to the fully microscopic picture deriving from M.D., analytical methods intend to recognize the physical

107

Th. Dorfmüller (ed.), Reactive and Flexible Molecules in Liquids, 107–139.
© *1989 by Kluwer Academic Publishers.*

basis of the conformational motions independently of the detailed form of the interactions between the molecules in solution. On the other hand, the analysis of Molecular Dynamics results often requires the use of models in order to rationalize the correlations between computed observables.

The models can be formulated at different levels of accuracy depending on the choice of the relevant degrees of freedom. Starting from the torsional angles which normally represent the reaction coordinates for the conformational transitions, more detailed representations can be generated by including the overall rotations and translations, the bending and stretching degrees of freedom, or relevant solvent interactions.

Several theoretical tools are required in this modelling procedure. In particular stochastic equations are used to represent the time evolution of the continuous distributions on the set of relevant variables. A common choice for them are the Fokker-Planck equations or, when the overdamped limit is attained, the diffusion-Smoluchowski equations [2,3]. The potential function which determines the Boltzmann distribution on the relevant coordinates, and the friction matrix which characterizes the viscous drag generated by the solvent, are the main ingredients of these equations. Careful modelling of both are required when dealing with specific molecular systems.

A much simpler picture is given in terms of conformer populations governed by master equations for discrete variables. This is a kinetic representation which specify the transition rates between conformers. A fundamental problem is the derivation of the master equation from the more general stochastic equations for continuous variables. This is strictly related to the Kramers theory of kinetic rates [4].

Finally, the experimentally accessible observables, like NMR T_1's, dielectric relaxation profiles etc, should be calculated within a given model. With complex systems like chain molecules with several degrees of freedom, this is not an easy task and it requires the implementation of specific algorithms.

An extensive discussion of all these topics is outside the objectives of this communication. We shall focus our attention to the Kramers theory and its generalizations, by reviewing some applications to the conformational dynamics in relation to our past works.

2. One-dimensional models

Kramers was the first to consider activated processes in the context of the Brownian motion theory [4]. He calculated the kinetic rate from the particle flux at the saddle point of a bistable potential in the presence of non-equilibrium distributions between the states at the minima of the potential energy [4]. The same method was applied in most subsequent works about the generalization of the Kramers theory (see ref. 5 for a recent review). We shall follow an alternative route based on the calculation of the singular eigenvalues of the time evolution operator, considered first by Widom even if in a rather different context [6]. In this way, the kinetic regime is identified with the long-time behaviour of the correlation functions which are often used to represent the dynamical observables in modern statistical mechanics.

We shall consider in this chapter the simplest conformational problem in relation to a molecule having one torsional degree of freedom for the relative rotations of two molecular fragments about a common axis. This could be the case of the biphenyl molecule or the normal butane when the bending and stretching degrees of freedom are frozen in correspondence of their equilibrium values. In all generality, the coupling of the overall translations and rotations with the torsional motion should be taken into account (see the next section). In order to deal with one-dimensional problems, we suppose that one molecular fragment is rigidly held in the space, so that the torsional angle θ represents the rotations with respect to a laboratory frame, of the mobile fragment only. This corresponds to the limit case of a molecule composed by subunits with very different sizes.

In the overdamped limit, the diffusion-Smoluchowski equation (SE) can be used to described the evolution of the probability density $P(\theta; t)$

$$\partial P(\theta; t)/\partial t = -\Gamma P(\theta; t) \tag{2.1}$$

The diffusion operator can be written in all generality as:

$$\Gamma = -(\partial/\partial\theta)D(\theta)P_{eq}(\partial/\partial\theta)P_{eq}^{-1} \tag{2.2}$$

where P_{eq} is the equilibrium Boltzmann distribution determined by the torsional potential $V(\theta)$

$$P_{eq} = \exp[-V(\theta)/k_B T]/Z \tag{2.3}$$

and D is the diffusion coefficient calculated from the friction ξ exerted by the solvent

$$D = k_B T/\xi \tag{2.4}$$

The friction coefficient could be a function of the torsional angle because of hydrodynamic interactions between the two molecular fragments. Usually, the torsional potential is mainly determined by intramolecular interactions, even if it includes mean field contributions generated by the solvent [7].

When the torsional potential have well defined minima separated by large energy barriers, one expects that the system can be described by a discrete set of conformer populations $P_j(t)$. The time evolution of the system will then be characterized by a set of transition rates collected in the matrix $\mathbf{W}$ of the master equation (ME)

$$\partial P_j(t)/\partial t = -\sum_m W_{jm} P_m(t) \tag{2.5}$$

Evidently ME model is much simpler than SE and so the calculation of time dependent properties becomes much easier.

Two main questions arise:
i) when the use of ME instead of SE is legitimate ?
ii) How to obtain ME from SE ?

In order to give precise answers, we shall consider the simplest diffusion equation with a constant D and the symmetric double minimum potential given by the cosine function

$$V = (\Delta k_B T/2)(1 - \cos 2\theta) \tag{2.6}$$

The parameter Δ determines the height in $k_B T$ units of the barriers located at $\theta = \pi/2$ and $\theta = 3\pi/2$ (see Fig. 2). The transition rate matrix is simply given as

$$\mathbf{W} = \begin{pmatrix} w & -w \\ -w & w \end{pmatrix} \tag{2.7}$$

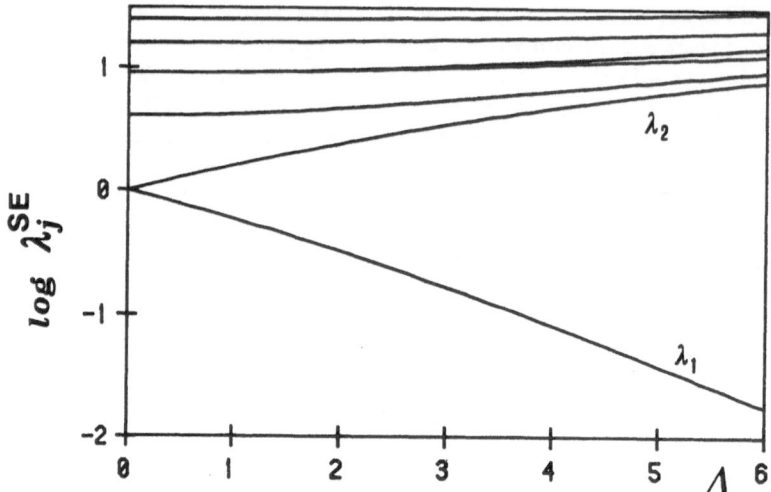

Figure 1. Barrier height dependence of the eigenvalues of the diffusion-Smoluchowski equation.

where w is the elementary rate for the interconversion of the two conformers.

It is convenient to compare the two models by looking at the corresponding eigenvalue problems. For the ME

$$\mathbf{W}\mathbf{x}_j = \lambda_j^{ME}\mathbf{x}_j \tag{2.8}$$

one readily obtains the two eigenvalues $\lambda_0^{ME} = 0$ and $\lambda_1^{ME} = 2w$. The eigenvalue problem for the diffusion equation is conveniently written in the following form

$$\Gamma\psi_j P_{eq} = \lambda_j^{SE}\psi_j P_{eq} \tag{2.9}$$

where the equilibrium distribution has been factorized from the eigenfunctions, with the orthogonality condition written as:

$$< \psi_j \mid P_{eq} \mid \psi_m > \equiv \int_0^{2\pi} d\theta \psi_j^* P_{eq}\psi_m = \delta_{jm} \tag{2.10}$$

No analytical solution of eq. (2.9) is available except for the stationary solution $\psi_0 = 1$ with $\lambda_0^{SE} = 0$.

The eigenvalues and the eigenvectors can be used to explicit the time dependence of a non-equilibrium distribution or of a generic correlation function. In this way, the fundamental difference between ME and SE will emerge in the decay to the equilibrium. While the master equation generates a mono-exponential relaxation, multi-exponential decay is found with the diffusion equation. Therefore, correlation functions calculated with ME should always be considered as approximations of the more general results obtained from SE. The conditions of validity of such approximation are derived from the dependence of λ_j^{SE} on the barrier height.

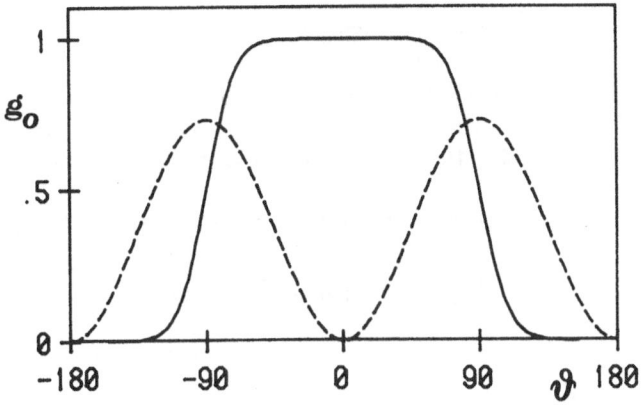

Figure 2. Localized function $g_0(\theta)$ for $\Delta = 8$ (continuous line). The potential function given by eq. (2.6) is also represented by a dashed line for comparison purposes.

In Fig. 1 we have represented the eigenvalues obtained from the numerical solution of eq. (2.9). For Δ large enough, the first positive eigenvalue λ_1^{SE} is well separated from the others. Correspondingly, the long-time behaviour of a non-equilibrium distribution is substantially described by a single exponential. Therefore for $\Delta \geq 2$ the master equation may be used to approximate the results of the diffusion equation under the condition of neglecting the multi-exponential initial decay.

As long as $\lambda_1^{SE} \ll \lambda_2^{SE}, \lambda_3^{SE} ...$, the eigenfunction ψ_1 represents a "slow mode" of the diffusion operator. The profile of ψ_1 clarifies its nature. In fact ψ_1 is almost constant, like the stationary solution ψ_0, with values ± 1 in correspondence of the two minima, and it changes sign abruptly near the location of the potential maxima. Therefore, ψ_1 is associated to the transition through the potential barriers in the presence of a quasi-equilibrium distribution inside the potential wells. Eigenfunctions ψ_j with $j \geq 2$ have a well-defined θ-dependence near the potential minima and they represent the fast modes which restore the equilibrium distribution inside the potential wells.

Let us now examine the second question about the derivation of ME from SE. Of course this problem is strictly related to the calculation of the transition rates from the diffusion equation, and it has been considered by several authors with different methods [8-13]. We shall apply the localized function method [15,16], which leads unambiguously to the definition of conformer populations. In essence, it is a projection procedure onto the subspace of slow modes including the stationary solution. A derivation which utilizes the exact eigenfunctions is presented here.

The localized functions $g_0(\theta)$ and $g_\pi(\theta)$ associated to the minima at $\theta = 0$ and $\theta = \pi$ respectively, are given as:

$$g_0 = (\psi_0 + \psi_1)/2 \qquad (2.11a)$$

$$g_\pi = (\psi_0 - \psi_1)/2 \qquad (2.11b)$$

Note that because of the translational symmetry of the potential, $g_\pi(\theta)$ is obtained from $g_0(\theta)$ by a π-shift of the argument θ. The function g_0 represented in Fig. 2 is nearly zero for $\theta = \pi$ and it steeply rises in correspondence of the potential maxima, almost reaching an unitary value for $\theta = 0$. By increasing the potential barrier, g_0 becomes even more sharply defined and in the limit $\Delta \rightarrow \infty$ it becomes a step function selecting the range $-\pi/2 \leq \theta \leq \pi/2$, that is the domain under the influence of the associated potential minimum.

Let us introduce the projection operator P onto the subspace of localized functions

$$P = \sum_m P_{eq} \mid g_m > Q_m^{-1} < g_m \mid \tag{2.12}$$

where

$$Q_m = < g_m \mid P_{eq} > \tag{2.13}$$

Because of eqs. (2.11a,b), such subspace can be defined alternatively as the linear combinations of the slow modes. The operator P can then be used to project the diffusion equation onto the subspace of slow modes, so obtaining the master equation [16], with the conformer populations and the transition rate matrix given as:

$$P_m(t) = < g_m \mid P(\theta; t) > \tag{2.14}$$

$$W_{jm} = < g_j \mid \Gamma P_{eq} \mid g_m > /Q_m = < \partial g_j/\partial\theta \mid D P_{eq} \mid \partial g_m/\partial\theta > /Q_m \tag{2.15}$$

The quantities Q_m should be identified with the equilibrium populations in agreement with eq. (2.14) and, in fact, they are the components of the stationary solution of the matrix $\mathbf{W}$ given in eq (2.15). In conclusion, the functions g_m transform the continuous distribution $P(\theta; t)$ in a discrete set of conformer populations in correspondence of localized states according to the shape of the potential.

In principle this procedure could be generalized to potentials with several minima and without particular symmetries by finding the suitable combinations of the slow modes. However, it would have a very limited utility, since it requires the knowledge in advance of the exact eigenfunctions of the diffusion equation. Direct derivation of approximate localized functions can be performed by modifying the method proposed by Larson and Kostin to calculate the singular eigenvalues of the diffusion equation [9]. Since the eigenvalues associated to the slow modes vanish exponentially with increasing potential barriers so that the right hand side of eq. (2.9) becomes negligible, the equation

$$\Gamma \psi_j P_{eq} = 0 \tag{2.16}$$

can be used to generate approximate eigenfunctions. The same equation can be applied to the localized functions since they are linear combinations of slow modes. Equation (2.16) has two independent solutions: the constant function, i.e. the stationary solution, and the function

$$f(\theta) = \int_{\theta_0}^{\theta} d\alpha [P_{eq}(\alpha) D(\alpha)]^{-1} \tag{2.17}$$

However, this function does not satisfy the periodicity condition when taken in the full range of the torsional angle. It can be used only to approximate locally the solutions of the diffusion equation. By properly matching functions f with different intial conditions, one can derive localized functions which mimic the behaviour of g_0 displayed in Fig. 1 [16].

The approximate localized function g_m associated to the minimum θ_m is null for θ outside the range $\theta_{m-1} \leq \theta \leq \theta_{m+1}$, while inside is given by the equation:

$$g_m(\theta) = \int_{\theta}^{\theta_{m\pm1}} d\alpha[D(\alpha)P_{eq}(\alpha)]^{-1} \bigg/ \int_{\theta_m}^{\theta_{m\pm1}} d\alpha[D(\alpha)P_{eq}(\alpha)]^{-1} \qquad (2.18)$$

with the positive (negative) sign in the upper integration limit taken for θ greater (less) than θ_m. One easily realizes that the localized functions have an error function shape in proximity of the saddle points θ_S with a slope determined by the the second derivative $V_S^{(2)}$ of the potential calculated at θ_S:

$$g_m \approx c_1 + c_2 \; \mathrm{erf}\left[(\theta - \theta_S)\sqrt{(|V_S^{(2)}|/2k_BT)^{1/2}}\right] \qquad (2.19)$$

By projecting the diffusion equation onto the subspace of approximate localized functions, the master equation is obtained with time dependent conformer populations, the transition rate matrix and the equilibrium populations given as before [16].

The Kramers result for the diffusion equation can now be recovered as the asymptotic results with respect to infinite energy barriers ($\Delta V/k_BT \to \infty$). This can be easily done by applying the Laplace method [17] to evaluate the integrals in eqs. (2.13) and (2.15) after substitution of eq. (2.18) for the localized functions. The resulting equilibrium populations are

$$Q_j = \exp(-E_j/k_BT)/\sum_m \exp(-E_m/k_BT) \qquad (2.20)$$

where E_j is the "free energy" of the j-th conformer, that is the local value $V(\theta_j)$ of the potential energy increased by an entropy term calculated from the potential curvature $V_j^{(2)} \equiv (\partial^2 V/\partial\theta^2)_{\theta=\theta_j}$ [18]

$$E_j = V(\theta_j) + (k_BT/2)\ln(|V_j^{(2)}|/2\pi k_BT) \qquad (2.21)$$

The elements of $\mathbf{W}$ vanish in correspondence of conformers separated by more than one saddle point, while the transition rate for nearest-neighbour minima is given as

$$W_{jm} = -(D_S |V_S^{(2)}|/2\pi k_BT) \exp[-(E_S - E_m)/k_BT] \qquad (2.22)$$

with the activation energy calculated as the difference between the energy of the saddle point defined in analogy to eq. (2.21), and the free energy of the starting conformer (note that $-W_{jm}$ is the kinetic rate for the transition $m \to j$). The values at the saddle point of the diffusion coefficient, $D_S \equiv D(\theta_S)$, and of the potential curvature determine the pre-exponential factor. The diagonal elements of $\mathbf{W}$ are derived by imposing the condition that its stationary solution is given by the equilibrium populations given in eq. (2.20).

The model diffusion equation with the cosine potential eq. (2.6) can be used to establish the range of validity of the approximations. In Fig. 3, the numerical value of λ_1^{SE} and its approximation derived from the master equation, are presented as function of the potential

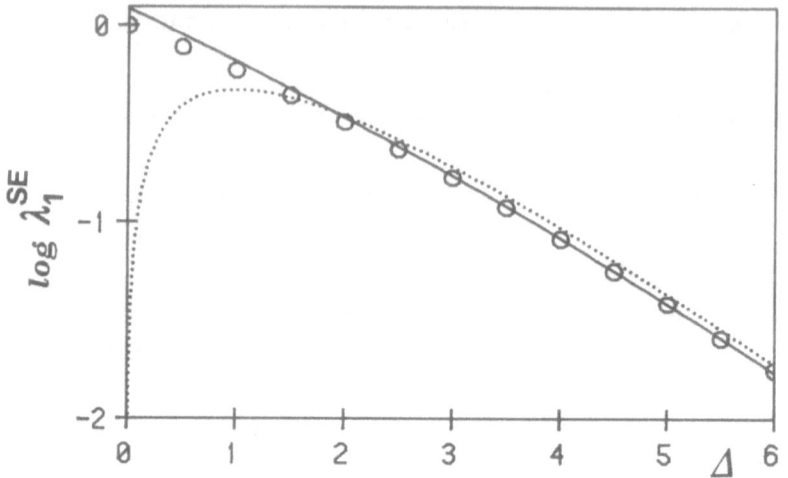

Figure 3. Slowest relaxation rate for the cosine potential as function of the barrier height. Circles: λ_1^{SE} from the numerical solution of the diffusion-Smoluchowski equation; continuous line: λ_1^{ME} with $\mathbf{W}$ computed by means of approximate localized functions; dotted line: λ_1^{ME} from the asymptotic kinetic rates.

barrier. Only for large potential barriers ($\Delta > 10$) the Kramers result is a reasonably good approximation (deviations less than 5%). The transition rate matrix calculated by numerical integration of the approximate localized functions given in eq. (2.18), extends the asymptotic approximation to intermediate values of the barrier height. In fact, for $\Delta > 2$ it approximates the exact eigenvalue within 5%. By increasing the values of Δ, the localized function results converge much faster than the asymptotic approximations. For example, the error with the former is less than 0.5% already for $\Delta = 5$. Therefore, when it is possible, calculation of transition rates with the localized function method should always be preferred to the use of asymptotic results.

When the condition of overdamped motion is removed, the more general Fokker-Planck equation (FP) should be used for the distribution $P(\theta, \dot{\theta}, ; t)$ on the torsional angle and torsional velocity $\dot{\theta} = d\theta/dt$ [2-4,11]

$$\partial P(\theta, \dot{\theta}; t)/\partial t = \left\{ -\dot{\theta} \frac{\partial}{\partial \theta} + I^{-1} \frac{\partial V}{\partial \theta} \frac{\partial}{\partial \dot{\theta}} + \frac{\xi k_B T}{I^2} \frac{\partial}{\partial \dot{\theta}} P_{eq} \frac{\partial}{\partial \dot{\theta}} P_{eq}^{-1} \right\} P(\theta, \dot{\theta}; t) \qquad (2.23)$$

The momentum of inertia I enters as new parameter and P_{eq} is now the full Boltzmann distribution on angle and velocity variables. The Fokker-Planck equation defines two characteristic frequencies: the streaming frequency $\omega_s = (k_B T/I)^{1/2}$ associated to the conservative streaming motion, and the collisional frequency $\omega_c = \xi/I$ describing the rate of angular velocity randomization. Different motional regimes are predicted by eq. (2.23) depending on the ratio ω_c/ω_s which is proportional to the friction ξ. This is illustrated in Fig. 4 where we have reported the first positive eigenvalue λ_1^{FP} computed numerically

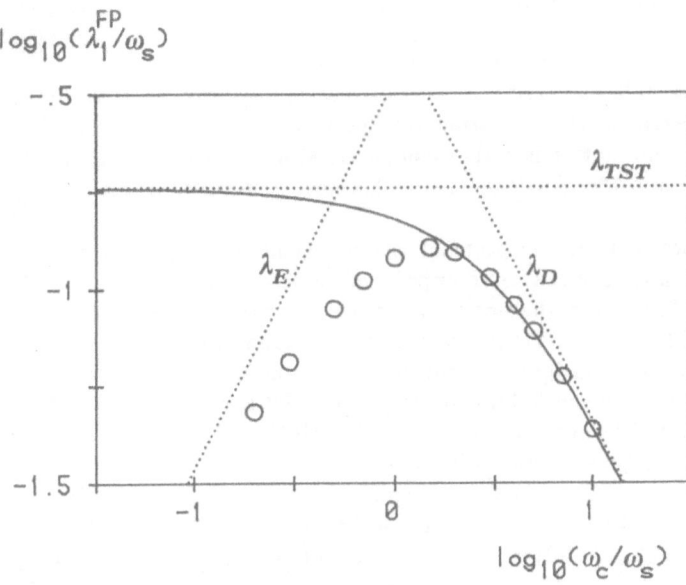

Figure 4. Lowest positive eigenvalue of the Fokker-Planck equation as function of the ratio ω_c/ω_s. The circles indicate the numerical results with a potential barrier of $2k_BT$ units. Continuous and dotted lines represent different approximations discussed in the text.

from the Fokker-Planck equation with a symmetric bistable potential given as the quartic polynomial of a cartesian coordinate [19]. In the Figure, λ_D, λ_E, λ_{TST} and λ_{ID} indicate different approximations described below.

The condition $\omega_c \gg \omega_s$ defines the overdamped limit when the angular velocity relaxes much faster than the coordinate. The interconversion process can then be described by a diffusion-Smoluchowski equation with the rate inversely proportional to the friction. Because of the logarithmic scales used in Fig. 4, the approximation $\lambda_D \equiv \lambda_1^{SE}$ appears as a straight line. The energy becomes the slow variable in the opposite limit of underdamped motion ($\omega_c \ll \omega_s$). So, a diffusion equation on the energy can be used to calculate the relaxation of the conformer populations [4], with a kinetic rate which is now proportional to the friction (approximation λ_E).

The transition between these two opposite limits is operating in the intermediate range of the friction, i.e. for $\omega_c \approx \omega_s$. The numerical values of λ_1^{FP} display the characteristic bell-shaped behaviour often called Kramers turnover, with the transition rate reaching the largest value for the given potential function. An upper bound to the kinetic rate can be derived by means of the Transition State Theory, that is by calculating the one-way particle flux at the saddle point in the presence of an overall equilibrium distribution (approximation λ_{TST}) [20].

A further approximation was derived by Kramers from the parabolic expansion of the potential at the saddle point, in order to connect the overdamped limit with the result of the Transition State Theory [4]. This asymptotic approximation can be extended to intermediate values of the potential barrier by means of the localized function method [19], so generating the approximation λ_{ID} in Fig. 4. In this way, substantial improvements with respect to the overdamped approximation are achieved with decreasing values of the friction. However, a gap exists between λ_E and λ_{ID} approximations. Recently, some efforts have been directed to the derivation of an unified kinetic rate covering all the friction range [21-25]. Still, the matter is rather controversial and it waits a definitive solution, even if the Mel'nikov and Meshkov equation [25] seems to be closer to the exact numerical results [19].

As a matter of fact, the problem is more general than the derivation of an analytical relation connecting the overdamped and underdamped solutions of the Fokker-Planck equation. In fact, when the potential function has several minima, qualitatively different behaviours of the population relaxation can be found in the two limits. While only transitions between nearest-neighbour conformers are recovered from the diffusion-Smoluchowski equation in the asymptotic limit of large potential barriers, multiple transitions between minima separated by more than one saddle point are obtained from the energy diffusion equation [26]. In general, multiple transitions should be expected when the energy relaxation is much slower than the free streaming motion, that is when the collisional processes, which deactivate the system during the saddle point crossing, occur in a time scale longer than the conservative motion over a sequence of potential barriers. Quenching of the multiple transitions is expected in the intermediate friction regime. Some evidence of multiple transitions has been found in theoretical calculations with the reactive flux formalism of the surface diffusion of atoms [27].

Multiple transitions might be important for the conformational dynamics of alkanes in low viscosity solvents or in the gas phase. Let us consider the normal butane whose torsional potential is well known [28-30], and characterized by three minima in correspondence of the trans (t) and the two equivalent gauche ($g_\pm$) states. Usually, only the $t \leftrightarrow g_\pm$ transitions are considered while direct transitions between gauche states are neglected in consideration of the large potential barrier separating them. However, this is strictly correct only in the overdamped limit. In the case of low friction, the multiple transition processes could activate the interconversion $g_- \leftrightarrow g_+$ with trajectories crossing both saddle points between gauche and trans states [1].

In conclusion, by passing from one-dimensional diffusion-Smoluchowski equations to the corresponding Fokker-Planck equations, a much more complex picture of the conformational dynamics is obtained. Even if the underlying physical phenomena are well understood, a complete formal treatment which derives accurate kinetic rates from the full Fokker-Planck equation without resorting to lengthy numerical procedures, is not yet available. Therefore, in the following sections concerning the coupling between different degrees of freedom, we shall confine the treatment to the diffusion-Smoluchowski level, so that one-dimensional results can be generalized without ambiguities.

3. Coupling of overall rotations with conformational dynamics

In this chapter, we shall consider the full roto-translational problem for a molecule having one torsional degree of freedom. In Fig. 5, the system is schematically represented as the ensemble of two rigid bodies with arbitrary shape. The line connecting the two centers O_1 and O_2 is the axis of internal rotation, with the angle θ determining the relative orientation

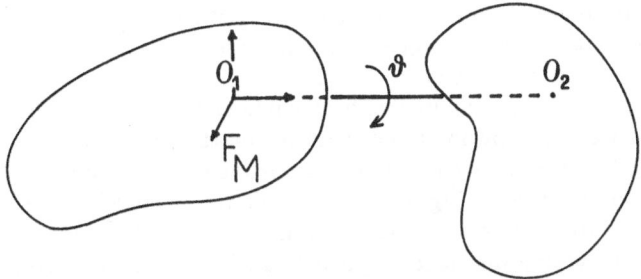

Figure 5. The model system with one internal degree of freedom.

of the two subunits.

Within a diffusion-Smoluchowski formalism, seven coordinates are required to represent the state of the system at a given time: the three components of the position vector $\mathbf{r}$ of the molecular frame F_M of the first subunit with respect to a laboratory frame F_L, the three Euler angles $\mathbf{\Omega} = (\alpha, \beta, \gamma)$ for the rotation from F_L to F_M, and the torsional angle θ. The diffusion equation for the probability density $P(\mathbf{r}, \mathbf{\Omega}, \theta; t)$ is written in all generality as [31]

$$\partial P(\mathbf{r}, \mathbf{\Omega}, \theta; t)/\partial\theta = -\nabla^\dagger \mathbf{D} P_{eq} \nabla P_{eq}^{-1} P(\mathbf{r}, \mathbf{\Omega}, \theta; t) \tag{3.1}$$

where P_{eq} indicates the equilibrium Boltzmann distribution specified by the torsional potential $V(\theta)$ according to eq. (2.3). We do not consider here the case of anisotropic environments, like nematic liquid crystals, which would require a potential with an explicit dependence on the orientation $\mathbf{\Omega}$. In eq. (3.1), ∇ is the gradient operator with respect to the seven degrees of freedom. Its translational (T), rotational (R) and internal (I) components can be written in the following form

$$\nabla = \begin{pmatrix} \nabla_T \\ \nabla_R \\ \nabla_I \end{pmatrix} = \begin{pmatrix} \mathbf{S}(F_M \leftarrow F_L)\partial/\partial\mathbf{r} \\ \partial/\partial\mathbf{\Phi} \\ \partial/\partial\theta \end{pmatrix} \tag{3.2}$$

where $\mathbf{S}(F_M \leftarrow F_L)$ is the Euler rotation matrix which generates the vector components in F_M from those expressed in F_L. The rotational gradient is written as the differential operator with respect to the components in the F_M frame of an infinitesimal rotation vector $\mathbf{\Phi}$.

The 7×7 diffusion matrix $\mathbf{D}$ can be partitioned in blocks with respect to the translational, rotational and internal components of the gradient operator

$$\mathbf{D}(\theta) = \begin{pmatrix} \mathbf{D}_{TT} & \mathbf{D}_{TR} & \mathbf{D}_{TI} \\ \mathbf{D}_{RT} & \mathbf{D}_{RR} & \mathbf{D}_{RI} \\ \mathbf{D}_{IT} & \mathbf{D}_{IR} & D_{II} \end{pmatrix} \tag{3.3}$$

Because of the choice of the translational and rotational gradient operators, the components of $\mathbf{D}$ are expressed in the molecular frame and, therefore, they depend only on the torsional angle θ. The diffusion matrix can be calculated, like in eq. (2.4), from the 7×7 friction matrix which determines the linear relations between the frictional force, the frictional torque and the frictional internal torque, on one hand, and the velocity, the angular velocity and the torsional velocity, on the other hand.

The calculation of the diffusion matrix for flexible molecules has been widely considered from the point of view of imposing the constraints on the equations of motion of independent subunits [32-37]. Equivalent results can be derived by mechanical composition of the frictional forces and torques acting on each subunit (see Appendix A of ref. 31). In this way, models for the diffusion matrix can be easily generated from the friction matrices of each subunit and by taking into account their hydrodynamical interactions.

Equation (3.1) can be used to describe the internal dynamics. Since the torsional potential and the diffusion matrix depend only on the θ variable, a reduced equation is obtained for the distribution $P(\theta; t)$ on the torsional angle

$$\partial P(\theta; t)/\partial t = (\partial/\partial\theta) D_{II}(\theta) P_{eq} (\partial/\partial\theta) P_{eq}^{-1} P(\theta; t) \tag{3.4}$$

It has the same form of the diffusion equation considered in the previous section and, therefore, the same analysis can be applied so deriving the master equation

$$\partial P_j(t)/\partial t = -\sum_m W_{jm} P_m(t) \tag{3.5}$$

Direct observation of the conformational dynamics, like in photochemical isomerization measurements [38], is rarely possible. With NMR or dielectric relaxation experiments, one investigates the relaxation of a tensorial property of a given subunit, which is modulated by both the overall rotation and the internal motion. The reduced probability density $P(\Omega, \theta; t)$ is required to calculate the corresponding correlation functions, and a simplified diffusion equation is obtained from eq. (3.1) by eliminating the translational components of the gradient operator. Then, one would like to simplify further the problem with a discrete representation of the torsional variable in terms of the probability density $P_j(\Omega; t)$ for the orientational distribution of the j-th conformer. The localized function method allows one to perform such reduction procedure [31], and so the following time evolution equation is obtained

$$\partial P_j(\Omega; t)/\partial t = -\frac{\partial}{\partial \Phi}^{\dagger} \mathbf{D}_j^R \frac{\partial}{\partial \Phi} P_j(\Omega; t) - \sum_m W_{jm} \mathcal{R}(\Xi_{jm}) P_m(\Omega; t) \tag{3.6}$$

The first term of the r.h.s. of eq. (3.6) describes the rotational motion of the j-th conformer in the absence of internal dynamics. In fact, the 3×3 matrix $\mathbf{D}_j^R$ represents the rotational diffusion tensor for the overall molecule with the torsional angle frozen at the value θ_j. The second term in the r.h.s. of eq. (3.6) takes into account the effects of conformational transitions with the kinetic rates already given in eq. (3.5). A new ingredient is the operator $\mathcal{R}(\Xi_{jm})$ which transforms the function $P_m(\Omega; t)$ by rotating the molecular frame F_M according to the Ξ_{jm} set of Euler angles. In order to understand the underlying physical phenomena, let first consider eq. (3.6) without this operator. Such equation would imply that for a given orientation Ω, $P_m(\Omega; t)$ contributes to $P_j(\Omega; t)$ during the $m \to j$ transition or, in other words, that the orientation of the F_M frame fixed on the first subunit does not change during this transition. In general, this cannot be true, since rotations of both fragments should compensate the change $\theta_m \to \theta_j$ of the torsional angle.

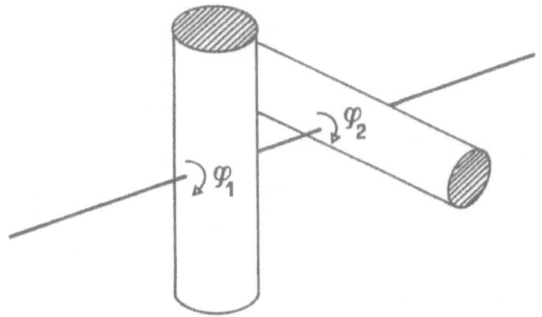

Figure 6. The model system of two coupled rotors.

Then, the first subunit together with the molecular frame F_M should reorient according to the set Ξ_{jm} of Euler angles, in the following called recoil rotation for the $m \to j$ transition.

In order to illustrate the effects of recoil rotations, it is useful to consider the simplified system of two coupled rotors constrained to rotate about a common axis. The recoil rotations of the first rotor are given now by the shifts β_{jm} of the corresponding angle φ_1, and so eq. (3.6) can be simplified in the following form

$$\partial P_j(\varphi_1; t)/\partial t = D_j^R \partial^2 P_j(\varphi_1; t)/\partial \varphi_1^2 - \sum_m W_{jm} P_m(\varphi_1 - \beta_{jm}; t) \qquad (3.7)$$

Let us indicate with $(\varphi_1^i, \varphi_2^i)$ and $(\varphi_1^f, \varphi_2^f)$ the angular positions of the two rotors before (i) and after (f) the transition $\theta_m \to \theta_j$. By definition, the quantity $\beta_{jm} \equiv \varphi_1^f - \varphi_1^i$ defines the recoil rotation of the first rod. During the given transition, the overall angular distribution is displaced, so that only $P_m(\varphi_1^i; t) = P_m(\varphi_1^f - \beta_{jm}; t)$ contributes to $P_j(\varphi_1^f; t)$. This is in agreement with eq. (3.7), if the angle φ_1 is identified with φ_1^f.

Instead of using these phenomenological arguments, eq. (3.7) can be derived in a more formal way by means of the localized function method, so supplying an explicit relation for the recoil rotations [39]:

$$\beta_{jm} = - \int_{\theta_j}^{\theta_m} d\theta\, D_{RI}(\theta)/D_{II}(\theta) \qquad (3.8)$$

The scalar quantity D_{RI} replaces the array $\mathbf{D}_{RI}$ of eq. (3.3), because one coordinate describes the overall rotation of the system instead of the three Euler angles. The generalization to the full rotational problem is given in terms of a differential equation [31]:

$$\delta \mathbf{\Psi} = -\mathbf{D}_{RI}(\theta) D_{II}(\theta)^{-1} \delta \theta \qquad (3.9)$$

where the vector $\delta \mathbf{\Psi}$ represents the infinitesimal rotation of F_M frame for the change $\theta \to \theta + \delta\theta$ of the torsional angle. The Euler angles Ξ_{jm} are derived by numerical integration of eq. (3.9) in the range $\theta_j \le \theta \le \theta_m$. Figure 7 is a pictorial representation of the recoil rotations for $g_{\pm} \leftrightarrow t$ transitions of the butane molecule. A bead model for the friction has been used [40,41]. The most evident effect is the opposite $60°$ rotations of the terminal $C - C$ bonds. At the same time, the central $C - C$ bond changes its orientation by about

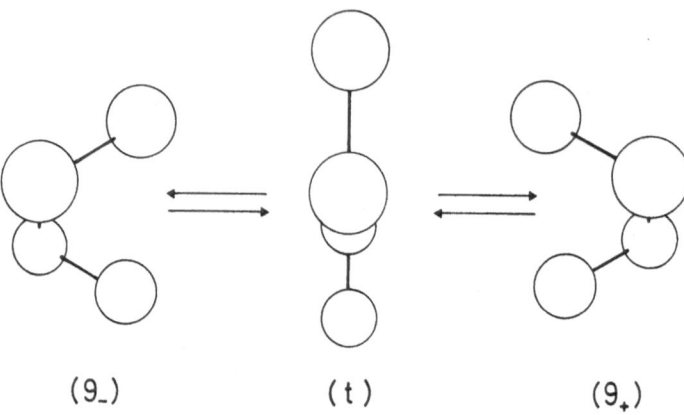

(9_-) (t) (9_+)

Figure 7. recoil rotations of the butane molecule.

16° in order to minimize the displacement of the centers of the ethyl fragments. With highly symmetric molecules, the recoil rotations can be anticipated without numerical solution of eq. (3.9). In the biphenyl molecule, for example, the change of the torsional angle should determine equal, but opposite rotations of the phenyl rings without reorientation of the long molecular axis.

A simplified form of eq. (3.6) can be used when the first segment bearing the F_M frame has a much larger size than the second fragment (the mobile fragment). Then, the rotational diffusion tensor becomes independent of the conformational state, since it is substantially determined by the shape of the first subunit only. Moreover, the operator $\mathcal{R}(\boxminus_j; m)$ can be neglected since the first segment suffers almost no recoil rotations, because only the mobile fragment rotates during the conformational transition. Under these conditions, the overall motion and the conformational transitions are decoupled, and the time evolution is determined by the simple superposition of these two motions. In this way, one recover the model introduced by Woessner to describe the NMR relaxation times of a rotating group [42]. The same kind of decoupling procedure has been used in other theoretical studies about NMR [43-47] or dielectric [48] relaxation of flexible molecules. The use of such models is justified only when the rotating groups have a very different size. On the contrary, equation (3.6) provides a much more general model without such an assumption. It should be emphasized that the model is invariant with respect to the choices of the molecular frame specifying the orientation Ω. A complementary equation can be written by using a F_M frame fixed on the second fragment. The recoil rotations are defined in such a way that the same behaviour of physical observables is derived from the two alternative equations [31].

4. Transition rates from multi-dimensional diffusion equations

The study of conformational dynamics in chain molecules like alkanes, requires the analysis of the coupling among several torsional degrees of freedom. The one-dimensional Kramers theory is no longer sufficient to calculate the transition rates. Generalization to multi-dimensional problems has been considered in the past from different point of view [49-55]. We shall present here an analysis of the diffusion-Smoluchowski equation by using the localized function method.

Let us consider N coordinates represented by the array $\mathbf{x}$ with a diffusion equation written in all generality as:

$$\partial P(\mathbf{x};t)/\partial t = -\Gamma P(\mathbf{x};t) \tag{4.1}$$

$$\Gamma = -\frac{\partial}{\partial \mathbf{x}}^{tr} \mathbf{D}(\mathbf{x})P_{eq}\frac{\partial}{\partial \mathbf{x}}P_{eq}^{-1} \tag{4.2}$$

with the equilibrium distribution determined by the potential $V(\mathbf{x})$ according to eq. (2.3), and the superscript tr indicating transposed arrays. Like in the one-dimensional problem, the master equation representation should be adequate if the potential have a well defined set of minima separated by large barriers. Moreover, the same projection procedure could be applied, if the complete set of localized functions are available. Again, the localized functions could be written as linear combination of slow modes of the diffusion operator, but only their approximations are useful in practice.

One difficulty immediately arises in the generalization of the localized functions. In the asymptotic limit of a very steep potential, they should resemble step functions selecting attraction domains for each minimum in the full phase space. While in one-dimensional problems the potential maxima naturally provides the boundaries of such domains, their derivation in multi-dimensional problems is not a trivial task and it requires the deterministic approximation of the diffusion equation [51].

With the probability density factorized with respect to the equilibrium distribution

$$P(\mathbf{x};t) = P_{eq}(\mathbf{x})P'(\mathbf{x};t) \tag{4.3}$$

the time evolution equation (4.1) can be rewritten as

$$\partial P'(\mathbf{x};t)/\partial t = \left[\frac{\partial}{\partial \mathbf{x}} - \frac{1}{k_B T}\frac{\partial V}{\partial \mathbf{x}}\right]^{tr} \mathbf{D}(\mathbf{x})\frac{\partial}{\partial \mathbf{x}}P'(\mathbf{x};t) \tag{4.4}$$

The asymptotic limit of very narrow distributions around the minima is reached by considering infinitely steep potentials or, for a given potential, by taking the low temperature limit. Then, the gradient operator between square brackets in eq. (4.4) can be neglected, so generating the following approximation:

$$\partial P'(\mathbf{x};t)/\partial t \approx \mathbf{b}(\mathbf{x})^{tr}\partial P'(\mathbf{x};t)/\partial \mathbf{x} \tag{4.5}$$

$$\mathbf{b}(\mathbf{x}) = -(1/k_B T)\mathbf{D}(\mathbf{x})\partial V/\partial \mathbf{x} \tag{4.6}$$

This is a first order partial differential equation similar to the Liouville equation in classical mechanics. Therefore, its solutions can be derived from the deterministic trajectories specified by a set of ordinary differential equations

$$d\mathbf{x}(t)/dt = \mathbf{b}(\mathbf{x}) \tag{4.7}$$

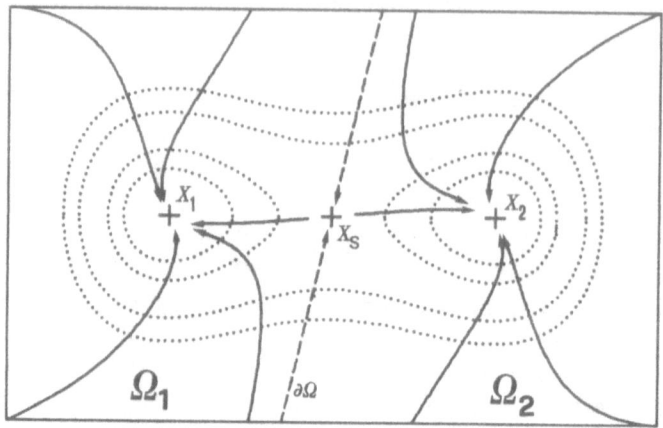

Figure 8. Deterministic trajectories for a two-dimensional problem. The equipotential lines are indicated by dots.

In Fig. 8, some deterministic trajectories are drawn for a two-dimensional diffusion equation with the potential having two minima, indicated as x_1 and x_2, separated by the saddle point x_S. One can partition the phase space according to the final stationary points of the trajectories. The domain of attraction Ω_1 denotes the points belonging to trajectories which fall into x_1 [51]. A similar definition is given for the domain of attraction Ω_2 of the minimum x_2. The separatrix $\partial\Omega$ is determined by the two trajectories pointing to the saddle point. This procedure can be easily generalized to any multi-dimensional problem.

The domains of attraction can be used to specify the main features of the localized functions. The localized function $g_j(x)$ associated to the minimum x_j should be

$$g_j(x) \approx \delta_{jm} \quad \text{for} \quad x \in \Omega_m \tag{4.8}$$

with a step-like behaviour at the boundary of Ω_j. Explicit forms of the localized functions in all the phase space are not available, even if an attempt was done with a two-dimensional problem describing the diffusional rotation of a dipole in a cubic field [56]. We shall consider here only a local approximation near the saddle points. This is justified when the transition rates are calculated in the asymptotic limit of very large potential barriers. In fact, after the projection onto the subspace of the localized functions, the elements of the transition rate matrix are calculated like in eq. (2.15). Because of the presence of gradients of localized functions, only a boundary layer of the separatrix contributes significantly in the integration. Moreover, the equilibrium distribution selects the region around the saddle point because it has the least potential energy.

Let us consider the generic saddle point x_S with the corresponding parabolic approximation of the potential

$$V(x) \approx V_S + \delta x^{tr} V_S^{(2)} \delta x/2 \tag{4.9}$$

where $\delta x = x - x_S$ is the displacement from the saddle point. The principal components of the curvature matrix

$$[V_S^{(2)}]_{ij} = (\partial^2 V/\partial x_i \partial x_j)_{x=x_S} \tag{4.10}$$

can be classified according to their sign. There is only one negative or reactive component in correspondence of the direction of steepest descent: the positive components correspond to non-reactive displacements which increase the potential energy. A simplified equation is written for the deterministic trajectories near x_S

$$d\delta x/dt = -(1/k_B T)D_S V_S^{(2)} \delta x \tag{4.11}$$

The "normal mode" analysis of the trajectories is done by solving the eigenvalue problem

$$(1/k_B T)D_S V_S^{(2)} u_i = \lambda_i u_i \tag{4.12}$$

with eigenvectors u_i defining the normal directions with characteristic frequencies λ_i [51]. Like the curvature matrix, eq. (4.12) has only one negative eigenvalue λ_1 with the unit vector $u \equiv u_1 / |u_1|$ defining the reactive normal mode, i.e. the direction of the trajectories leaving the saddle point to reach one of the two minima. Note that by considering the transposed matrix in eq. (4.12), a conjugated set of eigenvectors is obtained

$$u^i = D_S^{-1} u_i \tag{4.13}$$

A biorthogonality relation exists between the two set of vectors.

A two-dimensional problem can be used to illustrate the behaviour of the reactive normal mode [57]. It is convenient to choose the components of δx along the principal directions of the curvature matrix written as

$$V_S^{(2)} = \begin{pmatrix} V_r^{(2)} & 0 \\ 0 & V_{nr}^{(2)} \end{pmatrix} \tag{4.14}$$

with $V_r^{(2)}$ and $V_{nr}^{(2)}$ the reactive (negative) and the non-reactive (positive) components, respectively. The diffusion matrix is given in all generality as

$$D_S = T(\theta)^{\text{tr}} \begin{pmatrix} d_1 & 0 \\ 0 & d_2 \end{pmatrix} T(\theta) \tag{4.15}$$

where d_1 and d_2 are its principal values and $T(\theta)$ is the matrix which rotates the principal directions of D_S by an angle θ, as displayed in Fig. 9.

Different directions u of the reactive normal mode are obtained by varying θ. Only for $\theta = 0$ or multiples of $\pi/2$, u is the same of the reactive mode of the potential. In general the diffusion matrix couples the reactive and the non-reactive modes of the potential, so that the effective reactive mode u will be biased in the direction of the largest component of D_S. This corresponds to a smaller friction acting on the trajectory leaving the saddle point [57]. The anisotropies of the curvature matrix and of the diffusion matrix strongly influence the frictional coupling between reactive and non-reactive modes of the potential. Obviously it is absent in isotropic diffusion. Another case of vanishing coupling is that of a potential with a very large $V_{nr}^{(2)}$ component. The corresponding saddle point has very steep walls opposing to displacements orthogonal to the direction of steepest descent, so that deviations from this direction become energetically very unfavourable.

The previous analysis can be used in the calculation of the localized function g_j near the saddle point x_S lying on the boundary of Ω_j. Like in the one-dimensional case, equation

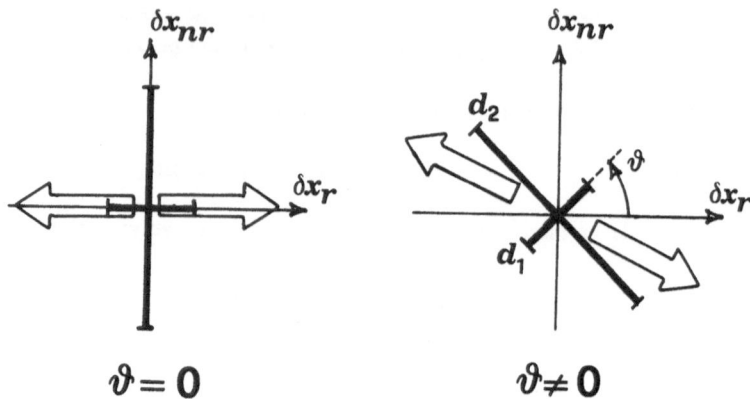

Figure 9. Effective reactive mode for a two dimensional problem. The cartesian axes are taken along the principal components of the curvature matrix. The large segments indicate the principal components of the diffusion matrix, while the large arrows indicate the direction of the trajectories leaving the saddle point.

(2.16) is taken as the generating equation of the approximate localized functions. However, we simplify further the problem by considering the expansion around the saddle point:

$$(\Gamma P_{eq} g_j)_{\mathbf{x} \approx \mathbf{x}_S} = - \left[\frac{\partial}{\partial \delta \mathbf{x}} - \frac{\mathbf{V}_S^{(2)} \delta \mathbf{x}}{k_B T} \right]^{tr} \mathbf{D}_S \frac{\partial}{\partial \delta \mathbf{x}} g_j(\delta \mathbf{x}) = 0 \qquad (4.16)$$

The differential equation is factorized by changing the variables to the displacements $y_i \equiv (\mathbf{u}^i)^{tr} \delta \mathbf{x}$ along the normal modes

$$\sum_i (\partial/\partial y_i - \lambda_i y_i)(\partial/\partial y_i) g_j = 0 \qquad (4.17)$$

The solution with the proper boundary conditions is [58]

$$g_j = \frac{1}{2} \pm \frac{1}{2} \mathrm{erf} \left(y_1 \sqrt{2/|\lambda_1|} \right) \qquad (4.18)$$

so that $g_j \approx 1$ ($g_j \approx 0$) for large positive (negative) values of y_1. The sign in eq. (4.18) is determined by the direction of $\mathbf{x}_j$ according to eq. (4.8). We remind that an error function shape was found from the more accurate localized functions in one-dimensional problems, when small displacements from the potential maxima were considered.

By projecting the diffusion operator Γ onto the subspace of localized functions, the master equation is finally recovered. By expanding the Boltzmann distribution around the minima, equation (2.20) for the equilibrium populations is derived with the following free energy of the j-th conformer

$$E_j = V(\mathbf{x}_j) + (k_B T/2) \ln |\mathrm{Det}(\mathbf{V}_j^{(2)}/2\pi k_B T)| \qquad (4.19)$$

with the curvature matrix $\mathbf{V}_j^{(2)}$ calculated like in eq. (4.10) in correspondence of the minimum $\mathbf{x}_j$. The transition rate for two minima $\mathbf{x}_j$ and $\mathbf{x}_m$ separated by the saddle point $\mathbf{x}_S$, is calculated like in eq. (2.15) as the matrix element of Γ between the corresponding localized functions. From the expansion around the saddle point, one obtains the following asymptotic result [58]

$$W_{jm} =< g_j \mid \Gamma P_{eq} \mid g_m > /Q_m \approx (\lambda_1/2\pi) \exp[-(E_S - E_m)/k_B T] \qquad (4.20)$$

which generalizes equation (2.22). The saddle point energy E_S is calculated like in eq. (4.19), while λ_1 is the negative eigenvalue calculated from eq. (4.12) by using the specific curvature and diffusion matrices whose values depend on the location of $\mathbf{x}_S$. Equation (4.20) agrees with the result of Langer [50], derived by considering the particle flux at the saddle point.

The negative eigenvalue determines the magnitude of the pre-exponential factor of the kinetic rates. The following question arises: how the frictional coupling between reactive and non-reactive modes influences λ_1 ? Let us specify λ_1 in terms of the direction $\mathbf{u}$ of the trajectories leaving the saddle point, by taking into account eqs. (4.12) and (4.13). The result obtained in this way is

$$\mid \lambda_1 \mid = -\frac{\mathbf{u}^{tr}\mathbf{V}_S^{(2)}\mathbf{u}}{\mathbf{u}^{tr}\xi_S\mathbf{u}} \qquad (4.21)$$

whit the friction matrix given as

$$\xi_S = k_B T \, \mathbf{D}_S^{-1} \qquad (4.22)$$

Therefore, λ_1 is the ratio between the contribution of the potential curvature and a purely frictional term. One can show by a linear expansion of $\mathbf{u}$, that the eigenvalue problem eq. (4.12) can alternatively be formulated as the maximization of the r.h.s. of eq. (4.21) with respect to all possible directions of $\mathbf{u}$. In other words, the direction of the reactive normal mode is that giving rise to the largest pre-exponential factor compatible with the r.h.s. of eq (4.21). The actual value of $\mathbf{u}$ is a compromise between two opposite conditions: the maximization of the potential term which favours the steepest descent direction, and the minimization of the effective friction.

In some problems, multiple transitions between states which are not connected by a well defined saddle point could be important. Our treatment based on the expansion about the saddle points cannot recover such processes. On the other hand, complete analyses of the multiple transitions are not available. Given the potential function and the diffusion matrix for a specific problem, one cannot anticipate the kinetic rate of multiple transitions even in the asymptotic approximation. At present, only qualitative considerations about their activation energy are possible by considering the potential profiles, like in the analysis of the crankshaft motion in polyethylene by Blomberg [53]. In this way one finds always that multiple transitions have an unfavourable activation energy, but this could be compensated by a small effective friction in the pre-exponential factor.

5. The chain of coupled rotors

Before to discuss the models for real chain molecules, a much simpler system is analysed here to illustrate the consequences of the transition rate theory. We shall examine the

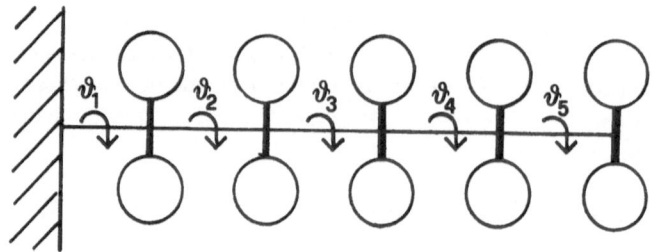

Figure 10. The model system of a chain of rotors.

chain of coupled rotors, which was already considered in the past to study collective dynamical properties [59,60]. In Fig. 10 the system is represented as an ensemble of dumbells constrained to rotate about a common axis.

The diffusion equation in eqs. (4.1) and (4.2) can be used to describe the system by identifying the variables $\mathbf{x}$ with the set of torsional angles $(\theta_1, \theta_2, ..., \theta_N)$. The coordinate θ_i denotes the orientation of the i-th rotor with respect to the previous one. The overall rotation motion is excluded by considering the chain bounded to a rigid wall. The potential is taken to be pairwise additive with respect to the interactions of nearest-neighbour rotors

$$V(\mathbf{x}) = \sum_{n=1}^{N} V_p(\theta_n) \tag{5.1}$$

with the pair potential V_p having the same symmetry of the cosine potential of eq. (2.6), so that each rotor have two equivalent stable positions for $\theta_n = 0$ and $\theta_n = \pi$. Note that potential V_p describes also the interaction between the wall and the first rotor. In this way all the conformational states have the same energy. Moreover, the saddle points are easily found and they also have the same energy. Therefore, the transition rates have the same activation energy and they differ only by the pre-exponential factor. In the following, the simplest diffusion matrix is used with an unique diffusion coefficient D_0 for the rotational motion of each rotor.

The calculation of λ_1 requires the diagonalization of matrices of the following type [61]

$$\mathbf{D}_S \mathbf{V}_S^{(2)} = D_0 V_{nr}^{(2)} \begin{pmatrix} 1 & -1 & & & & \\ -1 & 2 & -1 & & & \\ & -1 & 2 & \rho & & \\ & & -1 & -2\rho & -1 & \\ & & & \rho & 2 & -1 \\ & & & & -1 & 2 \end{pmatrix} \tag{5.2}$$

where ρ is the ratio

$$\rho = | V_r^{(2)} | / V_{nr}^{(2)} \tag{5.3}$$

between the reactive and the non-reactive mode determined by the second derivatives of the pair potential. The particular matrix written in eq. (5.2) describes the transition of

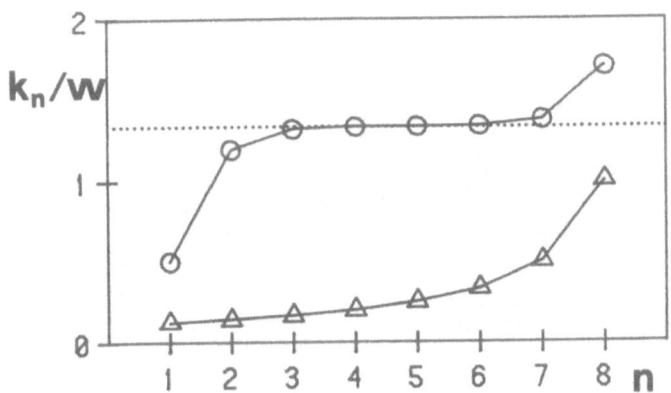

Figure 11. Position dependence of the bond-specific rates k_n calculated with $\rho = 1$ (circles) and $\rho = 0$ (triangles) in the eight rotor chain.

the fourth torsional angle. In the corresponding saddle point, the fourth and the following rotors have an orthogonal orientation with respect to the plane of Fig. 10.

With this simple model, the transition rate for a specific bond is an uniquely defined quantity since it does not depend on the state of other torsional angles. Its position dependence is determined by the values of λ_1 calculated by varying the ρ-dependent column in the matrix at the r.h.s of eq. (5.2). It is convenient to define an elementary frequency w

$$w = (D_0 \mid V_r^{(2)} \mid /2\pi k_B T) \exp(-\Delta E/k_B T) \tag{5.4}$$

which is the transition rate for a single rotor chain. The set of bond-specific transition rates k_n, when scaled by the elementary frequency w, depends only on the curvature ratio ρ. In Fig. 11, we have represented the rates k_n as function of the position n for two values of the curvature ratio.

The case $\rho = 0$ corresponds to an infinite $V_{nr}^{(2)}$, so that the coupling between the reactive and the non-reactive modes of the potential is quenched. The steepest descent trajectory is represented by a rotation of the terminal rotors as a rigid ensemble. The effective friction opposing their motion is proportional to the number of displaced rotors and, in fact, k_n is inversely proportional to $(N - n + 1)$.

With finite values of the curvature ratio, the frictional coupling becomes very important and it changes drastically the behaviour of the transition rates. Two major effects are recognized from Fig. 11: much faster transition rates are obtained with $\rho = 1$, and their values are almost position-independent except for a few terminal rotors. This behaviour is a consequence of the frictional coupling which generates a cooperative distortion of adjacent torsional angles, as shown by Skolnick and Helfand in a detailed study of the transition rates for alkyl chains [54,55]. These authors were the first to clarify the cooperative nature of the conformational dynamics in chain molecules, by showing that only a small number of bonds reorients significantly with respect to a laboratory frame, during the passage through the saddle point. This obviously explains the weak position dependence of the kinetic rates in Fig. 11. At the same time, the coupling between different torsional degrees of freedom reduces the effective friction so increasing the transition rates.

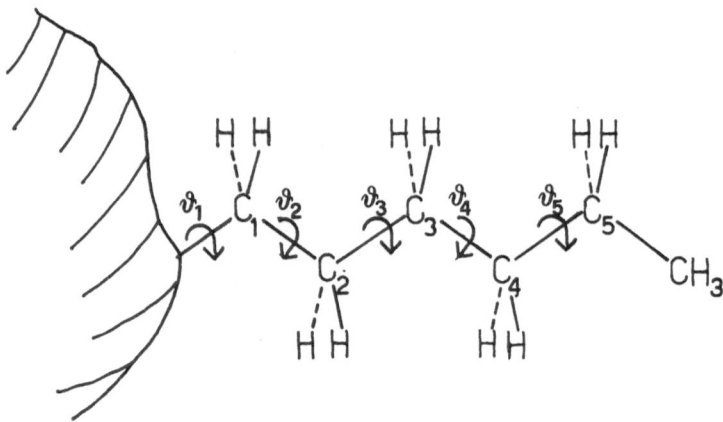

Figure 12. Torsional angles for an alkyl chain attached to a massive rigid core.

With infinitely long chains, a limit value k_∞ of the kinetic rates is obtained. An analytical expression has been derived for the rotor chain [61]

$$k_\infty = \frac{4\rho}{1+2\rho} w \tag{5.5}$$

The value for $\rho = 1$ is represented by a dotted line in Fig. 11. It appears that k_∞ describes quite well the transition rates of the central bonds in a relatively short chain.

Some comments about the multiple transitions are now in order. These processes are necessarily required in hypothetical chains with $\rho = 0$. Given the steady decrease of the kinetic rates, crankshaft-like transitions [53,62] involving only local displacements become important in spite of their unfavourable energetics. In our model system one can have, for example, reorientations of the n-th rotor alone, with an opposite change of the torsional angles θ_n and θ_{n+1}. A doubled activation energy could be anticipated for such transition, but with a small and position independent friction.

On the contrary, multiple transitions are expected to have only a secondary role in the presence of cooperative effects due to the frictional coupling. In this case, also the single bond transitions are characterized by an almost constant effective friction, which prevents their decrease below the rate of multiple transitions.

6. Alkyl chain dynamics

We have now all the ingredients to discuss the conformational dynamics of an alkyl chain like that represented in Fig. 12. We suppose that the chain is bound to a massive rigid core, so that the overall rotational motion and the internal motion are decoupled. The generalization of the theory presented in section 3 would be required in the absence of this condition.

The stable conformations can be classified in terms of gauche ($g_\pm$) or trans (t) states for each $C - C$ bond. A collective index $J = \{j_1, j_2, ..., j_N\}$ is introduced to describe the states $j_n = g_\pm, t$ of the N torsional degrees of freedom $\theta_1, \theta_2, ..., \theta_N$. Note that we do not examine the torsional dynamics of the terminal methyl group. Diffusion-type equations cannot be adequate to represent methyl rotations since one expects that inertial effects are important. When the torsional potential and the diffusion matrix are specified, the theory presented in section 4 can be applied to derive the following master equation

$$\partial P_J(t)/\partial t = \sum_M W_{JM} P_M(t) \tag{6.1}$$

We shall consider only single bond transitions, so that off-diagonal elements W_{JM} vanish except when conformations J and M differ only by the state of a specific bond.

The knowledge of the full torsional potential appears necessary to calculate equilibrium populations and transition rates. In principle one can derive it from the detailed intramolecular potential used in Brownian Molecular Dynamics simulations of chain molecules [35,41, 63-65]. We shall follow instead an alternative strategy based on the parametrization of the free energies like in the Rotational Isomeric State (RIS) model [66]. In this way we intend to isolate from the full torsional potential, the relevant features which directly influence the conformational dynamics. We have proposed a specific parametrization for alkyl tails in ref. 67, and we shall summarize here only the main ingredients in relation to the geometry, the energetics and the friction of the chain.

(i) Geometry. Starting with fixed bond lengths and bond angles, the geometry of stable conformations is obtained by imposing the values $\theta_n = 0°, \pm120°$ to each torsional angle. The geometry of the saddle points is derived by a $\pm60°$ rotation of one torsional angle.

(ii) Energetics. Like in the RIS model, two additive contributions are considered: the energy increment E_g of the gauche state with respect to the trans state, and the increment E_p due to the "pentane effect" [66]. The energy E_J of conformer J is then written as

$$E_J = n_g E_g + n_p E_p \tag{6.2}$$

where n_g is the number of gauche states and n_p is the number of $g_\pm g_\mp$ sequences. Excluded volume effects are taken into account by introducing a distance of closest approach between two carbon atoms [67,68]. Only one parameter ΔE_S is used to calculate the energy of the saddle point for the single bond transition connecting conformers J and M

$$E_S = \Delta E_S + \text{Max}(E_J, E_M) \tag{6.3}$$

The shape of the potential near the saddle points is parametrized in terms of two fixed curvatures $V_r^{(2)}$ and $V_{nr}^{(2)}$ for the reactive and non-reactive modes, respectively.

(iii) Friction. The simple bead model is used with each methylene generating a translational friction ξ_0. The friction matrix can then be written in the simple form [67,68]

$$\xi_{ik} = \xi_0 \sum_{j=\text{Max}(i,k)}^{N+1} [\vec{z}_{i-1} \times (\vec{r}_j - \vec{r}_i)] \cdot [\vec{z}_{k-1} \times (\vec{r}_j - \vec{r}_k)] \tag{6.4}$$

where $\vec{r}_j$ is the vector position of the j-th carbon atom and $\vec{z}_n$ the unit vector along the $C_n - C_{n+1}$ bond.

Note that eq. (6.4) is valid only in the limit of an infinitely massive rigid core, so that its contribution to the torsional friction can be neglected. More realistic models necessarily require some new parameters to describe the frictional coupling with the rigid core. The use of eq. (6.4) is justified by the requirement of dealing with the simplest model having the least number of parameters. On the other hand, preliminary calculations have shown that the main features of the conformational dynamics are not affected by considering a rigid core of finite but large size [69].

Like in the rotor chain, we introduce an elementary frequency w related to the $g \rightarrow t$ transition rate of the N=1 chain, i.e. the ethyl tail,

$$w = (\mid V_r^{(2)} \mid /2\pi \xi_0 d^2) \exp(-\Delta E_S/k_B T) \tag{6.5}$$

where d is the $C - C$ bond length. Then, only three parameters are required to specify the matrix $\mathbf{W}$ scaled by w: the curvature ratio ρ given in eq. (5.3), and the two energy parameters $e_g \equiv E_g/k_B T$ and $e_p \equiv E_p/k_B T$. The results of the calculations discussed below, are obtained with standard values of these parameters at room temperature: $\rho = 1$, $e_g = 0.84$ and $e_p = 3.0$.

An important feature of the present model is that all the torsional degrees of freedom are coupled because of the pentane effect and of the frictional coupling generated by eq. (6.4). Therefore, the transition rate at a given bond depends on the state of all other bonds. This feature should be present in any model based on a microscopic description of the alkyl chain dynamics [54,55,70]. On the contrary, most phenomenological models proposed in the past consider independent single-bond transitions. This is the basic assumption of the Wallach model for alkyl chain dynamics [44]. The same hypothesis is found in subsequent applications or generalizations [71-74]. The coupling generated by the pentane effect was examined only in few cases [75,76]. Wittebort and Szabo considered the master equation in the most general form, but again by resorting to a phenomenological approach [77]. It should be emphasized that the main difference between phenomenological and microscopic models, is that kinetic rates are free parameters in the former case, while they are derived from molecular properties in the latter.

The presence of many different kinetic rates for a specific bond transition complicates the analysis of microscopic models. One can rationalize this complexity by recognizing the effects of specific sequences of gauche and trans states, as done by Skolnick and Helfand [54,55]. The alternative is the calculation of correlation functions, so deriving single bond properties from the dynamical behaviour of the system.

Let us consider a function $f(\theta_1, \theta_2, ..., \theta_N)$ of the N torsional angles. A discrete representation is provided by an array $\mathbf{f}$ whose elements f_J are calculated by assigning to the torsional angles their values in the conformation J. The master equation (6.1) can be used to calculate the corresponding autocorrelation function $G(t) = \overline{f(t)^* f(0)}$. The spectral density associated to a particular spectroscopic observable is conveniently written as [67,68]

$$J(\omega) \equiv \int_0^\infty dt \, G(t) \exp(-i\omega t) = \mathbf{v}^\dagger (i\omega \mathbf{1} + \tilde{\mathbf{W}})^{-1} \mathbf{v} \tag{6.6}$$

where the array $\mathbf{v}$ has elements $v_K = f_K \sqrt{Q_K}$ and the transition rate matrix has been symmetrized in the following form

$$\tilde{W}_{KM} \equiv W_{KM} \sqrt{Q_M/Q_K} = \tilde{W}_{MK} \tag{6.7}$$

By diagonalizing $\tilde{\mathbf{W}}$, one readily obtains the full frequency profile of the spectral density. However, severe computational difficulties arise when the number of degrees of freedom is

Table. Matrix size as function of the number N of torsional degrees of freedom. The truncated basis is obtained by eliminating conformations with an equilibrium population less than that of all-trans configuration by a factor 10^{-3}.

N	Matrix size	
	full basis	truncated basis
1	3	3
4	71	69
7	1,551	953
10	34,161	9,741
13	752,529	106,225

increased: as shown in the table, the matrix size grows very quickly beyond the capabilities of standard diagonalization procedures [78].

The Lanczos algorithm, which takes into account the sparsity of the matrix [79,80], must be used to treat long chains. This method allows one to calculate the spectral density as a continued fraction expansion

$$J(\omega) = \cfrac{\mathbf{v}^\dagger \mathbf{v}}{i\omega + a_1 - \cfrac{b_2}{i\omega + a_2 - \cfrac{b_3}{i\omega + a_3 - \cfrac{b_4}{i\omega + a_4 \ldots}}}} \qquad (6.8)$$

with coefficients a_j and b_j derived by means of a recursive relation [81,82]. It should be evident the analogy with the Mori's continued fraction expansion of autocorrelation function [83]. In fact, it has been shown that the two methods are based on the same projection procedure [84,85].

Given the efficiency of the Lanczos algorithm, matrices of the order of several thousands are easily solved in a μVAX computer [67]. Larger problems have been treated with the IBM 3090 supercomputer, for example the 15-member chain representative of an alkyl tail of phospholipids in the membrane environment [86].

The relaxation of non-equilibrium distributions can be used to characterize the conformational dynamics at specific positions. For example, one can perturb the equilibrium distribution by introducing at $t = 0$ an excess of trans states at the n-th position

$$P_J(0) \propto Q_J + \epsilon Q_J \delta_{j_n,0} \qquad (6.9)$$

where $j_n = 0$ and $j_n = \pm 1$ indicates trans and $\pm$gauche states, respectively, of configuration J. From the solution of eq. (6.1) with this initial condition, one can calculate the time-dependent excess of trans states (et) at the n-th position as the deviation from the equilibrium distribution

$$G_{et}^{(n)}(t) = \sum_J \delta_{j_n,0}[P_J(t) - Q_J] / \sum_I \delta_{i_n,0}[P_I(0) - Q_I] \qquad (6.10)$$

132

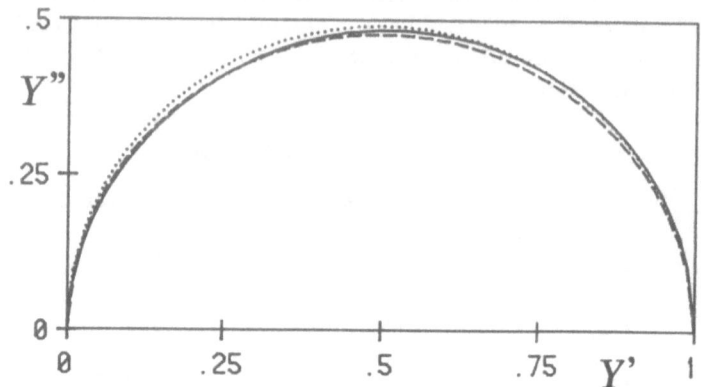

Figure 13. Cole-Cole plots for the real Y' and the imaginary Y'' parts of $Y = J_{et}^{(n)}(\omega)$ of the N=7 chain. The continuous, dashed and dotted lines present the results for $n = 1$, $n = 4$ and $n = 7$ positions, respectively.

with $G_{et}^{(n)}$ normalized at $t = 0$. Independent informations about the local dynamics are derived by considering an initial distribution with unbalanced gauche states at the n-th bond

$$P_J(0) \propto Q_J + \epsilon Q_J j_n \tag{6.11}$$

The time-dependent unbalanced gauche (ug) populations are calculated as

$$G_{ug}^{(n)}(t) = \sum_J j_n P_J(t) / \sum_I i_n P_I(0) \tag{6.12}$$

Both $G_{et}^{(n)}$ and $G_{ug}^{(n)}$ can be alternatively formulated as autocorrelation functions whose spectral density is calculated from eq. (6.6) [67].

Both these functions have a mono-exponential decay in the N=1 chain. On the contrary, multi-exponential behaviour is found when $N \neq 1$, because the matrix $\mathbf{W}$ couples the different degrees of freedom. However, their multi-exponential decay has a limited dispersion. This is clearly shown by the Cole-Cole plots derived from the imaginary and the real parts of the associated spectral densities $J_{et}^{(n)}(\omega)$ and $J_{ug}^{(n)}(\omega)$. Figure 13 is an example for the excess trans populations of a $N = 7$ chain. One can see only small deviations from the perfect semicircle. A similar behaviour is found with the unbalanced gauche populations. Under this condition, the effective rates defined as time integral of the correlation functions

$$1/k_{et}^{(n)} = \int_0^\infty dt\, G_{et}^{(n)}(t) \quad ; \quad 1/k_{ug}^{(n)} = \int_0^\infty dt\, G_{ug}^{(n)}(t) \tag{6.13}$$

well characterize the local conformational dynamics at specific positions.

We examine now the more complex problem of the rotational dynamics of methylene groups. NMR and dielectric relaxation measurements can be used to obtain informations at specific positions. We shall then consider the relaxation behaviour of the Wigner function $D_{q0}^p(\Omega)$, where Ω are the Euler angles for the rotation from the laboratory frame F_L to a local molecular frame $F_M^{(n)}$ in the n-th methylene group. Calculations of the corresponding correlation functions have been done under the condition of uncoupled overall rotational motion and internal dynamics [67]. Simplified expressions result when isotropic overall diffusion is considered with a diffusion coefficient D much smaller than the effective transition rates. The spectral density at zero frequency can then be written as [67,68]

$$J_p^{(n)}(0) = (2p+1)^{-1}\left[\frac{(S_p^{(n)})^2}{p(p+1)D} + \frac{1-(S_p^{(n)})^2}{k_p^{(n)}}\right] \tag{6.14}$$

The overall rotational motion contributes according to the first term in the square brackets, with a weight determined by the order parameters $S_p^{(n)}$ for the "spatial restriction of the internal motion" [87]. They are given by the following average to be evaluated with the equilibrium populations Q_J

$$S_p^{(n)} = \left[\sum_q | \overline{D_{q0}^p(\Omega^{(n)})} |^2\right]^{1/2} \tag{6.15}$$

where $\Omega^{(n)}$ is the set of Euler angles for the rotation from the molecular frame $F_M^{(0)}$ of the rigid core to the local frame $F_M^{(n)}$.

The second term in the square brackets of eq. (6.14), represents the contribution of the conformational transitions, with an effective relaxation rate $k_p^{(n)}$ depending on the rank p of the observed tensorial property

$$1/k_p^{(n)} = \left[1 - (S_p^{(n)})^2\right]^{-1} \int_0^\infty dt \sum_q \overline{\delta D_{q0}^p(\Omega_t^{(n)})^* \delta D_{q0}^p(\Omega_{t=0}^{(n)})} \tag{6.16}$$

In the previous equation, the correlation functions are defined for the deviations of the Wigner functions from their equilibrium average.

In isotropic phases, the order parameters $S_p^{(n)}$ are negligibly small, except for a few methylenes near the rigid core [67]. This implies that the conformational dynamics is mainly responsible of the rotational relaxation along the alkyl chain. An opposite behaviour is found instead with anisotropic environments, like in liquid crystals or lipid membranes, since mean field contributions generate finite order parameters [86].

In fig. 14, the previously defined effective rates are presented as function of the position. One can see that the rotational relaxation rates are comparable with the population relaxation rates, with a steady increase along the chain. Therefore, position-dependent relaxation times measured by NMR techniques can be used to characterize the local conformational dynamics as long as the effects of the overall motion can be neglected. But this is correct only as a zeroth order approximation. In fact, a closer inspection of fig. 14 reveals that the rotational rates increase along the chain faster than the population relaxation rates. This is explained by considering that the conformational transitions of the first n bonds have a cumulative effect on the rotational relaxation of the n-th methylene group.

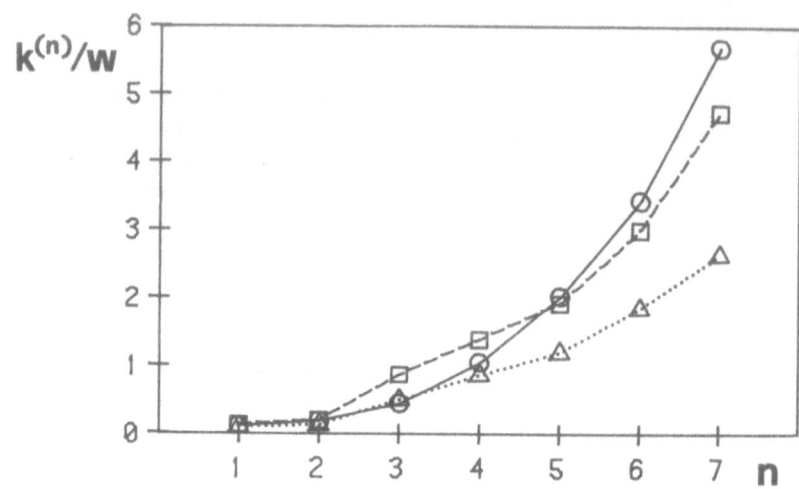

Figure 14. Effective rates as function of the position in the $N = 7$ chain. Circles: second rank rotational relaxation $k_2^{(n)}$; squares: relaxation of excess trans populations $k_{et}^{(n)}$; triangles: relaxation of unbalanced gauche populations $k_{ug}^{(n)}$.

The theoretical analysis presented in ref. 67 has been applied to interpret selective NMR measurements of ^{13}C relaxation times in tetralkylammonium salts $NR_4^+X^-$ with different solvents [88,89]. In this system, the decoupling between overall rotational motion and internal dynamics can be safely assumed. In fact, when examining the rotational motion of a methylene group in the alkyl chain R, one identifies the rigid core with the fragment NR_3^+ which has a larger size.

Simplified expressions like eq. (6.14) can be derived only for the zero-frequency spectral densities. A much more complex behaviour is found by analyzing their frequency dependence. In fig. 15, some Cole-Cole plots are presented for the spectral density $J_2^{(n)}(\omega)$ relative to the Wigner function $D_{00}^2(\Omega)$. The decay to the equilibrium is characterized by a wide distribution of relaxation times. In opposition to fig. 13, we have now large deviations from a semicircle, with both the overall rotational motion and the cumulative effects of the conformational transitions determining the dispersion of relaxation frequencies.

With the rotor chain model, it has been found that cooperative effects stabilize the transition rates of the central bonds, and this seems contradictory with the steady increase of the population relaxation rates displayed in fig. 14. As matter of fact, the presence of a plateau for the bond-specific transition rates depends strongly on the chain length, and it is clearly visible only for $N > 10$. In fig. 16, the transition rates $\overline{k}_{g \to t}^{(n)}$ are presented as function of the position in a chain with 15 methylenes. These kinetic rates are defined as the equilibrium average of the elements of matrix $\mathbf{W}$ leading to a transition $g_\pm \to t$ at the n-th bond, and they can be efficiently calculated in long chains by means of Monte Carlo sampling techniques [61]. In order to facilitate the comparison with fig. 11, also

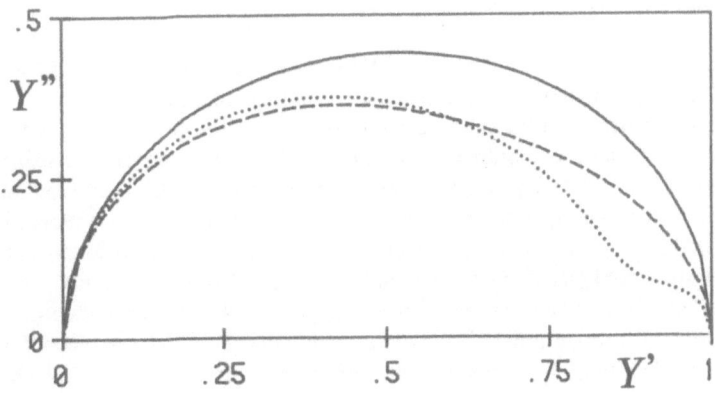

Figure 15. Cole-Cole plots for $Y = J_2^{(n)}(\omega)$ at positions $n = 1$ (continuous line), $n = 4$ (dashed line) and $n = 7$ (dotted line) of the $N = 7$ chain. A value $D = w/100$ has been used for the isotropic diffusion coefficient.

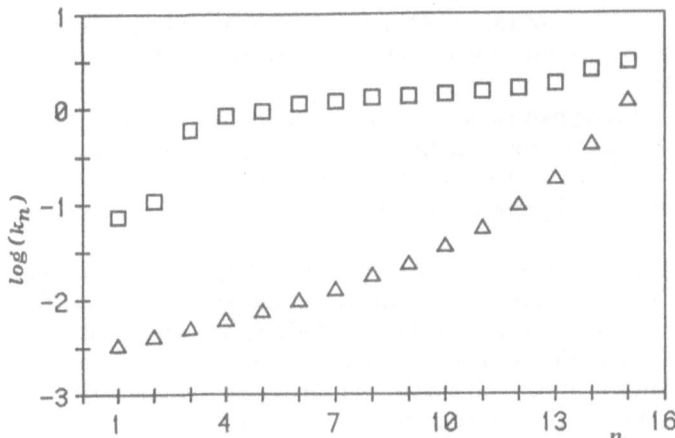

Figure 16. Mean kinetic rates $k_n \equiv \bar{k}_{g \rightarrow t}^{(n)}$ in the N=15 chain. Curvature ratios: $\rho = 1$ (squares) and $\rho = 0$ (triangles).

the rates calculated with a vanishing curvature ratio are reported. The frictional coupling between reactive and non-reactive modes generates the same effects found in the rotor chain: with a finite curvature ratio, much larger kinetic rates are obtained, with almost position-independent values in the central portion of the chain.

136

7. Conclusions

By means of the Kramers theory and its generalizations provided by the localized function method, the kinetic rates of conformational transitions can be obtained from the diffusion-Smoluchowski equation for the torsional variables, after generating a discrete representation based on the notion of conformational sites. By including the overall rotational motion, a realistic picture of the conformational processes and of their effects on spectroscopic observables has been given. Still, several problems requiring a refinement of our theoretical tools remain to be clarified. We mention here some objectives for future research.

(A) A complete analysis of the inertial effects has not been done as yet. Only partial improvements of overdamped results have been proposed. In particular, Skolnick and Helfand have generalized to multi-dimensional problems the Kramers result for intermediate and large friction [54,55]. Even if their equation is useful to describe the effects of under-damped motion of non-reactive coordinates like bond stretching and bending, it does not describe the inertial effects on the torsional motion when energy relaxation becomes important (see fig. 4 for the one-dimensional case). The inertial effects on the the transition rate is a topic relevant not only to the conformational dynamics, but also to the more general field of the chemical kinetics [90].

(B) Multiple transitions have been neglected in our treatment. This seems a resonable approximation. Brownian Molecular Dynamics simulations by Weber and Helfand have shown that these transitions contribute to a limited extent [91]. Numerical solutions of the diffusion equation of a chain with few rotors supports their conclusion [92]. However, a definitive solution of this problem requires accurate estimates of the rate of multiple transitions. At present, only a qualitative evaluation of their activation energy seems to be possible.

(C) When the rotating parts of a molecule have about the same size, the coupling of the overall rotation with the conformational dynamics should be taken into account. A theory is available only for molecules with one torsional degree of freedom. Its generalization to multidimensional problems is required to study the rotational relaxation of free linear alkanes.

(D) In one-dimensional problems, the presence of large barriers is sufficient to determine different time scales for the conformational transitions and the librational motions in the potential well. This could be not the case in long chain molecules, because the frequency of librational modes is reduced by cooperative displacements [59]. In this case, cooperative librational modes could represent an alternative pathway contributing to rotational relaxation. A more general theory including the conformational transitions and the cooperative librations would then be required.

Acknowledgements

This work has been supported by the Italian Ministry of Public Education and in part by the National Research Council through its Centro Studi sugli Stati Molecolari ed Eccitati. The authors are indebted to Prof. U. Segre and Dr. F. Coletta for many stimulating discussions.

References

1) Bellemans, A., this volume.
2) Hynes, J.T. and Deutch, J.M. (1975), in H. Eyring, D. Henderson and W. Jost (eds.), *Physical Chemistry*, Vol XIB, Academic Press, New York.
3) Hwang, L.P. and Freed, J.H. (1975), *J. chem. Phys.* **63**, 118.
4) Kramers, H.A. (1940), *Physica* **7**, 284.
5) Hanggi, P. (1980), *J. Stat. Phys.* **42**, 105.
6) Widom, B. (1971), *J. chem. Phys.* **55**, 44.
7) Rebertus, D.W., Berne, B.J. and Chandler, D. (1979), *J. chem. Phys.* **70**, 3395.
8) Tomita, H., Ito, A. and Kidachi, H. (1976), *Prog. Theor. Phys.* **56**, 786.
9) Larson, R.S. and Kostin, M.D. (1978), *J. chem. Phys.* **69**, 4821.
10) Caroli, B., Caroli, C. and Roulet, B. (1979), *J. Stat. Phys.* **21**, 415.
11) Risken, H. (1984) *The Fokker Planck Equation*, Springer, Berlin.
12) Blackmore, R. and Shizgal, B. (1985), *Phys. Rev.* **A31**, 1855.
13) Drozdov, A.N. and Zitserman, V.Yu. (1986), *Russ. J. Phys. Chem.* **60**, 35.
14) Bratos, S., this volume.
15) Moro, G. and Nordio, P.L. (1985), *Molec. Phys.* **56**, 255.
16) Moro, G. and Nordio, P.L. (1986), *Molec. Phys.* **57**, 947.
17) de Bruijn, N.G. (1981) *Asymptotic Methods in Analysis*, Dover, New York.
18) Karplus, M. and Kushick, J.N. (1981), *Macromolecules* **14**, 325.
19) Moro, G. and Polimeno, A. (1988), *Chem. Phys.* (in press).
20) Chandler, D. (1978), *J. chem. Phys.* **68**, 2959.
21) Carmeli, B. and Nitzan, A. (1983), *Phys. Rev. Lett.* **51**, 233.
22) Matkowsky, B.J., Schuss, Z. and Tier, C. (1984), *J. Stat. Phys.* **44**, 443.
23) Buttiker, M. and Landauer, R. (1984), *Phys. Rev.* **B30**, 1551.
24) Zawadzki, A.G. and Hynes, J.T. (1985), *Chem. Phys. Lett.* **113**, 476.
25) Mel'nikov, V.I. and Meskov, S.V. (1986), *J. chem. Phys.* **85**, 1018.
26) Polimeno, A., Nordio, P.L. and Moro, G. (1988), *Chem. Phys. Lett.* **144**, 357.
27) Doll, J.D. and Voter, A.F. (1987), *Ann. Rev. Phys. Chem.* **38**, 413.
28) Scott, R.A. and Scheraga, H.A. (1966), *J. chem. Phys.* **44**, 3054.
29) Ryckaert, J.P. and Bellemans, A. (1975), *Chem. Phys. Lett.* **30**, 123.
30) Steele, D. (1985), *J. Chem. Soc., Faraday Trans 2.* **81**, 1077.
31) Moro, G. (1987), *Chem. Phys.* **118**, 181.
32) Erpenbeck, J.J. and Kirkwood, J.G. (1958), *J. chem. Phys.* **29**, 909; (1963), *Ibid.*, **38**, 1023.
33) Fixman, M. and Kovac, J. (1974), *J. chem. Phys.* **61**, 4939.
34) Titulaer, U.M. and Deutch, J.M. (1975), *J. chem. Phys.* **63**, 4505.
35) Fixman, M. (1978), *J. chem. Phys.* **69**, 1527, 1538.
36) Weneger, W.A., Dowben, R.A. and Koester, V.J. (1980), *J. chem. Phys.* **73**, 4086.
37) Weneger, W.A. (1982), *J. chem. Phys.* **76**, 6425.
38) Ladanyi, B.M. and Evans, G.T. (1984), *J. chem. Phys.* **79**, 944.
39) Moro, G. (1987), *Chem. Phys.* **118**, 167.
40) Knauss, D.C. and Evans, G.T. (1980), *J. chem. Phys.* **72**, 1499.
41) Levy, R.M., Karplus, M. and McCammon, J.A. (1979), *Chem. Phys. Lett.* **65**, 4.
42) Woessner, D.E. (1965), *J. chem. Phys.* **42**, 1855.
43) Woessner, D.E., Snowden Jr., B.S. and Meyer, G.H. (1969), *J. chem. Phys.* **50**, 719.

44) Wallach, D.J. (1967), *J. chem. Phys.* **47**, 5258.
45) Tsutumi, A. (1979), *Molec. Phys.* **37**, 111.
46) Tropp, J. (1980), *J. chem. Phys.* **72**, 6035.
47) Bluhm, Th. (1982), *Molec. Phys.* **47**, 475.
48) Bottcher, C.J.F. and Burdewijck, P. (1975) *Theory of Electric Polarization*, Vol. 2, Elzevier, Amsterdam.
49) Landauer, R. and Swanson, J.A. (1961) *Phys. Rev.* **121**, 1668.
50) Langer, J.S. (1969), *Ann. Phys.* **54**, 258.
51) Schuss, Z. (1980) *Theory and Applications of Stochastic Differential Equations*, Wiley, New York.
52) Matkowsky, B.J. and Schuss, Z. (1981), *SIAM J. Appl. Math.* **40**, 242.
53) Blomberg, C. (1979), *Chem. Phys.* **37**, 219.
54) Skolnick, J. and Helfand, E. (1980), *J. chem. Phys.* **72**, 5489.
55) Helfand, E. and Skolnick, J. (1982), *J. chem. Phys.* **77**, 5714.
56) Moro, G. and Nordio, P.L. (1986), *Z. Phys. B* **64**, 217.
57) van der Zwan, G. and Hynes, J.T. (1982), *J. chem. Phys.* **77**, 1295.
58) Ferrarini, A., Moro, G. and Nordio, P.L. (1988), *Molec. Phys.* **63**, 285.
59) Shore, J.E. and Zwanzig, R. (1975), *J. chem. Phys.* **63**, 5445.
60) Zientara, G.P. and Freed, J.H. (1983), *J. chem. Phys.* **79**, 3077.
61) Ferrarini, A., Moro, G., Nordio, P.L. and Polimeno, A. (1988), *Chem. Phys. Lett.* (in press).
62) Helfand, E. (1971), *J. chem. Phys.* **54**, 4651.
63) Levy, R.M., Karplus, M. and Wolynes, P.G. (1981), *J. Am. chem. Soc.* **103**, 5938.
64) Brown, M.S., Grant, D.M., Horthon, W.J., Mayne , C.L. and Evans, G.T. (1985), *J. Am. chem. Soc.* **107**, 6698.
65) Pastor, R.W., Venable, R.M. and Karplus, M. (1988), *J. chem. Phys.* **89**, 1112.
66) Flory, P.J. (1969) *Statistical Mechanics of Chain Molecules*, Wiley, New York.
67) Ferrarini, A., Moro, G. and Nordio, P.L. (1988), *Molec. Phys.* **63**, 225.
68) Nordio, P.L., Ferrarini, A. and Moro, G. (1988), in M. Moreau and P. Turq (eds.), *Chemical Reactivity in Liquids, Fundamental Aspects*, Plenum, New York.
69) Polimeno, A. (1988), unpublished results.
70) Bahar, I. and Erman, B. (1987), *Macromolecules* **20**, 1386.
71) Levine, Y.K., Birdsall, N.J.M., Lee, A.G., Metcalfe, J.C., Partington, P. and Roberts, G.C.K. (1974), *J. chem. Phys.* **60**, 2890.
72) London, R.E. and Avitable, J. (1977), *J. Am. chem. Soc.* **99**, 7765.
73) Edholm, O. and Blomberg, C. (1979), *Chem. Phys.* **42**, 449.
74) Wittebort, R.J., Szabo, A. and Guard, F.R.N. (1980), *J. Am. chem. Soc.* **102**, 5723.
75) Levine, Y.K. (1973), *J. Mag. Res.* **11**, 421.
76) Chachaty, C. and Langlet, G. (1985), *J. Chim. Phys.* **82**, 613.
77) Wittebort, R.J. and Szabo, A. (1978), *J. chem. Phys.* **69**, 1722.
78) EISPACK, Argonne Code Center, Argonne National Laboratory.
79) Lanczos, C. (1950), *J. Res. Natn. Bur. Stand* **45**, 255; (1952), *Ibid.*, **49**, 33.
80) Cullum, J. and Willoughby, R.A. (1985) *Lanczos Algorithm for Large Symmetric Eigenvalue Computations*, Birkhauser, Basel.
81) Moro, G. and Freed, J.H (1981), *J. chem. Phys.* **74**, 3757.
82) Moro, G. and Freed, J.H. (1986), in J. Cullum and R.A. Willoughby (eds.),*Large Scale Eigenvalue Problems*, North-Holland, Amsterdam.

83) Mori, H. (1965), *Prog. Theor. Phys.* **34**, 399.

84) Moro, G. and Freed, J.H. (1981), *J. chem. Phys.* **75**, 3157.

85) Wassam Jr., W.A. and Torres-Vega, Go. (1988), *J. chem. Phys.* **88**, 1837.

86) Ferrarini, A. (1988), unpublished report.

87) Lipari, G. and Szabo, A. (1982), *J. Am. chem. Soc.* **104**, 4546, 4559.

88) Coletta, F., Moro, G. and Nordio, P.L. (1987), *Molec. Phys.* **61**, 1259.

89) Coletta, F., Ferrarini, A. and Nordio, P.L. (1988), *Chem. Phys.* **123**, 397.

90) Berne, B.J., Borkovec, M. and Straub, J.E. (1988), *J. phys. Chem.* **92**, 3711.

91) Weber, T.A. and Helfand, E. (1983), *J. phys. Chem.* **87**, 2811.

92) Ferrarini, A. (1987), unpublished results.

DIELECTRIC POLARIZATION OF FLUIDS COMPRISING RIGID DIPOLES AND
MOLECULES WITH INTERNAL ROTATIONAL DEGREES OF FREEDOM

W. SCHRÖER, D. LABRENZ, C. RYBARSCH
Department of Chemistry
University of Bremen
Leobener Str.
D-2800 Bremen 33
Germany (FRG)

ABSTRACT. From the molecular theory of dielectric polarization an
expression for the average moment is obtained, which formally agrees with
the results of the Onsager–Böttcher–Scholte–model. The reaction field and
the cavity field are replaced by the results of a molecular statistical
theory. A simple approximation is proposed, which is exact at low
densities, and in fair agreement with the results of the simulations of the
Stockmayer fluid at high densities. The approximation is applied in order
to explain the density and temperature dependence of the dielectric
permittivity of pure fluids (CH_3F, CF_3H, CF_3Cl, H_2O, CH_3CN), and mixtures
with non–polar solvents.

In order to describe the dielectric permittivity of molecules with
internal rotational degrees of freedom, such as $CO(OR)_2$, $PO(OR)_3$ (R =
CH_3, CH_2CH_3), an equilibrium between the conformeres is assumed. The
dipole moment and the energies of the conformeres are calculated by
semi–empirical and ab initio methods (MINDO/3, MNDO, STO–3G, 3–21G, 4–
31G, 4–31G*). The reaction field stabilizes the conformere with the higher
dipole moment, which is the less stable conformere. Reasonable agreement
between theoretical and experimental results is found for the variation of
the average dipole moment with temperature and concentration.

1. INTRODUCTION

The dielectric properties of liquids have been a challenge for researchers
in liquid state physics since the early days of statistical mechanics. The
purpose of the classical work was to extract microscopic properties, such
as the molecular dipole moment μ, or the polarizability α, from
macroscopic properties as the dielectric permittivity ε, the refractive
index n and the thermodynamic functions. This was achieved by
assuming:
a) effective values α^* and μ^* for the polarizability and the dipole
   moment, and

141

Th. Dorfmüller (ed.), Reactive and Flexible Molecules in Liquids, 141–180.
© *1989 by Kluwer Academic Publishers.*

b) a local field E_{loc} determining the induced moment and the orientational energy of a dipole in the liquid,
both based on models taken from continuum electrostatics.

In the classical models the averaged dipole moment $\langle p \rangle$, of a molecule in a polarized fluid of a pure substance is given by an expression of the form

$$\langle p \rangle = [\, \alpha^* + \beta \frac{\mu^{*2}}{3} g \,] E_{loc} \tag{1.1}$$

Consequently the macroscopic polarization P is

$$P = \frac{\varepsilon - 1}{4\pi} E = \rho \langle p \rangle \tag{1.2}$$

where ρ is the number density and E the electric field. In the Debye model [1] we have $\alpha^* = \alpha$, $\mu^* = \mu$, and the local field is the Lorentz field. In the Onsager-model [2] the local field is the Cavity-field and α^* and μ^* are

$$\alpha^* = \frac{\alpha}{1 - \alpha R} \qquad \text{and} \qquad \mu^* = \frac{\mu}{1 - \alpha R} \tag{1.3}$$

R is the reaction field tensor which is for a sphere in a continuum

$$R = \frac{2}{\sigma^3} \frac{\varepsilon - 1}{2\varepsilon + 1} \tag{1.4}$$

σ is the radius of the assumed cavity representing the molecule in the medium. Scholte has modified the Onsager model for ellipsoidal molecules [3]. Details on the classical work and applications to experiments may be found in [4,5].

The aim of the more recent work on the statistical mechanics of dielectric fluids is the converse of the classical work: Using the machinery of statistical mechanics it is attempted to calculate correlation functions determining the macroscopic properties such as ε, from the molecular properties assumed to be known.

The basis of this modern approach is the work of Kirkwood [6] and Fröhlich [4] who derived rigorous relations between the fluctuations of the dipole moment of macroscopic spherical samples in vacuum and in a medium of arbitrary dielectric permittivity. The corresponding expressions for ellipsoidal samples are also known [7].

In most of the work on the statistical mechanics of dielectric fluids only non-polarizable dipoles are considered. By this restriction the theory is largely simplified as the evaluation of many-body averages is not required. However a fundamental problem, not present e.g. in Lennard Jones fluids, had to be solved: as a consequence of the long-range dipole-dipole interactions the macroscopic boundary conditions influence the molecular statistical theory. It was not clear for some time, which type of macroscopic sample was considered in the standard molecular theories. Consequently it was not clear, how the mean square fluctuations of the dipole moment, calculated by means of an analytical theory or simulation methods, were related to ε. Concerning the analytical theory for rigid dipoles, the problem was clarified by the work of Nienhuis and Deutch [8], Ramshaw [9] and Hoye and Stell [10], while Perram et.al. [11], Alder [12,13] and Neumann [14,15] gave solution for the simulation methods. Quite extensive simulations have been carried out for the Stockmayer potential [12,13] at the temperature $T^* = kT / \varepsilon_{LJ} = 1,35$ at the reduced density $\varrho\sigma^3 = 0.8$, and hard sphere fluids [16–19] at the same density. ε_{LJ} is the energy parameter in the Lennard Jones potential, and σ the corresponding diameter or the hard-sphere diameter. Looking at the results (Fig. 2.1) one can say that the dielectric permittivities, calculated for the Stockmayer fluid and the hard-sphere dipole system, are almost identical. It would be interesting to see if this remains a true statement at lower densities and lower temperatures T^*.

We now turn to the discussion of analytical theories. A complete analytical solution of the Ornstein–Zernicke equation, which is the basis of any analytical theory, is not possible, even for the hard-sphere fluid. Therefore one cannot expect this for hard-sphere dipoles or even more realistic molecular models. One has to confine oneself to simplifications of the direct correlation function, determining the Ornstein–Zernicke (OZ) equation, and a numerical solution for a restricted set of equations, determining the rotational invariant expansion of the pair-correlation function. Only for the mean-spherical (MSM) model, an exact analytical solution can be obtained [20]. More complete approximations to the direct correlation function, such as the linearized (LHNC) [21] and the quadratic (QHNC) [22] hypernetted chain theory, require numerical iterative solutions of the Ornstein–Zernicke equation. The LHNC and the QHNC theories lead to values of ε which are far too high, while the values of the MSM model are too small. Considering the deviations from the simulation result the MSM model seems superior to the more refined approximations.

All three approximations are fundamentally in error at low densities, as the 2nd dielectric virial coefficient (2nd DVC), which was exactly calculated by Joslin and Buckingham [23], is neglected entirely. The 2nd DVC however is the first manifestation of the intermolecular interactions. It describes the deviations linear in the density from the Debye model, which is exact for the ideal gas.

The 2nd DVC is calculated rigorously from the Boltzman factor involving the true two-body potential without any further statistical mechanical formalism. Further developments of the theory of dielectric polarization taking into account the molecular shape [24], higher multipole moments [25,26], anisotropic polarizability and multipole polarizabilities [27], would definitely require the consideration of the 2nd DVC. Therefore Patey's solution of the hypernetted chain theory (HNC) for the hard-sphere dipole [28] and for the Stockmayer fluid [29] was a major advance in the development of the theory of dielectric polarization, as it contains exactly the 2nd DVC. It is comforting to see that the agreement with the simulation is rather good. For reviews of the modern developments of the statistical theories of dielectrics the articles [30–33] may be consulted. It should be noted however, that these have been written before Patey advanced to the solution of the HNC theory.

The theory of polarizable dipoles was pioneered by Wertheim [34,35]. From the graph-theoretical analysis of the partition function, it can be extracted that the MSM theory of hard-sphere dipoles may be adapted to the case of polarizable molecules, if $\beta\mu^2/3$ is replaced by $\beta\mu^{*2}/3 + \alpha^*$, where μ^* and α^* are renormalized for the one-terminal graphs, which can be interpreted as the reaction field. The theory becomes much more involved if, as in the hypernetted chain theory, a higher connectivity in the direct correlation function is taken into account.

In this work we present a simple numerical approximation to the mean spherical model, the HNC-theory and the generalization to polarizable molecules [33,36]. The theory, which is almost as simple to apply as the expression of the Onsager-model, is generalized for ellipsoidal molecules, molecules with internal degrees of freedom and mixtures. It is applied to the analysis of experimental data from dipolar fluids as functions of the density and composition. The applied approximation is exact at low densities, including the third dielectric virial coefficient, but it is also in reasonable agreement with the simulation results at high densities.

2. RIGID DIPOLES

2.1 General Theory

The most important model of a polar molecule is the rigid dipole. Because of the absence of the polarizability, many-body interactions are not present, and the properties of this model fluid can be calculated by means of standard methods of statistical mechanics, based on two-body interactions.

From Kirkwood's theory [6] the average moment $\langle\mu\rangle$ in a sample of regular shape, containing rigid dipoles interacting in vacuo with a uniform field $\mathbf{E},^\circ$ is to first order in the field, given by

$$\langle\mu\rangle = \frac{1}{3}\beta\left[\mu^2 + \frac{\rho}{(8\pi^2)^2}\int\mu(\Omega_1)\cdot\mu(\Omega_2)h^{(2)}(\Omega_1,\Omega_2,r_{12})d\Omega_1 d\Omega_2 dr_{12}\right]E^\circ$$

(2.1)

where $r_{12} = r_2 - r_1$. ρ is the number density. $h^{(2)}$ is the pair-correlation function which is in general angular-dependent. Using the Ornstein-Zernicke equation, equ.(1) can be reformulated into the self-consistent equation [9,10,37]

$$\langle\mu\rangle = \beta\frac{\mu^2}{3}\left[E^\circ + \frac{4\pi}{3}\rho\, C\langle\mu\rangle\right]$$

(2.2)

C is defined by the correlation integral (3) involving the direct correlation funktion $c(\Omega_1, \Omega_2, r_{12})$.

$$y\,\frac{\mu^2}{3}\,C = \beta\frac{\rho}{(8\pi^2)^2}\int\mu(\Omega_1)\mu(\Omega_2)c(\Omega_1,\Omega_2,r_{12})\,d\Omega_1 d\Omega_2 dr_{12}$$

(2.3)

where $\quad y = \dfrac{4\pi}{3}\,\rho\beta\,\dfrac{\mu^2}{3}$

Equ. (3) holds rigorously, provided the dipole axis is a symmetry axis [9b]. We aim to calculate the integral C on the basis of the cluster expansion of c in an approximate but systematic fashion. Denoting the Mayer-f-function by a line, the fixed particles 1 and 2 (called terminals) by open dots, and the other particles determining the direct correlation function (called field particles) by a dot, the graphical representation of c is [38-40]

(2.4)

where $\circ\!\!-\!\!\circ = f(1,2) = \exp[-\beta u(1,2)] - 1$ $\quad$ and $\quad \bullet = \dfrac{\rho}{8\pi^2}\int d\Omega\,dr$

The terms in the first row represent the Percus-Yevick-approximation. In the hypernetted chain approximation the terms in the second row are added, while the terms in the last row called 1,2-irreducible are neglected in standard analytical theories.

Since the liquid structure is determined to a large degree by radial symmetric interactions it is convenient to separate the intermolecular potential into two parts: the radial potential u_0 and the angle-dependent potential u_1, which is here the dipole-dipole interaction energy [41]. Denoting the corresponding parts of the f-function by f_0 and f_1, and noting that the low-densitiy limit g_0 of the radial distribution function is given by $g_0 = 1 + f_0$, the f-function may be represented as

$$f = f_0 + f_1\,(1 + f_0) \tag{2.5}$$

The HNC theory involves the full f-function and all graphs representing c, except the 1,2-irreducible graphs.

The mean spherical model [42,20] and the linearized hypernetted chain theory [21,22] involve only the high temperature limit of f_1, which is linear in β and just a single chain of $-g_0\ \beta\mu_1 T_{12}\mu_2$ bonds. Denoting $-g_0\beta\mu_1 T_{12}\mu_2 = \sim\!\!\sim$ and $f_0 = -$, the graphs of the mean spherical model are

$$(2.6)$$

In the LHNC-theory graphs as

$$(2.7)$$

are included [43] while the graphs required to give exactly the 2nd dielectric virial coefficient, are

(2.8)

The dotted line denotes $-\beta\mu_1 T_{12}\mu_2$.

The integral of the graph $\text{o}\!\sim\!\!\sim\!\text{o}$ added to E^0 gives the Lorentz field. Equation (2) can now rigorously be written as [9,10,37]

$$\frac{\varepsilon - 1}{\varepsilon + 2} = \frac{y}{1 + C^* y} \tag{2.9}$$

The asterix denotes that the long-range contribution $-g_0\beta\mu_1 T_{12}\mu_2$ to C is evaluated and not included in C^*. By neglecting C^* the Debye model is regained. The Debye model is the exact low-density limit for any theory of dielectric polarization. Eq. (9) ist just another formulation of the dielectric virial expansion. C^* however is a more fundamental function than the virial expansion: it contains only irreducible clusters. Eq. (9) shows, how the first terms of C^* contribute to the higher dielectric virial coefficients.

The higher contributions to eq. (6) may be interpreted as terms shielding the Lorentz correction $4\pi/3\ P$ to the electric field. In Wertheim's MSM- theory the many-body distribution functions, required to evaluate this shielded interaction, are obtained using the superposition approximation, involving the low-densitiy limit of the hard-sphere pair-distribution function. In the LHNC-theory the density-dependent hard-sphere pair-distribution function is used instead [33,36].

As mentioned in the introduction, the MSM-theory is superior to the LHNC theory. The reason for this is probably a fortuitous cancellation: the LHNC-theory includes terms like 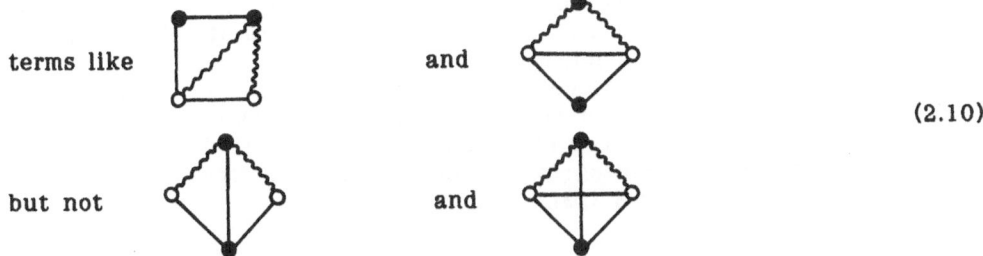 and

(2.10)

but not and

which are of similar size but opposite sign.

The hypernetted chain theory includes the uneven terms in equ. (8). The high-density contributions cause shielding of the dipolar interactions in the same manner as in equ. (6).

From this discussion we reach a simple interpretation of that part of c determining the dielectric polarization: it can be regarded as a Mayer-f-function involving a shielded dipolar interaction. Expanding the exponential, the integral of the linear term, which is considered in the MSM model or the LHNC theory, can be interpreted as the contribution to the local field, which corrects the Maxwell field E.

2.2 Numerical approximations
2.2.1 Hard-sphere dipoles

We now discuss the numerical estimates for the various contributions to C^*. The first contribution to the shielded dipolar interactions, which was calculated by Kirkwood, is $-15/16$ y [6]. If we considered just this term, then the shielding of the Lorentz factor would be greatly overestimated. One has to consider that the four-body term reduces this contribution and that in the five-body term the three-body graph

(2.11)

appears as a subgraph. We do this by replacing the field particle, which in the three particle graph in (6) has the weight y, by an effective field particle with the weight $x = (\varepsilon-1/\varepsilon+2)$. Just considering this modified three particle graph, we get a fairly good approximation to the mean spherical model, and a noticeable improvement in comparison to the Onsager model.

It is interesting to note that the Onsager model is obtained as follows: the field particles are given a weight x and their size is ignored. The series (6) is easily evaluated by Fourier transform techniques [44,45]. With this approximation we get exactly the terms for the series (6), which reduce the Lorentz field:

$$\frac{2x}{1+x} = \frac{2(\varepsilon - 1)}{2\varepsilon + 1}$$

(2.12)

This is the cavity field correction to the Lorentz field. The evaluation of the graphs, representing the MSM theory, in the approximation that the size of the field particles is ignored, gives the Onsager model. This allows for configurations of the particles, which are not possible in reality, because of the finite volume of the molecules. That is the reason why in the Onsager contiuum model the shielding is over-estimated.

The two-body contribution to C^* can be extracted from the 2nd dielectric virial coefficient B [23].

$$B\rho^2 = y^2 b$$

where

$$b = \sum_{n=1}^{\infty} \frac{12}{n} \left(\frac{\beta\mu^2}{\sigma^3} \right)^{2n} \left(\frac{2^n(n+1)}{(2n+3)!} \right)^2 \sum_{t=0}^{n} \frac{(3t-n)(2t)!}{t!^2} \tag{2.13}$$

b is always positive and therefore enhancing the value of the Clausius-Mossotti function. Its contribution is reduced at higher densities, because of the shielding. In a heuristic manner this is taken into account by multiplying b with x/y.

For the average moment we get on the basis of these two terms

$$\langle \mu \rangle = \frac{\beta\mu^2}{3} \left[E_L - \left(\frac{15}{16} x - b\frac{x}{y} \right) \frac{4\pi}{3} \rho \langle \mu \rangle \right] \tag{2.14}$$

Writing eq.(14) in the form of eq. (1.1) we find for the local field

$$E_{loc} = E_L - \frac{15}{16} x \cdot \frac{4\pi}{3} P \tag{2.15}$$

and for the correlation factor.

$$g = \frac{1}{1 - bx} \tag{2.16}$$

The Onsager model gives for the local field

$$E_{loc} = E_{cav} = E_L - \frac{2x}{1+x} \frac{4\pi}{3} P \tag{2.17}$$

but no estimates for the correlation factor.

Multiplying equ. (14) with $4\pi g/3$ and dividing by the Lorentz field E_L, we get the final equation to be used in the calculation of the dielectric permittivity

$$x = \frac{y}{1 - \frac{15}{16} xy - bx} \tag{2.18}$$

Fig. (1) shows the dielectric permittivity of the Stockmayer-fluid [12] and of the hard-sphere dipole system [16-18] as calculated by computer simulation. Those values are compared with the predictions of the Debye model, the Onsager model, and equ. (17). The figure also shows the prediction of ε by our approximation to the mean-spherical model, which means b = 0, and by a modification of our HNC approximation, in which the shielding of b is not taken into account.

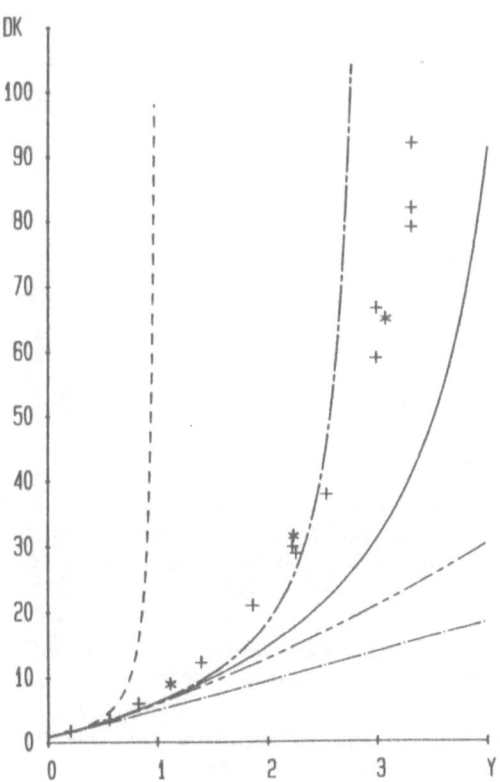

Fig. 2.1   Dielectric permittivity DK of a rigid spherical dipole fluid as function of y defined in equ. (2.3). Simulation results of the Stockmayer fluid (+) [12] and hard— sphere dipole fluids (*) [16] are compared with the predictions of
the Debye model (---),
the Onsager model (-·-·)
and our approximation to the HNC theory (——) (equ. 2.18),
to the mean spherical model (-···-···),   where b = 0,
and a modification in which the shielding of b is neglected (--·--).

It is quite remarkable that equ. (18) which is exact at low densities, is in fair agreement with the results of the simulations at high densities, and for the highest dipole moments which can be treated in the simulations.

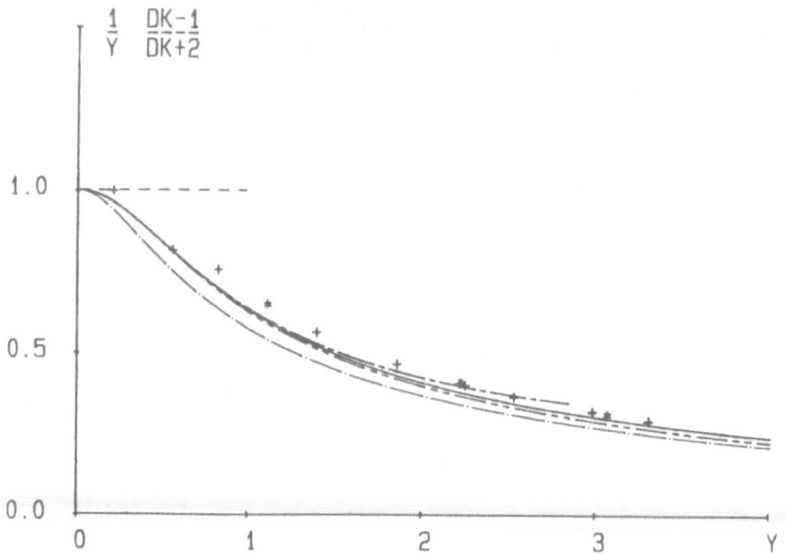

Fig. 2.2   Clausius–Mossotti function for the Stockmayer fluid and for the hard–sphere dipole fluid as function of y for various models explained in the legend of Fig. 2.1. In this representation, the models predicting large differences in the DK, give very similar graphs.

In Fig. 2 we show that in the plot of the Clausius–Mossotti function, the differences between the various models – with the exception of the Debye model – are very small.

In Fig. 3 the correction factors g, obtained for the various models, are drawn. The definition of these factors are

$$g_{Debye} = \frac{1}{y} \frac{\varepsilon - 1}{\varepsilon + 2} = \frac{x}{y}$$

$$g_{Onsager} = \frac{1}{y} \frac{(\varepsilon - 1)(2\varepsilon + 1)}{9\varepsilon}$$

$$\quad (2.19)$$

$$g_{MSM} = \frac{1}{y} \frac{x}{1 + \frac{15}{16}xy} = \frac{1}{1 - bx} g_{HNC}$$

g_{MSM} is the only factor which can be interpreted as describing the short-range correlation of the dipoles. g_{HNC} gives the deviation between the estimate of this factor in the theory, and the value of g_{MSM}.   The value deviates noticeably from unity. This deviation however is small if compared to g_D, g_O and g_{MSM}. From g_{MSM} we may conclude that hard-sphere dipoles are correlated in parallel orientation. The estimate of this correlation factor according to equ. (18) is slightly too small. The refinement of our approach would require a balanced consideration of quite a few more terms contributing to the third, fourth and fifth virial coefficient in the dielectric virial expansion. However, only the low-temperature contributions to the third virial coefficient are exactly known. [46]

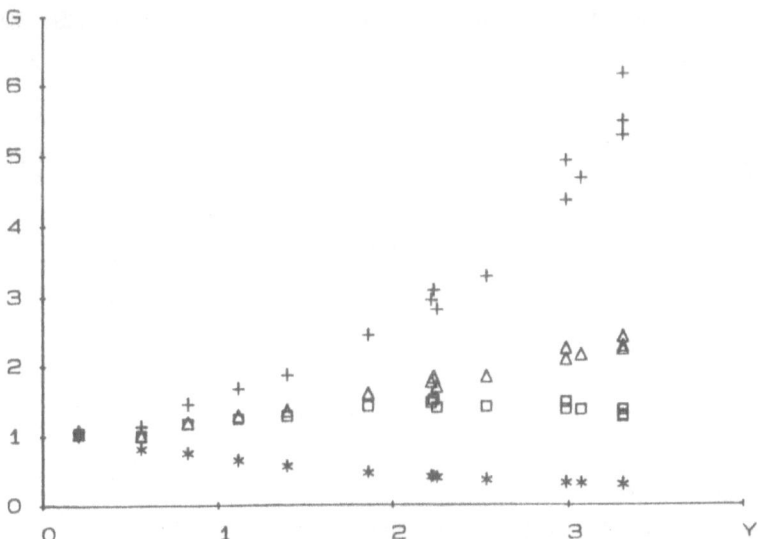

Fig. 2.3 Correction factors G of the Stockmayer fluid defined in equ. (2.19) calculated from the simulation results:
G_{Debye} (*), $G_{Onsager}$ (+), G_{MSM} (Δ), G_{HNC} ($\square$).
Only G_{MSM} has the physical meaning of a short-range correlation factor, G_{Debye} and $G_{Onsager}$ are largely determined by the short comings of the local field approximation, while G_{HNC} represents just the deviations of our HNC approximation to the simulation-results.

2.2.2 Ellipsoidal dipoles

The consideration of ellipsoidal molecules played an important role in the earlier macroscopic models, extending the Onsager-model [2] for spherical molecules to the more general case [3,9]. In respect to the application of the modern analytical and simulation methods, the consideration of ellipsoidal dipoles still seems to be complicated but is approached [47,50-52].

The consideration of non-spherical molecules is straightforward within the framework of our approach. A temperature expansion of the 2nd dielectric virial coefficient of ellipsoidal molecules is required. This would be used to replace b in equ. (18). A general analytical solution is not available, but a numerical evaluation of the 2nd DVC for ellipsoids with a central dipole pointing in the D_∞ symmetry axis of the molecule [53].

Analysing the temperature dependence of these results, it is found that the high-temperature limit can be calculated analytically and is determined by the shielding factor $S_\parallel$ of the ellipsoid, known from classical electrostatics [54].

$$b \simeq \Delta S \tag{2.20}$$

In the cases, when the direction of the dipole is parallel to the D_∞ axis

$$\Delta S = 3 S_\parallel - 1 \tag{2.21}$$

but, $\quad \Delta S = \frac{1}{2}(1 - 2S_\parallel) \tag{2.22}$

if the direction of the dipole is perpendicular to it.

$S_\parallel$ is

$$S_\parallel = \begin{array}{ll} (1 - \gamma^2)^{-1} + \gamma(\gamma^2 - 1)^{-3/2} \, \mathrm{arch}\, \gamma & \gamma \geq 1 \\ (1 - \gamma^2)^{-1} - \gamma(\gamma^2 - 1)^{-3/2} \, \mathrm{arcos}\, \gamma & \gamma \leq 1 \end{array} \tag{2.23}$$

where $\quad \gamma = a_\parallel / a_\perp$

$a_\parallel$ is the length of the D_∞ axis of the ellipsoid, $a_\perp$ the length of the axis perpendicular to it. In principle one would expect for the high temperature limit of b twice the value for a collision ellipsoid. The axes ratio for the collision ellipsoid is $(a_\parallel + \sigma)/(a_\perp + \sigma)$ where $\sigma = (a_\parallel \cdot a_\perp^2)^{1/3}$. This gives numerically the same value as for one collision ellipsoid with the axes ratio $a_\parallel/a_\perp$, which is assumed in the Onsager–Böttcher–Scholte model, where the cavity is the molecular cavity.

ΔS contributes to the local field in the same manner as the Lorentz factor, and does not vanish, even for the rare gas case. It is a residuum of the long-range integral and independent of the molecular size. For oblate molecules ($\gamma < 1$) the shape contribution to the local field is positive, and negative for prolate molecules ($\gamma > 1$). We emphasize that this contribution is a result of the shape of the molecules. It is obtained from the evaluation of $\multimap\kern-6pt\sim\kern-6pt\sim$, which is the high-temperature limit of C. The sign of ΔS is no indicator for parallel or antiparallel orientation. However, at low temperature for oblate molecules, b becomes more positive, while for prolate molecules the opposite is true, which indicates indeed a tendency of antiparallel orientation for prolate molecules and of parallel orientation for oblate molecules. This was predicted long ago by Buckingham [24]. The shielding of b is taken into account in the same manner as in equ. (18).

The equation evaluated is

$$x = \frac{y}{1 + \frac{15}{16} x\, y - \Delta S\, x}$$ (2.24)

which is a heuristic adaption of our approximation to the MSM-model for hard ellipsoidal dipoles. On the level of our approximation of the hypernetted chain theory, we would have to include the full second DVC. Equ. (18) would then be transformed to

$$x = \frac{y}{1 + \frac{15}{16} x\, y - b\, x}$$ (2.25)

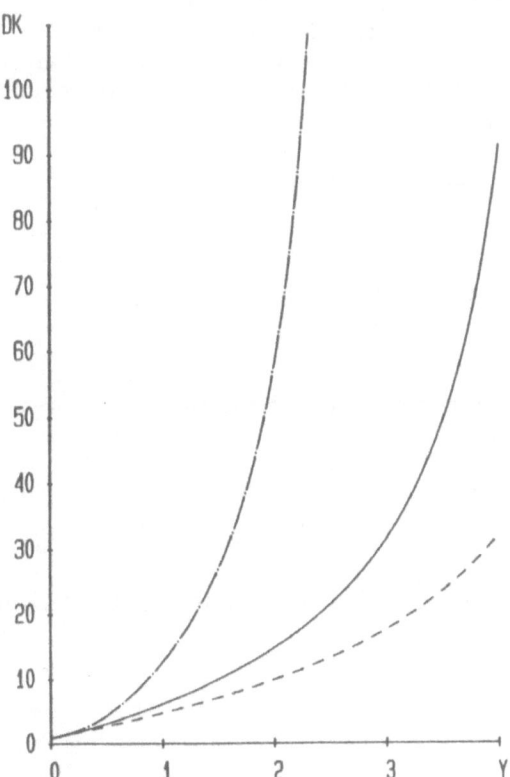

Fig. 2.4 Dielectric permittivity of ellipsoidal dipoles as function of y. The predictions for prolate molecules (– – –) with $a_\parallel/a_\perp = 2$ and for oblate molecules (–·–·) with $a_\parallel/a_\perp = \frac{1}{2}$, calculated from equ. (2.24), are compared to the predictions for spherical molecules (—), calculated from equ. (2.18)

Separating ΔS from b, as this is a contribution to the local field, the correlation factor estimated by this model would be

$$g_{MSM} = \frac{1}{1 - (b - \Delta S)x}$$ (2.26)

In Fig. (4) we show the influence of the shape correction on the dielectric permittivity, calculated on the level of our MSM approximation.

The effect of the shape is large. Unfortunately calculations of the third DVC, solutions of hypernetted chain theory or simulations for this molecular model are not yet available. Therefore we are not in a position to check the accuracy of this approximation. The statement, that the influence of the molecular shape on the dielectric permittivity is vast, can be made safely.

We compare our expression for the local field in a fluid of ellipsoidal dipoles with the cavity field of the ellipsoid [3]. In classical electrostatics, the field in a cavity is given by the electric field E, the field due to the surface polarization at the cavity, and the reaction field of this cavity [7].

$$E_{cav} = E + 4\pi S \cdot P - \frac{4\pi}{3} a_{\parallel} a_{\perp}^2 R \cdot P$$ (2.27)

where $$R = \frac{3}{a_{\parallel} a_{\perp}^2} S [(\varepsilon - 1)(1 - S)][1 + (\varepsilon - 1)(1 - S)]^{-1}$$

With the cavity field as the local field we get for the Clausius–Mossotti function of rigid dipoles

$$x = \frac{y}{1 + \frac{2xy}{1+x} - \frac{\Delta S x}{1+x}}$$ (2.28)

For small x equ. (28) agrees with our equ. (24) if the factor two is replaced by 15/16.

3. POLARIZABLE DIPOLES

3.1 General Theory

Following Kirkwood [6], we consider again a canonical ensemble of samples comprising N polarizable dipolar molecules interacting with a fixed uniform field E^o in vacuo. The potential energy H of the sample is the sum of the dipolar contributions and H_o, which represents the two-body short-range interactions.

$$H = H_o + \frac{1}{2} \sum_{i,k} \mu_i \cdot T_{ik} \cdot m_k - \sum_i m_i \cdot E^\circ - \frac{1}{2} \sum_i E^\circ \cdot A_i \cdot E^\circ \tag{3.1}$$

This Hamiltonian was given by Mandel and Mazur [55] and was also applied in the work of Wertheim [38]. A_i is a generalized polarizability of a molecule and is determined by the polarizabilities α_j of all molecules in the sample.

$$A_i = \alpha_i \left[1 - \sum_k T_{ik} \alpha_k + \sum_{k,k'} T_{ik} \alpha_k T_{kk'} \alpha_{k'} - \ldots \right] \tag{3.2}$$

Similarly, the moment m_1 is given by the permanent moment μ_1, corrected for the induced contributions.

$$m_i = \mu_i - \alpha_i \sum_k T_{ik} \mu_k + \alpha_i \sum_{k,k'} T_{ik} \alpha_k T_{kk'} \mu_{k'} - \ldots \tag{3.3}$$

The average moment $\langle p_1 \rangle$ is to first order in the applied field E°

$$\langle p_1 \rangle = \left[\beta \langle m_1 \sum_k m_k \rangle + \langle A_1 \rangle \right] \cdot E^\circ \tag{3.4}$$

The averages $\langle \ \rangle$ involve the distribution functions of the isotropic system.

In order to evaluate the square bracket, we expand the Boltzmann factor of the distribution function in the dipolar interactions. At first, we consider only the linear term. The dipole-dipole interaction can formally be written

$$\mu_1 T_{12} m_2 = \mu_1 \left[T_{12} - \sum T_{1k} \alpha_k T_{k2} + \ldots \right] \mu_2 \equiv \mu_1 T_{12}^* \mu_2 \tag{3.5}$$

Only the terms, where μ_1 and μ_2 represent different particles, contribute to the distribution functions determining the averages in equ. (4). The terms where $\mu_1 = \mu_2$, are parts of the graphs determining the free energy. In the distribution functions such terms occuring in the Boltzmann factor and the partition function cancel. The same is true for all even terms $(\beta \mu_1 T_{12}^* \mu_2)^{2n}$.

First we restrict our consideration to the linear terms of the f_1 function generalized for the many-body effect. The first term in equ. (4) can now be written.

$$\langle m_1 \sum_k m_k \rangle = \langle \mu_1 \mu_1 - \alpha_1 T_{1k}^* \mu_k \mu_k - \mu_1 \mu_1 T_{1k}^* \alpha_k - \mu_1 \mu_1 T_{1k}^* \mu_k \mu_k$$

$$+ \alpha_1 T_{1i}^* \mu_i \mu_i T_{ik}^* \alpha_k + \ldots \rangle_o \tag{3.6}$$

The distribution functions, applied to evaluate the many-body averages, are the many-body distribution functions of the reference system, determined by H_0.

The various interaction patterns, averaged in equ. (6) can be represented by Eulerian graphs [56]. Eulerian graphs are graphs which can be drawn without taking pen from paper. The lines represent a T-tensor, and the points may represent α or $\beta\mu^2$. We neglect for the moment all terms in which two particles are connected by three and more chains. With this simplification α and $\beta\mu^2$ can be treated in the same manner. A new interaction tensor τ can be introduced depending on $\alpha + \beta\mu^2$.

$$\tau_{12} = T_{1k}(\alpha_k + \beta\mu_k^2)T_{k2} - T_{1k}(\alpha_k + \beta\mu_k^2)T_{kk'}(\alpha_{k'} + \beta\mu_{k'}^2)T_{k'2} + \ldots (3.7)$$

A distinction must be made between the shielded interaction $\bar{\tau}_{12}$, where the particles labelled 1 and 2 are different and the reaction field $\overset{o}{\tau}_{11}$, where the start and the end of the interaction are at the same particle. We notice that any point in a graph, representing $\bar{\tau}$ or $\overset{o}{\tau}$, can be the root of all one terminal graphs. $\bar{\tau}$ is a chain decorated, by one-terminal graphs, and $\overset{o}{\tau}$ a cactus. $\bar{\tau}$ becomes a simple chain and $\overset{o}{\tau}$ a ring if the dots representing the polarizability and the dipole moments are renormalized for the reaction field.

$$\alpha' = \frac{\alpha}{1 - \alpha\overset{o}{\tau}} \qquad \text{and} \qquad \mu' = \frac{\mu}{1 - \alpha\overset{o}{\tau}} \qquad (3.8)$$

From this analysis, we can reformulate equation (4) in terms of α', μ' and τ

$$\langle p_1 \rangle = \langle \beta\mu_1'^2 + \alpha_1' - (\beta\mu_1'^2 + \alpha_1')(T_{12} - \bar{\tau}_{12})(\beta\mu_2'^2 + \alpha_2') \rangle \cdot E^o \qquad (3.9)$$

The averaging involves the distribution functions of the reference system. Many-body distribution functions are in general reducible and can always be written as sum of products of irreducible components. The three-particle distribution function contains $h^{(2)}$ and $h^{(3)}$ as irreducible components.

$$g(1,2,3) = 1 + [h^{(2)}(1,2) + h^{(2)}(1,3) + h^{(2)}(2,3)]$$
$$+ [h^{(2)}(1,2) \ h^{(2)}(2,3) + \ldots]$$
$$+ [h^{(2)}(1,2)h^{(2)}(2,3)h^{(2)}(1,3) + h^{(3)}(1,2,3)] \qquad (3.10)$$

As a result of this structure, the various terms, represented by equation (9), can also be written as sum of products of irreducible terms: Equ. (9) can now be rewritten as a chain of cut point-free averages $\langle T \rangle$ and $\langle\langle \bar{\tau} \rangle\rangle$, which means that the subgraphs representing $\langle\langle \bar{\tau} \rangle\rangle$ contain no point where the graph may be cut into two parts, each of which contains one terminal. The dots in $\langle\langle \bar{\tau} \rangle\rangle$, and between two averages $\langle\langle \bar{\tau} \rangle\rangle$ represent

$$y^* = \frac{4\pi}{3} \rho \left(\alpha^* + \beta \frac{\mu^{*2}}{3} \right) \tag{3.11}$$

where $\quad \alpha^* = \dfrac{\alpha}{1 - \alpha \langle\langle \overset{o}{\tau} \rangle\rangle} \quad$ and $\quad \mu^* = \dfrac{\mu}{1 - \alpha \langle\langle \overset{o}{\tau} \rangle\rangle} \tag{3.12}$

α^* and μ^* are the effective values of α and μ, obtained by correcting for the reaction field $R = \langle\langle \overset{o}{\tau} \rangle\rangle$. The interaction patterns representing $\langle\langle \overset{o}{\tau} \rangle\rangle$ are simple rings, whereas the dots represent y^*.

From this analysis, equ. (9) can be written in a self–consistent manner

$$\frac{4\pi}{3} \rho \langle p \rangle = y^* \left[E^o - (\langle T \rangle - \langle\langle \overline{\tau} \rangle\rangle) \frac{4\pi}{3} \rho \langle p \rangle \right] \tag{3.13}$$

Working out the long–range average $\langle T \rangle$, equ. (13) can be rewritten in terms of the Lorentz field. The Clausius–Mossotti function x is then

$$x = \frac{y^*}{1 - y^* \langle\langle \overline{\tau} \rangle\rangle} \tag{3.14}$$

3.2 Numerical Approximations
3.2.2 Spherical Dipoles

Neglecting the irreducible many–body terms in the distribution functions, and considering the low–density limit of the hard–sphere pair–correlation function, we get the generalization of the mean spherical model for polarizable molecules. Considering the true radial distribution function, the corresponding generalization of the linearized hypernetted chain theory is obtained.

As $\langle\langle \overline{\tau} \rangle\rangle$ is determined by the same graphs as the mean–spherical model, we adapt the result and approximate

$$\langle\langle \overline{\tau} \rangle\rangle \simeq - \frac{15}{16} x \tag{3.15}$$

The evaluation of $\langle\langle \overset{o}{\tau} \rangle\rangle$ gives

$$\langle\langle \overset{o}{\tau} \rangle\rangle \simeq \frac{2}{\sigma^3} x - \frac{1}{\sigma^3} \frac{15}{16} x^2 \tag{3.16}$$

In the Onsager model we have

$$\langle\langle \overline{\tau} \rangle\rangle = - \frac{2x}{1 + x} \tag{3.17}$$

and

$$\langle\langle \overset{o}{\tau} \rangle\rangle = \frac{2}{\sigma^3} \frac{x}{1 + x} \tag{3.18}$$

In our statistical approximation the leading term in the reaction field is the same as in the continuum model. The shielding, caused by a third particle, is overestimated in the continuum model, for the same reasons as discussed above (see equ. 2.12). Nevertheless, the difference between the estimate of $\langle\langle\overset{o}{\tau}\rangle\rangle$ in the Onsager model and the molecular approach is marginal, as α/σ^3 is a small quantity in organic liquids.

Now we want to develop an approximation, which is the analogue to the hypernetted chain approximation for hard-sphere dipole fluids. The leading term in the expansion of f_1 is $1/3!\ (\mu_1{}^*T_{12}{}^*\mu_2)^3$. This term vanishes in the angular average, except when multiplied by μ_1 and μ_2. Replacing μ by μ^* in the expression for b (equ. 2.13), we may write for the averaged moment

$$\langle p \rangle = \left[\alpha^* + \beta \frac{\mu^{*2}}{3} g \right] \left[E_L + \langle\langle \bar\tau \rangle\rangle \frac{4\pi}{3} \rho \langle p \rangle \right] \tag{3.19}$$

The value expected for the correlation factor is

$$g_{MSM} = \frac{1}{1 - bx} \tag{3.20}$$

As in equ. (2), the shielding of the dipolar interactions is taken into account heuristicly. The formula, finally used in the estimation of the dielectric permittivity, is

$$x = \frac{4\pi}{3} \rho\ (\alpha^* + \beta \frac{\mu^{*2}}{3} \frac{1}{1 - bx})(1 + \frac{15}{16} x y^*)^{-1} \tag{3.21}$$

As α/σ^3 is a very small quantity and since the leading term in the reaction field is the same in the Onsager model as in the molecular theory, the Onsager model may safely be used in the actual calculation of the reaction field. However, in variance to the common use of the Onsager model, the cavity representing the molecule is the collision volume, but not the molecular volume.

We now apply equ. 3.19 to analyse experimental data. Fig. 3.1 − 3.5 show the apparent correlation factors, obtained by evaluating the density dependence of the dielectric permittivity, using equ. 3.19 for the various models. The general feature is that $g_{Debye}\ \langle\ 1,\ g_{Onsager}\ \rangle\ 1$, while g_{MSM} and g_{HNC} are in most cases very similar with comparitively small deviations from unity. This shows that the estimate of the local field by the MSM model is realistic. For CClF$_3$ which has a small dipole moment of 0,5 D, all models give $g \approx 1$ (Fig. 3.1).

In the case of fluoroform the second DVC is known to be positive [48,49], while the third DVC is always negative. Therefore g_{Debye} must pass a maximum at intermediate densities. This maximum was not seen before for polar liquids. Earlier measurements consider either the low density region [48,49,82] or are carried out at high pressure [83]. Our measurements fill this gap nicely. As a consequence of the ellipsoidal shape the analysis using the MSM model gives an increase of g with the density for CHF$_3$ and a decrease for CH$_3$F.

For Water [88] and CH$_3$CN [83,90] the apparent g factors show very large deviations from unitiy: $g_{MSM}\ \rangle\rangle\ 1$ for water and $g_{MSM}\ \langle\langle\ 1$ for CH$_3$CN. This is to be expected from the 2nd dielectric virial coefficient. If the dielectric data are evaluated presuming a spherical shape for acetonitrile g_{HNC} is even smaller than g_{Debye}. This is so since the 2nd DVC of a sphere is positive, while it is negative for CH_3CN [91]. Therefore g_{MSM} is corrected in the wrong direction by equ. (3.20).

Table 3.1 Molecular parameters used in the calculations

	μ [D]	α [cm^3/mol]	σ [10^{-8} cm]	$a_{\parallel}/a_{\perp}$	Lit.
$CClF_3$	0.5	3.44	5.40	1.05	[49,95,96]
CHF_3	1.65	2.08	3.94	0.95	[48]
CH_3F	1.85	1.789	4.75	1.1	[85]
CH_3CN	3.96	3.90	4.45	1.3	[91]
H_2O	1.83	0.943	3.51	0.9	[89]
C_6H_6	–	6.38	5.30	–	[97]
C_6H_{12}	–	6.62	5.44	–	[97]
CCl_4	–	6.74	4.44	–	[97]
TEP	2.8	10.74	6.58	0.9	[76]

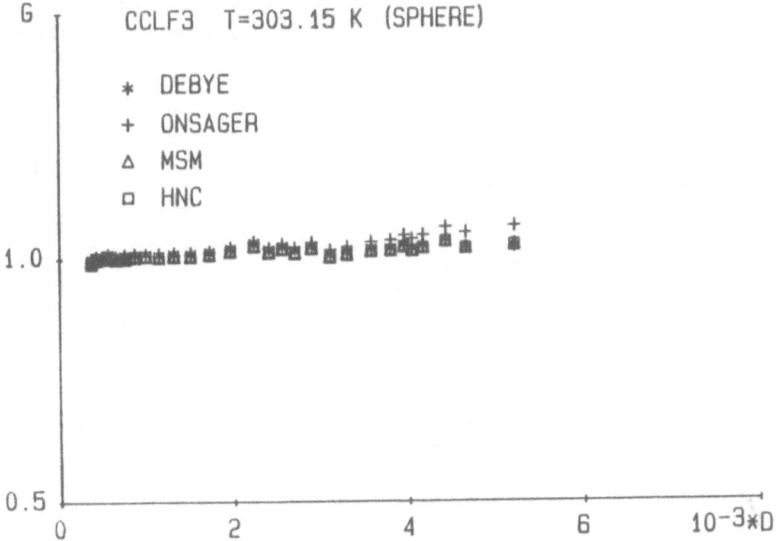

Fig. 3.1. Apparent correlation factors of $CClF_3$ [84] obtained from the density dependence of the dielectric permittivity assuming spherical molecular shape. D is the density [mole/cm^3].

160

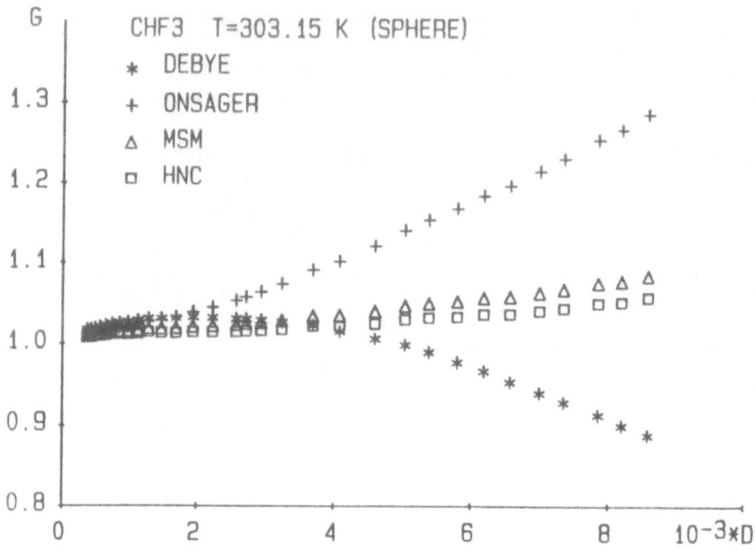

Fig. 3.2. Apparent correlation factors of CHF₃ [84] obtained from the
density dependence of the dielectric permittivity assuming spherical
molecular shape. D is the density [mole/cm³].

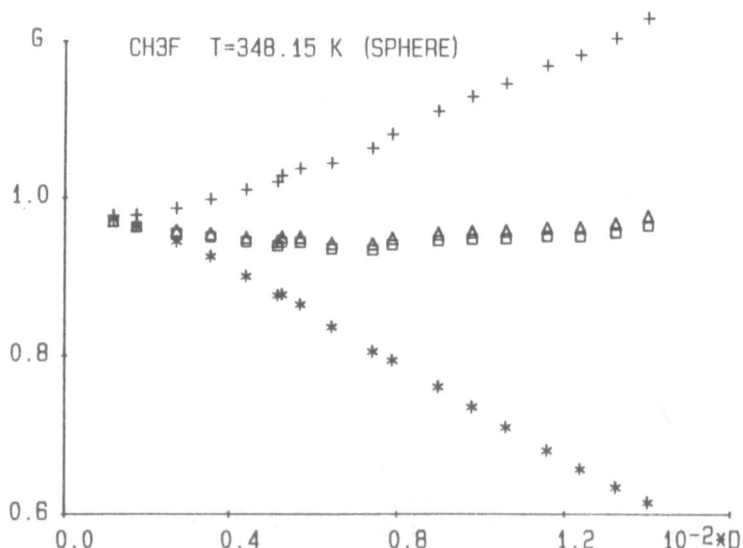

Fig. 3.3. Apparent correlation factors of CH₃F [85] obtained from the
density dependence of the dielectric permittivity assuming spherical
molecular shape. D is the density [mole/cm³].

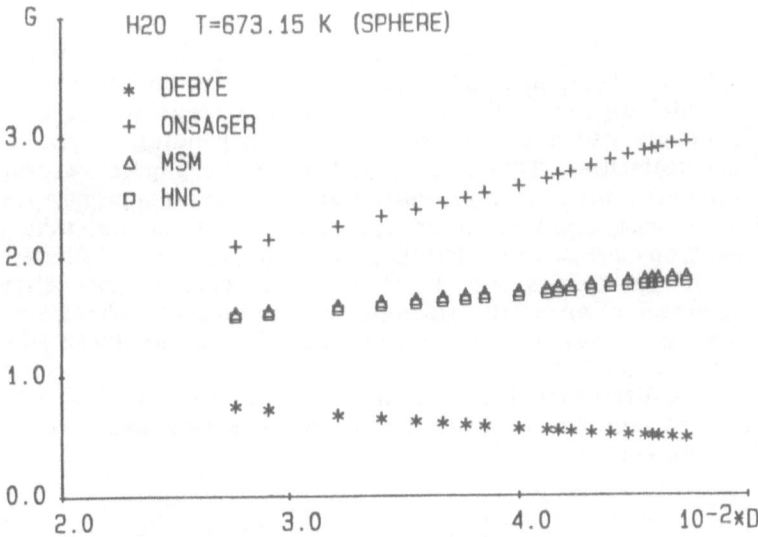

Fig. 3.4. Apparent correlation factors of H_2O [88] obtained from the density dependence of the dielectric permittivity assuming spherical molecular shape. D is the density [mole/cm³].

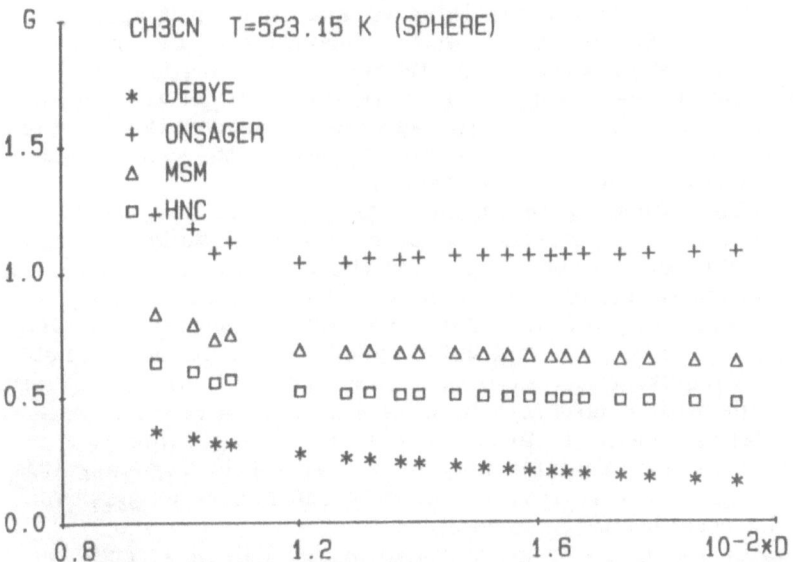

Fig. 3.5. Apparent correlation factors of CH_3CN [90] obtained from the density dependence of the dielectric permittivity assuming spherical molecular shape. D is the density [mole/cm³].

3.2.3 Ellipsoidal molecules

It is straightforward to introduce corrections for the ellipsoidal shape in the equations (19) and (21), analogous to the treatment for non-polarizable ellipsoidal dipoles. The evaluation of ⟨T⟩ leads in the case of non-spherical molecules, to a second-rank tensor, depending on the orientation of the molecules. This term vanishes in the angular average, except when contracted with another molecular second-rank property such as the second-rank component of μ^2 or α [54,57]. As we do not take into account the anisotropy of the polarizibility, the shape factor ΔS only contributes to the local field acting on the dipoles. This is quite different from Scholte's generalization of the Onsager model, in which the field in the ellipsoidal cavity is assumed to be the local field determining the orientation and the polarization.

Because of the difference between the orienting and the polarizing field the average moment ⟨p⟩ is given by the same expression as in the case of spherical molecules.

$$\langle p \rangle = \left[\alpha^* + \beta \frac{\mu^{*2}}{3} \frac{1}{1 - bx} \right] \left[E_L - \frac{15}{16} \times \frac{4\pi}{3} \rho \langle p \rangle \right] \qquad (3.22)$$

The expressions for b and the reaction field are those, evaluated for ellipsoidal molecules. In the same manner b is obtained from the second dielectric virial coefficient, as in equ. (2.13). It may be written as b = $\Delta S + \Delta b$. In the calculation on the level of the mean spherical model, Δb is assumed to be zero.

For the reaction field, determining α^* and μ^* we just apply the expression from the Scholte model. In the derivation of this reaction field expression, the model of a uniformly distributed polarization inside the molecule is assumed. We nevertheless apply the result of the continuum approach (equ. 2.27) here too, as the agreement between the continuum model and the molecular approach is rather good in the case of spherical molecules. The correction is in any case small.

In the same manner as in the last paragraph, we now use equ. (3.19) to calculate the apparent correlation factors. b is certainly not taken from the 2nd DVC for the sphere, but for the ellipsoid. For g_{Debye} and for g_{MSM} the high temperature limit of b is used. Fig. 3.6 and 3.7 show that for molecules like CH_3F and CHF_3 the high temperature limit suffices to make $g_{MSM} \approx 1$. For those molecules specific correlations are not present. This ist quite different from CH_3CN (Fig. 3.9), which is a small prolate molecule with a large dipole moment and definitely has a tendency for antiparallel alignment [61,91,93]. The value of b, obtained by extrapolation from Joslin's table [53] is −24. With this large negative value we find $g_{NC} \approx 1$ so that the density dependence of g_{MSM} for CH_3CN can be regarded to be understood.

For water the model of a prolate ellipsoidal dipole is still not sufficient to describe the density dependence of the dielectric permittivity.

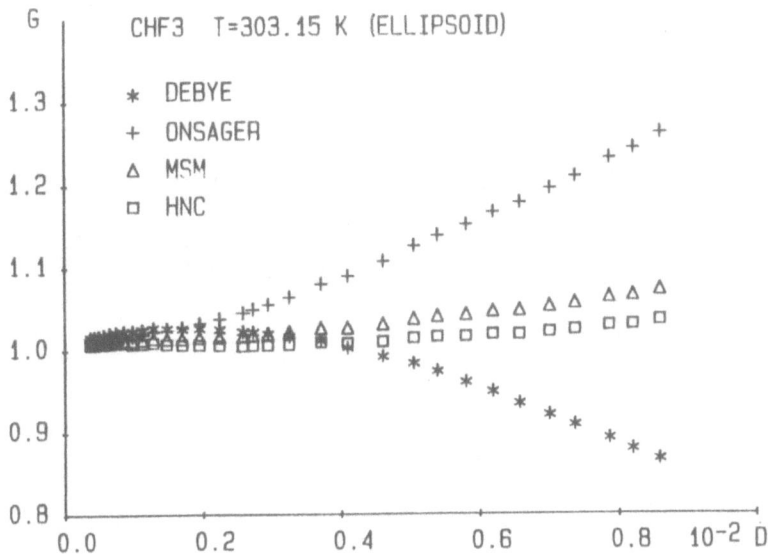

Fig. 3.6. Apparent correlation factors of CHF₃ [84] obtained from the density dependence of the dielectric permittivity assuming ellipsoidal molecular shape. D is the density [mole/cm³].

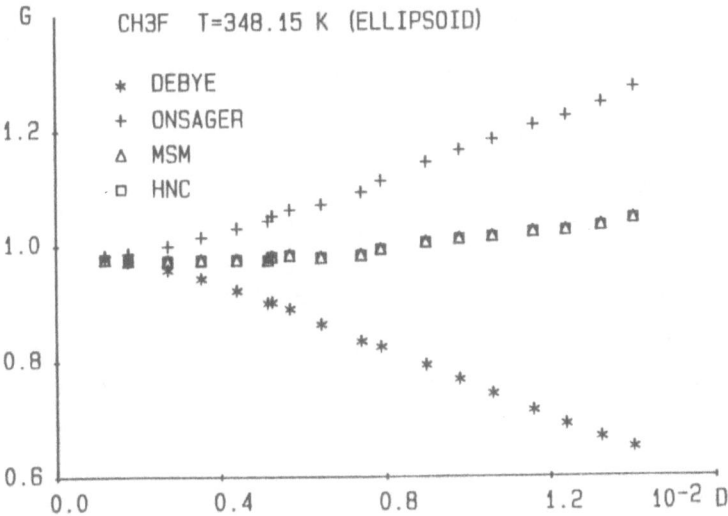

Fig. 3.7. Apparent correlation factors of CH₃F [85] obtained from the density dependence of the dielectric permittivity assuming ellipsoidal molecular shape. D is the density [mole/cm³].

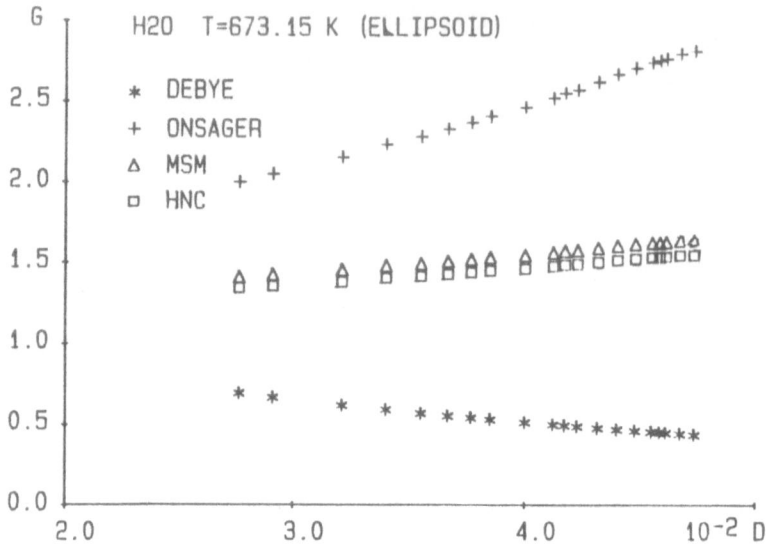

Fig. 3.8. Apparent correlation factors of H_2O [88] obtained from the density dependence of the dielectric permittivity assuming ellipsoidal molecular shape. D is the density [mole/cm³].

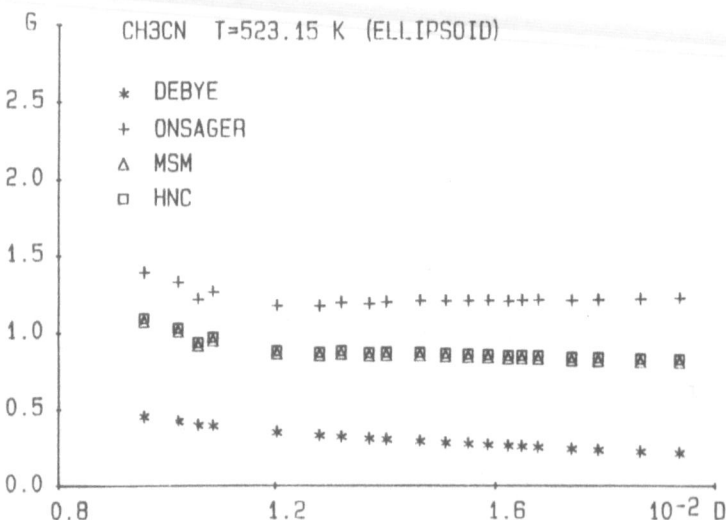

Fig. 3.9. Apparent correlation factors of CH_3CN [90] obtained from the density dependence of the dielectric permittivity assuming ellipsoidal molecular shape. D is the density [mole/cm³].

3.2.4 Mixtures

For mixtures the theory of dielectric polarization becomes very involved. The first correction to the Lorentz field gives only the famous factor 15/16, if all three particles of the cluster are of the same kind [44,45]. The collision volume determining the reaction field depends on the size and shape of the interacting molecules. In a binary mixture we have also to distinguish the different terms b_{11}, b_{12} and b_{22}, which form manifold ordered products, determining the correlation factors g_{11}, g_{12} and g_{22}. The analysis of this very complicated situation is beyond the scope of this paper.

The problem is considerably simplified in binary mixtures of a polar and a non-polar component. In this case, b concerns only the polar component. We furthermore adopt the local field approximation applied for pure fluids, which is correct if the molecules are of equal size. With these simplifications we get, instead of equ. (22)

$$\langle p \rangle = \left[x_s \alpha_s^* + x_p \alpha_p^* + x_p \beta \frac{\mu^{*2}}{3} \frac{1}{1 - x_p b_p x} \right] \left[E_L - \frac{15}{16} \times \frac{4\pi}{3} \rho \langle p \rangle \right] \qquad (3.23)$$

x_s and x_p denote the mole fraction of the non-polar and the polar component respectively.

One would expect different types of the reaction field tensors determining α_s^*, depending on the field particle considered. For convenience, we just take into account that the collision volume for the molecules is different, and varies with the composition.

$$\frac{1}{\sigma_s{}^3} = x_s \frac{1}{\sigma_{ss}{}^3} + x_p \frac{1}{\sigma_{sp}{}^3} \quad \text{and} \quad \frac{1}{\sigma_p{}^3} = x_p \frac{1}{\sigma_{pp}{}^3} + x_s \frac{1}{\sigma_{ps}{}^3} \qquad (3.24)$$

On the level of our approximation to the hypernetted chain theory equ. (23) is correct, only if the polar and the non-polar component are of the same size. Theoretical expressions for mixtures involving molecules of differing sizes are now investigated.

In the figures 3.10 - 3.13 we show the apparent correlation factors for mixtures of water with benzene [88] and acetonitrile with CCl_4 [94]. The apparent correlation factors are obtained by evaluating equ. (19) modified for mixtures. The figures just state the conclusions reached for the pure fluids. Taking into account the ellipsoidal shape of acetonitrile the concentration dependence of the dielectric permittivity is well described. The deviations from the predictions of our approximation to the HNC theory equ. (23) are small. For acetonitrile the Onsager model allows a fairly good description, which is quite exceptional. The reason for this is the cancellation of errors: the neglect of the very large negative second DVC and the overestimation of the shielding of the Lorentz field in the continuum approach.

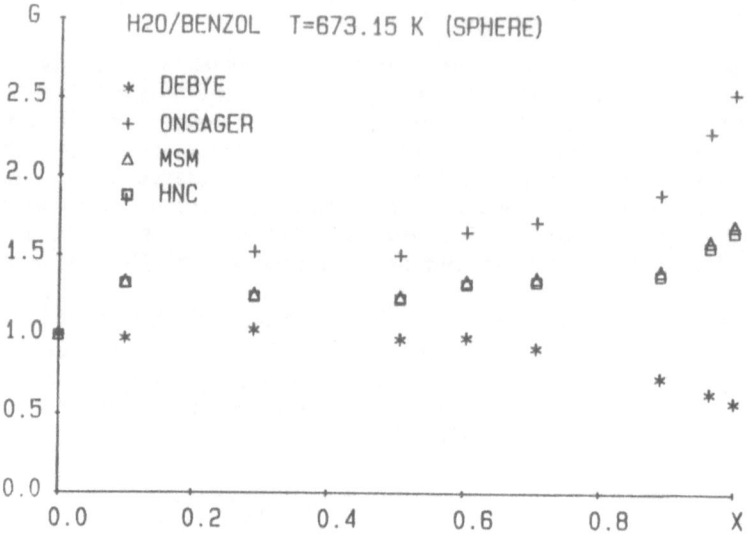

Fig. 3.10. Apparent correlation factors of H_2O/Benzol [88] obtained from the concentration dependence of the dielectric permittivity assuming spherical molecular shape. X is the mole fraction of H_2O.

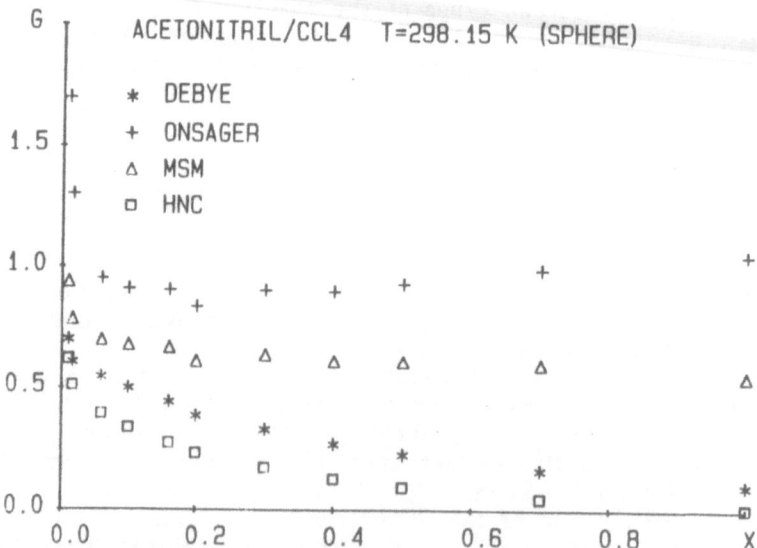

Fig. 3.11. Apparent correlation factors of CH_3CN / CCl_4 [94] obtained from the concentration dependence of the dielectric permittivity assuming spherical molecular shape. X is the mole fraction of CH_3CN.

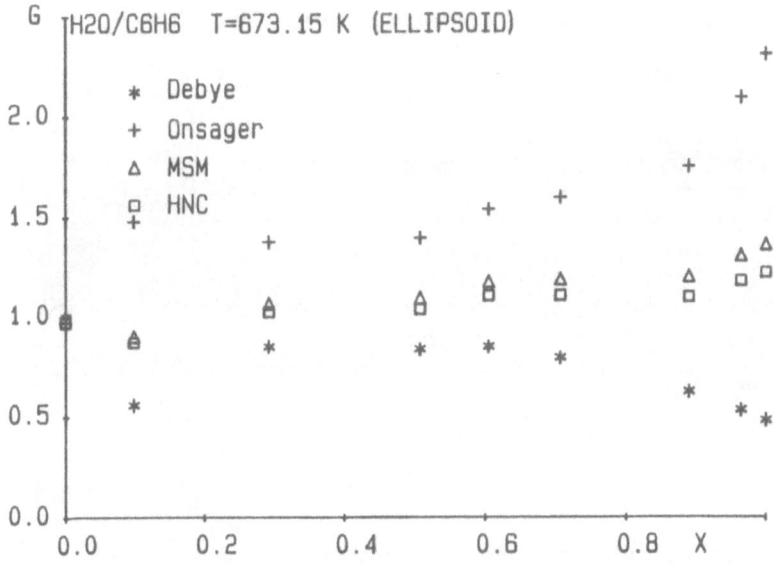

Fig. 3.12. Apparent correlation factors of H_2O / Benzol [88] obtained from the concentration dependence of the dielectric permittivity assuming ellipsoidal molecular shape. X is the mole fraction of H_2O.

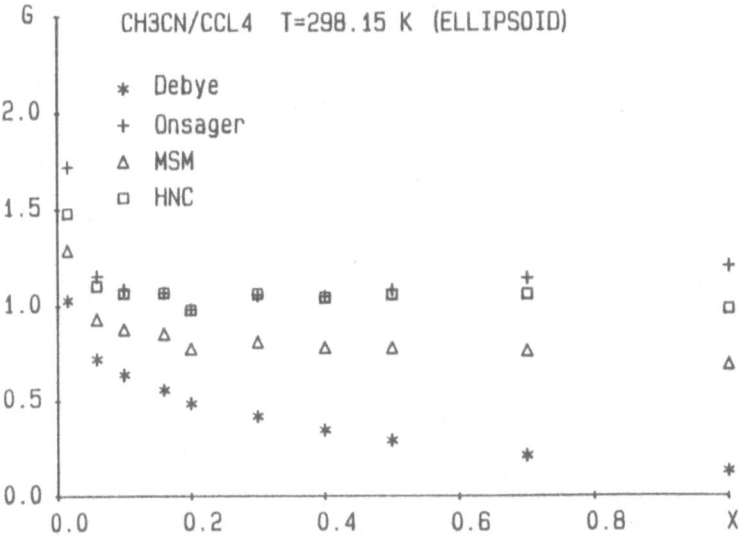

Fig. 3.13. Apparent correlation factors of CH_3CN / CCl_4 [94] obtained from the concentration dependence of the dielectric permittivity assuming ellipsoidal molecular shape. X is the mole fraction of CH_3CN

4. MOLECULES WITH INTERNAL ROTATIONAL DEGREES OF FREEDOM

4.1 General Considerations

Non-rigid molecules vary in their internal energy and entropy, their shape, their dipole moment and other multipole moments. Dielectric measurements have played a prominent part in the investigations of such systems with internal rotational degrees of freedom [58,59,60]. In principle, we may treat the liquid as a mixture of different species, which differ in their molecular properties, their interactions with the solvent, their reaction field and their local field. The correlations between the species should also be taken into account.

The distribution between the various species can be described by a Boltzmann distribution

$$\frac{N_i}{N} = \frac{e^{-\beta g_i}}{\sum\limits_j e^{-\beta g_j}}$$

$$(4.1)$$

g_i is the molecular free enthalpy of the species labelled i. It depends on the gas-phase free enthalpy, the potential energy due to the intermolecular interactions, and an entropy contribution arising from the change in the liquid structure, caused by inserting a particle of a certain size and shape into the liquid [61]. The latter effect can be estimated from the equation of state for hard convex bodies.

In this work, we just take into account the internal molecular energy and the reaction field [62,63] to model the intermolecular interactions. This is correct, provided that the internal and intermolecular entropy contributions are similar for the various species.

In many cases only a very limited number of conformeres is required in order to describe the liquid. In our investigation of esters of carbon acid and phosphoric acid only two conformeres (trans and gauche) need to be considered. The concentration ratio between the two conformeres is

$$r = \frac{N_g}{N_t} = \frac{N_g^o}{N_t^o} \exp\left[- \beta (E_g - E_t) + \frac{1}{2} \beta \left(\frac{\mu_g^2 R_g}{1 - \alpha R_g} - \frac{\mu_t^2 R_t}{1 - \alpha R_t} \right) \right]$$

$$(4.2)$$

N_g^o/N_t^o is the distribution of the conformeres in the high-temperature limit. The energy difference $\Delta E = E_g - E_t$ and the moments of the conformeres can be estimated by means of semi-empirical and ab initio quantum-chemical calculations or treated as experimental parameters. R is the reaction field of the Onsager-Scholte model. If the shape and the size of the conformeres are not very different, the difference in the reaction field factors may be neglected.

As said before, the general theory of dielectric polarization of a mixture of polarizable ellipsoids has not yet been developed. Important integrals, such as the second virial coefficient, describing the correlations between two different ellipsoidal dipoles, are not yet known. Therefore we just give the very simple result for spherical molecules of equal size on the level of the mean spherical approximation. The average moment $\langle p \rangle$ of a molecule is

$$\langle p \rangle = [\ \overline{\alpha^*} + x_p \beta \frac{\overline{\mu^{*2}}}{3}\ ][\ E_L - \frac{15}{16} \times \frac{4\pi}{3} \rho \langle p \rangle\ ]$$ (4.3)

The averages α^* and μ^{*2} are

$$\overline{\alpha} = (1 - x_p)\alpha_s + x_p \alpha_p$$ (4.4)

and

$$\overline{\mu^2} = \frac{1}{1+r}\mu_t^2 + \frac{r}{1+r}\mu_g^2$$ (4.5)

α_s is the polarizability of the non-polar solvent and x_p the molefraction of the polar component.

The Clausius–Mossotti function is

$$x = \frac{y^*}{1 + \frac{15}{16} \times y^*}$$ (4.6)

where $y^* = \frac{4\pi}{3}\rho\left[\overline{\alpha^*} + x_p \beta \frac{\overline{\mu^{*2}}}{3}\right] = y_\alpha^* + y_p^*$

Taking into account that the molecules have different shape, y_p^* may be separated into the contributions y^*_{pg} and y_{pt} from the gauche and the trans conformeres

$$y_p^* = \left[\frac{y_{pg}^*}{1 - \Delta S_{gg}\frac{x}{y^*}y_{pg}^*} + \frac{y_{pt}^*}{1 - \Delta S_{tt}\frac{x}{y^*}y_{pt}^*}\right]\frac{1}{1 - \Delta S_{gt}x}$$ (4.7)

ΔS_{gg}, ΔS_{tt}, ΔS_{gt} are the deviations of the electrostatic depolarization factors of the indicated collision volumes from the cases, when the collision volumes are spherical. As there are no reliable estimates for the distinct case ΔS_{AB}, we just apply equ. (3), which already involves three new parameters; if compared to rigid molecules; the energy difference and two dipole moments of the conformeres. In a refined treatment the entropy change, caused by a change in the molecular shape, should also be taken into account.

4.2 Applications
4.2.1 Dialkyl carbonates

In principle dialkyl carbonates allow for three conformeres:

trans-trans   trans-gauche   gauche-gauche

As a consequence of the conformere equilibrium, the dielectric permittivity shows a temperature dependence in the gas phase [66]. The IR-spectra also indicate the presence of distinct conformeres [78]. Using the Debye theory, which in this case applies, an apparent temperature-dependent dipole moment is obtained.

We have carried out a series of semi-empirical MINDO/3 [79] and MNDO [80] calculations using the MOPAC program package [64], and ab initio calculations with the Gaussian 80 program [65,81]. Starting from the three conformeres (trans-trans (tt), gauche-gauche (gg), gauche-trans (gt)) and optimizing the structure, we find in all calculations, that the tt conformation is not possible for steric reasons, and that with the exception of MINDO/3 the gg configuration is more stable than the gt conformation. The gg configuration also has the smallest dipole moment. This is in agreement with electrostatic reasoning, considering the interaction of the local dipole moments of the carbonyl and the alkoxy group.

In table (1) we give the values of the dipole moments μ_g and μ_t, and the energy difference $E_t - E_g$, for dimethyl and diethyl carbonate. These results are used in order to calculate the apparent values of the dipole moment in the gas and the liquid phase

$$\bar{\mu} = \left[\frac{1}{1+r} \mu_t^2 + \frac{r}{1+r} \mu_g^2 \right]^{1/2} \tag{4.8}$$

They are compared with values determined as the best fit parameters to describe the experiments [66,86].

Table 4.1: Energy and dipole moments of the conformeres
of dialkyl carbonates

Dimethylcarbonat [66]

	MOPAC		GAUSSIAN 80				
	MNDO	MINDO/3	STO-3G	3-21G	4-31G	4-31G*	exp.[2]
μ_{GG} [D]	0.62	1.33	0.38	0.14	0.01	0.3	0.8
μ_{GT} [D]	3.37	3.07	2.50	4.16	4.56	3.9	2.1
ΔE[kJ/ mole]	2.43	-9.25	10.13	6.99	8.74	15.44	11.7

Diethylcarbonat [86]

μ_{GG} [D]	0.87	2.06	0.15	0.26	0.43		1.0
μ_{TG} [D]	3.50	3.47	2.66	4.18	4.64		1.2
$\Delta E^{1)}$ [kJ/ mole	2.97	-8.30	10.38	6.82	8.83		5.6

[1] $\Delta E = E_{GT} - E_{GG}$
[2] The Debye model is applied for the evaluation of the gas
phase data

All calculations predict an increase in the average dipole moment with temperature. The minimal basis set (STO-3G) gives the best agreement with the values obtained from the experiments. As shown in Tab. 4.2 the agreement in the tendency and the order of magnitude is satisfactury. The amplitude estimated for the temperature dependence is too large.

Table (4.2) also shows the values of the apparent dipole moment of liquid dimethyl carbonate [87] as function of the temperature. The values calculated theoretically are obtained using the results of the STO-3G calculation for the energy and the moments of the conformeres, and the reaction field corrections. They are compared to the values of the apparent dipole moment obtained from the experimental data, presuming only one kind of rigid molecule. The agreement between the values obtained from the experiments and theoretical estimates is rather good for the ab initio calculations with the basis sets STO-3G and 4-31G*.

The reaction field is estimated for spherical molecules with the diameters $\sigma_g = 4.41 \cdot 10^{-8}$ and $\sigma_t = 4.08 \cdot 10^{-8}$ cm, which are obtained from the largest distances in the optimized structures, and from the van der Waals radii. For the molecular polarizability a value $\alpha = 5.06$ cm^3/mol is obtained from the refractive index n = 1.367 and the relation $\varepsilon_- = n^2 \cdot f$, where f = 1.05, which is a heuristic correction factor widely used in order to take into account atomic polarization [5].

Table 4.2: Temperature dependence of the apparent
dipole moment μ [Debye] of dimethyl carbonate
in the gas and liquid state

Gas [66]

T[K]	exp.	MNDO	STO-3G	3-21G	4-31G	4-31G*
479.35	1.00	2.47	0.98	2.11	1.95	0.87
412.25	0.94	2.42	0.85	1.90	1.68	0.69
350.05	0.89	2.36	0.71	1.64	1.37	0.55
328.15	0.86	2.31	0.66	1.53	1.25	0.51

Liquid [87]

T[K]	exp.	MNDO	STO-3G	3-21G	4-31G	4-31G*
278.15	0.79	3.04	0.73	3.32	3.48	0.78
298.15	0.85	3.03	0.82	3.30	3.49	0.86
323.15	0.93	3.01	0.90	3.29	3.51	0.97
353.15	1.02	2.99	0.95	3.28	3.53	1.09

[1] The gas phase data have been analysed using the Debye theory while equ. (3.19)
was used in the analysis of the data of the liquid. For the theoretical
estimation of the apparent dipole moment equ. (4.2) and the results given in
table 4.1 were used.

4.2.2 Trialkyl phosphates

As a second example we investigate trialkyl phosphates. In this case, we
have three groups, which allow in principle 2^3 conformeres. The trans-
gauche-gauche (tgg), and the gauche-trans-trans (gtt) conformeres, both,
represent three configurations. Spectra of electron-diffraction [69], NMR
[70], microwave [71,72] and IR [73-75] confirm the presence of different
conformeres in the gas phase and in the liquid.

The concentration dependence of the dielectric permittivity of
mixtures of triethyl phosphate (TEP) and cyclohexane [76] is analysed by
the same procedure applied for mixtures of rigid molecules (Fig. 4.1). An
apparent correlation factor is obtained, which cannot be explained by our
approximation to the HNC theory for spherical or ellipsoidal dipoles.
However, it can be explained by assuming an equilibrium between
conformeres and a mean dipole which varies with concentration. The
conformere with the higher dipole moment is stabilized with increasing
dielectric permittivity.

From the ab initio calculations, the conformations ggg and tgg are
obtained as the only stable conformeres.

(4.9)

gauche–gauche–gauche   trans–gauche–gauche

The energy difference between the conformeres is very small but the tgg conformere, which has the higher dipole moment, is predicted to be more stable. The results of the STO–3G calculations are given in table (4.3). They are compared to values obtained as the parameters of the best fit, to describe the concentration dependence of the dielectric permittivity of triethyl phosphate in cyclohexane [76]. While the calculations predict, in agreement with calculation for trimethyl phosphate [98], the tgg conformation to be slightly more stable than the ggg conformere in the gas phase, the analysis of the experiments leads to the opposite conclusion.

Table 4.3: Energy and dipole moments of the conformeres of triethyl phosphate

	MNDO	GAUSSIAN STO–3G	exp. [76]
μ_G [D]	1.55	0.39	0.4
μ_T [D]	3.09	2.84	3.65
ΔE [kJ/mol]	–0.04	–0.59	4.4

The disagreement in the sign of the energy is disappointing. Recent microwave measurements [71] indicate also that the less polar component is more stable, which supports our conclusion from the analysis of the experiments.

The reason for this disagreement is probably due to the fact that dispersion interactions cannot be calculated by the Hartree–Fock method. Through space dispersion interactions between the PO and the alkyl group are expected to stabilize the ggg configuration compared to the ggt configuration. From the group interaction model [77], used to describe thermodynamic functions of mixtures and pure fluids, a value of –4 to –10 kJ/mol is expected for this effect, which could explain the disagreement between the theoretical and the experimental results.

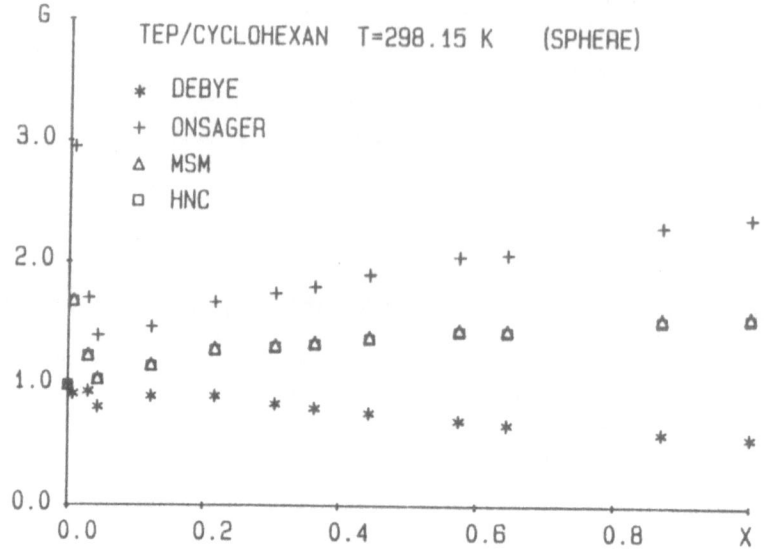

Fig. 4.1 Apparent correlationfactors of triethyl phosphate (TEP) in cyclohexane from the concentration dependence of the dielectric permittivity assuming <u>spherical</u> molecular shape. X ist the mole fraction For the dipole moment the apparent dipole moment at infinite dilution in cyclohexane is used (μ = 2,8 D).

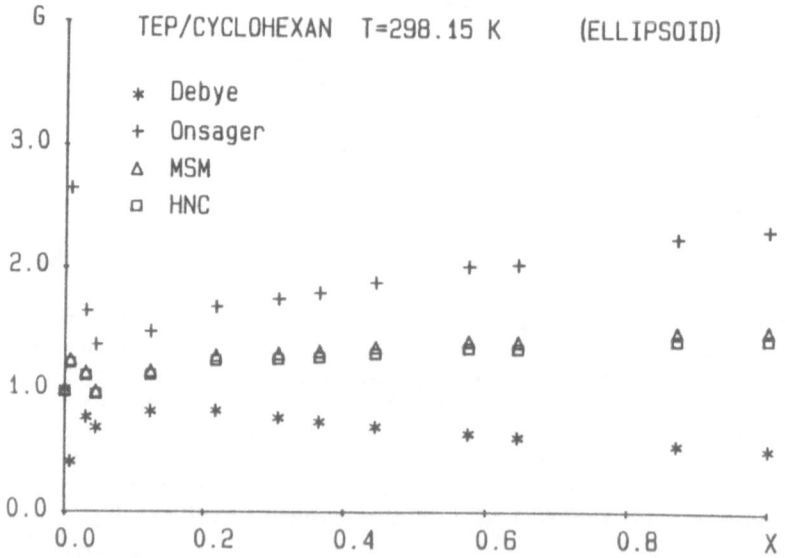

Fig. 4.2 Apparent correlationfactors of triethyl phosphate (TEP) in cyclohexane from the concentration dependence of the dielectric permittivity assuming <u>ellipsoidal</u> molecular shape. X ist the mole fraction. For the dipole moment the apparent dipole moment at infinite dilution in cyclohexane is used (μ = 2,8 D).

5. CONCLUSION

From the careful analysis of the structure of the distribution functions determining the dielectric polarization, approximate expressions have been obtained which hold exactly at low densities and are in fair agreement with the results of computer simulations for high densities. The formulae represent the first terms in a systematic renormalized self consistent cluster expansion. They reflect the most important contributions determining the self consistent process of dielectric polarization: the shielded dipolar interaction, the reaction field, and the low density two particle correlations. It is fairly straight forward to generalize the theory for mixtures and reacting fluids. Molecular properties such as higher moments and polarizabilities, as yet not considered can easily be taken into account provided second virial coefficients are known, and the theoretical estimates can be judged by comparison with the results of simulations and experiments on representative systems.

We like to thank Prof. E. U. Franck for giving access to the original partly unpublished data of the high pressure work of his group and Prof. W.-D. Stohrer for providing the library of quantum-chemical calculation programs. Support by the Deutsche Forschungsgemeinschaft, the research council of the University of Bremen and the Fonds der Chemischen Industrie ist gratefully acknowledged.

Literature

[1] Debye, P., Polare Molekeln, Hirzel, Leipzig (1929).

[2] Onsager, L., J. Am Chem. Soc. 58, 1468 (1936)

[3] Scholte, T.G., Physica 15, 437 (1949).

[4] Fröhlich, H., Theory of Dielectrics, 2nd ed., Oxford U.P. (1968).

[5] Böttcher, C.J., Theory of Electric Polarization, 2nd ed., Vol. 1 (1973); Vol. 2 (1978), Elsevier Amsterdam.

[6] Kirkwood, J.G., J. Chem. Phys. 7, 911 (1939).

[7] Schröer, W., Ber. Bunsenges. Phys. Chem. 86, 916 (1986).

[8] Nienhuis, G. and Deutch, J.M., J. Chem. Phys. 55, 4213 (1971).

[9] Ramshaw, J.D., a) J. Chem. Phys. 57, 2684 (1972);
    b) ibid 66, 3134 (1977);
    c) ibid 68, 4149 (1978).

[10] Hoye, J.S. and Stell, G., J. Chem. Phys. 61, 562 (1974);
     64, 1952 (1976).

[11] de Leeuw, S., Perram, J.S. and Smith, E.R. Proc. Roy. Soc., London, A373, 27, 57 (1980).

[12] Pollock, E.L. and Alder, B.G., Physica A102, 1 (1980).

[13] Alder, B.G., Alley, E. and Pollock, E.L.,
     Ber. Bunsenges. Phys. Chem. 85, 944 (1981) .

[14] Neumann, M., Mol. Phys. 50, 841 (1983).

[15] Neumann, M. and Steinhauser O., Mol. Phys. 39, 437 (1981).

[16] Levesque, D., Patey G.N. and Weis, J.J., Mol. Phys, 34, 1077 (1977).

[17] Patey, G.N., Levesque, D. and Weis, J.J., Mol. Phys. 45, 737 (1982).

[18] Neumann, M. and Steinhauser, O., Chem. Phys. Lett. 95, 417 (1983).

[19] Adams, D.J., Mol. Phys. 40, 1261 (1980).

[20] Wertheim, M.S., J. Chem. Phys. 55, 4291 (1971).

[21] Patey, G.N., Mol. Phys. 34, 427 (1977).

[22] Patey, G.N., Mol. Phys. 35, 1413 (1978).

[23] Buckingham, A.D. and Joslin, C.G., Mol. Phys. 40, 1513 (1980).

[24] Buckingham, A.D. and Pople J.A., Trans. Farad. Soc. 51, 1179 (1955).

[25] Buckingham, A.D., Adv. in Chem. Phys. 12, 107 (1967).

[26] Joslin, C.G., Mol. Phys. 47, 693 (1982).

[27] Buckingham, A.D. and Pople, J.D., Proc. Phys. Soc. 68A, 905 (1955).

[28] Fries, P.H. and Patey, G.N., J. Chem. Phys. 82, 429 (1985).

[29] Lee, L.Y., Fries, P.H. and Patey, G.N., Mol. Phys. 55, 751 (1985).

[30] Adelman, S.A. and Deutch, J.M., Adv. Chem. Phys. 31, 103 (1975).

[31] Stell, G., Patey, G.N. and Hoye, J.S., Adv. Chem. Phys. 48, 183 (1981).

[32] Madden, P. and Kivilson, D., Adv. Chem. Phys. 61, 411 (1984)

[33] W. Schröer, Adv. Chem. Phys. 63, 720 (1985).

[34] Wertheim, M.S. a) Mol. Phys. 26, 1425 (1973);
            b) ibid 33, 95 (1977);
            c) ibid 36, 1217 (1978).

[35] Wertheim, M.S., Am. Rev. Phys. Chem. 30, 471 (1979).

[36] Schröer, W., Thesis, Cambridge (1981).

[37] Schröer, W., Mol. Phys. 41, 239 (1980).

[38] Uhlenbeck, G.E. and Ford, G.W., Studies in Statistical Mechanics 1, 123, North Holland, Amsterdam (1962).

[39] Rushbrooke, G.S. in Physics of Simple Liquids
     Temperley, N.N.V, Rushbrooke, G.S. and Rowlinson J.S. eds. Elsevier, Amsterdam, p. 25 (1968).

[40] Stell, G. in The Equilibrium Theory of Classical Fluids
     Frisch, H.L. and Lebowitz J.L. eds. Benjamin, New York, p. II–171 (1964).

[41] Lebowitz, J.L., Stell G. and Baer, S., J. Math. Phys. 6, 1282 (1965).

[42] Lebowitz, J.L. and Percus, J.K., J. Math. Phys. 4, 248 (1961).

[43] Rushbrooke, G.S., Mol. Phys. 37, 761 (1979).

[44] Jepsen, D.W. and Friedman, H., J. Chem. Phys. 38, 846 (1963).

[45] Jepsen, D.W., J. Chem. Phys. **44**, 774 (1966).

[46] Joslin, C.G., Mol. Phys. **42**, 1507 (1981).

[47] Baron, V., Mol. Phys, **28**, 809 (1974).

[48] Barnes, A.N.H., Sutton, L.E. Trans Farad Soc. **67**, 2926 (1971)

[49] Sutter, H., Cole, R.H., a) J. Chem. Phys. **52**, 132 (1970)
         b) ibid **96**, 2014 (1967)

[50] Perram, J.W. and Wertheim, M.S., J. Comp. Phys. **58**, 409 (1985).

[51] Frenkel, D., Mulder, B.M., Mol. Phys. **55**, 1171, 1193 (1985).

[52] Perera, A., Kusalik, P.G. and Patey, G.N.,
         J. Chem. Phys. **87**, 1295 (1987).

[53] Joslin, C.G., Mol. Phys. **47**, 771 (1982).

[54] Schröer, W. and Rybarsch, C., Chem. Phys. Lett. **126**, 342 (1986).

[55] Mandel, M. and Mazur, P., Physica **24**, 116 (1958).

[56] Harary, F., Graph Theory Addison–Wesley, London (1969).

[57] Schröer, W. and Rybarsch, C., Chem. Phys. Lett. **131**, 500 (1986).

[58] Orville. Thomas, W.J. ed., Internal Rotation in Molecules,
         John Wiley, London (1974).

[59] Mizushima, S.I., Structure of Molecules and Internal Rotations,
         Academic Press, New York (1954).

[60] Smyth, C.P. in Ref. 58, p. 29.

[61] Lippert, E., Chatzidimitriou–Dreismann, C.A. and Naumann, K.H.,
         Adv. Chem. Phys. **57** 311 (1984)

[62] Abraham, R.J. and Bretschneider, E. in Ref. 58, p. 481.

[63] Ref. [5], p. 145

[64] QCPE Nr. 464

[65] QCPE Nr. 437

[66] Mizushima, S. and Kubo, M., Bull. Chem. Soc. Jap. **13**, 174 (1938).

[67] Yasumi, M., J. Chem. Soc. Jap. **60**, 1208 (1939).

[68] Thomson, G., J. Chem. Soc. 1228 (1939).

[69] Oberhammer, H., Z. Naturforsch. A28, 1140 (1973).

[70] Khetrapal, C.L., Govil, G. and Yeh, H.S.C.,
J. Mol. Struct. 116, 303 (1984).

[71] Stockhausen, M. and Wessels, V. priv. Comm. (1987)
Wessels, V., Diplom-Thesis, Münster (1987)

[72] Katolichenko, V.I., Egorov, Yu.P., Borovikov, Yu.,Ya. and
Semenic, V.Ya., Theoret.; Eksperm. Khim 10, 88 (1974).

[73] Mortimer, F.S., Spectrochimica Acta 9, 270 (1957).

[74] F. Marsault-Hérail, J. Chim. Phys. 68, 274 (1971).

[75] Katolichenko, V.I., Egerov, Yu.P., Borovikov, Yu.Ya. and Golik, G.A.,
Zh. Obshch. Khim. 43, 2490 (1973).

[76] Labrenz, D. (1984) Diplom-Thesis, Bremen

[77] Schröer, W., Naumann, K.D., Domke, W.D. and Lippert E. in Molecular
Liquids, Buckingham, A.D., Lippert, Bratos, E.S., eds. Wiley, New
York, p. 309 (1978).

[78] Katon, J.E., Cohen M.D., a) Can. J. Chem. 53, 1378 (1974);
b) ibid 54, 1994 (1975).

[79] Bingham, R.C., Dewar M.J.S., Lo D.H.,
J. Am. Chem. Soc. 97, 1285, 1294 (1975).

[80] Dewar, M.J.S., Thiel, W., J. Am. Chem. Soc. 99, 4899, 4907 (1977).

[81] Binkley, J.S., Pople J.A., Mehre, W.J.,
J. Am. Chem. Soc. 102, 939 (1980).

[82] T.G. Copland, R.H. Cole, J. Chem. Phys. 64, 1741 (1976).

[83] Franck, E.U. in: Organic Liquids, Buckingham, A.D., Lippert, E., Bratos,
S. eds. Wiley, New York, p. 181 (1978).

[84] Labrenz, D., Thesis Bremen 1988

[85] David, H.G., Hamann, S.D., Pearse, J.F.,
J. Chem. Phys. 20 969 (1952).

[86] Kubo, M., Sci. Papers Phys. Chem. Res., 29 169 (1936).

[87] Thiebault, J.M., Rivail, J.L., Grefe, J.L., J.Chem.Soc.Farad.II, 72 2024
(1976).

[88] Deul, R., Thesis, Karlsruhe 1984.

[89] Franks, F.: Water − A comprehensive Treatise, Wiley, London 1973

[90] Francesconi, A.Z., Franck, E.U., Lentz, H.,
Ber. Bunsenges. Phys. Chem. 79 897 (1975).

[91] Buckingham, A.D., Raab,R.E., J.Chem. Soc. 5511 (1961).

[92] Kratochwill, A., Weidner, J.U., Zimmermann, H.,
Ber. Bunsenges. Phys. Chem. 77 408 (1973).

[93] Bertagnolli, H., Ber. Bunsenges. Phys. Chem. 81 739 (1977).

[94] Arnold, R., Yarwood, J., Price, T.E., Mol. Phys. 48 451 (1983).

[95] DiGiacomo, A., Smyth, C.P., J. Am. Chem. Soc. 77, 774 (1955).

[96] Kong, C.L., Larson, J.W., J. Chem. Phys. 55, 3051 (1971).

[97] Kielich, S. im Dielectric and Related Molecular Processes, Davis, M.
ed., The Chemical Society, London, 1972, p. 192.

[98] J.R. van Wazer, C.S. Ewig, J. Am. Chem. Soc. 108, 4354 (1986).

DYNAMIC PROPERTIES OF SHORT STYRENE CHAINS IN BULK MODELLED BY COMPUTER SIMULATION

M. BUCHNER and TH. DORFMÜLLER
Faculty of Chemistry
University of Bielefeld
D−4800 Bielefeld
West Germany

ABSTRACT. Disordered structures of styrene oligomers with five to ten monomeric units are developed in a cubic box using periodic boundary conditions. By energy minimization these structures are brought to detailed mechanical equilibrium. Harmonic analysis shows that, to a certain extent, side group libration is decoupled from the other degrees of freedom suggesting a simplified dynamical model for the relaxational behaviour of the phenyl groups. Rotational Raman spectra are calculated for the two limits of harmonic chain motion and decoupled side group libration respectively. In the light of these results a common way of evaluating low frequency Raman spectra of amorphous polymers is critically analysed. The light scattering coupling coefficient, which is usually assumed to depend on frequency quadratically, is shown to be sensitive to the chemical structure of the sample. This leads to serious problems in the standard evaluation procedure for this kind of spectra.

1. Introduction

For a long time the computational treatment of polymeric systems was the domain of Monte Carlo methods [1–3]. Only in recent years the problem of detailed modelling of chain molecules in solution [4] or bulk was tackled, either with the aim to gain insight into molecular elementary processes [5–8] or to study the structural and elastic behaviour of these systems [9–11].

The aim of this article is to demonstrate the usefulness of rather simple computational methods for examining the low frequency spectroscopic behaviour of amorphous polymeric material as compared to corresponding experimental results. Polystyrene was chosen as objective mainly for technical reasons.
  - Its physical properties have been extensively studied and documented.
  - Because of the small polarity of styrene chains long range interactions do not have to be included in the potential function.
  - The side group configuration may be well described with only one soft degree of freedom.
  - The phenyl groups are by far the most active entities in light scattering experiments because of their high anisotropic polarizability.

The following section deals with the way the structures are built up and the coordinates they are described with. Normal coordinates are used for an approximate dynamical treatment and to characterize the motion of side groups. The spectroscopic behaviour is derived for two dynamical models and compared to experiment in section three.

Th. Dorfmüller (ed.), Reactive and Flexible Molecules in Liquids, 181–198.
© 1989 by Kluwer Academic Publishers.

2. Methods Employed to Simulate Disordered Chain Structures with Side Groups

2.1. STRUCTURE DEVELOPEMENT AND DEGREES OF FREEDOM TAKEN INTO ACCOUNT

Computationally, disordered polymeric structures are usually built up in two steps. In the first step a rough estimate of the final structure is obtained with the help of a random process either on a lattice or in continuous space [10]. Energy restrictions are taken into account by forbidding overlap or by introducing a Boltzmann weighting factor. In the second step relaxed configurations can be reached by energy minimization [10] or by different types of simulated annealing, solving the Langevin equation of motion and decreasing temperature at the same time [12] or combining standard molecular dynamics simulation with local minimization techniques [13], in order to avoid being trapped in a local energy minimum.

In the present case $N = 10$ non−overlapping isotactic styrene oligomers with $M = 5$ to 10 monomeric units are placed in a cubic section of a tetrahedral lattice using periodic boundary conditions. At this stage the chains are restricted to pure trans or gauche states, the space for side groups is occupied by cyclohexane rings. The recursively formulated algorithm is outlined in Figure 1. Figure 2 shows a generated structure. It is evident that at this stage the chains are quite inhomogeneously distributed in space.

<u>try to build up the rest of the chain :</u>

```
if chain is completed
then
        return 'successful'
else
        while any possible configuration for current monomeric unit left
        do
                check for overlap with other chains or units
                if no overlap
                then
                        try to build up the rest of the chain
                        if this is successful
                        then
                                return 'successful'
                        end if
                end if
        end while
        return 'not successful'
end if
```

Figure 1. Simplified depiction of the algorithm used to obtain first guess structures for non−overlapping oligomeric chains. For multichain samples the procedure is called N (= number of chains) times.

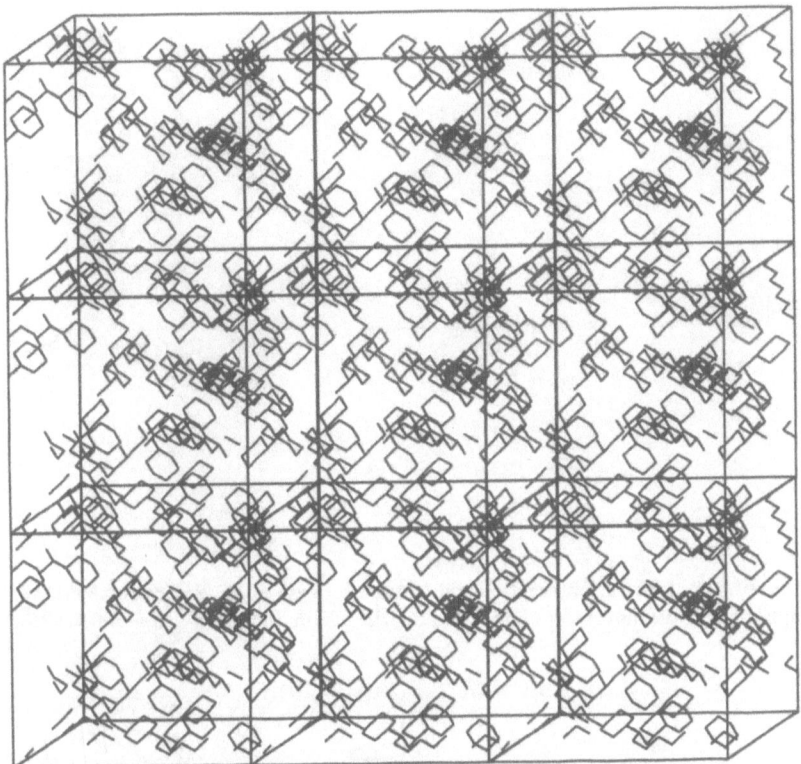

Figure 2. Random structure of styrene–like chains on a tetrahedral lattice.

The first guess lattice structure is converted to continuous space, at the same time the cyclohexane rings are flattened. The position and configuration of chain n, $1 \leq n \leq N$ is described with the cartesian coordinates (x_n, y_n, z_n) of an end–methyl group, the Eulerian angles $(\varphi_n, \theta_n, \psi_n)$ determining the orientation, and torsional angles of monomeric unit m, $1 \leq m \leq M$ along the chain's backbone $(\eta_{nm1}, \eta_{nm2},$ with $\eta_{n11} \equiv \psi_n)$ and the bonds to the phenyl groups (σ_{nm}) (Figure 3). Potential energy and its partial derivatives are expressed as functions of these coordinates. Bond lengths and bond angles are held constant. Carbon and hydrogen atoms are collapsed into common sites. Minimization of potential energy is performed involving the soft degrees of freedom mentioned above. Different sets of potential energy parameters are derived from two more detailed force fields [14,15]. The results presented in the course of this work turn out to be only weakly sensitive to the exact form of the force field. Detailed information can be found in [16]. A quasi–Newton algorithm is applied using first partial derivatives. Figure 4 shows a completely relaxed sample.

These and all subsequent calculations are performed for about fifty sample boxes in order to improve statistics.

184

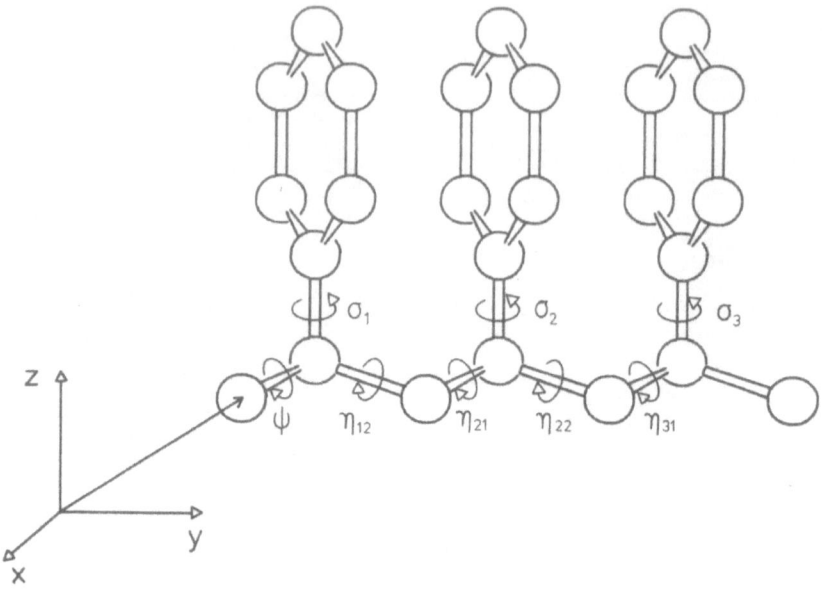

Figure 3. Sketch of the type of interaction centres and degrees of freedom taken into account. The spherical coordinates φ and θ and the chain index n are omitted for simplicity.

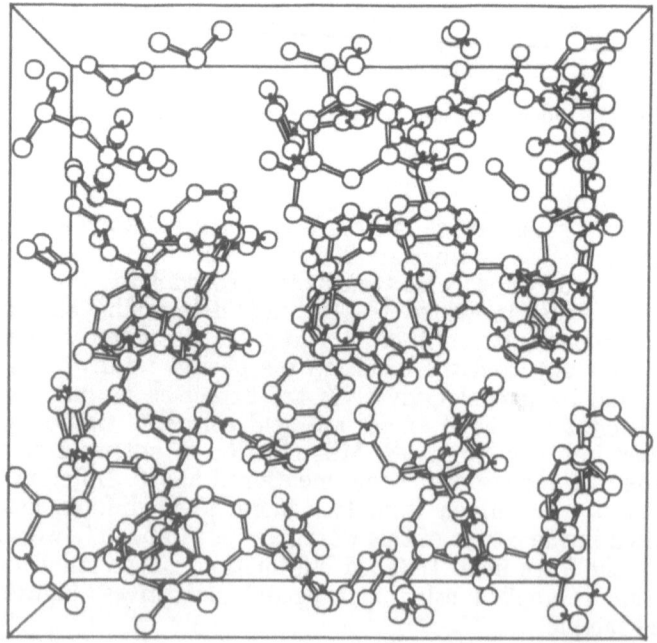

Figure 4. Plot of ten chains with five monomeric units in detailed energetic equilibrium.

2.2. NORMAL COORDINATE ANALYSIS AND DERIVATION OF A SIMPLIFIED DYNAMICAL MODEL

2.2.1. Main chain dynamics. Below the glass transition temperature extensive conformational interconversions are frozen in [8]. At even lower temperatures the short time dynamics of amorphous samples may be treated by including purely elastic forces only. Anharmonic effects and energy redistribution should be of minor significance on the timescale of the period of oscillation. Correspondingly the vibrational density of states is used as the key concept for describing the behaviour of inorganic glasses, e.g. of amorphous semiconductors [17]. Normal modes of disordered structures, especially those at higher frequency, are known to be highly localized [18] permitting in simulation the usage of moderately sized samples for harmonic analysis.

The technique of harmonic analysis of disordered polymers has been hitherto applied mainly to ensembles of isolated chains. Snyder and Strauss [19] performed a vibrational analysis of single disordered polymethylene–like chains including stretching, bending, and torsional degrees of freedom. They showed that in the region of $0 - 140$ cm^{-1} the density of states $g\,(\omega)$ is dominated by torsional modes. A gap in the histogram of $g\,(\omega)$ below 30 cm^{-1} introduced by excluding stretching and bending coordinates led to the conclusion that for very low frequencies all types of internal coordinates are essential.

The same set of coordinates used in the energy minimization step is taken into account to calculate normal modes. The treatment of generalized coordinates follows the procedure described in mechanics textbooks [20]. For the potential energy part, mixed derivatives with respect to positional, orientational, and torsional coordinates are computed by including intermolecular terms. The kinetic energy part is represented by the metric tensor T connecting generalized coordinates ξ_j and atomic or centre positions r_i and masses m_i. This results in a generalized eigenproblem.

$$V\,A = T\,A\,\lambda$$

V is the matrix of mixed derivatives of potential energy V with respect to every pair of generalized coordinates ξ_j, ξ_k.

$$V_{jk} = \frac{\partial^2 V}{\partial \xi_j\,\partial \xi_k}$$

$$T_{jk} = \sum_i m_i \frac{\partial r_i}{\partial \xi_j}^T \frac{\partial r_i}{\partial \xi_k}$$

$$\xi_j \in \{\, x_n,\, y_n,\, z_n,\, \varphi_n,\, \theta_n,\, \psi_n,\, \eta_{nm1},\, \eta_{nm2},\, \sigma_{nm}\,\}$$

$$1 \leq n \leq N\ ,\ 1 \leq m \leq M$$

λ is the diagonal matrix of eigenvalues and A is the matrix of the corresponding eigenvectors.

$$\lambda_{ij} = \delta_{ij}\,\omega_i{}^2$$

The symmetrical, positive definite matrix T is factored into the product of a lower triangular matrix and its transpose according to

$$T = L\,L^T$$

which leads to

$$V'\,A' = A'\,\lambda$$

with

$$V' = L^{-1}\,V\,(L^T)^{-1}$$

$$A' = L^T\,A$$

Eigenvectors and eigenvalues of this transformed eigenproblem are obtained with a standard Jacobi diagonalization procedure. The normal coordinates q_i are defined by

$$\xi = A\,q$$

Three of them are equivalent to purely translational coordinates of the whole simulation box with vanishing eigenvalues. The eigenvalues of the other modes turn out to be positive thereby confirming that the system is situated in a true local minimum of potential energy.

Figure 5 shows the computed density of states. There is a conspicuous maximum at thirty wavenumbers with considerable contributions at lower frequencies although stretching and bending coordinates, in contrast to the work of Snyder and Strauss, are not taken into account. Between 40 and 140 cm^{-1} $g\,(\omega)$ decreases almost linearly.

In Figure 6 the contributions of the different types of generalized coordinates to a normal mode are shown as functions of the normal mode frequency. The contributions are computed by summing up the squares of the transformed eigenvector components A_{jk}' corresponding to a given coordinate type. They therefore lie in the range of zero to one. At very low frequencies the normal modes can be characterized as a kind of tumbling motion involving whole molecules. At intermediate frequencies main chain motion takes place whereas at high frequencies librational motion of side groups dominates. A more detailed analysis shows that, on the one hand, all modes are strongly intermolecular in nature justifying the effort of the present approach of simulating polymers in bulk and not isolated. On the other hand, only a few torsional angles always contribute significantly to a given normal mode.

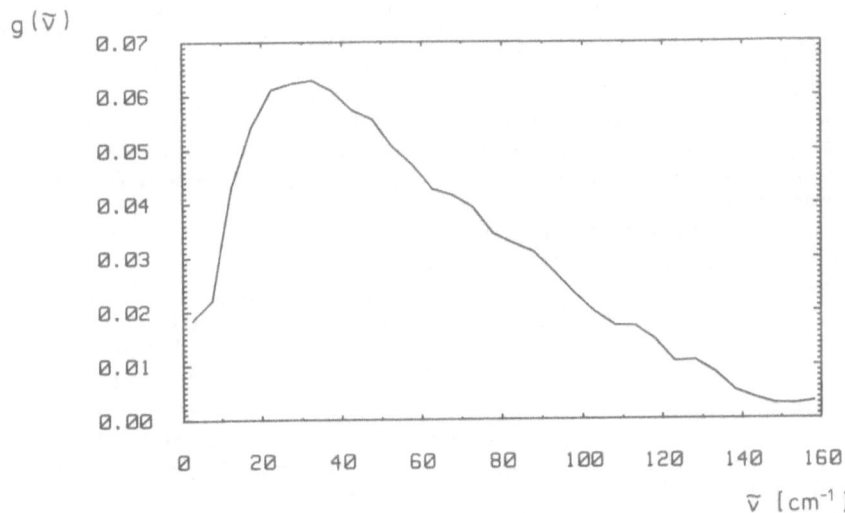

Figure 5. Computed density of states g $(\tilde{\nu})$. In this and the following figures frequencies are expressed in wavenumbers for consistency with experimental data. All figures showing data which are obtained by harmonic analysis are histograms with a channel width of 5 cm^{-1}.

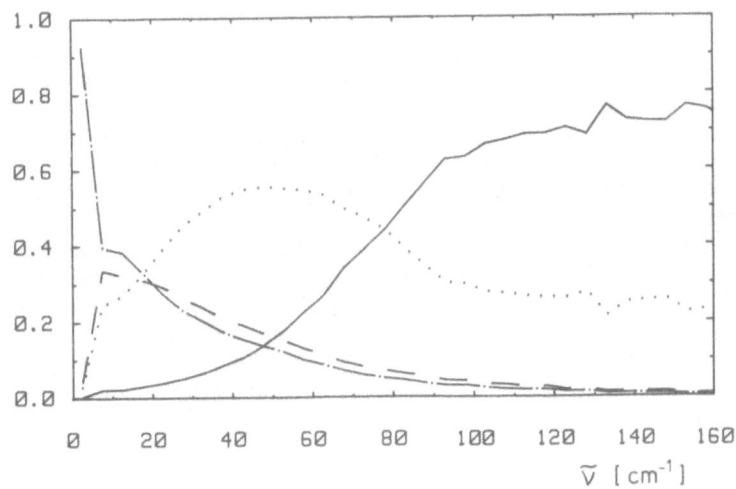

Figure 6. Contributions of different types of generalized coordinates to normal coordinates as a function of frequency. Different regions, though not sharply defined, where positional (x, y, z : dashed–dotted), orientational (φ, θ, ψ : dashed), internal main chain (η_{m1}, η_{m2} : dotted), or side group coordinates (σ_m : solid) dominate, can be distinguished.

2.2.2. *Side group dynamics.* The free rotation of phenyl groups is strongly hindered by high potential barriers, as was shown by several molecular mechanics calculations of isolated styrene chains [14,21,22]. The long rotational relaxation times discovered in high resolution Rayleigh [23] and Raman scattering [24] as well as in NMR experiments indicate a strongly cooperative type of motion. Thus an appropriate physical picture for styrene chains in solution or melt may be that of chain backbones slowly undergoing conformational transitions thereby inducing phenyl ring flips seemingly as a side effect.

In a solid polymer conformational changes can be considered as completely frozen in comparison to the fast librational motion of phenyl groups. This turns out to be confined to an angle of the order of twenty to thirty degrees. The question arises as to the nature of the main relaxation mechanisms for this oscillatory motion.

Two different channels for energy redistribution can be distinguished. The libration of the phenyl rings is coupled to the main chain motion through the connecting bond as well as to other chains and side groups by direct interaction. A rough measure for the type of coupling mechanism can be obtained by analysing the normal modes of the sample. Figure 6 shows that essentially two different types of normal modes can be observed. Below a frequency of about 50 cm^{-1} the normal modes involve almost exclusively main chain coordinates. At frequencies higher than 90 cm^{-1} a side group contribution of 60 to 80% per normal mode is found.

These circumstances suggest a rather simplified picture of dynamically coupled phenyl groups each experiencing an individual, insignificantly fluctuating potential field formed by the local chain configuration. The total energy of this model of anharmonic coupled librators can be formulated as

$$H = \sum_j \frac{1}{2} I \dot{\sigma}_j^2 + \sum_j v_j (\sigma_j) + \frac{1}{2} \sum_j \sum_k v_{jk} (\sigma_j, \sigma_k)$$

The first term is the kinetic energy with moment of inertia I. $\dot{\sigma}_j$ is the rotational velocity of side group j. The second term represents the potential energy of the side groups exerted by the, in this approximation, rigid main chains, whereas the third term describes the coupling between different phenyl rings. The chain configuration enters the potential energy only parametrically. Figure 7 elucidates this dynamical model.

The corresponding equations of motion are solved in the usual way yielding the temporal evolution of the phenyl group orientation. Figure 8 shows an example of a reorientational correlation function of a single ring. The motion of the side groups exhibits dynamical individuality by different vibrational frequencies as well as by a varying degree of coupling among two or more oscillators. The librational motion persists and is scarcely damped over several periods. Spectroscopic broadening can therefore be expected to be of inhomogeneous origin.

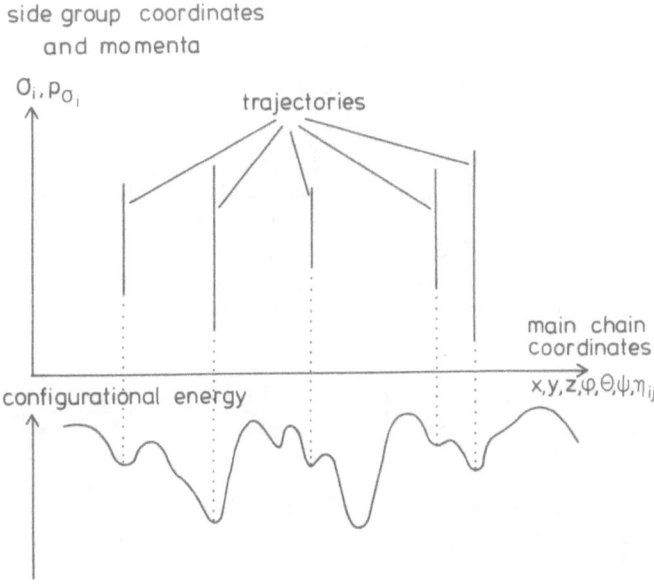

Figure 7. Sketch of phase space illustrating the model of anharmonically coupled, librating phenyl groups connected to stiff chain backbones.

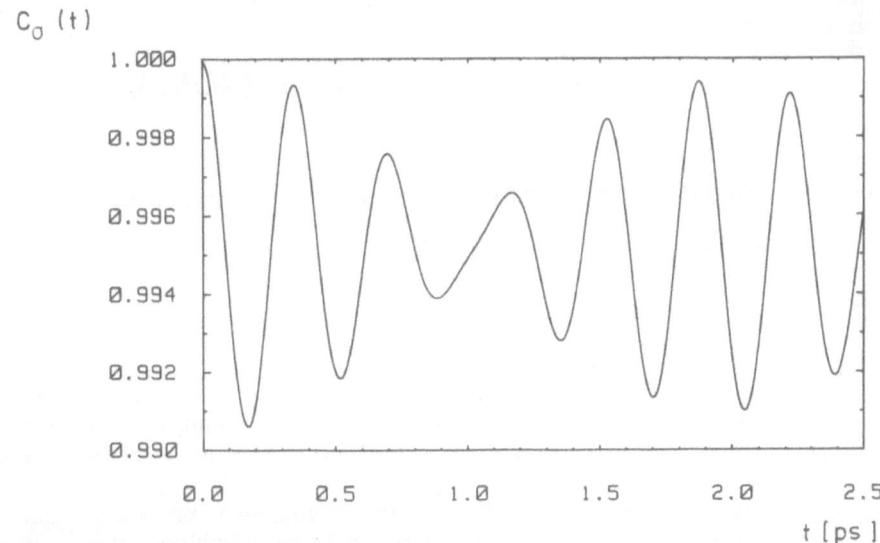

Figure 8. Typical reorientational correlation function of a single phenyl ring. Since the motion is hindered by high potential barriers the function retains a nonzero value at long times.

3. Application to low frequency Raman scattering

3.1. STANDARD EVALUATION PROCEDURE FOR LOW FREQUENCY RAMAN SPECTRA OF AMORPHOUS SUBSTANCES [25,26]

Experiments which strongly depend on the vibrational density of states and are therefore usually interpreted in connection to each other are the low frequency Raman spectrum of amorphous materials and the specific heat at low temperatures and constant volume. The specific heat of amorphous solids does not follow the T^3 behaviour predicted by the Debye theory for crystals but has much higher values [27]. In the harmonic approximation the constant volume specific heat C_V is given in terms of the density of states $g(\omega)$ by

$$C_v = k \cdot \int_0^\infty g(\omega) \cdot \left[\frac{\hbar\omega}{kT}\right]^2 \cdot \frac{\exp(-\hbar\omega/kT)}{[\exp(-\hbar\omega/kT) - 1]^2} \, d\omega$$

where the low frequency part of $g(\omega)$ is dominant.

Moreover light scattering intensities on the Stokes side of the spectrum are given by [28]

$$I(\omega) = \mathscr{C}(\omega) \cdot \frac{1}{\omega \cdot [1 - \exp(-\hbar\omega/kT)]} \cdot g(\omega)$$

$\mathscr{C}$ is the frequency dependent coupling coefficient, and the second factor accounts for the Bose–Einstein occupation density of the vibrational states.

In the usual analysis of Raman scattering spectra the reduced intensity I_{red} is first derived.

$$I_{red}(\omega) = I(\omega) \cdot \omega \cdot [1 - \exp(-\hbar\omega/kT)]$$

$$= \mathscr{C}(\omega) \cdot g(\omega)$$

Then it is assumed that in the low frequency part of the spectrum $\mathscr{C}$ varies as

$$\mathscr{C}(\omega) \sim \omega^2$$

so that I_{red}/ω^2 is proportional to the density of states.

$$g(\omega) \sim I_{red}/\omega^2$$

At last $g(\omega)$ is used to calculate the specific heat C_V which can be compared to low temperature measurements. This standard procedure of spectrum evaluation relies on several assumptions.
- The frequency dependence of the coupling coefficient is known, e.g. quadratic.
- The observed or the reduced spectrum can be reasonably extrapolated to low frequencies correcting for the Rayleigh wing.

These assumptions have been tested for the special case and within the limitations of the present model.

3.2. SPECTRA OBTAINED FROM THE NORMAL COORDINATE TREATMENT

The depolarized rotational Raman spectrum is determined by the time dependence of the total anisotropic polarizability $\beta(t)$. In the harmonic approximation the latter can be expressed in terms of the N normal coordinates q_i.

$$\beta(t) = \beta_0 + \sum_{i=1}^{N} \frac{\partial \beta}{\partial q_i} \cdot q_i(t)$$

The relevant correlation function is

$$< \mathrm{Tr}\, \beta(0) \cdot \beta(t) > = \mathrm{Tr}\, \beta_0^2 + \sum_{i=1}^{N} \mathrm{Tr} \left[\frac{\partial \beta}{\partial q_i}\right]^2 \cdot < q_i(0)\, q_i(t) >$$

The coupling coefficient at frequency ω_i is therefore given by

$$\mathscr{C}(\omega_i) = \mathrm{Tr} \left[\frac{\partial \beta}{\partial q_i}\right]^2$$

with

$$\frac{\partial \beta}{\partial q_i} = \sum_{j=1}^{N} A_{ij} \frac{\partial \beta}{\partial \xi_j}$$

For the present case the polarizability distribution of the molecules is approximated by anisotropic point polarizabilities at the side group centres.

In Figure 9 the coupling coefficient is shown as a function of normal mode frequency. It appears that even at frequencies below 50 cm^{-1} it can hardly be approximated by a quadratic function. Above 80 cm^{-1}, where side group libration dominates, the coupling coefficient is nearly constant. Comparison of Figures 9 and 6 reveals that the frequency dependence of the coupling coefficient resembles that of side group contribution to normal modes quite closely. Obviously the librational motion of the phenyl group around the bond connecting it to the main chain is the most efficient process with respect to scattering of electromagnetic radiation. The possibly strong influence of the detailed structure on coupling leads to serious problems in the standard evaluation procedure of low frequency Raman spectra.

192

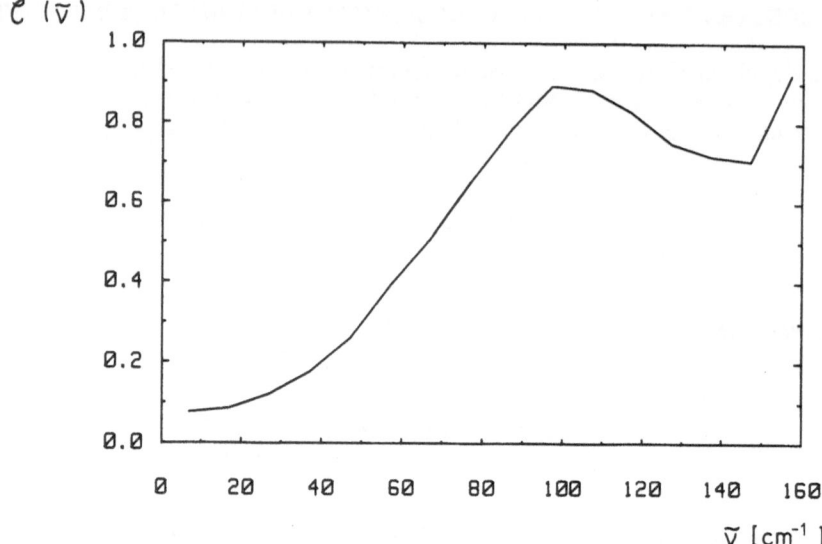

Figure 9. Coupling coefficient $\mathscr{C}$ of rotational Raman scattering as a function of frequency. The scattering at high frequencies is due to insufficient statistics because of low normal mode density (cf. Figure 5).

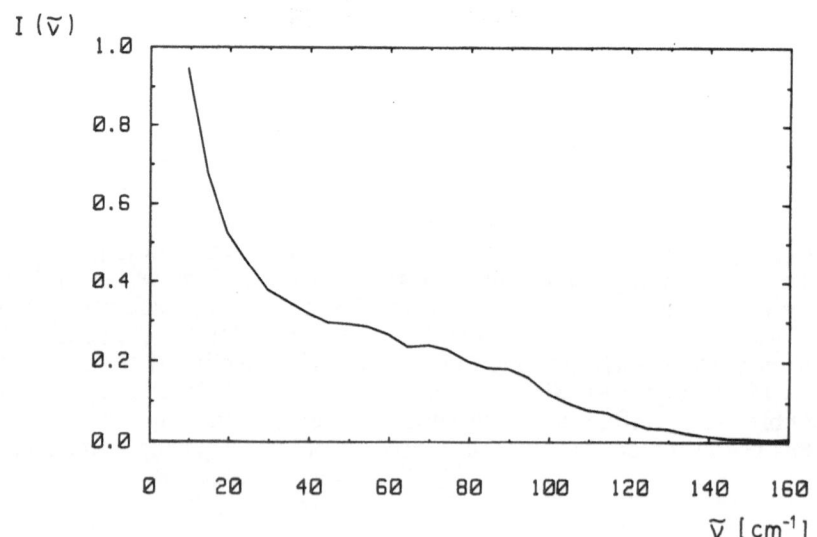

Figure 10. Simulated depolarized low frequency Raman spectrum I as a function of frequency in arbitrary units.

Figure 10 shows the computed depolarized low frequency Raman spectrum at 100 K. In addition to the well known shoulder between 30 and 100 cm⁻¹, which is mainly due to side group libration, there is an extra feature below 20 cm⁻¹. This strong band at low frequencies persists at rather low temperatures but it is suppressed in the reduced intensity spectrum I_{red} (Figure 11). Because of the overlapping strong Rayleigh wing this band could not be confirmed unambiguously as yet.

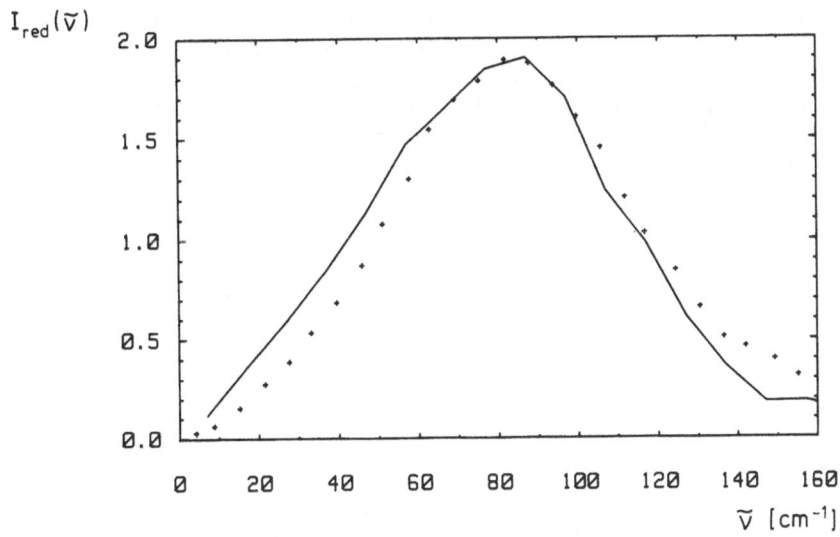

Figure 11. Comparison of experimental (crosses) and simulated (solid line) reduced intensity spectra. The computed spectrum is normalized so that the maximum intensities coincide.

In Figure 11 the computed reduced intensity spectrum is compared with experimental findings [29]. Both curves show the same broad feature peaked at about 85 cm⁻¹ . The experimental data are independent of temperature, which is in accordance with the present normal coordinate treatment. Closer inspection shows that in the simulated reduced intensity spectrum the low frequency region is slightly overstressed, an effect amplified in the intensity spectrum because of the statistical weighting factor. In view of the rough model used regarding degrees of freedom, potential energy, and interaction with radiation, simulation and experimental results agree unexpectedly well.

3.3. SPECTRA RESULTING FROM THE MODEL OF COUPLED SIDE GROUP OSCILLATION

Figure 12 shows the normalized correlation function of the fluctuating part of the total polarizability anisotropy as a function of time.

$$C_\beta(t) = \left\langle \frac{\mathrm{Tr}\ [\beta(0) - <\beta>]\ [\beta(t) - <\beta>]}{\mathrm{Tr}\ [\beta(0) - <\beta>]\ [\beta(0) - <\beta>]} \right\rangle$$

$<\beta>$ is the mean value of the total polarizability anisotropy. In spite of the scarcely damped oscillations observed for single phenyl groups (Figure 8) the correlation function of the total anisotropic polarizability reaches its long time value after 0.5 ps . This may be interpreted in terms of the phase relation of single phenyl group contributions due to their different fundamental frequencies. Because effects of coupling between pairs of oscillators are small on a short time scale, cross terms turn out to be of little importance.

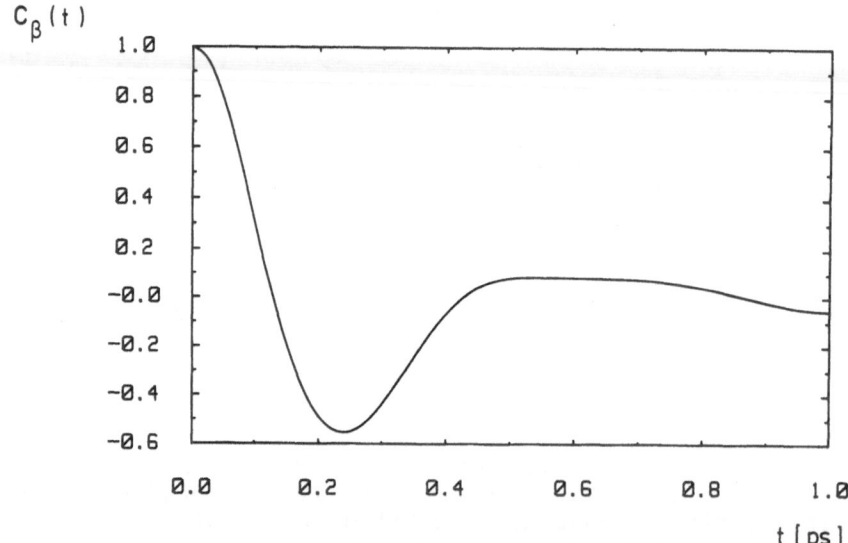

Figure 12. Normalized correlation function of the fluctuation of the total anisotropic polarizability as a function of time obtained from the model of stiff chains and librating side groups.

The power spectrum (Figure 13) obtained by Fourier transforming the correlation function lacks the low frequency contribution observed for the harmonic model (cf. Figure 10) because main chain motion is not taken into account. The reduced intensity function (Figure 14) peaks at the same frequency as the experimental curve but is narrower in shape.

I $(\widetilde{\nu})$

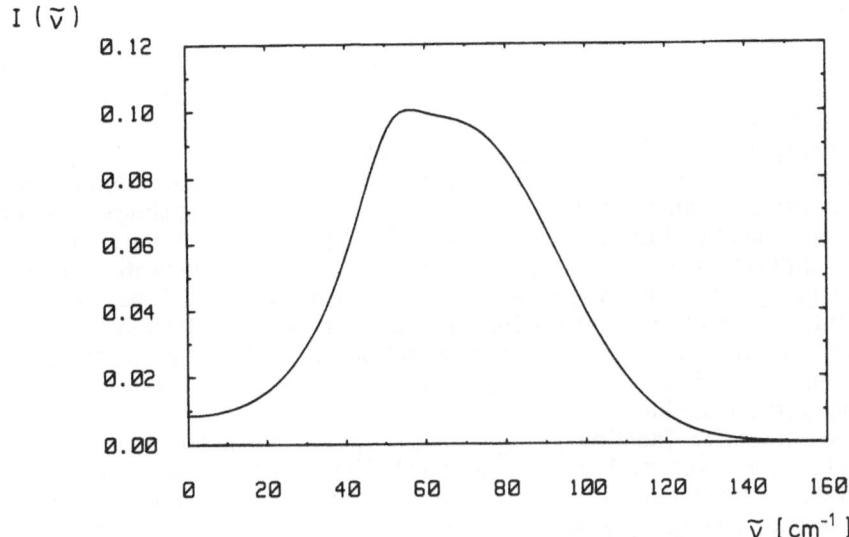

Figure 13. Power spectrum I $(\overset{\sim}{\nu})$ corresponding to the correlation function in Figure 12.

$I_{red}(\widetilde{\nu})$

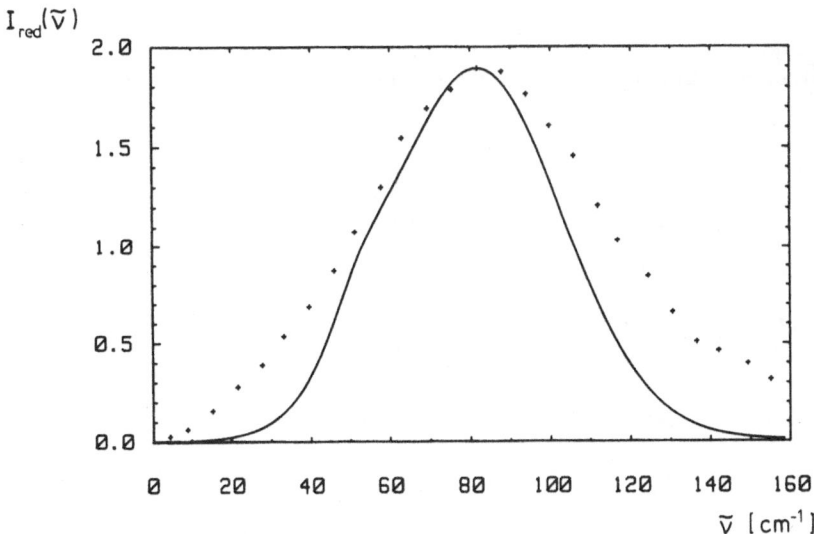

Figure 14. Reduced intensity spectrum I_{red} $(\overset{\sim}{\nu})$ obtained from the intensity spectrum in Figure 13 in comparison with experimental data analogous to Figure 11.

4. Conclusions

In the present article the simulation of dense systems of styrene oligomers aiming at modelling dynamical spectroscopic properties is performed in two steps. Firstly two mechanical models are specified differing in the degrees of freedom which are allowed to vary. Secondly the simplest model for interaction with radiation is applied, namely that of anisotropic point polarizabilities on the phenyl groups. Comparison of the reduced intensity functions of low frequency Raman scattering shows that the generally accepted model of harmonic expansion of the potential function around an equilibrium configuration turns out to be superior to that of anharmonically coupled, librating side groups. In the present case the mechanism of band broadening by anharmonic interaction of different oscillators appears to be less effective than the coupling of the librating rings to main chain vibration. Since different spectroscopic or thermodynamic experiments probe different dynamical properties, the relaxation model may be more successful in other cases.

The reduced intensity function is reproduced by the harmonic model within computational and experimental error (Figure 11). This function with a maximum at about 85 cm^{-1} turns out to factor into the density of states, which decreases linearly in this region (Figure 5), and the monotonically increasing coupling coefficient (Figure 9). Both factors seem to depend too much on sample structure to be amenable to simple approximations. Calculations using this model also predict appreciable contributions to rotational Raman spectra at low frequencies which lead to problems when extrapolating and trying to correct for the strong Rayleigh line. This explains the difficulties in correlating low frequency Raman spectra and low temperature specific heat measurements on a quantitative basis.

Acknowledgements :
The generous financial support of the 'Fonds der chemischen Industrie' is gratefully acknowledged.

References

[1] Baumgärtner, A. (1984) *'Simulation of polymer motion'*, Annual Review of Physical Chemistry 35, 419–435

[2] Baumgärtner, A. (1987) *'Simulation of polymer models'*, in Binder, K. (ed.), Applications of the Monte Carlo method in statistical physics, Springer, Berlin, pp. 145–180

[3] Bishop, M., Ceperly, D., Frisch, H.L., and Kalos, M.H. (1980) *'Investigations of static properties of model bulk polymer fluids'*, Journal of Chemical Physics 72, 3228–3235

[4] Fixman, M. (1986) *'Implicit algorithm for Brownian dynamics of polymers'*, Macromolecules 19, 1195–1204

[5] Ryckaert, J.-P. and Bellemans, A. (1978) *'Molecular dynamics of liquid alkanes'*, Faraday Discussions of the Chemical Society 66, 95–106

[6] Weber, T.A. (1979) *'Relaxation of a n-octane fluid'*, Journal of Chemical Physics 70.9, 4277–4284

[7] Edberg, R., Evans, D.J., and Morriss, G.P. (1986) *'Constrained molecular dynamics : Simulations of liquid alkanes with a new algorithm'*, Journal of Chemical Physics 84.12, 6933–6939

[8] Rigby, D. and Roe, R.-J. (1987) *'Molecular dynamics simulation of polymer liquid and glass. I. Glass transition'*, Journal of Chemical Physics 87.12, 7285–7292

[9] Weber, T.A. and Helfand, E. (1979) *'Molecular dynamics simulation of polymers. I. Structure'*, Journal of Chemical Physics 71.11, 4760–4762

[10] Theodorou, D.N. and Suter, U.W. (1985) *'Detailed molecular structure of a vinyl polymer glass'*, Macromolecules 18, 1467–1478

[11] Theodorou, D.N. and Suter, U.W. (1986) *'Local structure and the mechanism of response to elastic deformation in a glassy polymer'*, Macromolecules 19, 379–387

[12] Biswas, R. and Hamann, D.R. (1986) *'Simulated annealing of silicon atom clusters in Langevin molecular dynamics'*, Physical Review B 34.2, 895–901

[13] Sabochick, M.J. and Richlin, D.L. (1988) *'New global energy–minimization method'*, Physical Review B 37.18, 10846–10850

[14] Yoon, D.Y., Sundararajan, P.R., and Flory, P.J. (1975) *'Conformational characteristics of polystyrene'*, Macromolecules 8.6, 776–783

[15] Allinger, N.L. (1976) *'Calculation of molecular structure and energy by force–field methods'*, Advances in Physical Organic Chemistry 13, 1–82

[16] Buchner, M. (1989), Ph.D. Thesis, University of Bielefeld, West Germany

[17] Weary, D.L. (1981) *'The vibrational density of states of amorphous semiconductors'*, in Phillips, W.A. (ed.) Amorphous Solids, Springer, Berlin, pp. 13–26

[18] Böttger, H. (1983), Principles of the theory of lattice dynamics, Physik Verlag, Weinheim

[19] Snyder, R.G. and Strauss, H.L. (1987) *'Numerical studies of disordered–chain vibrations. I. Low–frequency modes of polymethylene–like skeletal chains'*, Journal of Chemical Physics 87.7, 3779–3788

[20] Goldstein, H. (1980), Classical Mechanics, 2. edition, Addison–Wesley, Reading

[21] Hägele, P.C. and Beck, L (1977) *'Calculation of phenyl group rotation in polystyrene by means of semiempirical potentials'*, Macromolecules 10.1, 213–215

[22]   Tanabe, Y. (1985) *'Phenyl–group rotation in polystyrene'*, Journal of Polymer Science, Polymer Physics Edition 23, 601–606

[23]   Bauer, D.R., Brauman, J.I. and Pecora, R. (1975) *'Depolarized Rayleigh spectroscopy studies of relaxation processes of polystyrene in solution'*, Macromolecules 8.4, 443–451

[24]   Samios, D. and Dorfmüller, T. (1985) *'A high–resolution Raman scattering study of reorientational correlation times in liquids'*, Chemical Physics Letters 117.2, 165–170

[25]   Spells, S.J. and Shepherd, I.W. (1977) *'Low temperature specific heat calculations for amorphous polystyrene'*, Chemical Physics Letters 45.3, 606–610

[26]   Viras, F. and King, T.A. (1984) *'Low frequency excitation in amorphous acrylic polymers'*, Polymer 25, 899–905

[27]   Pohl, R.O. (1981) *'Low temperature specific heat of glasses'*, in Phillips, W.A. (ed.) Amorphous Solids, Springer, Berlin, pp. 27–52

[28]   Shuker, R. and Gammon, R.W. (1970) *'Raman–scattering selection–rule breaking and the density of states in amorphous materials'*, Physical Review Letters 25.4, 222–225

[29]   Spells, S.J. and Shepherd, I.W. (1977) *'Low frequency Raman modes in solid amorphous polystyrene and polymethyl methacrylate'*, Journal of Chemical Physics 66.4, 1427–1433

DYNAMICS OF DENSE POLYMER SYSTEMS
DYNAMIC MONTE CARLO SIMULATION RESULTS AND ANALYTIC THEORY

JEFFREY SKOLNICK
Institute of Macromolecular Chemistry
Department of Chemistry
Washington University
St. Louis, MO 63130

Abstract

Dynamic Monte Carlo simulations of long chains confined to cubic and tetrahedral systems as a function of both volume fraction and chain length were employed to investigate the dynamics of entangled polymer melts. It is shown for a range of chain lengths there is a crossover from a much weaker degree of polymerization (n) dependence of the self-diffusion coefficient to a much stronger one, consistent with $D \sim n^{-2}$. Similarly, systems have been identified having a terminal relaxation time that varies as $n^{3.4}$. Since such scaling with molecular weight signals the onset of highly constrained dynamics, an analysis of the character of chain contour motion was performed. No evidence whatsoever was found for the existence of a well defined tube required by the reptation model. Lateral motions of the chain contour are remarkably large, and the motion appears to be essentially isotropic in the local coordinates. Results from this simulation indicate that the motion of a polymer chain is essentially Rouse-like, albeit, slowed down. Motivated by the simulation results, an analytic theory for the self-diffusion coefficient and the viscoelastic properties have been derived which is in qualitative agreement with both experimental data and the simulations.

I. Introduction

A long standing problem of polymer physics is the elucidation of the microscopic mechanism by which a given chain in an entangled polymer melt moves.[1] Although this problem has been a central one in polymer science for over the past forty years, the mechanism, in fact, remains unclear.[2] The answer to this question has practical applications to, among other areas, polymer flow rheology, polymer adhesion and polymer failure.[1]

Among the phenomenological properties that any microscopic theory must rationalize, in order to be considered a candidate for a correct

199

Th. Dorfmüller (ed.), Reactive and Flexible Molecules in Liquids, 199–220.
© 1989 by Kluwer Academic Publishers.

theory, are the following: Imagine a linear polymer composed of n bonds. The center of mass self-diffusion coefficient of the polymer, D, must scale like

$$D \sim n^{-1} \text{ when } n < n_c'$$
$$D \sim n^{-2} \text{ when } n > n_c' \tag{1}$$

n'_c is a crossover value of the degree of polymerization.[1] More recently, other values of the molecular weight dependence of D have been reported, and the situation is not quite as clear as it appeared a few years ago.[3] The degree of polymerization dependence of the zero frequency shear viscosity η is better characterized and is given by [4]

$$\eta \sim n \text{ if } n < n_c$$
$$\eta \sim n^{3.4} \text{ if } < n_c \tag{2}$$

One of the mysteries which remains unresolved is why n_c the crossover value for viscosity is smaller than that for self-diffusion. Whatever the microscopic mechanism giving rise to this behavior, however, it is clear that as the molecular weight of the chains increases, entanglements of some nature become important.

What are these entanglements? Computer simulations can be a particularly powerful technique for elucidating, at least in a qualitative sense, the microscopic mechanism of various physical processes and motion in polymers is no exception. This paper describes the results of simulations on melts of linear chains and describes a simple theory of melt dynamics that is suggested by those simulations.

One of the most remarkable, and in fact, surprising consequences of the experimental observations embodied in the dependence at low molecular weight of the self-diffusion coefficient and the viscosity is that the polymer chain in a melt behaves like a Rouse chain.[5] This is the simplest model for the properties of a polymer chain at infinite dilution. In a Rouse chain, which is just a bead-spring model of a polymer dissolved in a continuum solvent that exerts a high frequency, frictional force on the beads, the overall motion is locally, as well as globally, isotopic, and one can neglect hydrodynamic interactions between beads. Hydrodynamic interactions arise from the perturbation of the solvent flow about a bead due to the presence of the other beads. Thus, if the molecular weight of the polymer melt is sufficiently small, it behaves not only as if it is not entangled, but it behaves remarkably simpler than if it were at infinite dilution where hydrodynamic interactions must be considered.[6]

In subsequent discussion we will require various scaling properties of the Rouse model[7]. The mean square displacement of the center of mass

$$g_{cm}(t) = 6Dt. \tag{3}$$

In the long chain limit, the mean square displacement of a single bead,

$$g(t) \sim t^{\frac{1}{2}} \qquad t < \tau_{Rouse} \qquad (4)$$

with τ_{Rouse} the terminal or longest internal relaxation time; $\tau_{Rouse} \propto n^2$.

As n increases, the polymer changes its response from a viscous liquid to a viscoelastic liquid. Probably the most widely accepted model to describe the scaling behavior of D vs. n is the reptation model of de Gennes[8], which was later modified and elaborated on by Doi and Edwards.[9-12] Basically, one imagines that the surrounding matrix of linear chains that produces the entanglements remains static on the order of the relaxation time of the end to end vector. Thereby, an extremely complicated many body problem is reduced to an effective single particle picture. The matrix of chains is replaced by a tube, which is defined by the static entanglements. In the reptation picture, the only way the chain can move appreciable distances is by slithering out the ends of the tube; hence, the name reptation.

In the original reptation model, and in all its subsequent variants, the dominant long distance motion is longitudinal down the chain contour defined at zero time.[1] Lateral fluctuations are always of limited extent. Because of the torturous path that the chain must take with respect to the laboratory fixed frame, the longest internal relaxation time of the system τ_{Rep} turns out to be proportional to n^3. This is typically called the tube renewal time.

If one further assumes that there is a rubber-like elastic response of the polymer at short times[10], then $\eta \sim n^3$ as well. Observe that this is a slightly weaker molecular weight dependence than experiment indicates (see eq. 2). Similarly, the self-diffusion constant scales as n^{-2}. Thus, at face value it would seem that this simple model of polymer melt motion quite closely reproduces the experimentally observed behavior of the self-diffusion coefficient and the shear viscosity. In subsequent developments, however, it turns out that $D \sim n^{-2}$ is a ubiquitous property that is quite insensitive to the microscopic details of the motion, a point we will return to later.

Let us contrast the behavior of the single bead autocorrelation function g(t) for a reptating chain with that of a Rouse chain. For distances less than the tube diameter, g(t) scales like $t^{\frac{1}{2}}$. In this time regime the chain has absolutely no information about whether there is a tube or not, and so it will still behave like a pure Rouse chain. For somewhat longer times, the chain behaves like a Rouse chain confined to a tube which itself is a Gaussian random walk, and therefore, the $t^{\frac{1}{2}}$ behavior is further diminished so that g(t) is proportional to $t^{\frac{1}{4}}$ times for up to τ_{Rouse}. Next, one is left with the center of mass motion down the tube. Thus g(t) is proportional to $t^{\frac{1}{2}}$ for times less than the reptation time. Finally, in the free diffusion limit g(t) is proportional to t. Similar considerations indi-

cate that there is a range of times for which the mean square displacement of the center of mass scales not like t, but $t^{\frac{1}{2}}$ for times
less than the reptation time, but longer than τ_{Rouse}[7]. For $t > \tau_{rep}$,
$g_{cm}(t) \sim t$.

All this is extremely plausible for a regular gel for which, in
fact, reptation was originally derived.[8] It is not at all clear that
this picture holds for polymer melts where everything is moving on the
same time scale. The reality of a spatially fixed tube has been
previously questioned by Phillies[13] and Fujita and Einaga[14] for melts,
and by Fixman for concentrated solutions of rod-like polymers.[15]

The existence of reptation as the dominant mechanism in a polymer
melt forms the focus of the present paper, the outline of the remainder of which is as follows: We shall begin with the description
of the microscopic model of the polymer dynamics and then turn to the
dynamics of homopolymeric linear chains. Subsequently, we summarize
the results of an recent analytic theory which is in qualitative
agreement with both experiment and simulation.[16-17]

II. DESCRIPTION OF THE MODEL

Both diamond and cubic lattice models of a polymer melt have been
employed.[18-19] A lattice representation is used for a number of
reasons. First of all, it allows one to do the calculations in
integer arithmetic, thereby affording a factor of 10 to 100 speedup
over floating point calculations. Second of all, it allows one to
rigorously insure that no bond cutting occurs and thereby that the
excluded volume effect exerted by one chain upon another is rigorously
implemented. As in all simulations of this type, the lattice is
enclosed in a periodic box of volume L^3.[7] To avoid the problem of a
given polymer chain interacting with its image, we always chose L >
$<R^2>^{\frac{1}{2}}$ (the equilibrium root mean square end-to-end distance) to insure
that interactions of the chain with itself cannot occur. Each polymer
chain is assumed to occupy n consecutive lattice sites, and ϕ is the
volume fraction of occupied sites. In all cases, excluded volume is
implemented by prohibiting the multiple occupancy of any given lattice
site. The dynamic properties of homopolymeric diamond lattice
polymers were studied over a range of volume fractions ϕ from zero to
0.75 for chains up to n = 216.[18] The corresponding cubic lattice
polymers were studied at fixed ϕ = 0.5, but for n ranging from 64 to
800 for homopolymeric melts of linear chains.[19]

The first problem one faces in attempting to undertake such a
simulation is the construction of an equilibrated dense melt. Procedures for constructing such systems have been discussed in detail
elsewhere and need not concern us further.[20] The next problem encountered in studying the dynamics of the melt by Monte Carlo techniques
is that one must choose a set of local moves. One wants to make the
moves as local as possible, to avoid either distorting the time scale

of motion or somehow building in an artificial and unphysical dynamics.[7,18-19]

The crucial properties of the chain dynamics are as follows: One has to choose a set of moves on any given lattice that not only can diffuse local orientations down the chain but also has the possibility of locally introducing new random conformations into the chain. Otherwise, the dynamics of even an isolated chain would be non-physical. That is, if new chain orientations can only arise by diffusion from the free ends, an now artificial n^3 time scale is built into the algorithm which has absolutely nothing whatsoever to do with entangled dynamics.[21-22] For the case of both diamond[23] and cubic lattices,[19,21-22,24] the set of elementary jumps depicted in Figures 1 and 2 respectively satisfy the above requirements and are useful in that a reasonable fraction of the jumps are successful.

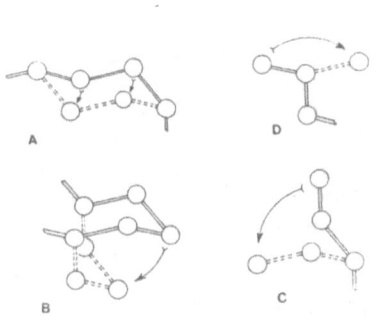

Figure 1. Elementary conformational jumps for tetrahedral polymers: (a) three-bond motion $g^{\pm} \to g^{\mp}$, (b) four-bond motion (with a random choice of the new orientation of the bonds), (c) two-bond motion of end units (with a random choice of the new orientation of the bonds), (d) one-bond motion of end units (with a random choice of the new orientation of the bonds).

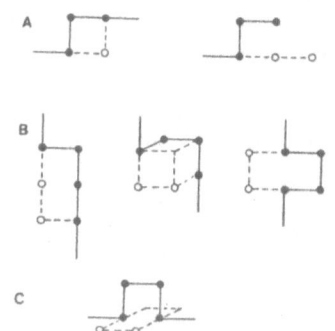

Figure 2. Elementary conformational jumps for cubic lattice polmers. (A) The normal bead motion and an example of chain end motion. (B) Examples of three-bond permutations. (C) The 90° -crankshaft motion of a U-shaped fragment of the chain.

The simulation proceeds as follows: A chain is picked at random and then a given bead is chosen. One then attempts to make each of the various kinds of moves per bead on average. The fundamental time unit is taken to be that required such that on average each of the lattice moves is attempted once per bead. Of course, in actual implementation, we randomly mixed the moves, and any move is rejected if such a move is not allowed. For example, in the case of a diamond lattice, if a three bond flip from one half of a cyclohexane ring to the other is attempted and the conformation of the selected three bonds is trans, as opposed to a gauche plus (g+) or gauche minus (g-) conformation, the move cannot occur. Similarly, the move will be rejected if the site(s) to which the jump is attempted is (are) already occupied. One of the nice things about doing simulations at high density is that while one may specify an *a priori* choice of any given type of move, the system itself due to excluded volume restrictions chooses the fraction of successful moves for a wide range of *a priori* probabilities.[18-20]

At this point, it is appropriate to review the relative advantages and disadvantages of using a lattice representation of a polymer melt. One uses a lattice as mentioned above because it allows one to simulate much longer polymers at much higher densities for longer times than corresponding off lattice systems. The disadvantage, of course, is that one has to demonstrate that the results obtained from such a simulation are physically meaningful and not an artifact of the lattice. While we cannot prove this, in fact, when comparisons can be made with off-lattice simulations,[25] all the qualitative conclusions are identical. Moreover, we have obtained identical results for both cubic and diamond lattices when corrections for differences in local persistence length and lattice coordination number are made.

III. SIMULATIONS ON HOMOPOLYMERIC LINEAR CHAINS

CENTER MASS MOTION AND THE LONGEST INTERNAL RELAXATION TIMES

The first problem that one faces when doing computer simulations is demonstrating that the scaling behavior of the self diffusion constant and terminal relaxation time are consistent with experiment. Thus we start by an examination of the center of mass motion, and in the Figure 3, plot, on a log-log scale, the mean square displacement of the center of mass, $g_{cm}(t)$ vs. t for homopolymeric cubic lattice systems at $\phi = 0.5$. Clearly, two regimes are evident. For distances such that $g_{cm}(t) \leq 2<S^2>$ ($<S^2>$ is the mean square radius of gyration), $g_{cm}(t) \sim t^a$ with the values of a monotonically decreasing from about 0.91 when n = 64 to 0.71 when n = 800. Thus, we conclude that these systems behave neither like a Rouse chain having a uniform friction constant where $g_{cm}(t)$ is always proportional to time, nor like reptating chains which have a $t^{\frac{1}{2}}$ regime. Qualitatively identical behavior is seen in diamond lattice simulations,[18] as well as in off-lattice simulations.[25] The existence of a t^a regime where a<1 indicates coupling between the center of mass of motion and

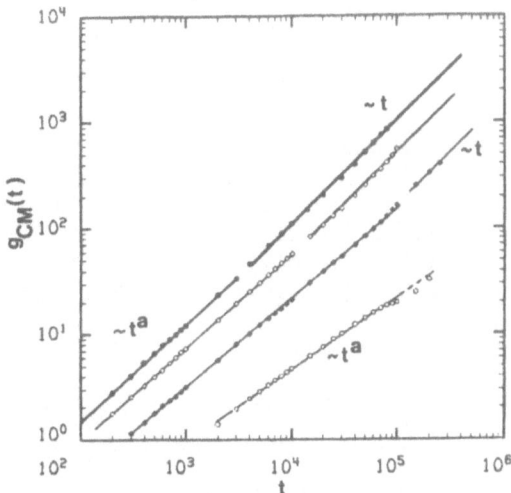

Figure 3. Log-log plots of the center-of-mass autocorrelation function $g_{cm}(t)$ *vs.* t for n = 64, 100, 216 and 800 melts reading from left to right (or top to bottom). ϕ = 0.5 in all cases. The shorter time t^a regime is to be distinguished from the long time diffusion regime.

the internal relaxation modes. This is also consistent with the fact that $2<S>^2$ is the maximum distance over which the internal modes of the chain relax to their equilibrium values if, in fact, one can treat the chains as statistically independent. The self-diffusion constant was obtained by fitting $g_{cm}(t)$ equal to the following functional form: $g_{cm}(t)$ = 6Dt + c, with c a small positive constant that arises from the fast motion of the center of mass at shorter times. If the data is fit over the regime from n = 64 to n = 216, then D ~ $n^{-1.52}$. Unfortunately, during the time these simulations were run we lacked the computer resources to run the n = 800 system out into the free diffusion limit. Thus, a number of extrapolation procedures were employed to obtain D for this system. These have been discussed in detail elsewhere,[19] but basically we feel that by n = 800, D is well into the n^{-2} regime.

We next examined the scaling of the longest relaxation time of the end-to-end vector obtained from the long-time decay of $<R(t) \cdot R(0)>$ where R(t) is the end-to-end vector at a time t. Following a period of very rapid initial relaxation we find that the autocorrelation function vector is rather well fit by a single exponential.

The scaling of D ~ $(n-1)^{-\alpha}$ and τ_R ~ $(n-1)^{\beta}$ on diamond and cubic lattice systems are summarized in Table I.

TABLE 1.   Chain Length Dependence of the Self-Diffusion Coefficient
$D \sim (n-1)^{-\alpha}$ and Terminal Relaxation Time $\tau_R \sim (n-1)^{\beta}$

ϕ	α	β
	Diamond Lattice	
Single Chain	1.154 (±0.010)[a]	2.349 (±0.018)[a]
0.25	1.372 (±0.021)[a]	2.563 (±0.061)[a]
0.50	1.567 (±0.017)[a]	2.677 (±0.035)[a]
0.75	2.055 (±0.016)[a]	3.364 (±0.082)[a]
	Cubic Lattice	
0.5[b]	1.52 (±0.06)[a]	2.63 (±0.04)[a]

[a] Standard deviation of the slope obtained from a linear
     least square fit of the log-log plots of D vs. n and
     τ_R vs. n, respectively.

[b] Fit over the n - 64 to 216 range, i.e., in the crossover
     region.

We point out that the diamond lattice system at ϕ - 0.75 is in
accord with the experimentally observed dependence of η on n if $\eta \sim \tau_R$. That is, η is proportional to the 3.4 power of n. We further
point out that a given $\phi \leq 0.5$ on the diamond lattice the chains are in
the crossover region. One would expect that on increasing n similar
values of the exponents α and β should be seen as in the ϕ =0.75 case.
One of the more interesting results that were obtained from this
series of simulations is that at all concentrations the product $D\tau_R$
scales like $n^{1.2}$ on a diamond lattice and $n^{1.1}$ on a cubic lattice
rather than the expected n^1. Based on elementary scaling considerations $D\tau_R$ should be on the order of the radius of gyration of the
chain, which in linear polymer melts is proportional to n. A possible
origin of this discrepancy might be the coincidence of statistical
uncertainties in the estimation of both α and β, which is approximately ± 0.05 for both exponents at high densities. A second explanation is that one is observing a crossover to $D\tau_R \sim$ n in the infinite
chain limit. If this were to be true, this would imply that the
experimentally determined scaling of the shear viscosity, $\eta \sim n^{3.4}$,
would be indicative of a crossover regime, and consist with the simulations it would be rather broad. Some recent experiments by Colby,
Fetters and Graessley are not inconsistent with this particular conclusion, but by no means demand it.[26] A final alternative explanation
is, in fact, that $D\tau_R$ always scales like $n^{1+\epsilon}$ with ϵ greater than zero.
That is to say, configurational relaxation is rapid relative to the

disentanglement between the chains. This idea has been put forth in a recent theory of Fixman.[27]

Having established that our simulations are on chains sufficiently large that the experimentally observed scaling with molecular weight of D vs. n obtains we next turn to a finer characterization of the dynamics, namely, single bead motion. In Figure 4A the log-log plots of the average mean square of the displacement per bead, g(t) vs. t for chains confined to a cubic lattice is displayed. Clearly, two different regimes of behavior are evident. The first regime, which extends once again up to $2<S^2>$, $g(t) \sim t^b$ with b decreasing gradually from the Rouse exponent of about 0.54 when n is equal to 64 to 0.48 when n is equal to 216. Thus, within the statistical error of the simulation one might conclude that these chains exhibit a Rouse-like dependence of g(t) on t. However, the n = 800 curve exhibits distinctly different behavior from all of the previous cases. There is a region where $g(t) \sim t^{0.36}$, indicative of more constrained dynamics in ‑hese chains.

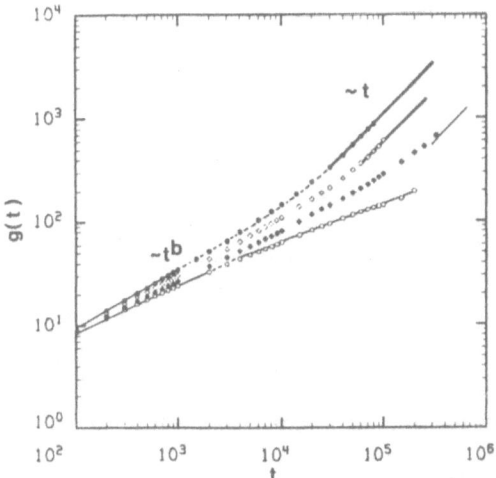

Figure 4A. Log-log plots of the single bead autocorrelation function g(t) vs. time t for n = 64, 100, 216 and 800 melts reading from top to bottom. The g(t) are averaged over all the beads in the system. $\phi = 0.5$ in all cases.

One's first response is to suspect the $t^{0.36}$ regime is similar to the $t^{\frac{1}{4}}$ regime predicted by reptation theory, and that these chains are reptating. However, getting slightly ahead of the story, microscopic examination of the motion demonstrates that the character of the chain motion is distinctly different. One might expect that the central beads of the chain, in fact, would cross over to the $t^{\frac{1}{4}}$ first. Thus, in Figures 4B and 4C, we plot present log-log plots of the average mean square displacement of the central five beads of the chain, $g_5(t)$, vs. t for n = 216 and n = 800. A $t^{\frac{1}{4}}$ regime is clearly evident (actually, $g_5(t) \sim t^{0.28}$). Thus, one is left with the question: Are these chains, in fact, reptating or is the character of their motion, in fact, different?

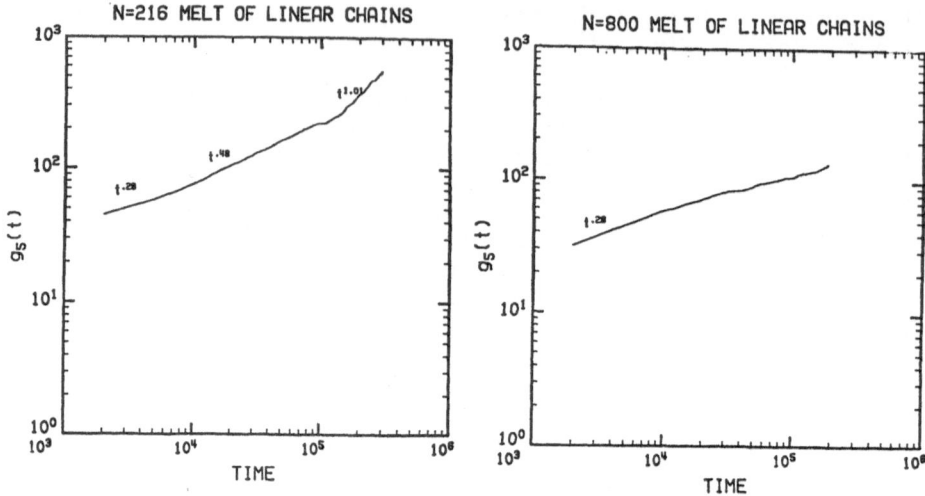

Figures 4B and 4C. Log-log plot of the single bead autocorrelation functions of the central five beads, $g_5(t)$ vs. t for n = 216 and n = 800 in Figures 4B and 4C respectively. In all cases $\phi = 0.5$.

EXAMINATION OF THE PRIMITIVE PATH DYNAMICS

In the classic treatment of polymer melt dynamics, assuming that reptation is the dominant mode of chain motion, Doi and Edwards used the idea of a primitive path.[9-12] This basically involves the replacement of the actual chain by an equivalent one in which all of the local fluctuations in chain contour irrelevant to the long distance motion are averaged out. Basically, one could imagine taking the chain and reeling in the slack and then looking at the resultant path. In what follows, we construct an equivalent chain which is quite close to the primitive path and follow its motion as a function time.

The basic outline of the procedure is as follows: Each bead in the original chain is replaced by a point on the equivalent path which is the center of mass of a subchain composed of n_b beads. Thus, one replaces the actual contour of the chain by a smooth path composed of these partially overlapping subchains which should be very close to the primitive path of Doi and Edwards, if n_b is close to the number of monomers between entanglements. At every time, we generate the equivalent path and look at the displacements down the original path defined at zero time. This corresponds to the reptation component.

What is left over is the non-reptation component which should be small if reptation is dominant. To determine whether, in fact, this is true we compute the average mean square displacement, $g_\parallel(t)$ down the original primitive path and perpendicular to the primitive path, $g_\perp(t)$. It is trivial to demonstrate that a reptating chain has a maximum value of $g_\perp(t)$ equal to one half the mean square tube radius for times less than the tube renewal time. Thus, the ratio $g_\perp(t)/g_\parallel(t)$ should monotonically decrease in time. On the other hand, if the motion is, in fact, globally isotropic and liquid-like with no memory whatsoever of a tube defined at zero time, $g_\perp(t)/g_\parallel(t)$ should monotonically increase with time. It is interesting to point out that reptation theory assumes a kind of glass transition has already occurred in the melt. That is to say, the motion of a chain perpendicular to the original path is essentially frozen out due to the existence of entanglements.

The question immediately arises whether, in fact, somehow we have not artificially suppressed reptation in our choice of moves, and if our particular criteria of the ratio of $g_\perp(t)/g_\parallel(t)$ vs. t would provide the signature of reptation when it exists. It has long been established that if a chain is in a fixed mesh and if it is sufficiently long, the chain should reptate.[25,28-29] Basically, the origin of the force that tends to keep the chain from having large lateral displacements is as follows: If a chain attempts to move between a set of fixed obstacles then the allowed number of configurations of the chain are reduced. In the asymptotic limit one could imagine that this forms a constrained ring whose configurational entropy is greatly reduced relative to the case when one takes the equivalent chain length and pokes it out the end of the tube. Thus, one would expect in the asymptotic limit that reptation should dominate.

To demonstrate that this is true for our simulation we have simulated a chain in a partially frozen environment.[18] Basically, what one does is take the original n = 216 diamond lattice polymer melt and freeze all the chains but a test chain. However, if one does this, since the tube is not very porous, one finds grid lock. Thus a partially frozen environment is used. We took every 18 beads in the matrix chains and pinned them. This provides for a set of local dynamics which is extremely close to that of the original polymer melt, but here all chains but the test chain of interest are constrained from moving appreciable distances. Looking at pictures of the primitive path, these chains reptate, and $g_\perp(t)/g_\parallel(t)$ vs. t monotonically decreases with increasing time. Thus, the signature of reptation is recovered, and one finds the presence of reptation when it is the dominant mechanism of long distance motion. Furthermore, since the chains do reptate when placed in a partially frozen environment, our choice of local Monte Carlo moves does not somehow artificially suppress reptation.

In Figure 5, we plot for the melt where everything moves $g_\perp(t)/g_\parallel(t)$ vs. t for n = 216 and 800 chains on a cubic lattice at a $\phi = 0.5$ for times below the tube renewal time of reptation theory. We have set $n_b = 17$ as well as 101 and find no qualitative difference.

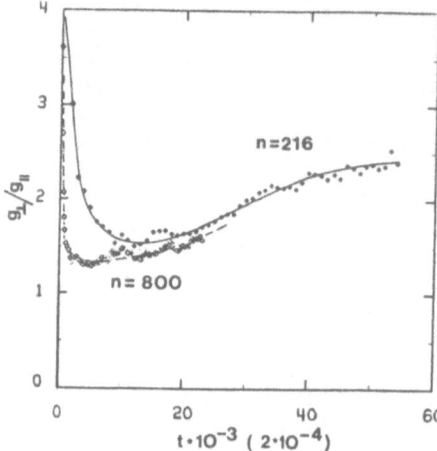

Figure 5. The plot of the ratio $g_\perp(t)/g_\parallel(t)$ vs. time for the $n = 216$ (upper curve) and $n = 800$ (lower curve) melt. See the text for more details.

The qualitative features displayed here are identical to those seen for shorter chains on a diamond lattice at higher density. At short times, transverse motion of a chain is preferred. This is an effect due to the nature of cooperative motions in high density systems whose origin is the following:[20] Imagine a chain has undergone a conformational rearrangement. Now the probability of the chain undergoing correlated motion is a product of two quantities. (1) The intrinsic probability that the chain is in a conformation such that it can undergo a jump. (2) The probability there are unoccupied sites the chain can jump into. For both down and cross chain motion, the intrinsic probabilities of undergoing a jump are identical. For cross chain motion, given that the chain has undergone a jump, there is now an unoccupied volume into which the neighboring chain can jump into and, therefore, the conditional probability that it can undergo a jump is now one. However, for down chain motion, this probability is roughly proportional (on a lattice) to $(1-\phi)$ raised to the power of the number of sites involved in the motional unit. Therefore, one would expect that with an increase in density, cross chain motion should dominate at short times, as is indeed observed.

Subsequent to the short time preference for transverse motion, there is a period when down chain motion becomes somewhat more important. This corresponds to, in fact, distances on the order of the excluded volume decay length. Basically what is happening is that the chain is starting to feel the effect of the environmental, topological constraints on this distance scale and has slowed down. There is a

certain incubation period before the collective motion giving rise to
the larger scale lateral motion takes over. Finally, at longer times
the reptation component becomes increasing less important and the
lateral component grows. In fact, it becomes increasingly difficult
to follow the original primitive path and project onto it. We point
out, however, that for the displacements shown, the maximum time is on
the order of a tenth of the terminal relaxation time, and the tube (if
it indeed exists) should have been well defined. An interesting
point observed on comparison of Figure 5 with Figure 4B is that the
minimum in $g_\perp(t)/g_\parallel(t)$ occurs when the chains are crossing out of the
$t^{1/4}$ regime in the $g_5(t)$ vs. t plot. Thus, if one were to merely look
at $g_\perp(t)/g_\parallel(t)$ for times just up to the end of the $t^{1/4}$ regime, but for
times and distances still short relative to the radius of gyration,
one would incorrectly conclude that reptation dominates. One still
has to go out further into the second $t^{1/2}$ regime where reptation, in
fact, becomes at most a minor component of the motion with respect to
the original primitive path defined at zero time. Thus, we conclude
that there is no tube confining chain of interest. We will return to
this point in subsequent discussion.

A more pictorial illustration of the character of the chain motion
is presented in Figures 6A-6C, where the trajectory of one of the n =
800 chains presented.

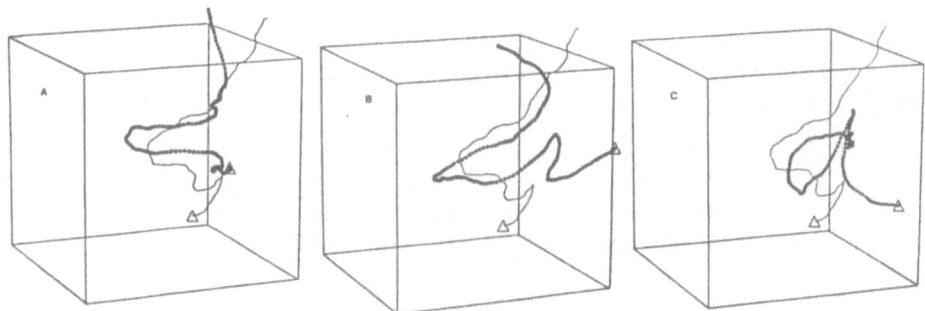

Figures 6A-C. Snapshot projections of the primitive path of a n =
800 chain with in the melt, at ϕ = 0.5. The thinner line corresponds to the conformation at the initial time. The thicker line
at a time t (indicated below) later. Triangles indicate one of
the chain ends. (A) The displacement after 6×10^4 steps; (B)
after 1.2×10^5 steps; C) after 2×10^5 steps. The equivalent chain
has been constructed as described in the text with n_b = 101.
Every bead in the plot corresponds to the center-of-mass of such
a blob, and for clarity only every fourth bead is plotted. The
density of the beads reflects the density of the chain beads.

The behavior of all of the other chains are qualitatively the same.
The thin curve is the initial configuration of the primitive path
defined at zero time and the thick solid curve, the path at times

$t = 6 \times 10^4$, 1.2×10^5 and 2×10^5 for Figures 6A-6C, respectively. For ease of visualization, $n_b = 101$ a very conservative value. It is entirely evident and consistent with the ratio of $g_\perp(t)/g_\parallel(t)$ that significant transverse fluctuations of the chain are observed, and the motion down the original path is basically insignificant. One is forced to conclude, based both on the examination of the primitive path and from the $g_\perp(t)/g_\parallel(t)$ analysis that the chains simply do not know they are confined to a fixed network or tube.

THE ORIGIN OF DYNAMIC ENTANGLEMENTS

Whatever their physical origin, in order to have an important effect on the long distance motion, entanglement constraints must live for times on the order of a terminal relaxation time. Otherwise, they can be subsumed into an effective molecular weight independent monomeric friction coefficient. Based on the simulations which behave much like slowed down, Rouse chains, one might conjecture that the slow down is due to dynamic entanglement contacts - that is, where one chain drags another chain for times on the order of the terminal relaxation time. Eventually, of course, we would expect these two chains to diffuse apart. However, this disengagement process would occur on a very long time scale. We next examine what the simulations have to say about this conjecture.

BEAD DISTRIBUTION PROFILES

Some further insight into the nature of dynamic entanglements emerges from Fig. 7, where we plot the time dependence of the average mean square displacement of the bead, $g_i(t)$, as a function of the position i along the chain. In the curves denoted by a through d, the time equals 3×10^4, 6.9×10^5, 1.35×10^6 and 2.1×10^6 time steps for the $\phi = 0.5$, n = 216 homopolymeric diamond lattice melt. The smooth curves through the data correspond to the values produced from the Rouse model obtained employing an apparent diffusion constant defined as $g_{cm}(t)/6t$.

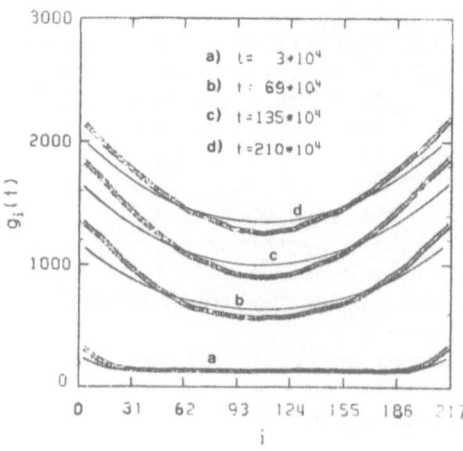

Figure 7. Plot for n = 216 chains in a melt having $\phi = 0.5$ of $g_i(t)$ vs. i, at times indicated in the figure, in the open circles, and calcualted assuming that the Rouse eigenvectors form a good basis set, in the solid lines. See the text for additional details.

The Rouse model over-estimates the mobility of the chain interior and somewhat underestimates the mobility of the ends. Overall, though, the bead distribution profiles are rather good. In fact, in general, it should be pointed out that even for a long isolated Rouse chain in the absence of any constraints the bead distribution profile is parabolic, with the ends moving more than the middle. This is a general result for linear chains that is independent of the particular model of dynamics.

NATURE OF THE DYNAMIC CONTACTS BETWEEN CHAINS

In order to examine the time evolution of contacts between chains the following procedure was employed. (1) We replaced each chain by a series of non-over-lapping blobs, each having $n_b = 18$ monomers; thus the resulting chain is analogous to a pearl necklace. (2) Next we searched for pairs of blobs belonging to different chains whose centers of mass are at a distance less than a distance $r_{min} = 5$ from each other at zero time (the length of a bond equals unity). (3) We count the number of such contacts. Let $n_c(t)$ be the fraction of such dynamic contacts that survive up to a time t later, given that the chains were at contact at time zero. In Figure 8, we plot $n_c(t)$ vs. t for the $n = 216$, $\phi = 0.5$ homopolymeric cubic lattice chains. $n_c(t)$ is found to be decomposable into a sum of three exponentials. While we realized that our three exponential decomposition is by no means unique, the time constants which are obtained, nevertheless, are highly suggestive. The majority of the

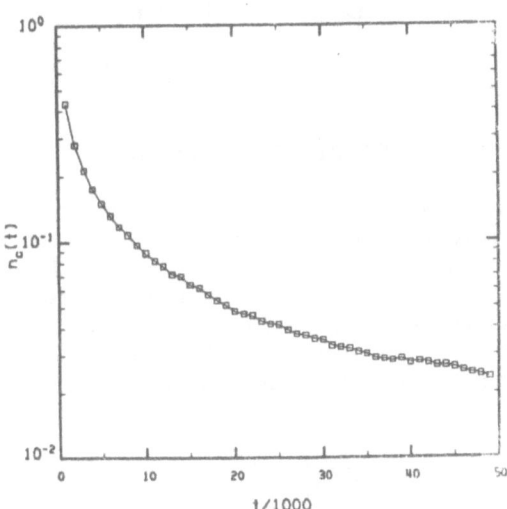

Figure 8. Log-log plot of the number of dynamic contacts that survive up to a time t, $n_c(t)$ vs. t for a $n = 216$ homopolymeric cubic lattice melt with $\phi = 0.5$, $n_B = 18$, and $r_{min} = 5$. See the text for further details.

contacts (64%) decay within 1% of τ_R, 91% of the contacts decay within 9% of τ_R, and the remaining 9% decay on the order of τ_R. Therefore, we conclude that dynamic entanglement events are very rare and approximately 10% of these contacts in an n = 216 chain are long lived. This translates into one dynamic entanglement for every 133 beads. This compares remarkably well with the estimate of the number of monomers between entanglements obtained from the self diffusion constant (see below, eq. 5) which equals 125. The mean lifetime of these contacts is consistent with the idea that contacts between polymers slow down the motion. Most local contacts, however, are rather short-lived, and apart from modifying the local friction constants have no effect on the long time dynamics. Thus, the first conclusion that emerges from these simulations is that the long-lived dynamic entanglement contacts have a distance scale which is order of a magnitude larger than the static screening length. In real polymer melts the static excluded volume screening length is on the order of a monomer unit or so, whereas, based on estimates of the plateau modulus, the mean number of monomers between dynamic entanglements is on the order of 100.[1] Moreover, we have also established that all entanglements are in fact moving with respect to the laboratory fixed frame. Thus, there is no fixed cage.

What then are the dynamic entanglements? Suppose that at zero time, a pair of chains are in a configuration where one chain forms a loop around the other chain - this is a necessary but not sufficient condition for an entanglement. They then subsequently have to move together in a direction that causes the contact to be long-lived. That is to say, one chain drags another chain. Currently we are in the process of examining the nature of the dynamic entanglement process in far more detail.

IV. RECENT ANALYTIC WORK

THE SELF-DIFFUSION CONSTANT

It is possible using the Mori projection operator treatment to calculate the effective friction constant of a polymer due to the presence of all of the other polymers. This was done in a recent elegant article by Hess.[30] If one then assumes that one need only consider two body terms, (consistent with the fact that the entanglements are rather dilute), and thus the propagator between collisions has a free Rouse component, then it is quite easy to show that[16]

$$D = \frac{d_o}{n + n^2/n_e} \tag{5}$$

with d_o the monomeric diffusion constant and n_e is the mean number of monomers between dynamic entanglements. Thus, based on these relatively benign assumptions we conclude that the n^{-2} dependence of D is very general and will, in fact, be independent of the microscopic details of the mechanism of motion. In other words $D \sim n^{-2}$ is not a unique signature of reptation.

What then must any successful theory say about the internal dynamics of polymer melts? It must rationalize the experimental molecular weight dependence of the self-diffusion coefficient and the shear viscosity (see eq. 1 and 2). It must also be consistent with the simulations which indicate that the motion appears to Rouse like, but slowed down and that there is no tube. $g(t)$ has a t^b regime with $b<\frac{1}{2}$; $g_{cm}(t)$ has a t^a regime with a less than one when $g_{cm}(t) < 2<S^2>$. The product $D\tau_R/n \sim n^\epsilon$ with ϵ between 0.1 and 0.2. Furthermore, it must rationalize the single bead, mean square displacement profiles which says that the ends are more mobile than the corresponding equivalent Rouse chain and the middle is less mobile.

We summarize below the features of a recent phenomenological theory that accounts for the above facts.[17] The following assumptions need to be made. (1) At short times a la Doi and Edwards,[10] we treat the response of a melt as identical to that of a rubber. We then focus on the motion of an average reporter chain. We are further going to assume that the long time relaxation behavior of a given chain in a polymer melt is adequately described by a Rouse model. However, due to the presence of dynamic entanglements there are some slow moving points.

Qualitatively, what might one expect from the crossover behavior of such a physical picture? Let us consider the behavior of a chain having a single dynamic entanglement. Physically one would expect that the longest lived dynamic entanglement contact to be located in the center of the chain. Clearly, the terminal relaxation time in this system does not change by much; basically, the slow moving point behaves like a local defect. However, in the absence of the entanglement, the self-diffusion is constant is do/n and in the presence of the entanglement it is do/2n. Therefore, one would expect the crossover regions of D and τ_R to be different. Moreover, the center of mass and the center of resistance are not identical. Basically, the center of mass motion couples into the internal coordinates and this gives rise to $g_{cm}(t) \sim t^a$ regime with a<1 for distances less than $2<S^2>^{\frac{1}{2}}$. Similarly, $g(t) \sim t^b$ with $b<\frac{1}{2}$. One can show is that as n goes to infinity, the dynamic properties behave as if the monomer friction constant can be replaced by the average friction constant per bead equal to $\zeta_0 (1+n/ne)$, with ζ_0 the friction constant in the absense of chain connectivity. The shear viscosity equals 4/15 of the Doi and Edwards value[31] and ultimately scales as n^3. Thus, we would predict that $\eta \sim n^{3+\delta}$ where δ goes to zero as n goes to infinity. And finally, the product of the plateau modulus times the shear compliance equals 10/7. Doi and Edward theory gives a value of 6/5 and experiments are in the range from $2\frac{1}{2}$ to 3.[31]

As shown in Figure 9 where the circles denote the calculated values and the dashed line is the fit through the points giving $\eta \sim n^{3.44}$ the crossover value of $\eta \sim n^3$ should be around 40 to 50 entanglements. This is not inconsistent with the recent work of Colby, Fetters and Graessley[26] although by no means demanded by it. We find in the present theory that the value of n'_c, in eq. 1 equals about $4.5 \, n_c$. Thus, the resulting regime where η depends on the 3.4 power of the molecular weight occurs at smaller chain lengths than for diffusion. Finally, $D\tau_R/\langle S^2 \rangle$ has an n^ϵ regime with $\epsilon = 0.1$ or 0.2 regime, depending on the particular distribution of friction constants.

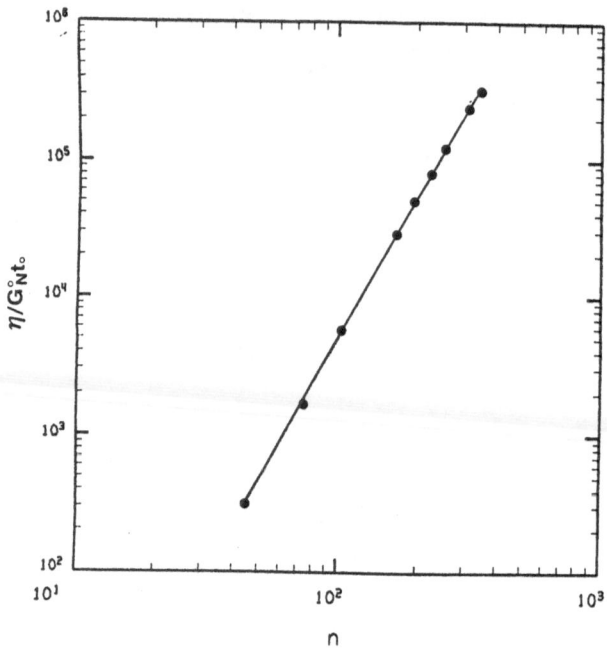

Figure 9. Plot of the shear viscosity in reduced units $vs.$ as n. The circles are the calculated values and the solid line is the least square fit through the data giving $\eta \sim n^{3.44}$. G_N^0 is the plateau modulus and t_0 is the time it takes a monomer to diffuse a distance equal to a bond length.

Finally, in Figure 10, log-log plots of $g_{cm}(t)$ and $g(t)$ $vs.$ t are presented for the case of a chain having n = 255 with a mean distance between entanglements of 15; $\tau_R = 1.88 \times 10^5$. We point out that τ_R of a Rouse chain of corresponding molecular weight will be 1.3×10^4. In the top solid curve, there are $t^{\frac{1}{2}}$, $t^{\frac{1}{4}}$ and $t^{\frac{1}{2}}$ regimes in $g(t)$. Clearly, however, these chains do not reptate. Thus, we have demonstrated that the existence of a $t^{\frac{1}{4}}$ regime in $g(t)$ is indicative of some kind of slow down in the dynamics of this system, but no means demands the existence of reptation as the dominant mechanism of polymer motion.

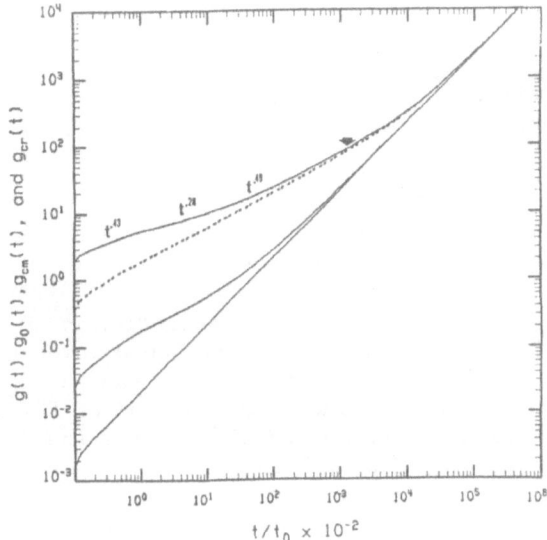

Figure 10. Going from top to bottom in reduced units, log-log plots of the mean square displacement per bead $g(t)$, the mean square displacement per bead if the friction constant were smeared uniformly over the entire chain, $g_0(t)$, the mean square displacement of the center of mass $g_{cm}(t)$, and the mean square displacement of the center of frictional resistance, $g_{cr}(t)$. $n = 255$ and $n_e = 15$. The arrow denotes $2<S^2>$.

V. SUMMARY

The computer simulations provide no evidence whatsoever that reptation is the dominant mode of polymer melt motion. There is no spatially fixed tube; rather to a very good approximation the character of the long-distance motion is essentially isotropic. Furthermore, there are two relevant distance scales in polymer melts. One is associated with the distance over which static excluded volume effects are screened out and a second, much longer distance, is associated with the mean distance between dynamic entanglements. These dynamic entanglements appear to be rare, and are the result of topological constraints. From an analytical viewpoint, the motion of the chain can be phenomenologically treated as that of a Rouse chain having a few less mobile points, corresponding to the long dynamic entanglements. Theory further indicates that the scaling $D \sim n^{-2}$ is due to the onset of some kind of constrained dynamics and nothing more.

Perhaps in the asymptotic limit chains reptate. If so, the transition to reptation behavior will be transparent to experiment. We have found cases in the simulation where D is proportional to n^{-2} and τ_R is proportional to $n^{3.4}$ and yet these chains do not reptate. Clearly, however, this is by no means a solved problem. One has to now examine

the nature of dynamic entanglements more closely and establish
whether, in fact, ultimately there is some kind of moving tube where
the long-lived entanglements somehow disengage by a reptating or
slipping mechanism of one chain past the other. Refinements of the
analytic theory are also required.

Acknowledgement

The author wishes to acknowledge the collaboration of Professors A.
Kolinski and R. Yaris, without which the work would never have been
done. This research was supported in part by a grant by the Polymer
Program of the National Science Foundation (No. DMR 85-20789) and the
National Institutes Health (Grant No. GM 37408) from the Division of
General Medical Science, U.S. Public Health Science.

References

1. Graessley, W.W. (1982) "Entangled linear, branched, and network
   polymer systems - molecular theories" *Adv. Poly. Sci.* 47, 67-
   117.

2. Bueche, F. (1962) *Physical properties of polymers.* John Wiley
   & Sons, Inc., New York.

3. Yu, H. (1988) "Polymer self-diffusion and tracer diffusion in
   condensed systems", in M. Nagasawa (ed), Molecular Conformation
   and Dynamics of Macromolecules in Condensed Systems. Studies in
   Polymer Science, Vol. 2. Elsevier, Amsterdam (1988) 107-181.

4. Berry, G.C. and Fox T.B. (1968) "The viscosity of polymers and
   their concentrated solutions" *Adv. Poly. Sci.* 5, 261-357.

5. Rouse, P.E. (1953) "A theory of the linear viscoelastic properties of dilute solutions of coiling polymers" *J.Chem. Phys.* 21,
   1272-1280.

6. Yamakawa, H. (1968) "Modern theory of polymer solutions"
   Harper and Row, New York.

7. Baumgartner, A. (1984) "Simulation of polymer motion" *Ann. Rev.
   Phys. Chem.* 35, 419-435.

8. De Gennes, P.G. (1971) "Reptation of a polymer chain in the presence of fixed obstacles", *J. Chem. Phys.*, 55, 572-578.

9. Doi, M. and Edwards, S.F. (1978) "Dynamics of concentrated polymer systems, part I. Brownian motion in the equilibrium state"
   J. Chem. Soc. Faraday Trans, 74, 1789-1801.

10. Doi, M. and Edwards, S.F. (1978) "Dynamics of concentrated polymer systems, part II. Molecular motion under flow" *J. Chem. Soc. Faraday Trans*, **74**, 1802-1817.

11. Doi, M. and Edwards, S.F. (1978) "Dynamics of concentrated polymer systems, part III. The constitutive equation" *J. Chem. Soc. Faraday Trans*, **74**, 1818-1832.

12. Doi, M. and Edwards, S.F. (1978) "Dynamics of concentrated polymer systems, part IV. Rheological properties" *J. Chem. Soc. Faraday Trans*, **75**, 38-54.

13. Phillies, G.D.J. (1986) "Universal scaling equation for self-diffusion by macromolecules in solution" *Macromolecules*, **19**, 2367-2376.

14. Fujita, H. and Einaga, Y. (1985) "Self diffusion and visco-elasticity in entangled systems. I. Self diffusion coefficients" *Poly. J.* (Tokyo) **17**, 1131-1139.

15. Fixman, M. (1985) "Dynamics of semidilute polymer rods: An alternative to cages" *Phys Rev. Lett.* **55**, 2429-2432.

16. Skolnick, J., Yaris, R. and Kolinski, A. (1988) "Phenomenological theory of the dynamics of polymer melts. I. Analytic treatment of self diffusion" *J. Chem. Phys.* **88**, 1407-1417.

17. Skolnick, J. and Yaris, R. (1988) "Phenomenological theory of the dynamics of polymer melts. II. Viscoelastic properties" *J. Chem. Phys.* **88**, 1418-1442.

18. Kolinski, A., Skolnick, J., and Yaris, R. (1987) "Does reptation describe the dynamics of entangled, finite length model systems? A model simulation" *J.Chem. Phys.* **86**, 1567-1585.

19. Kolinksi, A., Skolnick, J. and Yaris, R. (1987) "Monte Carlo studies on the long time dynamic properties of dense polymer systems. I. The homopolymeric melt" *J. Chem. Phys.* **86**, 7164-7173.

20. Kolinski, A., Skolnick, J. and Yaris, R. (1986) "On the short time dynamics of dense polymer systems and the origin of the glass transition. A model system." *J. Chem. Phys.* **84**, 1922-1931.

21. Hilhorst, H.J. and Deutch, J.M. (1975) "Analysis of Monte Carlo results on the kinetics of lattice polymer chains with excluded volume". *J. Chem. Phys.* **63**, 5153-5161.

220

22. Boots, H. and Deutch, J.M. (1977) "Analysis of the model dependence of Monte Carlo results for the relaxation of the end to end distance of polymer chains" *J. Chem. Phys.* **67**, 4608-4610.

23. Valeur, B., Jarry, J.P., Geny, F. and Monnerie, L. (1975) "Dynamics of macromolecular chains. I. Theory of motions on a tetrahedral lattice" *J. Poly. Sci., Poly. Phys. Ed.*, **13**, 667-677.

24. Gurler, M.T., Crabb, C.C., Dahlin, D.M. and Kovac, J. (1983) "Effect of bead movement rules on the relaxation of cubic lattice models of polymer chains" *Macromolecules* **36**, 398-403.

25. Bishop, M., Ceperley, D., Frisch, H.L. and Kalos, M.M. (1982) "Investigations of model polymers. Dynamics of melts and statics of a long chain in a dilute melt of shorter chains" *J. Chem. Phys.* **76**, 1557-1563.

26. Colby, R.H., Fetters, L.J. and Graessley, W.W. (1987) "Melt viscosity - molecular weight relationship for linear polymers". *Macromolecules* **20**, 2226-2237.

27. Fixman, M.M. (1988) "Chain entanglements. I. Theory" *J. Chem. Phys.* **89**, 3892-3911.

28. Evans, K.E. and Edwards, S.F. (1981) "Computer simulations of the dynamics of highly entangled polymers. Part 1. Equilibrium dynamics." *J. Chem. Soc. Faraday Trans* 2, 1891-1912.

29. Baumgartner, A. and Binder, K. (1981) "Dynamics of entangled polymer melts. A computer simulation" *J. Chem. Phys.* **75**, 2994-3005.

30. Hess, W. (1986) "Self diffusion and reptation in semi-dilute polymer systems" *Macromolecules* **19**, 1395-1403.

31. Graessley, W.W. (1980) "Some phenomenological consequences of the Doi-Edwards theory of viscoelasticity" *J. Poly. Sci., Poly. Phys. Ed.* **18**, 27-34.

ANOMALOUS DIFFUSION OF POLYMERS
IN DISORDERED MEDIA

A. BAUMGÄRTNER

Institut für Festkörperforschung

Kernforschungsanlage Jülich

D-5170 Jülich

Fed. Rep. Germany

Abstract : The effects on the dynamics of polymers resulting from diffusion in disordered solid media are discussed in the present article. The results of computer simulations are reviewed and discussed with respect to effects due the randomness of the media and due to entropic barriers separating cavities in the medium.

1. Introduction

Many experimental conditions, such as gel permeation chromatography, gel electrophoresis, enhanced oil recovery, and ultrafiltration, are controled by transport properties of polymers in porous materials /1/.

The diffusion of a single particle in porous media has been studied extensively using the "ant in a labyrinth" model /2-7/. On length scales over which the pore space is fractal one expects to observe anomalous diffusion. Related crossover behavior and corresponding exponents have been explained in terms of the fractal and spectral dimensions of the medium /3-5/.

The analogous problem of a "polymer in a labyrinth" is more complex, since the effects of entanglements /8,9/ of long chains with the constituents of the media may be in competition with disorder effects. Theoretical investigations of the dynamical properties of macromolecules trapped inside a porous medium have been started only very recently /10-15/.

Th. Dorfmüller (ed.), Reactive and Flexible Molecules in Liquids, 221–237.

In this article we are mainly concerned with the dynamics of polymers, whereas the discussion of configurational properties of a polymer in disordered media are mentioned only briefly. For more detailed discussions of this point we refer to refs./10,16-19/.

The outline of the present article is as following. In the next paragraph we present the results from computer simulations of gaussian (i.e. without self-excluded volume) and self-avoiding polymers in three- and two-dimensional disordered media. The third paragraph is concerned briefly with a crossover scaling assumption, and finally the most important aspects are summerized and discussed in section four.

2. Monte Carlo Simulations : Models and Results

2.1 GAUSSIAN POLYMERS IN 3D POROUS MEDIA /10/

The model for a porous solid in three dimensions consists of small, hard cubes distributed at random with probability $1 - p$ on the cubic lattice with mesh size a (Fig.1).

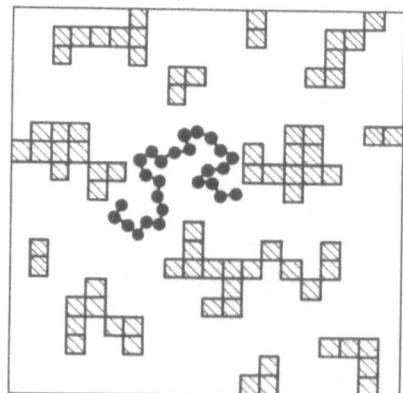

Fig.1 : Two-dimensional representation of a model porous media with a "pearl-necklace" polymer.

For porosities $p \leq 0.6883$ ($= 1 - p_c$, p_c is the site percolation threshold /20/) the solid is percolated. The polymer is modelled by the freely-jointed chain consisting of $N - 1$ rigid links of length $l/a=0.6$ freely jointed together. The dynamics is associated to the model by the conventional kink-jump technique /21,22/ : a chain configuration is changed locally by trying rotations of two successive links around the axis joining their end points, by an angle ϕ chosen randomly from the interval $(-\Delta\phi, \Delta\phi)$. The parameter $\Delta\phi$ is chosen arbitrarily. If an end point of the chain is chosen the terminal link is rotated to a near position by specifying two randomly chosen angles (ϕ, θ) in three dimensions, with $cos\theta$ being equally distributed in the interval $-1 < cos\theta < 1$. We rejected all rotations which lead to an overlap with the solid cubes constituting the disordered medium. When rejection takes place, we retain the old configuration and count it as the new one.

The diffusion constant has been estimated as usual from the slope of the mean square displacement of the center of mass $R^2(t)$ of the chain,

$$R^2(t) =< [\mathbf{R}_{cm}(0) - \mathbf{R}_{cm}(t)]^2 > \tag{1}$$

according to

$$D = \lim_{t \to \infty} R^2(t)/6tl^2. \tag{2}$$

Estimates of the mean square displacement of a bead relative to the center of mass of the chain have been obtained according to

$$r^2(t) =< [\mathbf{r}_k(0) - \mathbf{R}_{cm}(0) - \mathbf{r}_k(t) + \mathbf{R}_{cm}(t)]^2 >, \tag{3}$$

where $\mathbf{r}_k(t)$ is the position vector of the k-th bead at time t.

We have estimated the configurational correlationtime of the polymer τ_p. This has been done in two ways. Firstly, τ_p has been identified with the time for the onset of time-independent behavior of $r^2(t)$, i.e. $r^2(t)$=const. for $t \geq \tau_p$, and secondly from

$$\tau_p = \int_0^\infty dt \, \phi(t) \tag{4}$$

where

$$\phi(t) = [< S^2(0)S^2(t) > - < S^2 >^2]/[< S^4 > - < S^2 >^2] \tag{5}$$

is the configurational correlationfunction of the square radius of gyration at time t, $S^2(t)$. We took ensemble averages over the chain configurations as well as over different realizations of the porous media.

Typical results of $R^2(t)$ and $r^2(t)$ for a gaussian polymer chain are presented in Fig.2 for various chain lengths N and at a porosity $p = 0.6$.

According to Fig.2 the displacements for a particular bead of the chain is very close to the Rouse dynamics /23/

$$r^2(t) \sim t^{1/2} \tag{6}$$

and the corresponding configurational correlationtime is also that of a free-draining chain, i.e. $\tau_p \sim N^2$ (Table I).

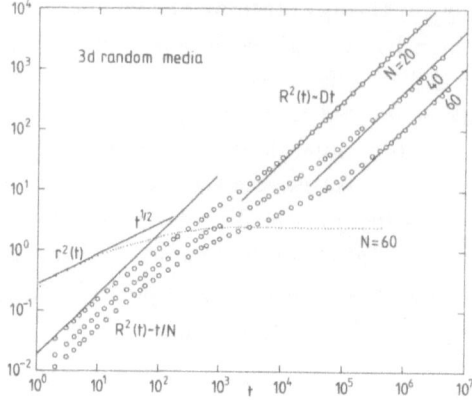

Fig.2: Log-log plot of the mean square displacements $R^2(t)$ and $r^2(t)$ of the center of mass and one particular bead of the chain respectively versus time for various chain lengths N at porosity $p = 0.6$ for a gaussian polymer /10/.

Table I : Normalized configurational correlationtime of a gaussian polymer τ_p/N^2 for various p and chain lengths N. (error $\leq 10\%$).

$N =$	20	40	60	80	100
$p = 0.9$	0.20	0.22	0.22	0.21	0.20
0.8	0.25	0.28	0.23	0.34	0.30
0.7	0.38	0.32	0.42	0.38	0.36
0.6	0.40	0.36	0.40	0.40	0.35
0.5	0.38	0.32	0.33	....	....

Table II : Normalized terminal relaxationtime of the polymer τ/N^2 for various p and chain lengths N. (error $\leq 10 - 20\%$).

$N =$	20	40	60	80	100
$p = 0.9$	0.4	0.9	1.4	....	2.0
0.8	4.0	10.0	....	20.0	24.0
0.7	12.5	37.5	39.8	....	170.0
0.6	25.0	130.0	180.0	....	....
0.5	100.0	312.0	440.0	....	....

The latter finding is quite important since the longest relaxationtime τ defined as the time characterizing the center-of-mass' crossover from anomalous to classical diffusion

$$R^2(t) \sim Dt \qquad for \qquad t > \tau \tag{7}$$

is orders of magnitude larger than τ_p. For comparison, Table II contains the Monte Carlo data of τ/N^2 for various porosities p and chain lengths N.

It has been argued that τ could follow either a power law /10/

$$\tau/N^2 \sim (1-p)^4 N^2 \qquad for \qquad (1-p)\sqrt{N} \gg 1 \tag{8a}$$

or an exponential law of Arrhenius type due to entropic barriers /24/

$$\tau/N^2 \sim \exp(AN) \tag{8b}$$

where A is a constant. Corresponding interpretations of Monte Carlo data for the diffusion coefficient D have been also propounded /10,12/, following either a power law

$$DN \sim (1-p)^{-4} N^{-2} \qquad for \qquad (1-p)\sqrt{N} \gg 1 \tag{9a}$$

or an exponential law of Arrhenius type due to entropic barriers /12/

$$DN \sim \exp(-BN) \tag{9b}$$

where B is a constant.

The dynamics of gaussian polymer chains in disorderd media has an additional complication : the equilibrium configuration of such a chain for $(1-p)\sqrt{N} \gg 1$ is strongly collapsed ("localized random walk") and the mean square radius of gyration is independent of chain length.

$$< S^2 > \sim (1-p)^{-2} \qquad for \qquad (1-p)\sqrt{N} \gg 1 \tag{10}$$

This has been discovered first by Monte Carlo simulations /10/, and discussed intensively later /16-19/. It has been argued /18/ that the chain might also be localized in space and diffusion is essentially suppressed.But this limiting case is hardly accessible by simulations.

2.2 GAUSSIAN POLYMERS IN 2D POROUS MEDIA

Similar findings have been found during corresponding simulations of a gaussian polymer in two dimensions moving among hard squares of width a /11/. The squares are distributed at random with probability $1-p$ $(0 \le p \le 1)$ on the square lattice with mesh size a. Here the polymer chain is modelled by a slightly different freely jointed chain consisting of $N-1$ rigid links freely jointed together. Dynamics is associated to the model by a "bead-spring" technique especially suitable for polymer motions in two dimensions. It consists in randomly selecting a new position $\mathbf{r}_k$ of the k-th bead under the constraint $|r_k - r_{k+1}| = |r_k - r_{k+1}| + |r_k - r_{k-1}|$, i.e. all possible new positions r_k are located on an ellipse defined by r_{k-1}, r_{k+1} and $|r_k - r_{k+1}| + |r_k - r_{k-1}| = $ const. If an end point of the chain is selected, the terminal link is rotated to a new position by specifying a randomly chosen angle and keeping the length of the link fixed. This type of "freely jointed" polymer model has a fixed contour length $<\ell> (N-1)$ with an average length $<\ell>$ of the link which has been chosen as $<\ell>/a=0.6$. (Of course, it has been verified by Monte Carlo methods that the present model exhibits in the case of absence of obstacles the desired properties of Gaussian statistics and of Rouse dynamics). During the simulations all attempts were rejected which led to an intersection of the polymer and the obstacles. Simulations have been performed for $20 \leq N \leq 100$ and $p = 0.75$. The results are presented in Fig.3. The first crossover of $R^2(t)$ from Rouse to anomalous diffusion appears at $t \simeq \tau_o$, which is according to our data approximately constant, i.e. independent of N.

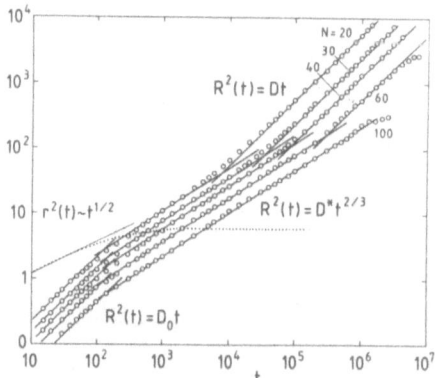

Fig.3: Log-log plot of the mean square displacements $R^2(t)$ and $r^2(t)$ of the center of mass and one particular bead of the chain respectively versus time for various chain lengths N at porosity $p = 0.6$ for a gaussian polymer in two dimension /11/.

In the intermediate time regime, the motion of the chain is restricted by the presence of the obstacles, and undergoes anomalous diffusion $R^2(t) = D^* t^{2k}$, where $k = 0.33 \pm 0.03$ and the anomalous diffusion coefficient $D^* \sim N^{-1.0\pm0.1}$ (Fig.4). The third regime of ordinary diffusion, where $R^2(t) \sim Dt$ and $D \sim N^{-2.0\pm0.2}$ (Fig.4), emerges for times $t > \tau$, where τ is the corresponding termination time of anomalous diffusion. This time is estimated from the Monte Carlo data to $\tau \sim N^{3.0\pm 0.4}$ (Fig.5). This is in variance with Monte Carlo results in three dimensions (sect. 2.1).

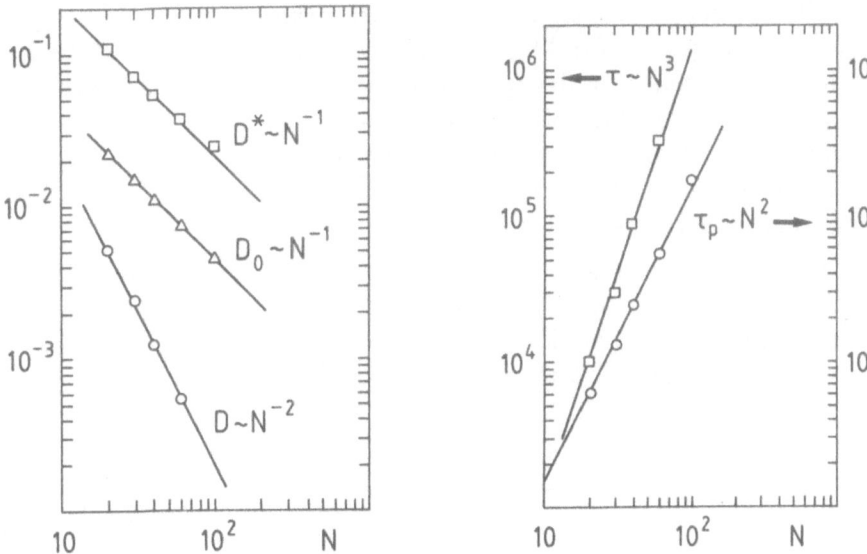

Fig.4 : Log-log plot of the anomalous diffusion coefficient D^* and the ordinary diffusion coefficients D and D_0 vs. N.

Fig.5 : Log-log plot of the polymer configurational relaxationtime τ_p and the longest relaxationtime τ vs N.

It is important to note that the configurational correlationtime of the polymer itself is still that of a Rouse chain, $\tau_p \sim N^2$, characterizing the onset of the time-independence of $r^2(t) \sim N$ for $t > \tau_p$ (Fig.5). Some possible consequences of the difference between τ and τ_p will be discussed in sec.5.

According to $r^2(t)$, presented in Fig.3, there are no indications for a "defect"-like motion leading to $r^2(t) \sim t^{1/4}$ as expected according to the reptation model /8,9/. Rather the dynamics is closer to the Rouse model $r^2(t) \sim t^{1/2}$. Of course, since one expects entanglements between the chain and hard squares, the appearance of reptation for much longer chains as an additional dynamical mechanism beside effects due solely to the randomness of the medium cannot be strictly excluded.

2.3 SELF-AVOIDING POLYMERS IN 3D MEDIA

The model for a porous solid is the same as defined in sec.2.1, i.e the percolation model (Fig.1).

The polymer is modelled by the "pearl-necklace" chain consisting of N hard spheres of diameter $h/l = 0.5$ which are freely jointed together by $N - 1$ rigid links of length $l/a=0.6$. The dynamics is associated to the model by the kink-jump technique . We rejected all rotations which lead to an overlap with any other sphere of the chain (self-excluded volume) or with the solid cubes forming the disordered medium. When rejection takes place, we retain the old configuration and count it as the new one.

If the medium is organized in such a way that the small cubes form a regular periodic network of infinitely long rods on the cubic lattice (yielding a porosity of $p = 1/2$), then the dynamics of the chain is that of the reptation model (Fig.6), i.e. $r^2(t) \sim t^{1/4}$ for $t < \tau$, $D \sim N^{-2}$, and $\tau \sim \tau_p \sim N^3$ /8,9/.

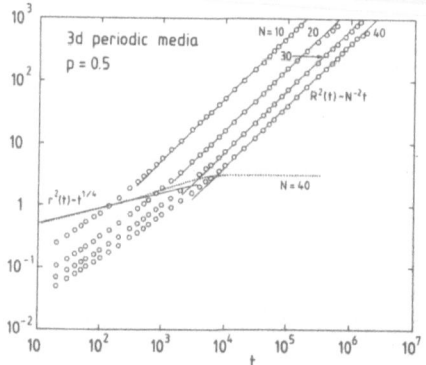

Fig.6 : Log-log plot of the displacements $R^2(t)$ and $r^2(t)$ (Eq.(1) and (3)) vs time t for a "pearl-necklace" chain moving among a periodic array of long rods.

However, if the medium consists of randomly distributed small cubes, the dynamics (Fig.7a-c) is quite different from reptation, rather similar to the results presented above for gaussian chains, but with essentially two important exceptions . (1) The chains are not collapsed but still self-avoiding, i.e. $< S^2 > \sim N^{6/5}$. (2) The configurational relaxation τ_p has a similar N-dependence as the terminal relaxation time τ (Tables III and IV), although they differ considerably with respect to their amplitudes, $\tau >> \tau_p$.

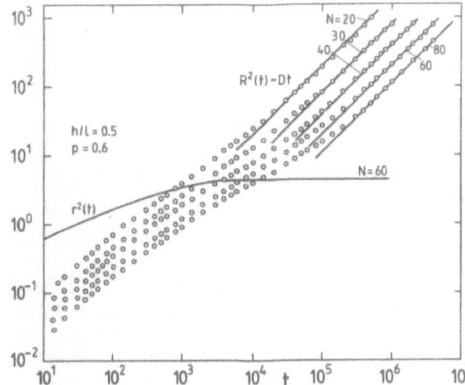

Fig. 7a : Log-log plot of the mean square displacements $R^2(t)$ and $r^2(t)$ of the center of mass and one particular bead of the chain respectively versus time for various chain lengths N at porosity $p = 0.6$ /12/.

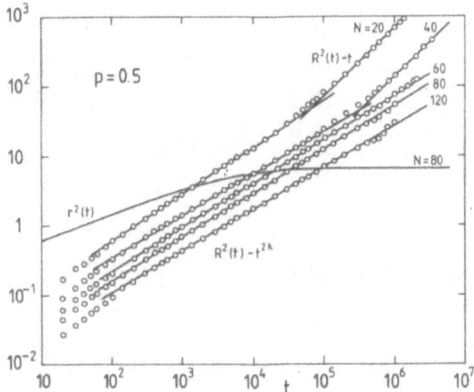

Fig. 7b : The same as in Fig.7a, but at porosity $p = 0.5$.

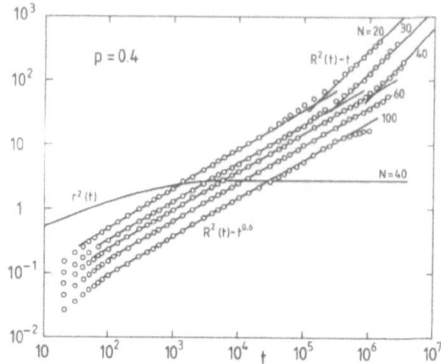

Fig.7c : The same as in Fig.7a , but at porosity $p = 0.4$.

Table III : Normalized correlationtime of the polymer configuration $10^3 \times \tau_p/N^3$ for various p and chain lengths N. (error $\leq 10\%$).

$N =$	20	30	40	60	80	120
$p = 0.8$	7.5	...	5.3	5.1	5.3	...
0.7	8.1	7.8	6.9	6.5	5.9	...
0.6	9.4	9.3	10.0	10.6	9.4	13.0
0.5	12.5	...	14.8	13.0	14.5	19.1
0.4	13.7	...	20.0	50.0	...	...

Table IV : Normalized terminal relaxationtime of the polymer τ/N^3 for various p and chain lengths N. (error $\leq 20\%$).

$N =$	20	30	40	60	80
$p = 0.8$	0.5	0.7	0.8	0.7	0.8
0.7	1.2	1.5	1.5	1.4	1.6
0.6	2.5	2.6	2.8	2.8	3.0
0.5	12.5	...	13.0	...	...

Table V : Normalized diffusion coefficient DN^2 for various porosities p and chain lengths N. (error $\leq 20\%$).

$N =$	20	30	40	60	80
$p = 0.8$	1.11	1.20	1.33	1.51	1.60
0.7	0.67	0.74	0.74	0.72	0.76
0.6	0.35	0.37	0.31	0.33	0.29
0.5	0.13	...	0.11	...	...

At intermediate times, the obstacles restrict the motion of the chain, which undergoes anomalous diffusion $R^2(t) \sim t^{2k} D^*$ where $2k = 0.60 \pm 0.05$ and the anomalous diffusion coefficient $D^* \sim N^{-1.0 \pm 0.2}$ (Fig.7b,c). Finally, the chain resumes ordinary diffusion $R^2(t) \sim tD$ at long times (Fig.7b,c) and $D \sim N^{-2.0 \pm 0.3}$ (Table V).

2.4 SELF-AVOIDING POLYMERS IN 2D MEDIA

We assume that a self-avoiding chain can move among infinitely long parallel hard rods of diameter a. The chain can never cut through these "obstacles" of one dimension. Such an anisotropic arrangement of rod my be realized in nematic fluids or in uniaxial strechted networks. In our model the rods lie in the z direction and are distributed in the xy-plane (compare Fig.1) at random with probability $1 - p$ on a square lattice with mesh size a. For $p \leq p_c$ ($p_c = 0.59$) the material is percolated /20/ in the x and y directions for the medium. We used periodic boundary conditions in the x and y directions. The linear dimension of the basic cell is $L/a \leq 350$, which is much larger than the distances travelled by the polymer chain during the simulations. The "pearl-necklace" chain is the model for the polymer molecule /21,22/. This consists of N hard spheres of diameter $h/a=0.45$ freely jointed together by $N - 1$ rigid links of length $\ell/a=0.5$. The dynamics is that of the conventional kink-jump technique. We rejected all rotations which brought the center of the chosen sphere inside any rod (entanglements) or led to an overlap with any other sphere of the chain (self-excluded volume). Simulations for $10 \leq N \leq 200$ and $p = 0.75$ and $p = 0.65$ have been performed /13/.

The static properties of the polymer do not significantly differ from those of the unperturbed, isotropic case ($p = 1$) and the mean square radius of gyration exhibit well-known self-avoiding characteristics $< S^2 > \sim N^{1.2}$.

The chain dynamics parallel to the rods do not differ significantly from ordinary Rouse dynamics. However, we observe strikingly different dynamics for the motion of the chain in the xy-plane, perpendicular to the rods. The perpendicular component of the mean square displacement of the center of mass $R^2(t)$ of the chain (1) appears in Fig.8 as a function of time for various chain lengths N. In Fig.8 we also include the xy-component of the mean square displacement of a bead relative to the center of mass of the chain $r^2(t)$ (Eq.3). There are three regimes of center of mass behavior. At short times, one observes the Rouse behavior $R^2(t) \sim t N^{-1}$. This follows from the unhindered displacement over distances smaller than the average separation of the rods, $\sim p^{-1/2} a \simeq 1.1 a$. At intermediate times, the rods restrict the motion of the chains, which undergoes anomalous diffusion $R^2(t) \sim t^{2k} D^*$ where $2k = 0.70 \pm 0.05$ (see insert in Fig.8) and the anomalous diffusion coefficient $D^* \sim N^{-0.85 \pm 0.2}$ (Fig.9). Finally, the chain resumes ordinary diffusion $R^2(t) \sim tD$ at long times and $D \sim N^{-2.0 \pm 0.2}$ (Fig.9).

Local motions of the chain are described approximately by the Rouse model; i.e. the mean square displacement of one particular sphere of the chain is close to $r^2(t) \sim t^{1/2}$ (Fig.8). There is no evidence for "defect"-like motions $r^2(t) \sim t^{1/4}$ according to the reptation model. We have estimated the configurational relaxation time of the polymer τ_p. Both estimates exhibit the same power law behavior $\tau_p \sim N^{3.0 \pm 0.3}$, which is shown in Fig.10. This result has to be compared with the time τ for the termination of the anomalous center of mass diffusion, which has been estimated from the data of $R^2(t)$. The results for τ are included in Fig.10 and exhibit a power law $\tau \sim N^{3.3 \pm 0.6}$ which is different from that of τ_p. Of course, with regard to the error bars one cannot exclude that both τ and τ_p actually follow the same power law. But we emphasize that τ is almost three orders of magnitude larger than τ_p, i.e. anomalous diffusion takes place in the present model over three decades of time. We expect that the range of anomalous diffusion becomes larger as p approaches p_c.

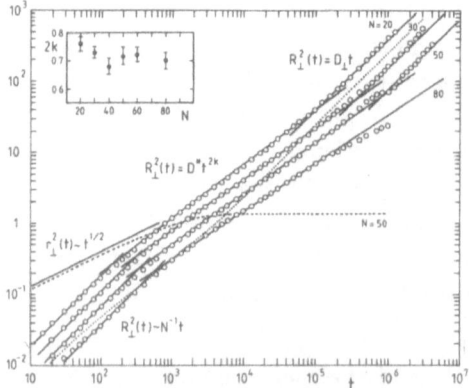

Fig.8 : Log-log plot of the mean square displacements $R^2(t)$ and $r^2(t)$ of the center of mass and one particular bead of the chain respectively vs time for various chain lengths N at porosity $p = 0.75$ in two dimensions.

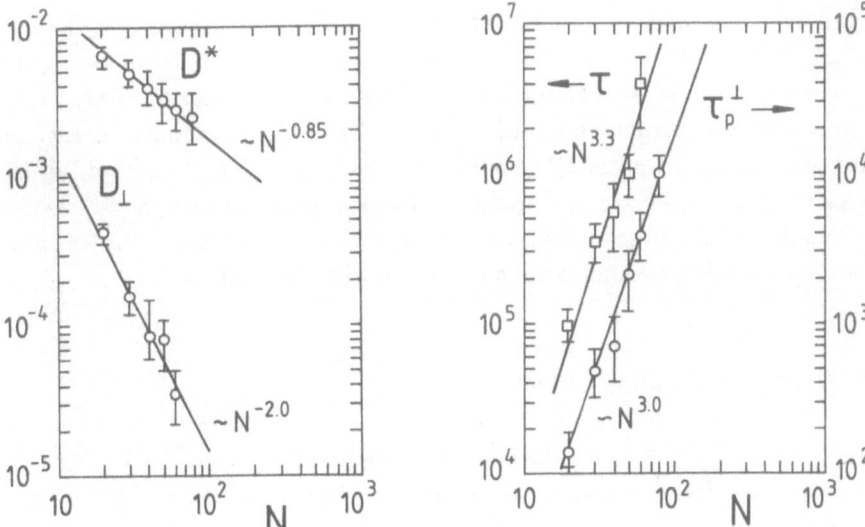

Fig.9 : Log-log plot of the anomalous diffusion coefficient D^* and the ordinary diffusion coefficients D and D_0 *vs.* N at porosity $p = 0.75$.

Fig.10 : Log-log plot of the polymer configurational relaxationtime τ_p and the longest relaxationtime τ *vs.* N at porosity $p = 0.75$.

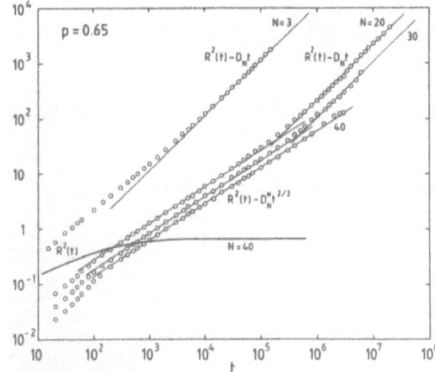

Fig.11 : Log-log plot of the mean square displacements $R^2(t)$ and $r^2(t)$ of the center of mass and one particular bead of the chain respectively vs time for various chain lengths N at porosity $p = 0.65$ in two dimensions.

Simulations at a much lower porosity $p = 0.65$, closer to the percolationthreshold $p_c = 0.59$, are presented in Fig.11. The results for $R^2(t)$ confirm the appearance of anomalous diffusion found also for $p = 0.75$. The corresponding anomalous exponent, which is estimated to $2k = 0.67\pm0.06$, is in variance within the statistical error with the exponent found for $p = 0.75$. The estimate of the anomalous diffusion exponent gives $D^* \sim N^{0.9\pm0.2}$. Estimates for D and τ from the results presented in Fig.11 are not possible. The crossover to ordinary Fickian diffusion (7) could not be observed because Monte Carlo times much larger than 10^7 are hardly accessible by simulations of chains with $N > 40$.

3. Crossover Scaling Analysis

The results from the simulations presented above, seem to be consistent with the crossover formula from anomalous to ordinary diffusion of the center-of-mass displacements

$$R^2(t) = D^* t^{2k} f(t/\tau) \tag{11}$$

with $f(s) \sim s^{1-2k}$ for $s \gg 1$, and $f(s) = const.$ for $s \ll 1$. Using the definitions for anomalous diffusion coefficient $D^* \sim N^{-x}$, $terminal relaxation time \tau \sim N^z$ and Fickian diffusion coefficient $D \sim N^{-w}$, we have the scaling relation

$$w = x + z(1 - 2k) \tag{12}$$

According to Monte Carlo results presented above the estimates of the various exponents,

(1) $x = 0.9\pm0.2$, $2k = 0.66\pm0.05$, $z = 3.3\pm0.6$ and $w = 2.0\pm0.2$ for self-avoiding chains in 2D media (sec.2.4),

(2) $x = 1.0\pm0.2$, $2k = 0.6\pm0.05$, $z = 3.0\pm0.7$ and $w = 2.0\pm0.3$ for self-avoiding chains in 3D media (sec.2.3),

(3) $x = 1.0\pm0.1$, $2k = 0.66\pm0.06$, $z = 3.0\pm0.4$ and $w = 2.0\pm0.2$ for gaussian chains in 2D media (sec.2.2),

are in reasonable agreement with the scaling relation (12).

4. Discussions

The reasons for the appearance of the anomalous diffusion of a polymer are not well understood. Therefore it is important to note that anomalous diffusion has *not* been observed for *periodically* distributed rods (sec. 2.3 and /10/). This obviously supports the idea that the disorder of the medium plays a dominant role in the appearance of anomalous diffusion by a polymer chain. With respect to this fact and remembering the well-known case of anomalous diffusion of a single particle on percolating clusters (/6/ and references therein), one is tempted to attribute the anomalous polymer diffusion to the fractal structure of the media.

In fact, estimates given above for the anomalous exponent $2k$ for polymers are in reasonable agreement with estimates for the "ant" /6/ : 0.69 in d=2 and 0.57 in d=3. The particular difference between an "ant in a labyrinth" and a "polymer in a labyrinth" is due to the chain length dependence. The time interval during which anomalous diffusion is observed is the larger the longer the molecule is. In fact, diffusion of an ant in those media, which have been considered above for polymer dynamics, i.e. $0.4 \leq p \leq 1$ for 3D media and $0.65 \leq p \leq 1$ for 2D media, exhibit only weak diffusional anomalies, because the percolation correlationlength is still quite small.

So the question remains why polymers exhibit anomalous diffusion over a much larger regime than single particles do. A simple possible explanation is the following. The displacement for the "ant" is $R^2 \simeq \sigma^2 t^{2k}$, where σ is the local jump distance. Assuming that the center of mass of the polymer chain follows essentially the same trajectory as the "ant", but needs, according to the Rouse model, a time of the order of $\tau_o \simeq \sigma^2 N$, in order to move the distance σ, leads to the displacement for the center of mass $R^2(t) \sim N^{-1} t^{2k}$, which is in agreement with (11).

However, this important effect of the disorder of the medium on the polymer dynamics can strongly be modified or even be changed if the chainlength becomes much larger than those which have been investigated so far. One effect might be due to the presence of randomly distributed bottlenecks ("entropic barriers") in the medium which is the the more pronounced the longer the chain or the lower the porosity is. The other effect might be due to entanglements leading eventually to reptation, especially if the medium consists of long rods as described in sec. 2.4.

In fact, the coincidence of diffusivity and relaxationtime exponents, $w \approx 2$ and $z \approx 3$, with the corresponding exponents from reptation theory is surprising, despite the fact that local reptational correlations $r^2(t) \sim t^{1/4}$ are not observed, in contrast to the case of a periodic network (sec.2.3). But this coincidence can only be fortuitous, because in contrast to reptation the dynamics of a polymer in a random media has two different timescales, the configurational correlationtime τ_p and the terminal relaxationtime $\tau >> \tau_p$, characterizing the onset of isotropic motion of the chain over large distances. Among others there are two questions : (1) What are the reasons for $\tau >> \tau_p$? (2) Why is for gaussian chains $\tau \sim N^3$ and $\tau_p \sim N^2$, but for self-avoiding chains $\tau \sim \tau_p \sim N^3$, but still $\tau >> \tau_p$? Of course, since our estimates of the corresonding exponents of τ and τ_p are not very precise, it might be that even for self-avoiding chains the two exponents are different. In the special case of a polymer moving among randomly distributed rods as described in sec.2.4, or between hard squares as described in sec. 2.3, it is conceivable that for very long chains entangled with rods or squares, reptation is an additional effect beside effects solely due to the randomness of the medium. Eventually reptation might dominate the dynamics.

Entropic barriers have probably a very important influence on the polymer dynamics in disordered solids /12/. There the chain is forced to squeeze through narrow channels in order to move to regions which are less occupied by obstacles and hence entropically more favorable. The motion of the chain between these different "entropic traps" is slowed down significantly by the bottlenecks connecting these regions.Assuming that there exist an average domain size C_1 and an average size C for the bottleneck, one has a diffusion coefficient /24/

$$D/D_o = \exp\{-N[f(\frac{1}{C})^{1/\nu} + (\frac{1-f}{z} - 1)(\frac{1}{C_1})^{1/\nu}]\} \tag{13}$$

where D_o is the diffusion coefficient of the free-draining chain, f is the fraction of the monomers inside the bottelneck and z is the average number of domains per gate available for the remainder of the chain. ν is the usual Flory exponent of the radius of gyration of the chain. f is a crossover function /24/. This formula is in very good agreement with the data presented in sec.2.3 and 2.1 for polymers in d=3. Eq.(13) has not yet been tested for chains in two dimensions as discussed in sec.2.2 and 2.4. Of course, the concept of entropic barriers is not useful in explaining the anomalous diffusion reported above, rather it is an *additional* effect of random media on the dynamics of polymers.

References

1. F.A.L.Dullien (1979) "Porous Media, Fluid Transport and Pore Structure", Academic Press, New York

2. P.G. de Gennes (1976) La Recherche **7**, 919

3. S.Alexander, R.Orbach (1982) J.Physique Lett. **43**, L625

4. Y.Gefen, A.Aharony, S.Alexander (1983) Phys.Rev.Lett. **50**, 77

5. A.Aharony (1985) in "Scaling Phenomena in Disordered Systems", Eds. R.Pynn and A.Skjeltrop, Plenum Press, New York

6. S.Havlin, D.Ben-Avraham (1987) Adv.Phys. **36**, 695

7. J.W.Haus, K.W.Kehr (1987) Phys.Rep. **150**, 263

8. P.G. de Gennes (1971) J.Chem.Phys. **55**, 572

9. M.Doi, S.F.Edwards (1986) *The Theory of Polymer Dynamics*, Clarendon Press, Oxford

10. A.Baumgärtner, M.Muthukumar (1987) J.Chem.Phys.**87**,3082

11. A.Baumgärtner (1987) Europhys.Lett. **4**, 1221

12. M.Muthukumar, A.Baumgärtner (1989) Macromolecules (in press), "Diffusion of a polymer in random media".

13. A.Baumgärtner, M.Moon (1988) "Anomalous diffusion between long rods".

14. R.Loring (1988) J.Chem.Phys. **88**, 6631

15. G.C.Martinez-Mekler, M.A.Moore (1981) J.Physique **42**, L413

16. S.F.Edwards, M.Muthukumar (1988) J.Chem.Phys. **89**, 2435

17. S.F.Edwards, Y.Chen (1988) J.Phys.A **89**, 2963

18. M.E.Cates, R.C.Ball (preprint 1988)

19. A.Baumgärtner (1988) in *11th Taniguchi Symposium on Macromolecular Fluids*, Springer, "Polymers in disordered solid media".

20. D.Stauffer (1985) *Introduction to Percolation Theory*, Taylor and Francis, London

21. A.Baumgärtner (1984) Ann.Rev.Phys.Chem. **35**, 419

23. A.Baumgärtner, K.Binder (1979) J.Chem.Phys. **71**, 2541

23. P.E.Rouse, J.Chem.Phys. (1953) **21**, 1273

24. M.Muthukumar, A.Baumgärtner (1989) Macromolecules (in press), "Effects of entropic barriers on polymer dynamics".

Optical Anisotropy of Macromolecular Systems by Depolarized Rayleigh Scattering

G. Floudas and G. Fytas
Foundation for Research and Technology - Hellas
P.O. Box 1527, 711 10 Heraklion, Crete, Greece

ABSTRACT. Depolarized Rayleigh spectra for various macromolecular systems composed of optically anisotropic segments have been measured in solution at 25°C. The effective optical anisotropy $<\gamma^2>/x$ per monomer unit is a sensitive index of the chain conformation. Examples of applications are flexible atactic polymer chains, polymers with different microstructure and AB di- and BAB triblock copolymers in dilute solutions.

Introduction

Polarized Rayleigh scattering has been routinely used to study polymers in solution. The isotropic Rayleigh component is usually measured to obtain both static and dynamic properties, such as molecular weights, radii of gyration, virial coefficients of polymers and diffusion coefficients [1-3]. On the other hand, the depolarized Rayleigh scattering (DRS) has also been identified as a rich source of dynamical and structural information for macromolecules in solution however, less widely used. The DRS which suffers from the much weaker intensity compared to polarized scattering, is very sensitive to the degree of orientational order in liquids consisting of long-chain molecules [4]. The quantity which is sensitive to the orientational order and yet to conformation of optically anisotropic units, is the configurational average optical anisotropy $<\gamma^2>$, determined by DRS. Rotational isomeric-state calculations (RIS) can be used to obtain the "intrinsic" anisotropy of the polymer under investigation [5]. Thereby, the configurational average optical anisotropy can be used to elucidate the conformation of chain molecules by using RIS calculations. Calculations reported in the past demonstrated the possibility of using $<\gamma^2>$, as a conformational probe in polystyrene (PS) [6], polycarbonate (PC) [6,7] and poly(methyl methacrylate) (PMMA) [8]. Although there is no strong evidence of intermolecular correlations in the bulk polymers, these intensities were probably affected by collisionally induced scattering. Collision induced phenomena were first introduced in gaseous Ar and Kr [9] and soon identified as an important mechanism of light scattering in liquids. The depolarized spectrum of ccl_4 [10-12] and other spherically symmetric molecules is greatly affected by collision induced scattering. Attempts have been made in the past to separate this unwanted intensity from the intrinsic

239

Th. Dorfmüller (ed.), Reactive and Flexible Molecules in Liquids, 239–247.

optical anisotropy using two narrow bandpass interference filters of different widths [13-14]. Instead of using the above method, the depolarized intensity can be obtained from the spectrum $I(\omega)$ measured with a Fabry-Perot interferometer of a proper free spectral range. Such a method allows us to estimate the anisotropic part of Rayleigh intensity as the area under the measured Lorentzian, as well as the collision induced contribution which appears as a flat background [15].

Fabry-Perot depolarized Rayleigh spectra have been reported for PS solutions [16-17], bulk poly(propylene glycol) [18], poly(phenyl methyl siloxane) (PPMS) [19-20], linear and star polyisoprenes (PI) [21] and oligomers of poly(p-oxybenzoate) [22]. This article presents applications of DRS to the study of a) conformational properties of the flexible polymers PPMS, PI and PC, b) PPMS of different tacticities and c) conformation of diblock PS-PMMA and triblock PMMA-PS-PMMA copolymer in dilute solution.

Depolarized Rayleigh Scattering

Local fluctuations in the anisotropic part of the polarizability tensor give rise to a depolarized component of the scattered light. The optical anisotropy of a molecule is embodied in the traceless part $\hat{\underline{\alpha}}$ of the polarizability tensor $\underline{\alpha}$. The intensity of the DRS yields the invariant

$$\langle \gamma^2 \rangle = \frac{3}{2} \, \text{tr} \, \langle \hat{\underline{\alpha}} \, \hat{\underline{\alpha}} \rangle . \tag{1}$$

Treatment of this traceless tensor $\hat{\underline{\alpha}}$ as a constitutive property, where $\hat{\underline{\alpha}}$ is taken to be the sum of contributions of bonds comprising the molecule has been attempted repeatedly. Although this model is proved to be inadequate for determining polarizability anisotropies of chain alkanes [23] and other polyatomic molecules in the gas phase, we propose that it can be used successfully to study polymers in solution where a Fabry-Perot interferometer is used to separate the broad component of the depolarized spectrum from the desired intrinsic narrow component.

Pecora has formulated general expressions for the depolarized Rayleigh spectral distribution from dilute solutions of macromolecules composed of optically anisotropic monomer units in solvents consisting of optically isotropic molecules. The spectrum is given by the Fourier transform of the correlation function $S_{VH}(t)$

$$S_{VH}(t) = \sum_{i,j}^{x} \langle \alpha_{yz}^{i}(t) \, \alpha_{yz}^{j}(0) \, \exp \{ iq \cdot (r^i(t) - r^j(0)) \} \rangle \tag{2}$$

where x is the number of monomer units on a single chain, $\alpha_{yz}^{i}(t)$ is the yz component of the laboratory – fixed polarizability tensor of monomer i at time t, q is the scattering vector, and $r^i(t)$ is the position of the ith monomer unit. In this expression i and j may belong to different chains (i.e. both intra and interchain correlations are important) or to the same chain, (i.e. interchain correlations are ignored). Then, the depolarized intensity I_{VH} is given by :

$$I_{VH} = A\varrho^* \sum_{ij}^{x} <\alpha_{yi}{}^i(0) \; \alpha_{yz}{}^j(0)> = A\varrho^* \; <\gamma^2> \tag{3}$$

where A is a constant and ϱ^* is the number density of the polymer in solution. We can also express the depolarized intensity in terms of the optical anisotropy β of the monomer unit and the quantity $1+F$, which is a measure of the internal pair correlations of the polymer chain (when the polymer is dilute enough) :

$$1+F = \frac{<\gamma^2>}{x \; \beta^2} \tag{4}$$

From equations (3) and (4) the total integrated depolarized intensity is given by:

$$I_{VH} = A\varrho \; (1+F)\beta^2 \tag{5}$$

where ϱ is the number density of monomer units.

The quantity $1+F$ can be measured experimentally, using the measured values of $<\gamma^2>/x$ in the polymer chain and β^2 of the monomer unit. It is also possible to calculate the ratio $1+F$ using RIS theory[5,24]. The effective optical anisotropy $<\gamma^2>/x$ is finally given by :

$$<\gamma^2>/x = 15 \; (\frac{\lambda_o}{2\pi})^4 n^2 \; f(n)^{-1} \; (\frac{R_{VH}}{\varrho}) \; _{\varrho \to 0} \tag{6}$$

where λ_o is the wavelength of light in vacuo, n is the refractive index of the solution, R_{VH} is the Rayleigh ratio of the solute molecules and $f(n)$ is the local field correction. The exact form for the latter is still unknown and we use the approximation $f(n) = [(n^2+2)/3]^2$.

Results and Discussion

(a) Flexible polymers

A typical depolarized spectrum for a dilute solution of PPMS in ccl_4, at 25°C, is shown in Fig.1. For comparison, the spectra of the solvent ccl_4 and of neat benzene used as standard are also shown in Fig.1. The area under the measured Lorentzian line represents the depolarized intensity I_{yH} and the collision-induced contribution appears as a flat background in the free spectral range used ($40cm^{-1}$). Details on the experimental set-up, can be found elsewhere [22]. Values of $<\gamma^2>/x$ in dilute solutions of PPMS in ccl_4 and PC in dioxane at 25°C are plotted versus the degree of polymerization in Fig.2. In the same figure RIS calculations for PS [25] are included for comparison. According to these calculations for flexible coils in dilute solutions, intrachain correlations are relatively short and the ratio $<\gamma^2>/x$ rapidly attains its asymptotic value for large values of x. For PS and PPMS the asymptotic value is approached for x~50 and for PC even from x~20. RIS theory was also used to calculate the characteristic ratio $c_\infty = <r^2>/nl^2$, where $<r^2>$ is the mean

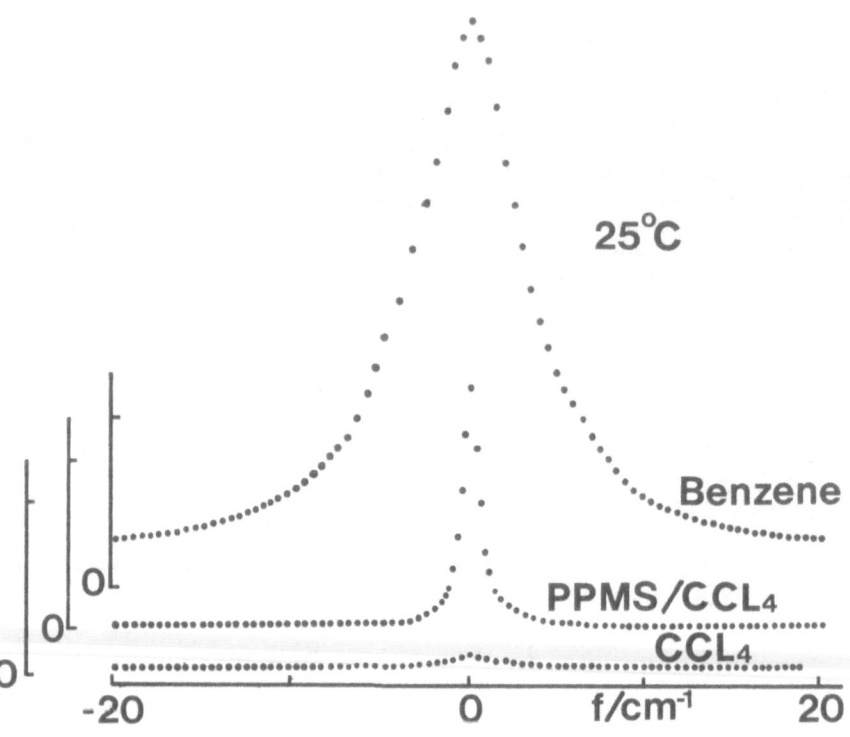

Figure 1. Depolarized Rayleigh spectra of 1% PPMS in ccl$_4$, of ccl$_4$ and of neat benzene at 25°C.

Table I : Depolarized Rayleigh Intensities, Optical Anisotropies and Intramolecular correlations of polymers in solution at 25°C.

Sample	$\dfrac{R_{VH}}{\varrho}$ /10^{-28}cm^2	β^2/Å^6	$\dfrac{\langle\gamma^2\rangle}{x}$ /Å^6	F_{intra}
PPMS/ccl$_4$	11.1	19.1[a,b]	28.0	~ 0.5
PDMS/ccl$_4$	0.038	0.08[a]	0.12	~ 0.5
PC/dioxane	39.4	58.0[a,b]	117.0	~ 1.0
PI/ccl$_4$	3.7	11.9[a]	10.5	~ 0

a : calculated values for structures very similar to the repeating unit.

b : measured values.

square end-to-end distance, n is the number of bonds and l is the bond length. This value resembles the ratio $<\gamma^2>/x\beta^2$ in that it is determined only by bond angles and internal potentials. For atactic cis-PI, PPMS and PC, this conformational property amounts to 4.5 [26], 5.8 [27] and 29.9 [28] respectively. These values correspond to 1.3, 2.0 and 2.2 monomer units per statistical segment in accord with the present experimental data.

To estimate the intramolecular correlation factor 1+F, the anisotropy of the monomer β^2 is needed. The latter can be either measured or computed from bond polarizabilities. In Table I, the absolute values of R_{VH}/ϱ, β^2 and $<\gamma^2>/x$ are given, for four polymers in solution. From these values it becomes evident that intrachain correlations are very weak in favor of the random coil conformation. That the random coil conformation is the one favored by these polymers in bulk and solution, has been tested by performing a concentration-dependent study of the R_{VH} intensity, where only a weak variation with solute concentration was observed [20].

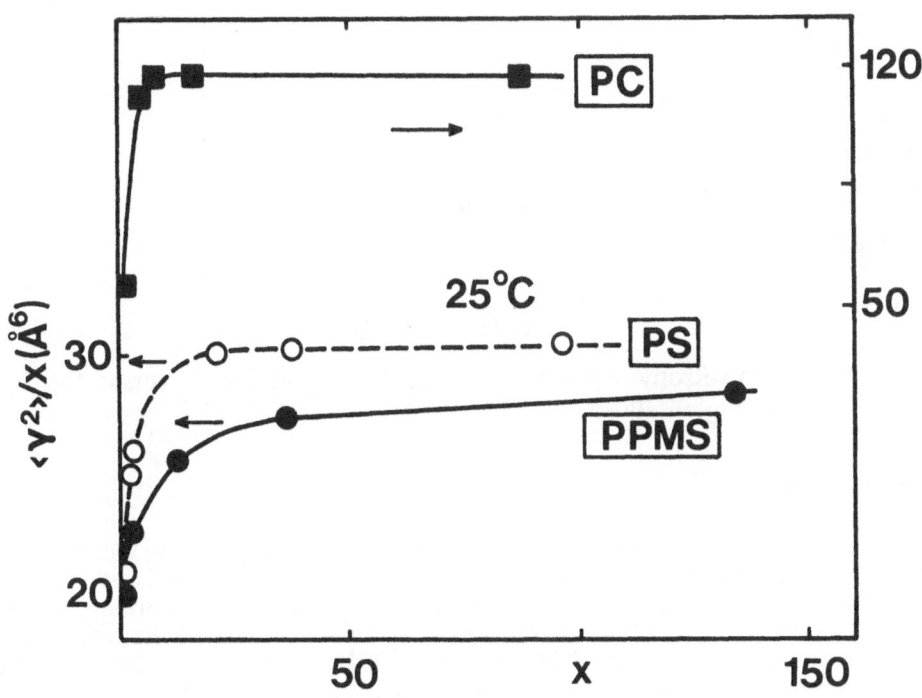

Figure 2. Effective optical anisotropy $<\gamma^2>/x$ per monomer unit of PPMS in ccl$_4$ and PC in dioxane, at 25°C. Dash line denotes the results of RIS calculations on PS.

244

b) PPMS of various tacticities

Poly(phenyl methyl siloxane) is an important polymer whose conformational properties have been studied in the past. The dependence of the characteristic ratio c_∞ and the temperature coefficient of the unperturbed dimensions on the chain microstructure, has been theoretically studied by Mark and Ko [27]. On the experimental side, strain birefrigence [29] and optical anisotropy measurements were reported only for atactic PPMS [20]. Very recently PPMS samples with different fraction W_m of mesodyads were synthesized. Results of the DRS measurements on the optical anisotropy as a function of W_m are shown in Fig.3. The optical anisotropy is sensitive to the chain microstructure and attains its lower value for nearly atactic chains but towards the region rich with isotactic chains (i.e. W_m=1). RIS calculations of the $<\gamma^2>/x$ for PPMS with W_m varying from syndiotactic (W_m=0) to pure isotactic chains are now in progress.

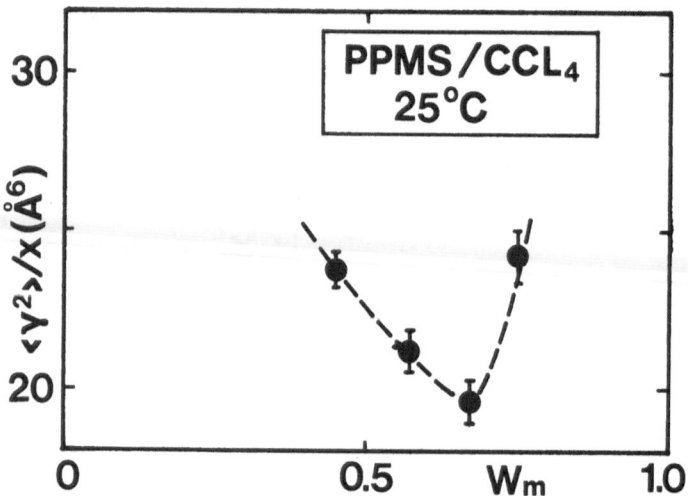

Figure 3. Optical anisotropy $<\gamma^2>/x$ of PPMS in ccl$_4$ at 25°C, versus the fraction W_m of meso dyads.

(c) Block copolymers

Conformational properties of PS-PMMA block copolymer and PMMA-PS-PMMA block copolymer were examined in the past by various methods. The radius of gyration of the PMMA block was obtained using isorefractive solvents; i.e. solvents having zero refractive index increment for one of the two blocks, say the PMMA block [30]. Chain dimensions was the result of a simulation on a self-avoiding type lattice-walk model [30,31]. Finally, both static light scattering and small angle neutron scattering were employed to examine the conformation of PS and PMMA in the diblock, respectively, using toluene as the solvent [32]. While the chain dimension of PS in the PS-PMMA block copolymer is only slightly bigger than that of its precursor, the PMMA block is substantially contracted in a near theta configuration. The picture inferred from the LS and SANS measurements is that of a partially segregated PMMA core and an expanded PS shell. A quantitative description is better given in terms of the distance between the center of mass of the two blocks [33]. In the case of the triblock PMMA-PS-PMMA copolymer, the central PS block is more expanded than the PS chain of the diblock, when compared at common composition and molecular weight. The same was observed for the overall dimension of the triblock.

In this article we demonstrate, for the first time, the application of DRS to the study of conformation of block copolymers in dilute solution. In contrast to the situation with polarized light scattering experiments, where an isorefractive solvent is needed, DRS is a selective technique in the sense that only optically anisotropic units will scatter light in the VH configuration (where V denotes polarization of the incident light and H the polarization of the scattered component). For an A-B block copolymer where A-block is highly anisotropic (PS) and B-block is predominantly isotropic (PMMA), the depolarized intensity I_{VH} should be attributed to the anisotropic block. Hence, the quantity $\langle\gamma^2\rangle/x$, is characteristic of the chain conformation of the A-block. Depolarized intensities and optical anisotropies calculated from eq.6 (after weight fraction correction), assuming the same local field correction, for diblock and triblock copolymers in dioxane are given in Table II. The copolymer

Table II : Depolarized Rayleigh Intensities and Optical Anisotropies of Block Copolymers in solution at 25°C.

Sample	M_w	f_{PS}	$\dfrac{R_{VH}}{\varrho}$ /10^{-27}cm^2	$\dfrac{\langle\gamma^2\rangle}{x}$ /Å^6
PS in ccl$_4$	3×10^5	–	1.3 ± 0.1	38 ± 2
PS-PMMA in ccl$_4$	8.3×10^4	0.68	1.35 ± 0.05	39 ± 1
PS-PMMA in dioxane	8.3×10^4	0.68	0.7 ± 0.1	22 ± 2
PMMA-PS-PMMA in dioxane	1.04×10^5	0.59	2.0 ± 0.1	61 ± 2

composition was also checked by NMR measurements. DRS measurements were carried out in the low concentration region ($\sim1\%$) and the obtained $\langle\gamma^2\rangle/x$ value was virtually independent of concentration. It is worthnoticing that $\langle\gamma^2\rangle/x$ for atactic PS ($M_w=3\times10^5$) in ccl$_4$ is 38 ± 2 Å^6 in agreement with the RIS result [25]. The much larger value of $\langle\gamma^2\rangle/x$ for the atactic PS in the triblock is compatible with the more extended PS chain increasing the intramolecular order (F). The anisotropic scattering of the PMMA block is very weak ($R_{VH} = 0.04\times10^{-27}$ cm^2 at 25°C), and therefore the PMMA block is nearly "invisible".

On the other hand, the reason for the much lower $\langle\gamma^2\rangle/x$ of PS in the diblock, which is remarkably close to the calculated value for isotactic PS chain, is not certain. Phenomenologically, low optical anisotropy is expected for PS if the chain favors more (tg) conformations [25]. The optical anisotropy of the PS block is a sensitive index of the copolymer chain conformation as indicated by the value of $\langle\gamma^2\rangle/x$ for the diblock dilute solution in ccl$_4$. The latter is a good and marginal solvent for the PS and PMMA block respectively, and the copolymer is organized in micelles where the PMMA block forms the interior core. The PS block which exudes out, has similar random coil conformation with that of the homopolymer PS as inferred from the values of $\langle\gamma^2\rangle/x$ listed in Table II. The microstructure of PS, examined by high resolution NMR, was the same in both homopolymer and copolymer.

Acknowledgement

We thank A. Lappas for his assistance in the measurements. The financial support of the Research Center of Crete is gratefully acknowledged.

References

1. Huglin, M.B. (1972) "Light Scattering from Polymer Solutions", Academic Press, New York.

2. Chu, B. (1974) "Laser Light Scattering", Academic Press, New York.

3. Berne, B.J. and Pecora, R. (1976) "Dynamic Light Scattering", Wiley-Interscience, New York.

4. Fischer, E.W., Strobl, G., Dettenmaier, M., Stamm, M., and Steidle, N. (1979) Faraday Disc. Chem. Soc. 68, 26.

5. Flory, P.J. (1969) "Statistical Mechanics of Chain Molecules", Wiley-Interscience, New York.

6. Dettenmaier, M., Fischer, E.W. (1975) Makromol. Chem. 177, 1185.

7. Dettenmaier, M., Kausch, H.H. (1981) Colloid Polym. Sci. 259, 209.

8. Fischer, E.W., Dettenmaier, M. (1978) J. Non-Cryst.Solids 31, 181.

9. McTague, J.P., Birnbaum, G. (1968) Phys. Rev. Lett. 21, 661.

10. Bucaro, J.A., Litovitz, T.A. (1970) J. Chem. Phys. 55, 3585.

11. Stevens, J.R., Patterson, G.D., Carroll, P.J. and Alms, G.R. (1982) J. Chem. Phys. 76, 5203.

12. Shin, S., Ishigame, M. (1988) J. Chem. Phys. 89, 1892.

13. Patterson, G.D., Flory, P.J. (1972), J.C.S. Faraday II, 68, 1098.

14. Carlson, C.W., Flory, P.J. (1977) J.C.S. Faraday II, 73, 1505.

15. Burnham, A.K., Alms, G.R. and Flygare, W.H. (1974) J. Chem. Phys. 62, 3289.

16. Bauer, D.R. Brauman, J.I., Pecora, R. (1975) Macromolecules 8, 443.

17. Alms, G.R., Patterson, G.D., Stevens, J.R. (1979) J. Chem. Phys. 70, 2145.

18. Jones, D., Wang, C.H. (1976) J. Chem. Phys. 65, 1835.

19. Lin, Y.H., Fytas, G., Chu, B. (1981) J. Chem. Phys. 75, 2091.

20. Fytas, G., Patkowski, A., Meier, G., Fischer, E.W. (1988) Macromolecules, 21, 000.

21. Fytas, G., Floudas, G., Hadjichristidis, N. (1988) Polym. Communic. 29, 322.

22. Floudas, G., Patkowski, A., Fytas, G., Ballauff, M., in preparation.

23. Ladanyi, B.M., Keyes, T. (1978) Molec. Phys. 37, 1809.

24. Abe, Y., Tonelli, A.E., Flory, P.J. (1970) Macromolecules 3, 295.

25. Suter, U.W., Flory, P.J. (1977) J.C.S. Faraday II, 73, 1521.

26. Abe, Y., Flory, P.J. (1971) Macromolecules, 4, 219.

27. Mark, J.E., Ko, J.H. (1975) J.Polym.Sci. Polym.Phys. Ed. 13, 2221.

28. Gawrisch, W., Brereton, M.G., Fischer, E.W. (1981) Polym. Bull. 4, 687.

29. Llorente, M.A., de Pierola, I., Saiz, E. (1985) Macromolecules, 18, 2663.

30. Tanaka, T., Kotaka, T., Inagaki, H. (1976) Macromolecules, 9, 561.

31. Birshtein, T.M. Skvortsov, A.M., Sariban, A.A. (1976) Macromolecules, 9, 888.

32. Han, C.C., Mozer, B., (1976) Macromolecules, 10, 44.

33. Tanaka, T., Kotaka, T., Ban, K., Hattori, M., Inagaki, H., (1977), Macromolecules, 10, 960.

APPLICATION OF PHOTON CORRELATION SPECTROSCOPY TO THE STUDY OF DIFFUSIONAL DYNAMICS IN COMPATIBLE POLYMER BLENDS

J. KANETAKIS, A. RIZOS and G. FYTAS
Foundation for Research and Technology - Hellas
and Chemistry Department, University of Crete
P.O. Box 1527, 711 10 Heraklion, Crete, Greece

ABSTRACT. Photon correlation spectroscopy has been employed to measure dynamic and static structure factor in compatible polymer blends. For the poly(ethylene oxide)(PEO)/poly(propylene oxide)(PPO) blend containing a volume fraction $\varphi=0.87$ of PEO and with constant molecular weight of PPO, the mutual diffusion and transport coefficients, D and D^o, decrease with the degree of polymerization N_{PEO} of the PEO component. The D^o coefficient exhibits a weak N_{PEO} dependence in accord with the "fast mode theory" of mutual diffusion. For the PPO/polystyrene (PS) mixture at extremely low concentrations of PS we have been able to detect for the first time to our knowledge two distinct concentration dependences of the coefficient D.

1. Introduction

Photon correlation spectroscopy (PCS) has been extensively utilized to study diffusion constants in polymer solutions [1] and relaxation processes in bulk amorphous polymers near the glass transition temperature T_g [2]. Relatively few photon correlation spectroscopic studies have, however, been performed on polymer mixtures. PCS is a very convenient technique to study the diffusional dynamics of such systems due to its wide time range ($10 - 10^{-6}$ s) which satisfies the diffusion times of a large number of polymer pairs. A prerequisite for the application of the technique is the sufficient refractive index difference between the two components. Unlike other techniques such as infrared densitometry and forward recoil spectrometry which require a built-in concentration gradient, PCS probes diffusion of the system at equilibrium. The signal intensity is due to local thermal concentration fluctuations whose autocorrelation function C(q,t) at a scattering vector q and time t decays exponentially as predicted theoretically [3] and proved experimentally [4]. Such experiments can be readily performed if certain precautions in the preparation of the samples are taken and careful attention is given to some experimental pitfalls.

In this report, we present mutual diffusion measurements on two groups of compatible polymer blends : poly(ethylene oxide)(PEO)/poly

Th. Dorfmüller (ed.), Reactive and Flexible Molecules in Liquids, 249–256.
© *1989 by Kluwer Academic Publishers.*

(propylene oxide) (PPO) at finite compositions and poly(styrene)(PS)/PPO at low PS concentrations. The main objective is to discriminate between the different theories of mutual diffusivities.

2. Theoretical Background

Theory for the diffusional dynamics in polymer blends has been developed in the last few years. For incompressible systems, the mean field theory expression for the dynamic structure factor $S(q,t)$ reads [3]

$$S(q,t) = <\varphi_q(t)\varphi_{-q}(0)> = S(q)\exp(-Dq^2t) \tag{1}$$

where $\varphi_q(t)$ is the magnitude of the Fourier transformed thermal concentration fluctuations and $S(q)=<|\varphi_q(0)|^2>$ is the static structure factor, which in the $q\to0$ limit assumes the form

$$S^{-1}(0) = 2(\chi_s - \chi_F) \tag{2}$$

with χ_s being the Flory–Huggins interaction parameter χ_F at the spinodal. The mutual diffusion coefficient D in eq.1 is a composite quantity; it comprises bare chain mobilities and thermodynamic interactions in the form

$$D = 2D^o (\chi_s - \chi_F)\varphi(1-\varphi) \tag{3}$$

where φ is the average volume fraction of one component. D^o in eq.3, termed dynamic transport coefficient, is a weighted average of the tracer diffusivities D_i^o's (i=A,B) of the blend components and is in dispute between the different theoretical approaches. Mean field theory [5] expresses D^o as :

$$(D^o)^{-1} = \varphi_B(D_A^oN_A)^{-1} + \varphi_A(D_B^oN_B)^{-1} \tag{4}$$

whereas within the frame of phenomenological treatment of mutual diffusion [6,7], D^o is defined as

$$D^o = \varphi_BD_A^oN_A + \varphi_AD_B^oN_B \tag{5}$$

where N_i is the degree of polymerization. There is no unanimity either experimentally; some experiments favor the "slow mode" (eq.4) and other the "fast mode" (eq.5) theory.

From eq.4 and 5 it is obvious that if $D_A^oN_A \neq D_B^oN_B$ one could discriminate between the two theories. In the PEO/PPO blend the degree of polymerization of PPO, N_{PPO}, is kept constant and N_{PEO} is varied at constant φ from the Rouse to the well entangled PEO regime. Any complications of the data analysis arising from a composition dependent $T_g(\varphi)$ may be considered negligible since the T_g's of the separate components are similar and the experiments are performed well above the blend T_g. Another possible way to verify either of the two theories is to keep $D_A^oN_A \approx D_B^oN_B$ while changing the composition of the blend. These conditions were applied to the PPO/PS blend.

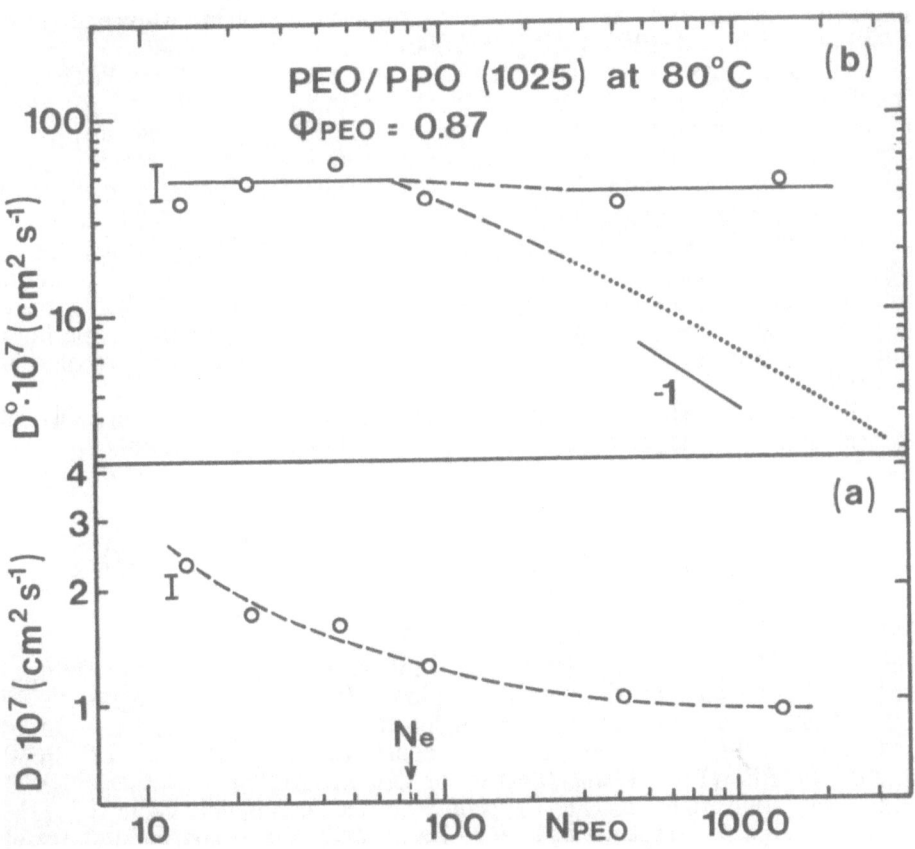

Figure 1. (a) Measured mutual diffusion coefficient D plotted as a function of degree of polymerization N_{PEO} of the PEO component at 80 °C with constant N_{PPO} and φ_{PEO}. N_e denotes the entanglement degree of polymerization of PEO. (b) Transport coefficient D^o versus N_{PEO}. For $N_{PEO}<N_e$ the solid line is an average D^o of the experimental results. For $N_{PEO}>4N_e$ the dotted and solid lines represent eq.4,5, respectively. For $N_e<N_{PEO}<4N_e$ (dashed line) the reptation model by itself can not describe diffusion in the entangled PEO matrix.

3. Molecular Weight Dependence

In eq.3 the diffusion coefficient D comprises a mobility factor D^o and a thermodynamic factor represented by $(\chi_s - \chi_F)$. The term Ω, given by

$$\Omega = D^o \varphi (1-\varphi) \qquad (6)$$

is the Onsager transport coefficient. For unentangled Rouse chains, that is when $D_i^o \sim N_i^{-1}$ (i=A,B), Ω is expected (eq.4,5) to be independent of molecular weight. This prediction was experimentally confirmed [8] for PEO/PPO blends with $\varphi_{PEO}=0.87$, PPO molecular weight 1025 and various molecular weights of PEO (600,1000,2000).

In this report we are mainly concerned with PEO/PPO blends consisting of PEO chains above their entanglement chain length N_e (Figure 1) and having constant molecular weight of PPO(1025) and $\varphi_{PEO}=0.87$. Figure 1a shows that the mutual diffusion coefficient D decreases with the degree of polymerization N_{PEO} of PEO, which can be attributed to the increase in the static structure factor S(o) [8,9] since, as Fig.1b shows, the mobility term D^o shows no significant decrease with N_{PEO}. From the two expressions for D^o, eq.4 predicts that the slower moving component controls interdiffusion at constant φ, while eq.5 favors the faster moving chain. We will try to compare these predictions with the experimental values of D^o in Fig.1 in the regime $N_{PEO} > N_e$ in which PEO chains are entangled.

For the entangled PEO($\equiv$A) and unentangled PPO($\equiv$B) chains, we have used the following expressions for the tracer diffusion coefficients

$$D_A^o = \frac{4}{15} W_A \frac{N_e}{N_A^2} \quad , \quad D_B^o = \frac{W_B}{N_B} \qquad (7)$$

where W_A, W_B are the microscopic mobilities. For the PEO chains we assumed the validity of the reptation model for molecular weights of PEO larger than four times its entanglement molecular weight. In the calculations, W_A and W_B are set equal to the value of D^o in the unentangled PEO/PPO blend (degree of polymerization less than N_e). The dotted and solid lines in Fig.1b represent the computed value of D^o by means of eq.4,5, respectively. For $N_{PEO} > 4N_e$ the experimental results obviously are in good aggreement with the "fast mode" theory (eq.5). The transport coefficient D^o is determined by the fast diffusing PPO chains in the entangled PEO matrix. In favor of the "fast mode" theory has been also a recent forward recoil spectrometry study [10] of mutual diffusion in the miscible polymer blend of deuterated polystyrene (d-PS)/poly(xylenyl ether) (PXE) at $\varphi_{d-PS}=0.55$ in which interdiffusion has been investigated as a function of the faster moving d-PS chains while keeping the degree of polymerization of PXE constant. Both experiments, however, have been

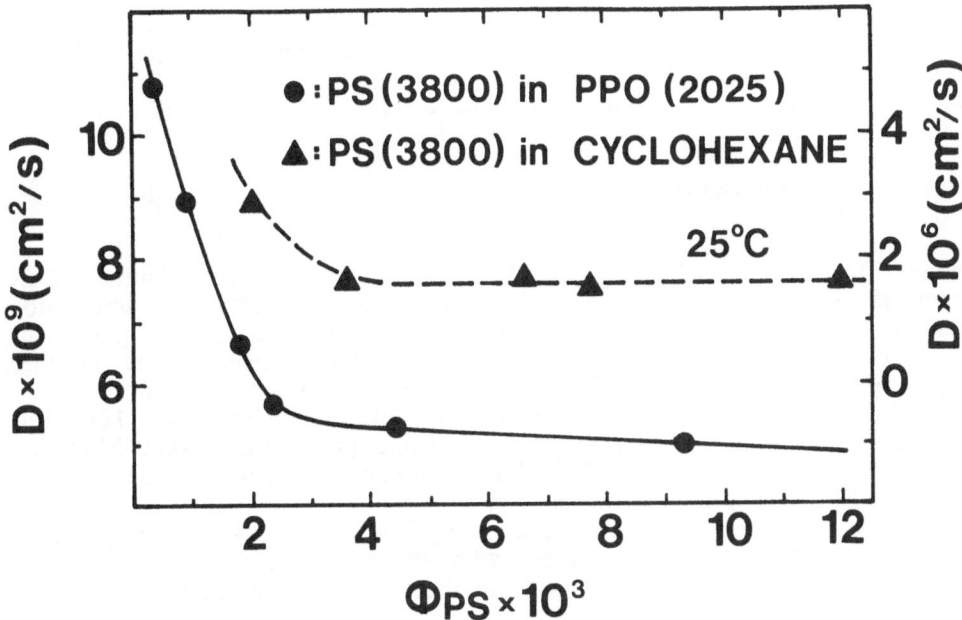

Figure 2. Measured diffusion coefficients D of PS/PPO and PS/cyclohexane versus the volume fraction φ_{PS}.

conducted at constant blend composition and can not comment therefore on the concentration dependence predicted by eq.5. A concentration dependent study in entangled PEO/PPO blends is however impossible since the phase diagram is shifted towards high temperatures (>120 ºC) at which PEO decomposes. On the other hand, a thorough composition dependent study has been performed for an unentangled PEO/PPO compatible blend [11].

4. Dilute polystyrene/poly(propylene oxide) mixtures

The diffusion coefficient in polymer mixtures at extreme compositions ($\varphi<0.01$) is not affected by the additional concentration dependence arising from $T_g(\varphi)$. The main objective is the evaluation of the concentration dependence of the mutual diffusion D and the tracer diffusion $D^0(PS)$ at sufficiently dilute PS compositions. We report preliminary results for the PS(3800)/PPO(2025) mixture at different volume fractions of PS in the range from $4\cdot10^{-4}$ to $93\cdot10^{-4}$ at 25 ºC. Owing to the relatively large difference between the refractive indices of PS and PPO it was possible to achieve very low concentrations.

Figure 2 shows the variation of D with φ_{PS} at 25 ºC; a weak and a remarkably strong composition dependence at very low φ's is clearly depicted. At the same φ range the static structure factor (Fig.3) exhibits a monotonic increase with PS concentration and yields a slightly positive second virial coefficient $A_2=7.1\cdot10^{-4}$ cm^3 mol g^{-2}. For polymer–solvent systems, the mutual diffusion D in dilute solutions is usually represented by

$$D(c) = D_0(1+kc) \tag{8}$$

where D_0 is the diffusion coefficient in the zero–concentration limit, and the concentration independent constant k represents hydrodynamic interactions among the various units. The sign of k depends on the solvent quality and is negative for θ-solvents. For the two highest concentrations in Fig.2 $D_0=5.5\cdot10^{-9}$ cm^2/s and k=-9 cm^3/g. For comparison, $D_0=1.9\cdot10^{-6}$ cm^2/s and k=-3.9 cm^3/g for PS(4000) in cyclohexane at 34.5 °C [12]. This difference in the k values may be due to the polymeric solvent PPO. The much slower diffusion D for the PS/PPO mixture is mainly attributed to the solvent (PPO) viscosity.

The novel feature of the data in Fig.2 is the significant change of the slope k for the PS/PPO mixtures with PS volume fractions smaller than about $2\cdot10^{-3}$. Despite the abundance in literature of diffusion data in dilute

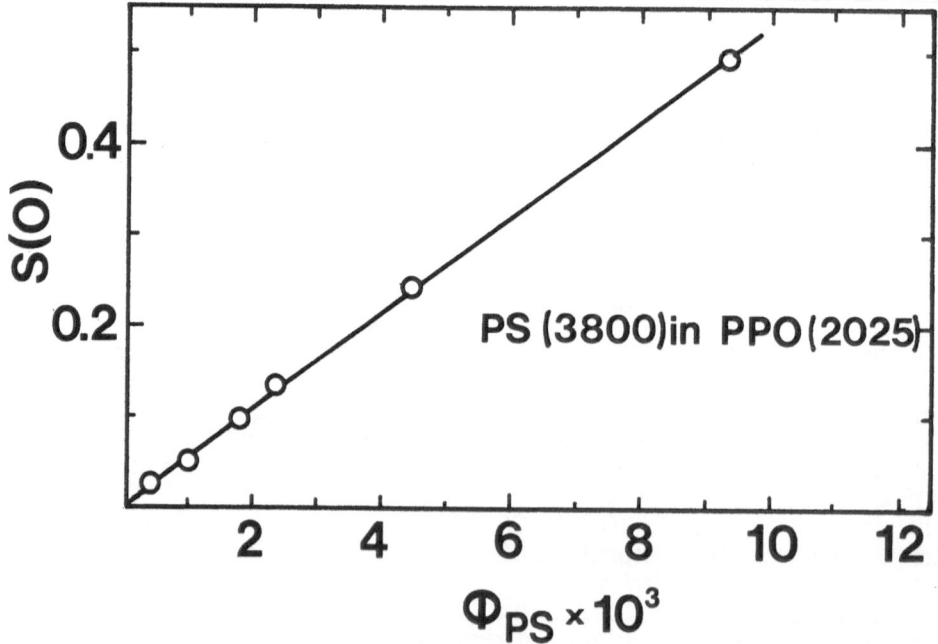

Figure 3. Variation of the structure factor S(o) with the volume fraction at 25 °C.

solutions the composition dependence of D shown in Fig.2 is being reported for the first time to our knowledge. To verify whether this behaviour is a solvent specific feature, we have measured the mutual diffusion D in PS(3800)/cyclohexane solutions at a scattering angle of 45°. Due to the much larger diffusion constants involved, the characteristic times fall in the upper frequency limit of the photon correlator. In addition, the scattering intensity arising from concentration fluctuations in the limit $\varphi \to 0$ is very low and therefore the accumulation time extends to a few hours. These are probably some reasons why such measurements were precluded so far. Figure 2 shows clearly an enhanced diffusivity below $\varphi \sim 2 \cdot 10^{-3}$ for the PS(3800)/cyclohexane dilute solutions at 25 $^\circ$C in close similarity with the data for PS(3800)/PPO(2025) dilute polymer mixtures. These preliminary experimental data make questionable the extrapolation of the translational diffusion coefficients at finite concentrations to obtain D_0 (eq.8) for size determination.

The measured diffusion coefficient at extremely dilute conditions is probably the tracer diffusion $D^o(PS)$ of the PS chain. Regrettably, independent $D^o(PS)$ measurements in the present mixture are not known for a direct comparison. However, it is known for few systems that self diffusion is always reduced with increasing concentration of the diffusant molecule [13]. In particular for the PEO-H_2O dilute solutions, $D^o(PEO)$ measured by pulsed-gradient-NMR technique follows the scaling law $D^o \sim c^{-0.35}$ [14]. A similar presentation of our data below $\varphi \sim 2.10^{-3}$ yields $D \sim c^{-0.27}$. Finally, the change in the rate of decrease in D (Fig.2) might also be interpreted as being caused by different amounts of free-draining versus non draining interactions; D should decrease as N^{-a} with a=1 for ther former and a<1 for the latter case. To provide a unique explanation of the concentration dependence of D, knowledge of the tracer diffusion D^o is needed.

Acknowledgements

The financial support of the Research Center of Crete is gratefully acknowledged.

References

1. Schaefer, D.W. and Han, C.C. (1985) "Quasielastic Light Scattering from Dilute and Semidilute Polymer Solutions", in R. Pecora (ed.), Dynamic Light Scattering, Plenum Press, New York, pp. 181-243.

2. Ngai, K.L., Mashimo, S. and Fytas, G. (1988), Macromolecules 21, 3030.

3. Brochard, F. and de Gennes, P.G. (1983), Physica A 118, 289.

4. Murschall, U., Fischer, E.W., Herkt-Maetzky, Ch. and Fytas, G. (1986), J. Polym. Sci. Polym. Letts. 24, 191.

5. Binder, K. (1983), J. Chem. Phys. 79, 6387.

6. Kramer, E.J., Green, P. and Palmstrom, C.J. (1984), Polymer 25, 473.

7. Sillescu, H. (1984), Makromol. Chem. Rapid Commun. 5, 519.

8. Kanetakis, J. and Fytas, G. (1987), J. Chem. Phys. 87, 5048.

9. Brereton, M.G., Fischer, E.W., Fytas, G. and Murschall, U. (1987), J. Chem. Phys. 86, 5174.

10. Composto, R.J., Kramer, E.J. and White, D.M. (1988), Macromolecules 21, 2580.

11. Kanetakis, J. and Fytas, G. submitted to Macromolecules.

12. Huber, K., Bantle, S., Lutz, P. and Burchard, W. (1985), Macromolecules 18, 1461.

13. Akcasu, A.Z. (1981), Polymer 22, 1169.

14. Brown, W. (1984), Polymer 25, 680.

Dynamics of DNA Double Helices

Dietmar PORSCHKE
Max-Planck-Institut für biophysikalische Chemie
P.O. Box 2841
D-3400 Göttingen, FRG

ABSTRACT. First, a short survey is given on the time scale of various 'elementary' processes observed in nucleic acids. Then, the 'self-organisation' of long viral DNA molecules from voluminous, disordered wormlike coils to compact well ordered toroids is discussed with respect to its dynamics. Two limit cases of the reaction mechanism for the spermine induced toroid formation have been demonstrated by stopped flow and electric field jump measurements:

1) At low spermine concentrations the rate of toroid formation is limited by the rate of spermine association to the DNA chain. The observed induction periods and the overall reaction rates indicate that spermine molecules move along the DNA with a rate of ~ 200 residues/s.

2) In the limit of high spermine concentrations winding of DNA strands into the toroidal form is reflected by a spectrum of time constants ranging from $25\mu s$ to $2ms$. This high rate is quite remarkable and suggests the analogy of a spring, which is kept under tension by electrostatic repulsion and which collapses immediately, when the repulsion is reduced by ligand binding.

Bending of DNA double helices has been analyzed in further detail by electrooptical measurements on restriction fragments. Bending amplitudes and bending time constants have been assigned as a function of the DNA chain length. An orientation function for weakly bent rods has been developed, which serves to explain the stationary dichroism and bending amplitudes. Most of the data are consistent with simple thermal bending, but deviations under special conditions suggest the existence of inherent curvature.

Finally, a simple model of toroid formation is proposed.

Introduction

Because of their biological function, DNA double helices are among the most intensively studied macromolecules. During the 35 years since Watson and Crick found their fundamental model, virtually all available techniques have been used to learn about these molecules and an enormous amount of information has been accumulated. In most of these investigations, the structure of DNA has been the main subject,[1] whereas the dynamics did not receive as

Th. Dorfmüller (ed.), Reactive and Flexible Molecules in Liquids, 257–268.
© *1989 by Kluwer Academic Publishers.*

much attention. Nevertheless, the literature on DNA dynamics is quite extensive and thus it is difficult to give a comprehensive review. In the present contribution a short survey will be given on various approaches to DNA dynamics, before some results obtained by general kinetic and special electrooptical procedures are described in more detail.

Standard figures showing DNA double helices suggest that these helices are quite rigid structures, which cannot be deformed easily. This impression is correct to a rather large extent, such that relatively short DNA fragments may be described as rigid rods at a sufficient accuracy. However, due to the enormous length of natural DNA molecules small deformations at the local structure level add up to large deformations of the long range structure with a very high probability.[2] Attempts to characterize the fluctuations of DNA double helices may be classified in several ways. First of all, information may be collected on the mean long range structure, which may be described by persistence lengths. This should be complemented by the corresponding information on the dynamics of the long range structure. The problem may also be approached from the opposite direction by an analysis of fluctuations of the local structure, which may then be used to construct models on the long range structure.

Certainly the most powerful method for investigations of local structures and of their fluctuations in solution is NMR spectroscopy. A major advantage of this method is that the properties of individual residues can be studied selectively. Various processes have been detected for double helical nucleic acids. For example, measurements of spin-spin and spin-lattice relaxation[3] suggest 'large amplitude out of plane motions' in the time range of 1 to $100ns$ and 'large amplitude local torsional motions' with time constants of 0.1 to 1.5 ns. Compared to these fluctuations the motions required, for example, to open internal base pairs for hydrogen exchange[4] take much more time and are associated with time constants of about 1 s.

Another method, which has been applied frequently, uses fluorescence depolarisation of intercalated dyes – usually ethidium – measured at ns-time resolution. Because intercalation distorts the native structures, the results must be interpreted with caution. The fluorescence anisotropy decays in a rather broad time range from $0.1ns$ to $150ns$, which has been attributed to 'wobble' of the chromophore and torsional motions of the DNA.[5] A corresponding method using triplet anisotropy decay[6] has been applied to study motions up to a longer time scale including slow overall rotational diffusion. Conclusions on local motions of DNA residues with time constants around $1ns$ are supported also by spin label measurements.[7] Finally, dynamic light scattering has been used to obtain information on the internal mobility of DNA over a relatively broad time range. However, it is not a trivial task to assign this information to specific modes of DNA internal mobility.[8,9]

The brief survey given above refers to small amplitude motions of DNA around its equilibrium structure. Motions of increasing amplitude may result in 'conformation changes' of the original structure to a different one. Since these conformation changes are particularly important for the biological function, much information has been accumulated on their thermodynamics and kinetics. Most conformation changes of nucleic acids are relatively fast reactions.[10] The elementary steps of base stacking and of hydrogen bonding are observed in the ns time range. However, due to cooperative coupling the time required to dissociate long double helices into single strands can be extremely long. A short summary of various reaction steps encountered in nucleic acids on various levels of organization is given in Tab. I.

Stacking
monomer base stacking [11]
CpC [12]
poly (C) [12]
poly (A) poly (dA) [13]

Hydrogen Bonding
AU pairs [14] (in unpolar solvents)

Ion Binding
monovalent ions [15]
Mg^{2+} inner spheres [16]
motion of spermine along DNA [17]

Base Pairing
fraying of AU ends [18]
sliding of AU chains [19]
8 AU pairs [10]
3 GC pairs [a]
A$_3$ GCU$_3$ [20]
6 GC pairs [a]
8 GC pairs [a]

DNA Modulation
twisting [3]
bending 95 bp [21]
bending 256 bp [21]
toroid formation ($5*10^4$ bp) [17]

Order Transitions
B - A transition [b]
B - Z transition [22]

(s)
10^{-9}
10^{-6}
10^{-3}
1
10^3
10^6
hour →
day →
month →

Table I. Logarithmic time scale of nucleic acid processes. The processes are characterized by their relaxation time constants τ; in the case of bimolecular reactions τ is given at zero concentration ($\tau \equiv 1/k_D$, corresponding to the lifetime of e.g. double helices, 20°C).

a average values from ref. 20

b not measured, but estimated to be between $1\mu s$ and $1ms$ according to unpublished data (Porschke).

Selforganization of long DNA molecules in toroids

Under standard conditions long DNA molecules – e.g. of viral origin – are found in solution as wormlike coils. This is the form expected for long polymers with a limited bending stiffness due to the very high number of different bent forms, which are favored by a large entropy term. When multivalent ions like spermine, spermidine or $[Co(NH_3)_6]^{3+}$ are added to long DNA molecules, however, the wormlike coils are converted to a completely different structure.[23-27] By electron microscopy of many DNA samples in different laboratories [24,25] it has been demonstrated that under these conditions the extremely long DNA threads are wrapped into toroidal forms as if on a spool. This spontaneous transition from a disordered, voluminous state to a well-ordered, compact one raises several questions. From a thermodynamic point of view it is not clear, how the large configurational entropy, the electrostatic repulsion between negatively charged phosphates and the bending energy, which are all against this transition, can be overcompensated. Obviously binding of positively charged ions to the DNA reduces electrostatic repulsion, but this alone does not explain the transition and the attractive interactions remain to be identified.

Another set of interesting problems arises with respect to the dynamics: How fast is this transition, what is the mechanism and how can entanglements of the extremely long DNA thread be avoided? Some answers to the latter set of questions were derived recently from kinetic investigations.[28,29] It has been shown that there are two limit cases of the reaction mechanism:

1. At low concentrations of added ligand, the rate of toroid formation is determined by the rate of ligand binding. This has been demonstrated by stopped flow experiments,[28] which show a reaction in the ms time range with a characteristic induction period. Since the reaction has been followed by measurements of the light scattering intensity, binding of ligands cannot be detected directly, but only indirectly via the formation of toroids. It is known from equilibrium measurements that the toroids are formed in a cooperative reaction as soon as the degree of ligand binding exceeds a theshold value.[27,28] In the case of spermine ions, which have been mainly used for stopped flow experiments, the threshold is observed at 86% saturation of the binding sites on the DNA double helix.[28] Upon mixing of DNA with spermine ions, the DNA binding sites are occupied in the first part of the reaction without any formation of toroids and thus without any change of the light scattering intensity. By this special coupling of a ligand binding reaction and a conformation change, the induction period observed in the stopped flow experiments can be explained without problems. A more quantitative analysis of the coupled reaction, however, proves to be rather complex. Each spermine molecule occupies more than a single lattice site, which leads to 'exclusion' of binding sites.[30] At high degrees of binding, exclusion of sites results in a very complex binding kinetics, which cannot be described anymore by analytical equations. Monte Carlo simulations [28,31] demonstrate that at high degrees of ligand binding the rate of further ligand binding is strongly dependent on the mobility of ligands along the lattice.

Comparison of experimental and simulated data[28] indicates that spermine molecules move by one nucleotide residue with a rate constant of $\sim 200~s^{-1}$.

2. The second limit case is observed at high ligand concentrations. Under these conditions ligand binding is fast compared to the 'conformation change' of the DNA chain. Due to the limited time resolution, this part of the reaction could not be studied anymore by stopped flow techniques, but had to be analyzed by the electric field jump technique.[29] Electric field pulses are used to induce dissociation of spermine from the DNA – presumably by a dissociation field effect. The dissociation of spermine molecules leads to decondensation of toroidal DNA, which is indicated by a strong decrease of the light scattering intensity. Upon termination of the electric field pulse the opposite reaction is observed: spermine binding to DNA is followed by toroid formation. Since the time resolution of the electric field jump technique extends to the ns time range, the time course of toroid formation can be studied up to very high ligand concentrations. In this limit a spectrum of relaxation time constants [28,29] is observed, which extends from $25\mu s$ to $2ms$.

Apparently this spectrum of time constants reflects the spectrum of internal mobilities of the long wormlike DNA coil. It should be expected that these segment motions are important for the transition of the wormlike coil to the well-ordered toroidal structure. However, it is quite surprising that this transition apparently is not limited by anything else but segment motion. The transition seems to be analogous to that of a spring, which is kept under tension by electrostatic repulsion, and a sudden release of this tension upon switching off the repulsion leads to an immediate collaps to a compact torus.

This remarkable reaction also has some biological implications. Long DNA molecules are exposed to many environmental factors, which may lead to destruction and loss of the genetic information. Obviously these molecules have to be protected. However, protection by some standard packing reaction implies that the genetic information is not as easily and quickly available as required for survival in a changing environment. Thus, a special form of packing has to be found, which is compatible with both demands: protection and fast accessibility of the genetic information. Packing of long DNA molecules into toroids appears to be ideal for this purpose. Fast packing or unpacking of the DNA may be induced by small changes of ligand concentration, which may be adjusted by simple regulation mechanisms.

Modulation of DNA structure by ligand binding

The peculiar transition of long natural DNA molecules from wormlike coils to toroids in the presence of multivalent counterions suggests that binding of these ions induces some change of the DNA structure. Because the circular dichroism of the DNA remains unchanged during the transition, it has been concluded that the local structure corresponds to the classical B-double helix in both forms. Probably the change of the structure is too subtle to be detected by circular dichroism measurements and can only be detected by more sensitive methods at the level of the long range structure.

Small changes of the long range structure can be analyzed at an unusually high sensitivity by measurements of rotational diffusion. The rotational time constants of rodlike molecules, for example, increase with the third power of their length[32] and can be measured very accurately by electrooptical techniques.[33] Due to the high sensitivity and accuracy it is not a problem to detect e.g. insertion of a single intercalator in a DNA double helix of 100 base pairs by measurements of dichroism decay time constants.[34]

Table II. Dichroism decay time constants of DNA restriction fragments as a function of the spermine concentration (at 20°C in 1m NaCl, 1mM Na-cacodylate pH 7.0, 200μM EDTA).

Spermine (μM)	0.0	1.5	3.0	4.0
69 bp fragment	317	298	289	261
84 bp fragment	525	460	454	423

This technique has been used to study changes of the DNA long range structure upon addition of multivalent counterions.[21] As shown by some examples given in Table II the rotational relaxation time constants decrease considerably upon addition of these ions. Since it has been shown that the local structure is hardly affected by these ions, the decrease of the time constants must be attributed to changes of the long range structure. Apparently ion binding induces bending of the double helices.

Bending of polymers like DNA is usually described in terms of persistence lengths. In the present case, the weakly bending rod model[35] developed by Hearst provides an accurate quantitative description of the data. The decrease of the rotation time constants observed for a set of 6 DNA restriction fragments of different chain lengths in the range from 43 to 256 base pairs is reflected by a clear decrease of the persistence length (cf. Table III).

Table III. Persistence lengths and Stokes' diameters in Å of DNA evaluated from dichroism decay time constants according to the weakly bending rod model of Hearst[35] at different spermine concentrations (20°C, buffer as in Table II).

Spermine (μM)	Stokes diameter	Persistence length
0.0	26.3	700
1.5	25.3	380
3.0	25.4	330
4.0	24.3	300

The weakly bending rod model as well as other corresponding models imply that bending results from thermal fluctuations of an inherently straight structure. For a rather long time it has been believed that this model is correct for DNA double helices. However, in recent years evidence has been accumulated that DNA double helices are not always inherently straight but are inherently curved at certain base sequences. Such inherent curvature is, of course, also reflected by reduced persistence lengths and cannot be easily distinguished from thermal bending. Because of the potential biological importance of inherent curvature, methods should be developed which are appropriate to characterize the different forms of bending in solution quantitatively.

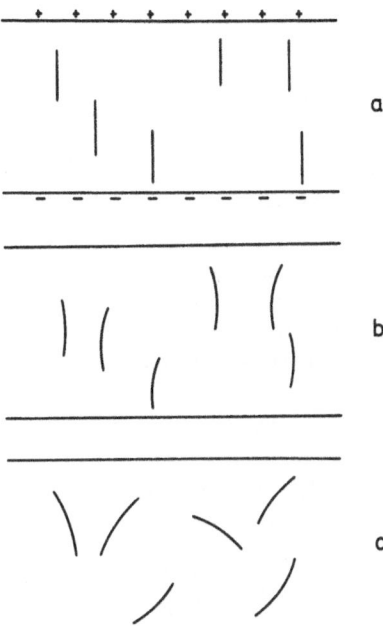

Figure 1. Scheme of the states of DNA molecules after termination of electric field pulses. After orientation and stretching of DNA under the electric field (a), the transition to the distribution of bent DNA's (a → b) is faster than overall distribution in space (b → c).

Bending and stretching of double helices

Electrooptical experiments can be used to obtain another information on bending, because electric fields do not only induce alignment of molecules parallel to the field vector but also stretching.[21] From the scheme shown in Fig. 1 it is apparent that upon termination of an electric field pulse the reverse process of bending should appear on the time scale clearly separated from overall tumbling. This expectation has been verified by measurements of the dichroism decay for DNA fragments with chain lengths around the persistence length. For example, bending of DNA fragments with 179 bp is associated with a bending amplitude of 30% of the total dichroism decay amplitude and a time constant of 180ns.

Field induced stretching has not been commonly accepted as a major contribution to electric field effects of biopolymers and thus an attempt to describe this process quantitatively has not been given until recently.[36] The description has been based on a relatively simple arc model, which is a reasonable approximation for dichroism amplitudes of DNA fragments with chain lengths not much beyond the persistence length.

The main steps of the evaluation may be summarized as follows:

First the dichroism is calculated for circular arcs,[21] which are aligned with their chord parallel to the field vector

$$\frac{\Delta\varepsilon^\infty}{\varepsilon} = \frac{3}{2}\left\{3\cdot\frac{1}{\varphi}\left[cos^2\kappa\left(\frac{1}{2}sin\varphi\cdot cos\varphi+\frac{\varphi}{2}\right)+\frac{1}{2}(1-cos^2\kappa)\left(-\frac{1}{2}sin\varphi\cdot cos\varphi+\frac{\varphi}{2}\right)\right]-1\right\}$$

where 2φ is the total bending angle, and κ is the angle of the transition dipole moments of the base pairs with respect to the helix axis.

Then, the 'electric energy' associated wih bending of DNA rods in the presence of an electric field is evaluated. Since DNA molecules are polyelectrolytes, an evaluation based on a proper molecular basis is very difficult. The same difficulty arises in the quantitative description of field induced alignment, which has always been described by simple dipole models. Thus, it is appropriate to describe stretching according to the same model by excess charges

$$q_E = \frac{\mu}{L}$$

at the ends of the molecules of length L, which corresponds to the dipole moment μ deduced from the alignment. In this case the electric free energy of bending is given by

$$\Delta G_e = L\cdot\left(1-\frac{sin\varphi}{\varphi}\right)\cdot q_E\cdot E$$

which has to be provided in addition to the mechanical energy of bending

$$\Delta G_m = \frac{\sigma}{2L}(2\varphi)^2$$

σ is the mechanical bending constant, which is related to the persistence length p according to

$$\sigma = kT\cdot p$$

The total energy required for bending is given by

$$\Delta G_b = \Delta G_m + \Delta G_e$$

At any given angle of the DNA axis or chord with respect to the field vector there is a distribution of bent molecules, which is associated with an average dichroism

$$\overline{\frac{\Delta\varepsilon}{\varepsilon}}(\vartheta = \frac{\int\limits_{-\infty}^{\infty}\int\limits_{-\infty}^{\infty} exp(-\Delta G_b^\vartheta/kT)\cdot\frac{\Delta\varepsilon}{\varepsilon}(\Theta)\cdot d\Theta_1 d\Theta_2}{\int\limits_{-\infty}^{\infty}\int\limits_{-\infty}^{\infty} exp(-\Delta G_b^\vartheta/kT)d\Theta_1 d\Theta_2}$$

Finally, these average values of the dichroism have to be integrated over the whole space by

$$\frac{\Delta\varepsilon}{\varepsilon} = \frac{\int\limits_0^\pi e^{-U/kT}\cdot\overline{\frac{\Delta\varepsilon}{\varepsilon}}(\vartheta)\cdot\frac{3cos^2\vartheta-1}{2}\cdot 2\pi sin\vartheta d\vartheta}{\int\limits_0^\pi e^{-U/kT}\cdot 2\pi sin\vartheta d\vartheta}$$

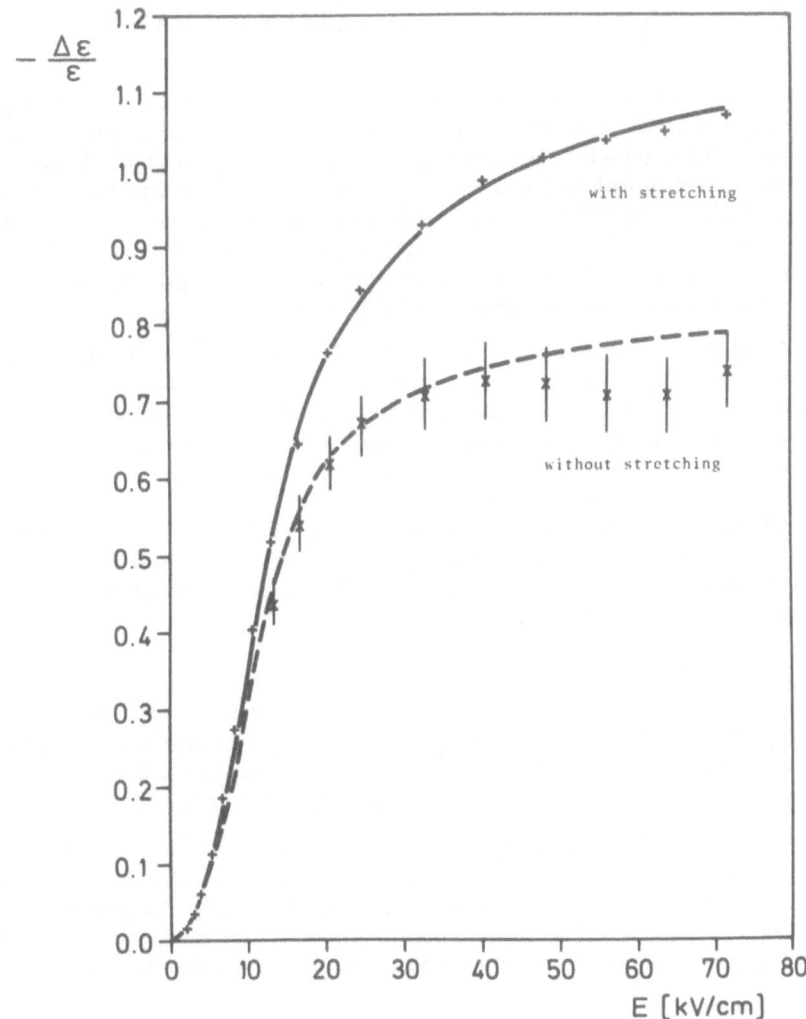

Figure 2. Reduced dichroism $\Delta\varepsilon/\varepsilon$ as a function of the electric field strength E for a 179 bp DNA in 1mM NaCl, 1mM Na-cacodylate pH 7.0, 100 μM Mg Cl$_2$ at 20°C. (+) measured stationary dichroism; (x) partial dichroism without contribution from stretching, derived from the amplitude of the slow dichroism decay; — fit of the measured stationary dichroism by the orientation function for weakly bent rods <u>with</u> stretching: polarisability $\alpha = 2.6 \cdot 10^{-32} Cm^2V^{-1}$, saturation field strength E$_0$ = 8.72 kV/cm; angle of transition dipole moments with respect to the long DNA axis 76.1°(evaluated with a persistence length of 103bp and a rise per base pair 3.4 Å); - - - calculated partial dichroism without stretching.

These components lead to an orientation function with the following variables: 1) the dipole moment μ, 2) the angle of the transition dipole moments with respect to the helix axis κ and 3) the persistence length p.

Since p is known from independent measurements of the rotational diffusion,[21] only two remaining parameters have to be fitted to the stationary values of the dichroism measured as a function of the field strength. An example given in Fig. 2 shows that both the stationary values of the dichroism with and without field induced stretching are represented by the model at a high accuracy. The bending amplitudes, which correspond to the differences between the dichroisms with and without stretching, are represented accurately by the model both with respect to their field strength and their chain length dependences, that have been observed in buffers containing multivalent ions.

It seems that these data support the conclusion that thermal bending is the only important mode of DNA bending. However, this conclusion is not yet justified, because the influence of inherent curvature may be hidden as a contribution to the persistence length p, which is then transmitted by the model given above into bending amplitudes. It should also be mentioned that the experimental data obtained in a buffer of low salt concentration are not consistent with the predictions of the model. This may be due to several reasons, among which inherent curvature is the most interesting possibility. Further experiments are required – in particular on inherently curved DNA fragments, which seem to be associated with quite unusual properties.[37] For a quantitative assignment of these properties bending and stretching processes certainly have to be considered.

Models of toroid formation

The remarkable transition of long DNA molecules from voluminous, disordered wormlike coils to compact well ordered toroids has not be described yet by a generally accepted model. Manning[38] has attributed the transition to a mechanical instability of DNA segments on the basis of "Euler buckling". However, this model still is very much under discussion.

The most attractive model for the explanation of toroid formation and its dynamics appears to be a spring model, which implies the existence of a DNA superhelix in solution. Such superhelices could be contracted to toroids at very high rates and entanglement of DNA segments could be avoided without much problem. Although superhelix structures of DNA are increasingly popular, it has not been possible yet, to give a definite proof for their existence.

In the absence of such a proof another model of toroid formation has to be considered. This model simply asks for a mode of DNA packing, which provides an optimum of interaction energy between DNA strands at a minimal expense of other energies. It may be assumed that there is given free energy of DNA contact per segment length. The DNA segments may come into contact with each other by sharp bends with parallel alignment of straight DNA segments or by smooth bending with circular folding of the DNA chain into toroids. For simplicity, the following discussion is restricted to these limit forms (cf. Fig. 3). In the first case, free energy of unstacking has to be provided at each sharp bent, whereas in the second case energy is required for smooth bending, which may be calculated from the experimental persistence length. A comparison of these energies reveals that smooth bending up to a limited degree of curvature as found in toroids is more favourable than sharp bends for parallel alignment of linear DNA segments (D. Porschke, in preparation). Thus, at least a driving force for the formation of toroids is available without need to postulate the existence of a special DNA structure in solution as a precursor.

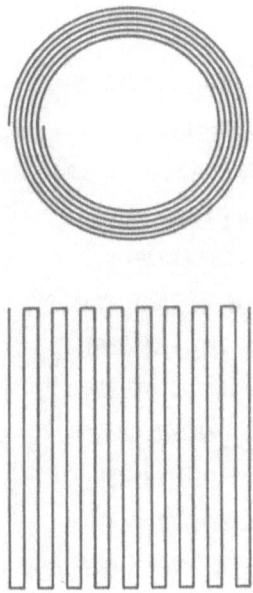

Figure 3. Scheme of potential arrangements of DNA chain for maximal segment contacts (cf. text).

If the main assumptions of this model are correct, curved DNA segments are not required to explain the thermodynamics of toroid formation. However, curved DNA segments could still be useful to support nucleation of toroids and to explain – at least to some extent – the high rate of toroid formation. According to sequence analysis curved DNA segments should exist in λ-DNA, which has been mainly used for the kinetic experiments described above. It is very likely that corresponding segments occur in other natural DNA samples. The curved segments should favour initial molecular contacts for a toroidal organisation and thus reduce the nucleation barrier.

References

1. W. Saenger, Principles of Nucleic Acid Structure, Springer, Berlin (1984)

2. V.A. Bloomfield, D. Crothers and I. Tinoco, jr., Physical Chemistry of Nucleic Acids, Harper & Row, NewYork (1974)

3. P.A. Mirau, R.W. Behling and D.R. Kearns, Biochemistry 24, 6200 - 6211 (1985)

4. D.R. Kearns, CRC Crit. Rev. Biochem. 15, 237 - 290 (1984)

5. J.H. Shibata, B.S. Fujimoto and J.M. Schurr, Biopolymers 24, 1909 - 1930 (1985)

6. M. Hogan, J. Le Grange and B. Austin, Nature 304, 752 - 754 (1983)

268

7. S.C. Kao and A.M. Bobst, Biochemistry 24, 5465-5469 (1985)

8. T. Dorfmüller et al., this volume

9. J. Langowski, Biophys. Chem. 27. 263 - 271 (1987)

10 D. Porschke, Mol. Biol., Biochem. Biophys. 24, 191 - 218 (1977)

11. D. Porschke and F. Eggers, Eur. J. Biochem. 26, 490 - 498 (1972)

12. D. Porschke, Biochemistry 15, 1495 - 1499 (1976)

13. D. Porschke, Biopolymers 17, 315 - 323 (1978)

14. G.G. Hammes and A.C. Park, J. Amer. Chem. Soc. 90, 4151-4156 (1968)

15. D. Porschke, Biophys. Chem. 22, 237 - 247 (1985)

16. D. Porschke, Nucleic Acids Res. 6, 883 - 898 (1979)

17. D. Porschke, Biochemistry 23, 4821 - 4828 (1984)

18. D. Porschke, Biophys. Chem. 2, 97 - 101 (1974)

19. D. Porschke, Biophys. Chem. 72, 83 - 96 (1974)

20. D. Porschke, O.C. Uhlenbeck and F.H. Martin, Biopolymers 12, 1313 - 1335 (1973)

21. D. Porschke, J. Biomol. Structure and Dynamics 4, 373 - 389 (1986)

22. F.M. Pohl and T.M. Jovin, J. Mol. Biol. 67, 375 - 396 (1972)

23. L.C. Gosule and J.A. Schellman, J. Mol. Biol. 121, 311 - 326 (1978)

24. D.K. Chattoray, L.C. Gosule and J.A. Schellman, J. Mol. Biol. 121, 327 - 337 (1978)

25. K.A. Marx and G.R. Ruben, Nucleic Acids Res. 11, 1839 - 1854 (1983)

26. J. Widom and R.L. Baldwin, J. Mol. Biol. 144, 431 - 453 (1980)

27. R.N. Wilson and V.A. Bloomfield, Biochemistry 18, 2192 - 2196 (1979)

28. D. Porschke, Biochemistry 23, 4821 - 4828 (1984)

29. D. Porschke, Biopolymers 24, 1981 - 1993 (1985)

30. J.D. McGhee and P.H. von Hippel, J. Mol. Biol. 86, 469 - 489 (1974)

31. I.R. Epstein, Biopolymers 18, 2037 - 2050 (1979)

32. S. Diekmann, W. Hillen, B. Morgeneyer, R.D. Wells and D. Porschke, Biophys. Chem. 15, 263 - 270 (1982)

33. S. Broersma, J. Chem. Phys. 32 1626 - 1631 (1960)

34. D. Porschke, N. Geissler and W. Hillen, Nucleic Acids Res. 10, 3791 - 3802 (1982)

35. J.E. Hearst, J. Chem. Phys. 38, 1062 - 1065 and personal communication

36. D. Porschke, Biopolymers, in press

37. S. Diekmann and D. Porschke, Biophys. Chem. 26, 207 - 216 (1987)

38. G. Manning, Cell Biophysics 7, 57 - 89 (1985)

FREQUENCY SPECTRUM OF F–ACTIN AND F–ACTIN/FILAMIN COMPLEXES AS STUDIED BY PHOTON CORRELATION SPECTROSCOPY

J. SEILS *and* TH. DORFMÜLLER
Faculty of Chemistry
University of Bielefeld
P.O. Box 8640
4800 Bielefeld, FRG

ABSTRACT. The frequency spectrum of entangled rigid rods with an exponential length distribution was calculated and compared to a spectrum obtained by an inverse Laplace transform of the first order autocorrelation function from dynamic light scattering data for F–actin, a polymerized muscle protein usually considered to be rather rigid. The experimental and model frequency distributions disagree substantially. In view of this discrepancy the effect of factors like flexibility, deviation from an exponential length distribution, interference with chemical reactions, and translational–rotational coupling as possible explanations were calculated. By comparison of the data to the model calculations or qualitative estimates all these could be excluded. Therefore we conclude that the frequency spectrum of F–actin is due to the formation of an infinite network structure. Further support to this interpretation is provided by the frequency spectra of actin/ filamin networks, that are very similar in shape, but show a shift of the main peak of the spectra to lower frequencies with increasing cross–link density. High frequency modes, which result from an increase of the elasticity constants of the network with progressive cross–link density are obviously more efficiently damped than the low ones, leading to a predominance of the slow motions in the spectra.

1. Introduction

Many processes of living materials are guaranteed by the flexibility and the reactivity of a number of molecules, among which the muscle–protein G–actin is one of the most frequent ones [1]. Intensive studies on this molecule during the last years have shown its high reactivity towards itself, a number of modulating proteins, ions and other biologically important molecules [2]. The self–reactivity of G–actin for instance causes the formation of rigid rodlike filaments, the F–actin, which is the basis of the constitution and the motility of the cell plasma [3]. Reactions of F–actin with modulating proteins lead to modifications such as shortening, bundling and cross–linking of the filaments. The latter, lead to rather flexible network structures [4,5,6]. In the past the attention has mainly been directed towards an elucidation of the formed structures with electron microscopy techniques [5,7]. Information about the flexibility of different structures were deduced from the shape of the molecules, as derived from static measurements in solid state. Determination of viscoelastic properties by viscosity and mechanical measurements in solution completed these studies [8,9]. A more direct approach to the flexibility of macromolecules in solution is provided by the technique of photon correlation spectroscopy [10]. A number of difficulties concerning the interpretation of the data from such measurements have

269

Th. Dorfmüller (ed.), Reactive and Flexible Molecules in Liquids, 269–283.

been overcome in the last years. Especially a priori assumptions about the number of relaxation times or the form of the frequency distribution are no longer necessary, since powerful algorithms for the inverse Laplace transforms are now available [11,12,13,14]. The resulting frequency spectra from such analyzes of the correlation functions give a detailed description of the dynamics of macromolecular systems. On the other hand, correlation functions or frequency distributions have been calculated for a number of models including rigid [15] and flexible [16] single particles, highly cross–linked gels [17] and entangled molecules [18]. Therefore the comparison of experimental frequency spectra with those models enables us to acquire a better insight and understanding of even complex biological systems. In order to achieve this we must first delimit the dominant factors which will enable us to describe the basic features observed in F–actin. To make things as simple as possible we begin with the single particle properties and then introduce polydispersity and particle interaction. After this some refinements must be made concerning internal flexibility, coupling of motions and interference with chemical reactions. Finally, after comparison with experimental spectra, we will present some results and a more general discussion on cross–linked structures.

2. Calculation of the frequency spectrum for an appropriate model of F–actin

2.1. SINGLE PARTICLE PROPERTIES

Since the actin filaments have persistence lengths of 6 μm with a mean contour length of 1–6 μm [7], a rigid rod model seems to be appropriate for our calculations and has been used by other groups for this molecule. The frequency distribution for a rigid rod model consists of a number of discrete frequencies Γ_n which, according to Pecora [15], depend only on the translational, D_T, and the rotational diffusion coefficient D_R (eq.1)

$$\sum_{n=0}^{\infty} \Gamma_n = \sum_{i=0}^{\infty} D_T q^2 + i(i+1) D_R \qquad (i = 2n) \qquad (1)$$

The relative amplitudes of these frequencies $B_i(qL)$ are given by eq.2.

$$B_i (qL) = (2i+1) [2/qL \int_0^{qL/2} j_i(x) \, dx]^2 \qquad (2)$$

They are functions of the product of the rod length L and the scattering vector q, as defined by eq. 3:

$$q = \frac{4 \pi n}{\lambda} \sin (\theta/2) \qquad (3)$$

$j_i(x)$ is the spherical Bessel function of order i. The sum in eq. 1 can be calculated since, depending on the value of the product qL, only a finite number of frequencies have significant amplitudes. The translational and rotational diffusion coefficients for rigid rods have been derived by Tirado and de la Torre [19,20] replacing the rod by an assembly of hydrodynamic spheres and taking additional friction for the ends into account. The translational diffusion coefficient at infinite dilution is an appropriate average of the longitudinal $D_{\shortparallel}$ and transversal diffusion coefficient $D_{\perp}$ according to eqs. 4–6.

$$D_{\shortparallel} = \frac{K_B T}{4\pi\eta L} (\ln \rho + \nu_{\shortparallel}) \qquad (4)$$

$$D_{\perp} = \frac{K_B T}{2\pi\eta L} (\ln \rho + \nu_{\perp}) \qquad (5)$$

$$D_0 = (D_{\shortparallel} + D_{\perp})\, 4/9 \qquad (6)$$

K_B is the Boltzmann constant, T the temperature in Kelvin, η the solvent viscosity and $\rho = L/d$, where d is the diameter of the rod. A similar equation can be derived for the rotational diffusion

$$D_R = \frac{3K_B T}{\pi\eta L^3} (\ln \rho + \delta_{\perp}) \qquad (7)$$

$\nu_{\shortparallel}$, $\nu_{\perp}$ and δ are the end effect corrections, that can be expressed as polynomials in ρ

$$\nu_{\perp} = 0.839 + 0.185/\rho + 0.239/\rho^2 \qquad (8)$$

$$\nu_{\shortparallel} = -0.207 + 0.980/\rho - 0.133/\rho^2 \qquad (9)$$

$$\delta_{\perp} = -0.662 + 0.917/\rho - 0.050/\rho^2 \qquad (10)$$

2.2.   POLYDISPERSITY

According to kinetic models for the polymerization reaction of G–actin F–actin should posses an exponential distribution of filament lengths [21]. This result is supported by the reported experimental length distributions from electron micrographs [5,22,]. The light scattering intensity of one species with degree of polymerization p is then proportional to the weight fraction $f_w(p)$ of this molecule, which can be calculated [23] according to eq. 11.

$$f_w(p) = p/<p> f_n(p) \qquad (11)$$

where $<p>$ is the mean degree of polymerization and $f_n(p)$ is the number fraction

$$<p> = 1./(1.-\nu) \tag{12}$$

$$f_n(p) = (1.-\nu) \, \nu^{(p-1.)} \tag{13}$$

ν is the fraction of bonds formed during polymerization.

2.3.   ENTANGLEMENT OF RODS

The large size of the monomer units (l_0 = 2.7nm) [2] and the high degree of polymerization ($<p>$ = 500–2000) [5,22] for actin lead to rather long filaments. For this reason the condition for entanglement of the rods ($cL^3 > 1$) is met at rather low concentrations leading to a hindrance of motions. According to the Doi–Edwards approach [18] the transversal translational diffusion for a probe rod is frozen and the rotation is severely restricted by a cage of neighboring rods to small–angle rotations of the probe. The rigidity of the cage assumed by Doi–Edwards, however, is rather questionable, so that Fixman suggests also to allow for small angle rotations of "caging"–rods [24]. From Brownian simulations of such a model it follows that in the regime of $24 < cL^3 < 120$ the rotational friction coefficient β_0 at infinite dilution must be corrected giving the rotational friction coefficient β at concentration c according to eqs. 14–16.

$$\beta/\beta_0 = 1 + (-\frac{cL^3}{24\,\pi\,R})\,Q(f)\,[\,f\ln\frac{1-f}{f}\,] \tag{14}$$

$$Q(f) = 1 + \pi/2\ cL^3\ d/L\ (1 - 2f)^2 \tag{15}$$

$$R = \frac{\beta_0\ D_{\shortparallel}}{L^2\ K_B\ T} \tag{16}$$

The parameter f in the equations gives the probability that a neighboring rod lies in an interval of a given size. Maximizing $\beta(f)$ with respect to f yields the rotational friction constant at a concentration c_j for a rod with diameter d_i and length L_i. The maximization problem can be solved by a Newton–algorithm. The lower bound of the validity of eqs 14–16 is only due to the limited precision of the calculations and not to fundamental limitations. For lower values of cL^3 it should therefore be possible to extrapolate the results of the Brownian simulations. The presence of different lengths in a polydisperse sample causes some problems in calculating cL^3. This parameter must be evaluated for all rods with length L_i separately, assuming that the overall concentration can be expressed as the sum of all concentrations c_j weighted by the length ratio L_j/L_i of rod j and rod i according to eq.17.

$$cL_i^3 = L_i^3 \sum_{j-1}^{N} c_j\ L_j/L_i \tag{17}$$

The assumption seems to be appropriate, since the concentration of filaments is in the dilute regime.

3.   Details of the calculations

In the previous chapter all equations that are necessary to calculate the frequency spectrum of an exponential length distributed, entangled, rigid rod model have been presented. Calculations have been carried out for different mean lengths of the rods in such a way, that for the lengths and also the frequencies a cutoff was defined.

With respect to the lengths those with weight fraction below a factor of 1×10^{-4} of the total length distribution were neglected. For a rod with a length of L_i a weighted mean frequency was calculated starting with the translational frequency and subsequently adding rotational modes. When these modes did not change the mean frequency by more than 0.1 per mille accumulation for the frequencies was stopped. If cL^3 for this length exceeded 1, the transverse translational diffusion was frozen and all rotational modes were corrected according to eqs. 14–16. The frequencies were then sampled in a histogram of 40 classes, where the mean frequency of one class and its height represent the coordinates of one grid–point of the frequency spectrum. All spectra were calculated in a concentration region where the theory of Fixman is applicable ($cL^3 < 120$) and which is easily accessible experimentally.

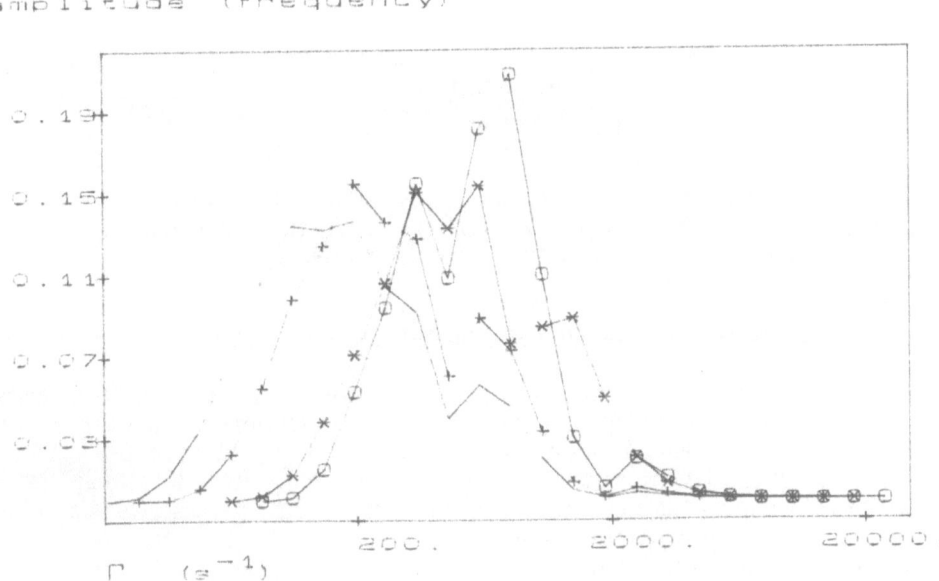

Fig.1   Frequency spectra calculated for a rigid rod model including entanglement. The spectra were calculated for different mean lengths of the rods ($o = 0.54$ μm ; $* = 1.35$ μm ; $+ = 1.80$ μm ; $\cdot = 3.00$ μm) and a scattering angle of 60^0, wavelength is 514.5 nm.

Typical frequency spectra for different mean lengths are shown in Fig.1 whereas the dependence of the mean frequency on this length is displayed in Fig. 2.

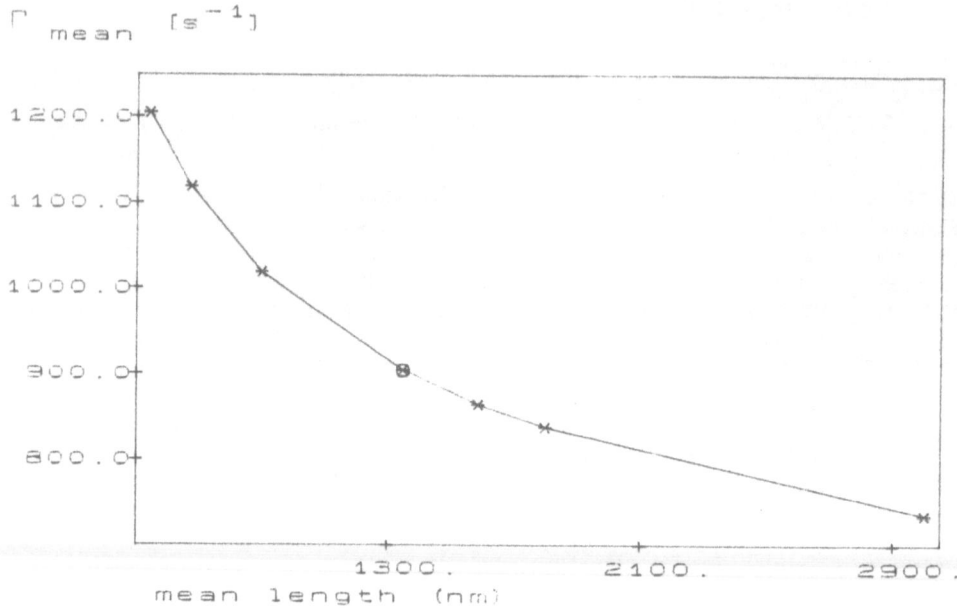

Fig.2   Mean frequency as a function of mean lengths for the discussed model. The mean frequency of the experimental correlation function is marked with o leading to a mean length of 1.35 μm at a scattering angle of 60^0.

In case of a fairly good description of the dynamic of F–actin by the model, the mean length of the filaments could be estimated from the mean frequency. But, as we will see, this requirement is not fulfilled in the present case.

4.   Comparison of experimental and modeled frequency spectra

With photon correlation spectroscopy the second order or intensity autocorrelation function $g_2(q,t)$ is measured [25]. If the Siegert [26] relationship is valid, i.e. the electric field is a gaussian random variable, $g_2(q,t)$ is simply the first order or electric field autocorrelation function $g_1(q,t)$ squared. Just for completeness, fluctuations Δ_1 due to the incoming beam and due to dust particles or air bubbles diffusing through the scattering volume Δ_2 have to be included so that the correlation function can be expressed as

$$g_2(q,t) = A(1+ \Delta_1 + \gamma^2 [g_1(q,t) + \Delta_2]^2) \qquad (18)$$

The parameters A and γ represent amplitudes, which depend on experimental conditions. Since the laser is stabilized Δ_1 is ignored in the data analysis. To calculate the frequency distribution function $G(\Gamma)$ from $g_2(q,t)$ the function must be normalized according to eq. 19 and the inverse Laplace transform has to be performed (eq.20)

$$y_1 = \sqrt{g_2(q,t)/A-1}$$ (19)

$$y_1 = \gamma \int_{\Gamma_{min}}^{\Gamma_{max}} G(\Gamma) \exp(-\Gamma t) + \Delta_2 + \varphi.$$ (20)

The parameter φ arises from noise in the data with the consequence that recovering $G(\Gamma)$ from the experimental function becomes an ill–posed and ill–conditioned problem. Nevertheless, with statistical a priori knowledge, the principle of parsimony and constraining the solutions on one side [11,12] or the principle of maximum entropy [14] on the other, algorithms have been developed that give reliable results on this problem. We have used the constrained algorithm CONTIN written by S.W. Provencher to analyze the correlation functions [11,12,27]. A typical experimental frequency distribution thus obtained is displayed in Fig. 3, where it has been compared to the model function.

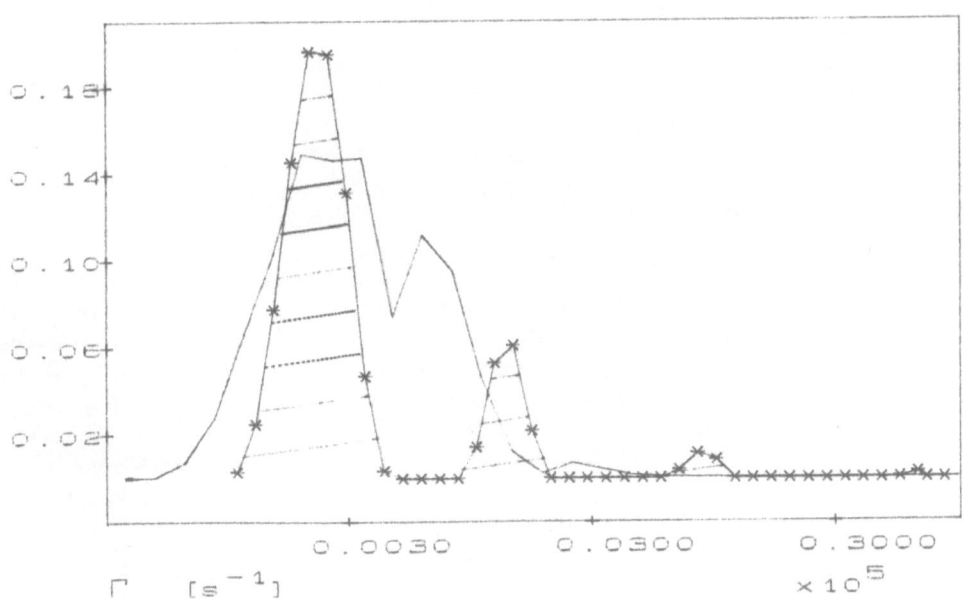

Fig.3   Comparison between experimental (*, hatched) and modeled (–) frequency distribution at a scattering angle of 45^0.

It is obvious that the experimental function strongly deviates from the function calculated for the model. There exist two main reasons for this discrepancy which need some discussion. First one might object that recovering $G(\Gamma)$ from $g_2(q,t)$ is not such a safe procedure allowing detailed comparison between experimental and modeled frequency spectra. However, this possibility can be ruled out rather easily by two experiments. In the first $g_2(q,t)$ was simulated according to the presented rod model and $G(\Gamma)$ obtained by the inverse Laplace transform compared to the modeled frequency spectrum. Apart from the relative contributions of different relaxation frequencies the agreement is good as presented in Fig. 4. In the second experiment the experimental frequency distribution was simulated by assuming four Poisson distributions. The Poisson distributions which were recovered with the inverse Laplace transform were excellent even if different amounts of noise were added (Fig.5). The only conclusions that can be drawn from these experiments is that experimental and modeled frequency spectra for F–actin indeed are different and F–actin in solution can not be viewed as an exponential length distributed ensemble of entangled rigid rods. In the following we will therefore try to answer the question, how the frequency spectrum will change if the described assumptions are not valid.

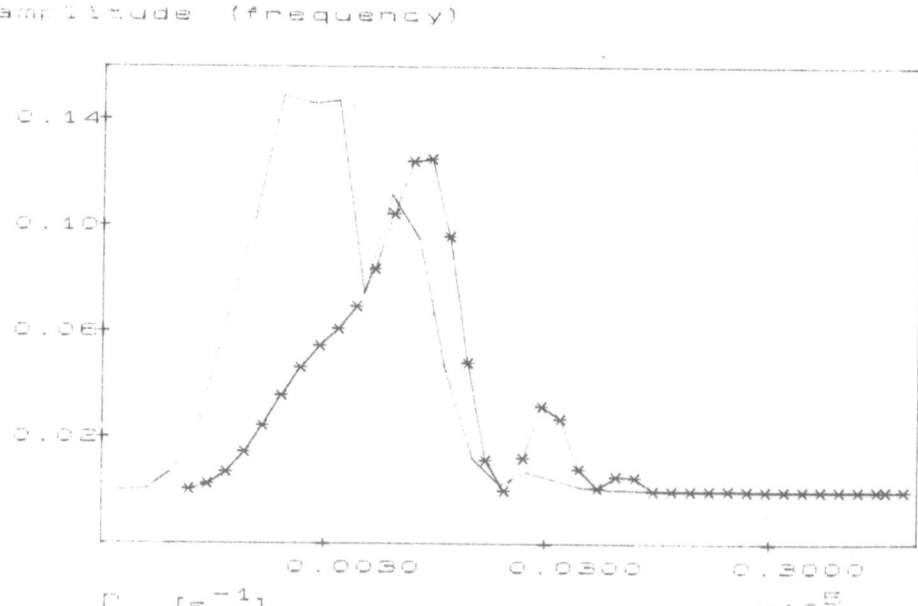

Fig.4   Recovery of the modeled frequency distribution (−) by the inverse Laplace transform CONTIN (*) [11,12,27].

amplitude (frequency)

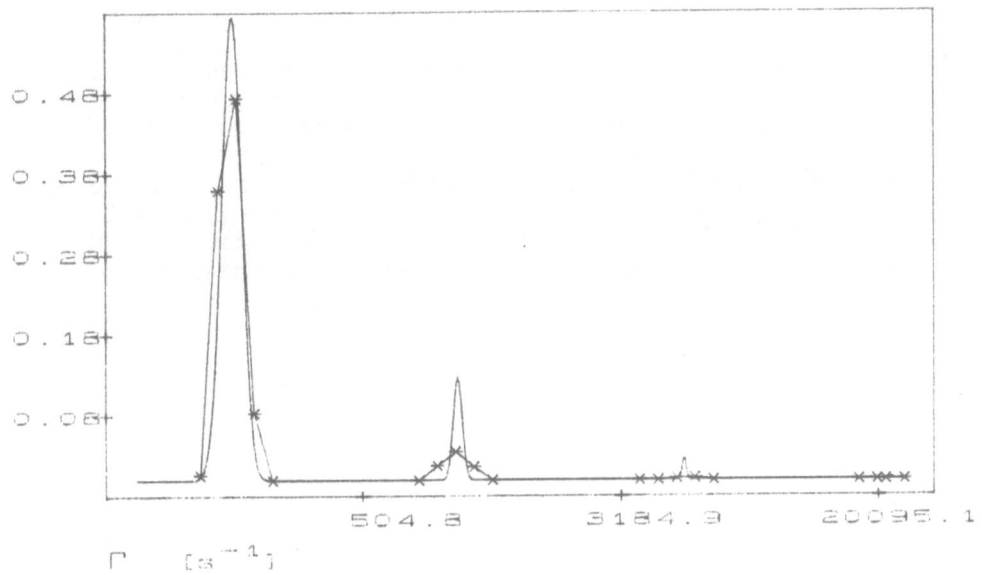

Γ [s⁻¹]

Fig.5   Recovery of four well separated Poisson–distributions (–) by the inverse Laplace transform CONTIN (*). The frequencies simulated were 200, 1000, 5000 and 22000 s⁻¹. Their areas were 20, 10, 5 and 1.5 respectively. The distribution displayed was analyzed without added noise leading to frequencies of 198, 998, 5027 and 22422 s⁻¹. When typical noise for a photon correlation experiment of $3.e^{-4}$ (baseline 1x10⁷) was added (not shown) the frequencies amount to be 204, 1164 and 5888 s⁻¹. The last frequency which was only 7.5% of the area of the first one, was not recovered.

5.   Discussion on the reliability of the model

The dominant influence on the frequency distribution should come from the polydispersity of the sample. A length distribution different from an exponential one should severely alter the measured correlation function. Since the typical feature of the experimental frequency distribution is the appearance of at least three rather narrow peaks we would expect them to be caused by three distinct narrow distributed populations of filament lengths. For instance distinct Poisson distributed lengths would fit the data (not shown). However, such length distributions are rather unlikely in view of the reported histograms of filament lengths [5,22] and also with respect to the proposed reaction mechanism for the polymerization reaction for G–actin [21]. Actually, the length distribution is very broad, even broader than an exponential distribution leading also to even broader frequency spectra in contradiction to the experiment.

A certain flexibility which has been neglected in the model might also contribute to the frequency spectrum. Additional bending and torsional modes are expected in

278

such a case. The calculation of correlation functions and their inversion for a slightly bendable rod model according to the theory of Maeda and Fujime [28] indeed shows high frequency terms appearing in the spectrum with increasing flexibility (Fig. 6). These terms also show up in the high frequency domain of the experimental spectra. However, a dominant influence of bending modes on the dynamics of F—actin would conflict with the reported persistence length for actin. Taking the lowest reported value of 6 μm [7] for this length in an ensemble of filaments with a mean length of 1.35 μm only seven percent by weight of the filaments are banded. As mean length of the filaments the value from Fig. 1. which would give the same mean frequency as experimentally determined was used. It is well inside the range of reported values for this parameter [5,22]. From the arguments above it follows that flexibility of the molecule should only contribute with minor corrections to the frequency spectrum.

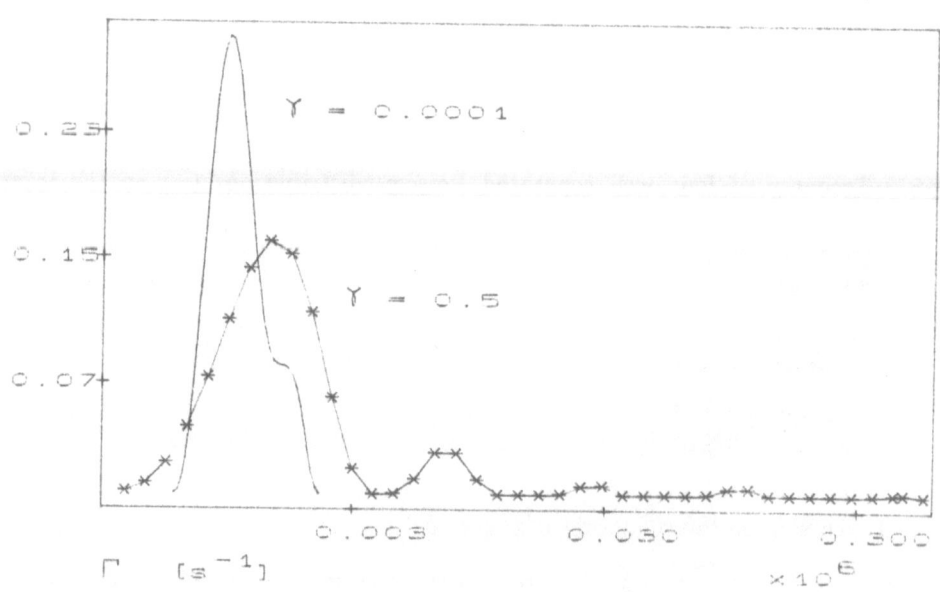

Fig.6 Frequency distributions for a slightly bendable rod model [28] with different degrees of flexibility ($\gamma = 0.0001$ rigid limit).

A more severe assumption of the presented model might be the fact that translational rotational coupling has been ignored. Although this coupling can in principle be included by the theory of Maeda and Fujime [29], for polydisperse samples these calculations are extremely time consuming. Moreover, coupling should rather lead to additional frequencies in the spectra than to a splitting and narrowing of peaks compared to the uncoupled situation. Therefore, translational rotational coupling, although it might have an important contribution to the frequency distribution of F—actin, can be excluded as a cause of the differences between experimental and modeled spectra.

All refinements of the rodlike polymer model that can be imagined fail to represent the main features of the experimental frequency distribution for F–actin. Obviously the dynamics of F–actin can not be described by such a model. There are only two possibilities left to explain the experimental data. Either the steady state kinetics of F–actin totally dominates the correlation function or structural particularities have not yet been taken into account.

The first possibility is rather unlikely in view of the fact that most of the reactions taking place in F–actin solutions under steady state conditions lie out of the time range of the correlator [2]. The dissociation of a monomer from a filament, although it is in the time range of seconds, would not alter the polarizability of the molecule as much as is needed for a dominant effect on the correlation function. Therefore, it only possesses a negligible amplitude in the correlation function.

With respect to polarizability the breaking and reannealing reactions should be better candidates to influence the frequency distributions. However, since the rate of these reactions is proportional to the number of ends [30] they can be calculated to be significantly slower than the time range of the experiment. A large effect of chemical reactions on the intensity autocorrelation function of F–actin can also be excluded by the more general argument, that in such a case additional frequencies would show up in the frequency spectra.

The arguments above force us to search for a structure totally different to rigid rods underlying the dynamic behavior of F–actin in solution. We suggest F–actin to form an infinite network structure giving rise to the measured frequency distributions. Below we present frequency spectra for F–actin/filamin complexes where network formation have been proven by the occurrence of T–junctions in electron micrographs [31]. These data together with theoretical predictions for the dynamics of networks strongly support our interpretation of the frequency distributions of F–actin solutions.

6. Dynamics of F–actin/filamin networks

For highly cross–linked gels the correlation function should only show one fundamental relaxation frequency which is connected to the elastic properties of the gel. Such a dynamical behavior is predicted according to Eq.21 by the theory of Tanaka which has also been demonstrated experimentally [17].

$$g_1(q,t) = \frac{I_0}{c} \left(\frac{\omega_0}{c}\right)^4 \frac{\sin^2 \phi}{4\pi R^2} \left(\frac{\delta\epsilon}{\delta c}\right)_T^2 c^2 \frac{V}{K+4/3\mu} K_B T \times \exp\left[-\frac{(k+4/3\mu)q^2 t}{\xi}\right] \qquad (21)$$

The relaxation frequency is governed by the bulk modulus K, the shear modulus μ, and a friction constant ξ that is assumed to be proportional to the relative velocity of the gel and the solvent. Properties that enter into the amplitude are the intensity of the incoming beam I_0, the velocity of light in vacuum c, the angular frequency of light ω_0, the distance between the scattering volume of length V and the detector, and the concentration c. ϵ are the diagonal elements of the dielectric tensor, so that $d\epsilon/dc$ can be determined by difference refractometry. Finally, ϕ is the angle of polarization planes of the incoming and the scattered beam.

In the case of actin/filamin networks, however, the mean distance between subsequent junctions can be estimated to lie in all cases above the inverse of the scattering vector. Therefore, the gel must be regarded as slightly cross–linked. Indeed, the frequency spectra for the gels with different cross–link densities reveal quite a large number of relaxation processes that are rather similar to those for the pure F–actin spectrum (Fig. 7). Whereas the overall shape of the frequency spectrum is not changed increasing the cross–link density it is evident that the low frequency peak, which is also the main one, is markedly shifted to lower values with increasing cross–link density. In a more functional form these findings are presented in Fig. 8 and Fig. 9 where the mean frequency of this peak is plotted as a function of the cross–link density which is monitored as the molar ratio of filamin/actin and simultaneously as the mean distance between subsequent junctions. Obviously mainly the slow part of the correlation function is affected by introducing more and more cross–links into the network. Therefore, the first cumulant should not be a sensitive focus when exploring the influence of cross–link density on the dynamics of networks. Nevertheless, a decrease of the first cumulant with increasing cross–link density has also been reported previously for synthetic polymers [32].

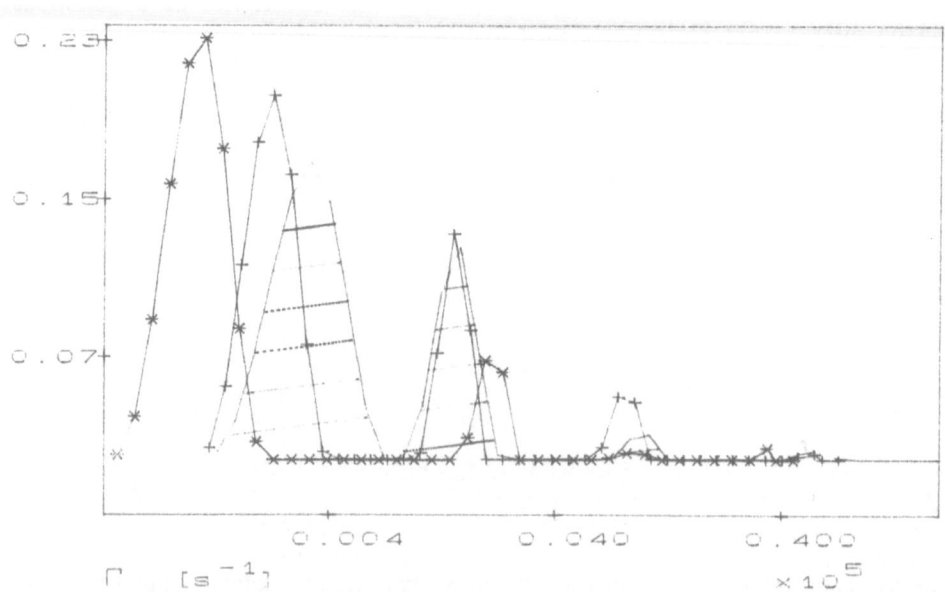

Fig.7   Comparison between frequency spectra for F–actin (– , hatched) and F–actin cross–linked with different amounts of filamin at a scattering angle of 60⁰. The F–actin concentration was always kept at 50 μg/ml whereas the filamin concentration was changed from 50 μg/ml (+) to 100 μg/ml (*) leading to molar ratios of F–actin/filamin of 48, 95 : 1. Not shown is the frequency distribution for the molar ratio of 71 : 1, which has also been measured.

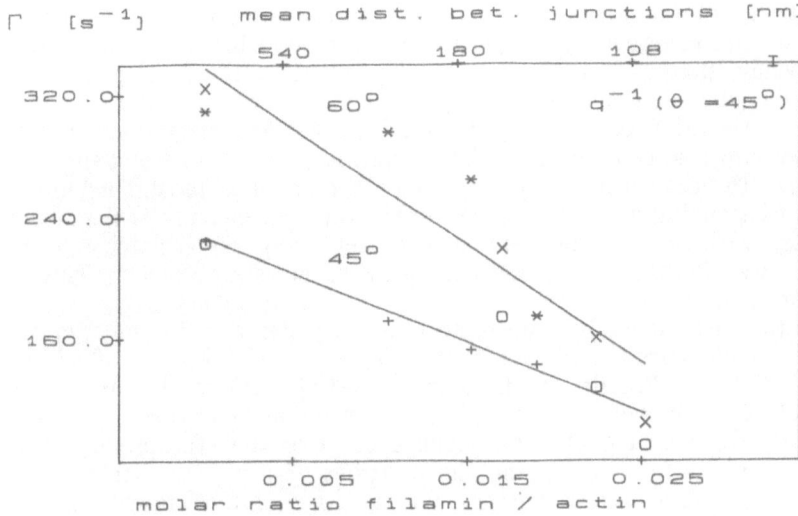

Fig.8   Main frequency as a function of cross–link density measured at scattering angles of 60⁰ (*,x) and 45⁰ (+,o). The different symbols for one angle indicate two different preparations. The points were fitted to a straight line.

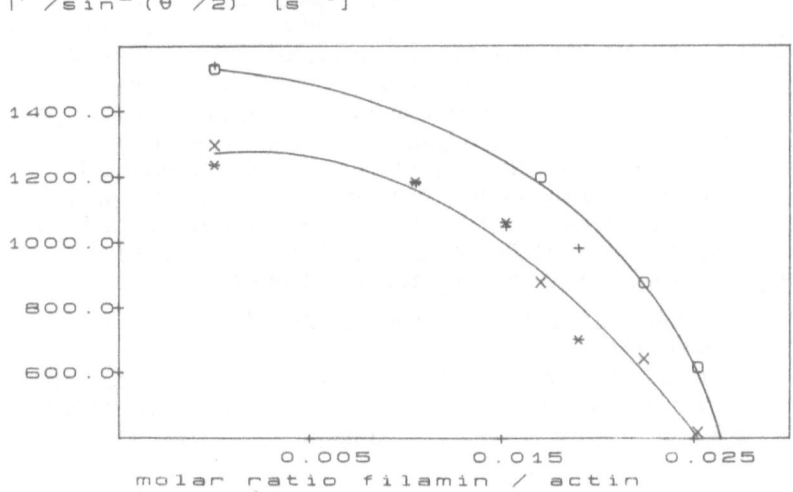

Fig.9   Apparent diffusion coefficient calculated from the main peak of the frequency distribution as a function of cross–link density. The symbols have the same meaning as in Fig. 8. The lines were drawn by hand as a guide for the eyes.

The sensitivity of the low frequency peak to the number of cross—links suggests that it is mainly governed by the gel frequency mode which is given by Eq. 21. We therefore would predict it to become q^2—dependent in the high cross—link density limit, which is obviously not reached in the present networks. Otherwise, the curves of different scattering vectors in Fig. 9 would merge. Assuming that similar terms as in eq. 21 would also contribute to the relaxation process in case of slightly cross—linked gels, i.e. the peak can be represented by a number of modes including an elasticity and a frictional damping term, we would expect an increase of the mean relaxation frequency with higher cross—link density regarding the elasticity term alone. This term is well known to increase with growing cross—link density giving rise to faster motions [33].

Obviously the fast modes become more efficiently damped by the friction which is rather reasonable since the friction should depend on the relative velocities of gel segments and solvent. Finally, in the high cross—link density limit only the slow gel mode should survive leading to a single exponential decay of the correlation function which is described by eq. 21. From such a point of view the picture of a gradual freezing of motions by constraining the positions of segments with the introduction of cross—links immediately emerges. On the basis of such an interpretation of the variation of the frequency distribution with the amount of cross—linking the similarity between the shapes of pure F—actin and filamin/actin spectra is very likely due to cross—links already present between F—actin filaments. Since a part of the molecular motions are frozen in the networks the frequency distributions of actin filaments show rather narrow and distinct peaks compared to the spectra expected for an entangled rigid rod model.

Certainly, the frequency spectra for actin filaments in solution do not give a direct proof of the presence of a network structure. However, an understanding of the dynamics of this system on the basis of an entangled rigid rod model can definitely be ruled out from these experiments and calculations. Moreover, all improvements of such a model by including flexibility of the filaments , anisotropy of motions or different approaches to the entanglement problem seem to point in a wrong direction with respect to an explanation of the observed dynamics. On the other hand the similarities between the profiles of frequency spectra for F—actin and actin/filamin complexes are so striking and the given arguments about their variation with cross—link density sound rather plausible, that we would support the presence of cross—linked actin filaments as they are also claimed by Sato et al. who found a finite storage modulus down to frequencies of 2×10^{-4} Hz from frequency dependent mechanical studies [34].

Concluding, we hope to have demonstrated that photon correlation spectroscopy in connection with calculations based on polymer models can be an important source to our knowledge of the dynamics of flexible and reactive polymers in solution.

Acknowledgements

The authors appreciate the critical comments and suggestions provided by Prof. M. Fixman. We gratefully acknowledge the financial support of the DFG, SFB—project 223, and thank J. Drögemeier for the preparation of filamin, J. Hattesohl for the preparation of actin and Prof. H. Hinssen for electron microscopic analysis.

7. References

1. Stossel, T.P., Chaponnier, C., Ezzel, R.M., Hartwig, J.H. Janmey, P.A., Kwiatkowski, D.L., Lind, S.E., Smith, D.B., Southwick, F.S., Yin, H.L. & Zaner, K.S. (1985) Ann.Rev.Cell.Biol. $\underline{1}$, 353–402.
2. Pollard Th. D., Cooper J.A. (1986), Ann. Rev. Biochem. $\underline{55}$, 987–1035.
3. Jockusch, B.M. (1983) Mol.Cell.Endocrinol. $\underline{29}$, 1–9
4. Wang K., Singer S.J. (1977), Proc.Natl.Acad.Sci.USA $\underline{74}$, 2021–2025
5. Hartwig J.H., Stossel T.P. (1979), J. Mol. Biol. $\underline{134}$, 539–553.
6. Jockusch B.M., Isenberg G. (1981), Proc. Natl. Acad. Sci. USA $\underline{78}$, 3005–3009.
7. Takebayashi T., Morita Y., Oosawa F. (1977), Biochim. Biophys. Acta $\underline{492}$, 357–363.
8. McLean–Fletcher S.D., Pollard T.D. (1980), J. Cell Biology $\underline{85}$, 414–428.
9. Jen C.J., McIntire L.V., Bryan J. (1982, Arch.Biochem.Biophys. $\underline{216}$, 126–132
10. Berne, B., Pecora, R. "Dynamic Light Scattering", J. Wiley, N.Y. (1976)
11. Provencher S.W. (1980), Comp. Phys. Commun. $\underline{27}$, 213–227.
12. Provencher S.W. (1980), Comp. Phys. Commun. $\underline{27}$, 229–242.
13. Bertero M., Boccacci P., Pike, E.R. (1984), Proc.R.
14. Bott, S.E., "Polyddispersity Analysis of Dynamic Light Scattering", Dissertation 1988
15. Pecora, R. (1968), J.Chem.Phys. $\underline{48}$, 4126–4128
16. Pecora, R. (1968), J.Chem.Phys. $\underline{49}$, 1032–1038
17. Tanaka T., Hocker L.O., Benedek G.B. (1973), J. Chem. Phys. $\underline{59}$, 5151–5159
18. Doi M., Edwards S.F. (1978), J.Chem.Soc.Faraday Trans. 2 $\underline{74}$, 560–570
19. Tirado M. M., de la Torre J.G. (1980), J. Chem. Phys. $\underline{73}$, 1986–1993.
20. Tirado M. M., de la Torre J.G. (1984), J. Chem. Phys. $\underline{81}$, 2047–2052.
21. Oosawa F. (1970), J. Theor. Biol. $\underline{27}$, 69–86.
22. Nunnally M., Powell L., Craig S. (1981), J. Biol. Chem. $\underline{256}$, 2083–2086.
23. Flory P.J. (1953), "Principles of Polymer Chemistry" Cornell University Press, Ithaca N.Y.
24. Fixman M. (1985), Phys.Rev.Lett. $\underline{55}$, 2429–2332
25. Cummins H.Z., Knable N., Yeh Y. (1964), J.Chem.Phys., 1033–1038
26. Wang M.C., Uhlenbeck G.E., Rev.Mod.Phys. $\underline{17}$, 323
27. Provencher S.W. (1984), CONTIN, Reference Manual.
28. Maeda T., Fujime S. (1981), Macromolecules $\underline{14}$, 809–818.
29. Maeda T., Fujime S. (1984), Macromolecules $\underline{17}$, 1157–1167
30. Murphy D.B., Gray R.O., Grasser W.A., Pollard Th. D. (1988), J. Cell Biol. $\underline{106}$, 1947–1954
31. Niederman R., Amrein P.C., Hartwig J.H. (1983), J. Cell Biol. $\underline{96}$, 1400–1413.
32. Mutin, P.H., Guenet J.M., Hirsch E., Candau S.J. (1988), Polymer $\underline{29}$, 30–36
33. Schanus E., Booth S., Hallaway B., Rosenberg A. (1985), J. Biol. Chem. $\underline{260}$, 3724–3730.
34. Sato M., Leimbach G., Schwarz W., Pollard Th. D. (1985), J. Biol. Chem. $\underline{260}$, 8585–8592.

ELECTRON SPIN RESONANCE STUDIES OF REORIENTATIONAL DYNAMICS IN LIPID SYSTEMS

L.J. KORSTANJE, E.E. VAN FAASSEN AND Y.K. LEVINE
Department of Molecular Biophysics
Buys Ballot Laboratory, University of Utrecht
P.O. Box 80.000
3508 TA Utrecht
The Netherlands

ABSTRACT. The orientational order and reorientational dynamics of lipid bilayer systems have been studied by ESR techniques. The rigid nitroxide spin label cholestane has been used as a probe molecule. We show that the observed ESR spectra must be interpreted within the formalism of the stochastic Liouville equation, as the rates of the reorientational motions are slow on the ESR timescale. Furthermore, we find that the picture of the orientational order and the rotational dynamics obtained with ESR techniques agrees well with that obtained from fluorescence depolarization experiments. This indicates that the probe molecules do not perturb the local structure of the lipid matrix to a large extent, so that their behaviour reflects the intrinsic order and dynamics of the system. We show that the orientational order decreases and the rates of the rotational motions increase with hydration and temperature. In marked contrast, increasing unsaturation of the lipid chains induces a reduction in the orientational order only. Furthermore we find that the curvature of the bilayer in vesicle systems reduces the orientational order, but has little effect on the reorientational dynamics. These observations are at odds with the correlation between the orientational order and the reorientational dynamics postulated by current membrane models.

1. INTRODUCTION

Lipid molecules form the backbone of the structure of biological membranes and act as an anisotropic solvent for the functional protein and polysaccharide molecules. The lipid matrix is commonly visualized as a bimolecular structure, about 5 nm thick, in which the flexible fatty acid chains are sandwiched between two layers of hydrated headgroups, fig. 1a. The lipid molecules are known to lie with their long axes preferentially along the normal to the bilayer surface and to undergo rapid lateral diffusion [1]. Moreover, the flexibility of the

Th. Dorfmüller (ed.), Reactive and Flexible Molecules in Liquids, 285–307.
© 1989 by Kluwer Academic Publishers.

lipid chains confers an anisotropic orientation gradient across the thickness of the bilayer. The lipid lamellum thus possesses the essential properties of an orientationally anisotropic liquid and it now becomes necessary to include both the orientational order and rotational motions of the molecules in any description of the physical properties of the bilayer system.

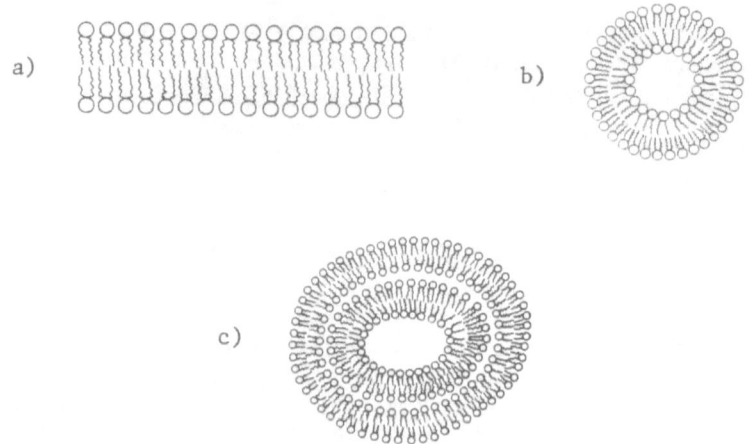

a)

b)

c)

Figure 1. The macromolecular organization of aqueous dispersions of lipid molecules: a) planar bilayer; b) vesicle; c) liposome.

On dispersing lipid molecules (fig. 2a) in water at physiological temperatures, macroscopic lamellar structures are formed. These structures, liposomes, consist of large, closed aggregates of stacked lipid bilayers (fig. 1c). No structural correlations exist in the liposome between adjacent bilayers. The stacking of the lamellae can be disrupted by exposing the liposomes to ultrasound, in which case closed single bilayer-walled vesicles are formed (fig. 1b). The vesicles have a diameter of around 25 nm, so that the bilayer assumes a considerable curvature. Dispersions of lipids in water are inherently isotropic on a macroscopic scale. Macroscopically aligned samples can also be prepared simply by rubbing an aliquot of hydrated lipids between two glass plates. The lipid bilayers align with their planes parallel to the glass plates, to form a planar multibilayer stack possessing an axis of uniaxial symmetry along the common normal to the bilayer planes. The disadvantage of using multibilayer systems is the limited hydration of the lipid headgroups.

The dynamic structure of lipid bilayers has been studied intensively during the past decade using ^{2}H-NMR, ^{13}C-NMR, ESR and Fluorescence Depolarization (FD) techniques. The latter two techniques, however, monitor the behaviour of extraneous probe molecules embedded in the lipid matrix at small concentrations. The behaviour of the probe molecules is taken to reflect that of the surrounding lipids. These probe techniques are particularly useful tools not only because of their great sensitivity, but also because the specific labelling of complex molecules with ^{2}H or ^{13}C atoms is not a trivial task. In this work we use the spin label 3-doxyl 5α-cholestane (CSL, fig. 2b) which is anchored with the nitroxide group at the headgroup layers of the lipid bilayers. The nitroxide group is fixed to the rigid steroid nucleus, so that the experimental ESR spectra yield information about the overall orientation and rigid body motion of the probe molecules.

a)

b)

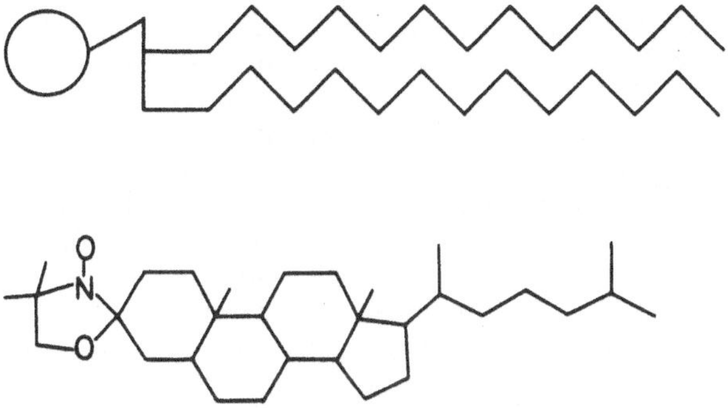

Figure 2. Schematic representations of the chemical structure of a) a lipid molecule and b) a cholestane spin label (CSL).

An important aspect which determines the information content of the experimental technique is its intrinsic timescale τ_{int}. In magnetic resonance techniques this is determined by the anisotropy of the magnetic interactions ($\approx 10^{-5}$ s for ^{2}H-NMR, $\approx 10^{-9}$ s for ESR), and in FD experiments by the lifetime of the excited state ($\approx 10^{-9}$ s). Optimally, the rotational motions, characterized by a rotational correlation time τ_R, should be studied by a technique for which $\tau_R \approx \tau_{int}$. It turns out that in the lipid systems $\tau_R \approx 10^{-10}$-10^{-8} s , so that ESR and FD techniques are particularly suitable for studying the dynamics of the lipid molecules.

Since $\tau_R \approx \tau_{int}$ for ESR, the interpretation of experimental data requires a theoretical approach different from that for NMR, where $\tau_R \ll \tau_{int}$. Consequently, the NMR experiments can be analyzed in the framework of the Redfield Motional Narrowing Approximation [2]. In contrast, the ESR spectral lineshapes must be described in terms of the stochastic Liouville equation (SLE) formalism [3-6] as the motions fall in the slow motion ESR regime ($10^{-9} < \tau_R < 10^{-7}$ s). Unfortunately, the SLE approach to the analysis of the ESR lineshapes is cumbersome and requires the application of complex numerical spectral simulation techniques. Consequently most ESR studies have made use of results based on the assumption of motional narrowing. This assumption, as will be shown here, is not generally valid for lipid bilayer systems.

The detailed description of the reorientational motions in the lipid matrix revealed by ^{2}H-NMR experiments seem to differ substantially from those obtained from ESR studies. As a result, the fidelity of the intrusive probe techniques has been often questioned on the grounds that as these molecules perturb the structure of their surroundings to such an extent, they do not monitor the intrinsic properties of the lipid system. However, it has been shown that the two techniques yield a consistent picture of the orientational order and rotational dynamics in the bilayer system, provided the differences in the approach to the analysis is recognized [6,7]. Furthermore, we shall show here that the picture of the orientational order and the rotational dynamics obtained with ESR techniques agrees well with that obtained from fluorescence depolarization experiments. This indicates that the probe molecules do not perturb the local structure of the lipid matrix to a large extent, so that their behaviour reflects the intrinsic order and dynamics of the system.

One of the main difficulties in the study of the reorientational motions in lipid systems is that the dynamic information cannot be obtained from the experiments in a straightforward way. In orientationally anisotropic liquids the observed dynamics is determined by both the static orientational order and the orientational fluctuations of the molecules. This implies that the interpretation of the experimental data may only be valid within the context of the model used in the spectral simulations. It now becomes imperative to validate of the theoretical description used in the analysis.

Dispersions of lipid molecules in water in the form of vesicles or liposomes afford the most convenient experimental systems. The random orientation of the dispersed bilayers, however, clearly does not provide the optimal sensitivity for a detailed analysis of the observed spectra. Yet, such an analysis is a prerequisite for a discrimination between the various models for the dynamic behaviour of the molecules. The required sensitivity can be enhanced by using macroscopically aligned lipid bilayer systems. This point immediately addresses the main question of our paper, whether the experimental results from vesicles, liposomes and aligned multibilayers give a mutually consistent picture of the bilayer dynamics.

We shall here demonstrate that results from lipid multibilayer systems indeed provide accurate information on the effects of temperature, hydration and unsaturation on the reorientational motions. The results from the dispersed systems fit both qualitatively and quantitatively into the same framework provided the differing water contents of the systems is taken into account.

Interestingly, this consistent behaviour fails to confirm the correlation between orientational order and rotational dynamics postulated by current membrane models.

2. THEORY

Our starting point in the description of the experimental data will be a simple mean molecular field model. In this model we shall take each probe molecule in the system to experience a mean time-dependent orienting torque $\vec{T}(t)$ due to its interactions with the lipid molecules. The orienting torque will be expressed as

$$\vec{T}(t) = <\vec{T}> + \vec{T}'(t) \qquad (1)$$

where $<....>$ represents a time average and $\vec{T}'(t)$ gives the isotropic fluctuating part of the torque. Furthermore we shall assume that $<\vec{T}>$ can be derived from a mean potential $U(\Omega)$ as

$$<\vec{T}> = i \underline{M} U (\Omega) \qquad (2)$$

where $\underline{M}$ is the quantum angular momentum operator and Ω denotes the set of Euler angles $\{ \alpha \ \beta \ \gamma \}$ characterizing the orientation of the probe molecule. Thus each probe molecule is subjected to the orienting potential $U(\Omega)$. It follows now from eq. (1) that in anisotropic liquids the rotational motions of the molecules are determined by both the orienting potential and the torque fluctuations in the system. This is in marked contrast to isotropic liquids where $<\vec{T}>\equiv0$.

We shall now consider the orientational order of the probe molecules in the bilayer system in greater detail before discussing suitable models for the rotational motions.

2.1. ORIENTATION OF MOLECULES IN MEMBRANE SYSTEMS

We shall consider here the orientational behaviour of molecules relative to the normal to the lamellar surface, the local director. We shall furthermore restrict the discussion to the case of a uniaxial bilayer system containing molecules with an effective cylindrically symmetric form. The latter restriction can in fact be verified experimentally for aligned multibilayers, by checking that the same response is observed on rotating the sample by an arbitrary angle about the macroscopic director. It can be justified in general on the grounds of the fast molecular lateral diffusion and the absence of structural correlations between the bilayers in a stack. We note that the bilayer also has a horizontal plane of symmetry.

The orientation of a molecule in our system is now specified by a single angle, β, between the molecular symmetry axis and the local director, fig. 3. The orientational distribution of an average molecule in the system is then characterized by a probability distribution $f_0(\beta)$. This distribution can be expressed as a series expansion of Legendre polynomials $P_L(\cos\beta)$, each of which is weighted by an order parameter $<P_L>$ which is the ensemble average of the corresponding term

$$f_0(\beta) = 1/2 \sum_{L=0}^{\infty} (2 L + 1) < P_L > P_L (\cos\beta) \quad ; \text{ L even} \tag{3}$$

$$f_0(\beta) \geq 0.$$

This expansion defines the order parameters $<P_2>$ and $<P_4>$ as

$$< P_2 > = 1/2 < 3 \cos^2 \beta -1 > \tag{4a}$$

$$< P_4 > = 1/8 < 35 \cos^4 \beta - 30 \cos^2 \beta + 3 > \tag{4b}$$

The orientational distribution function $f_0(\beta)$ is fully characterized if all the order parameters $<P_L>$ are known. In practice, however, only $<P_2>$ and $<P_4>$ are accessible experimentally.

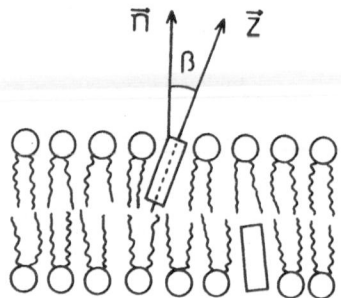

Figure 3. The orientation of a probe molecule in a lipid bilayer. β denotes the angle between the long molecular axis $\vec{z}$ and the normal (the director) $\vec{n}$ to the bilayer surface.

The main difficulty now is obtaining an objective and realistic estimate of the form of $f_0(\beta)$ from a limited knowledge of its moments. This may be accomplished by an information theory approach, in which the most probable values of the missing order parameters are calculated under the assumption that the informational entropy of $f_0(\beta)$ is a maximum under the constraints of the known values of $<P_2>$ and $<P_4>$. The essential point is that this affords the construction of the smoothest possible distribution function consistent with the given information. The resulting distribution function is of the form

$$f_0(\beta) = A \exp[\lambda_2 P_2(\cos\beta) + \lambda_4 P_4(\cos\beta)] \tag{5}$$

where A is a normalization constant and λ_2 and λ_4 are determined from the given values of $<P_2>$ and $<P_4>$. If only the order parameter $<P_2>$ is known, the distribution function takes the form of eq. (5), but with $\lambda_4=0$. We note here that eq.(5) has the form of a Boltzmann distribution with an angle-dependent orienting potential $U(\beta)= -kT \{ \lambda_2 P_2(\cos\beta) + \lambda_4 P_4(\cos\beta) \}$.

It can be seen from eq.(5) that if only $<P_2>$ is known, $f_0(\beta)$ either has a minimum at $\beta=0$ and decreases monotonically to a minimum at $\beta= \pi/2$ or vice versa. Knowledge of $<P_4>$ is required to establish the existence of a maximum in the distribution function at an angle intermediate between 0 and $\pi/2$, corresponding to a collective tilt of the lipid molecules in the structure. Similarly, $<P_4>$ is needed for obtaining an indication of the superposition of two or more independent populations of molecules. Such a superposition may be inferred from the observation of a minimum in the distribution function for $0 \le \beta \le \pi/2$. We also note here that a distribution function possessing a maximum at $\beta=0$ and decreasing monotonically to a minimum at $\beta=\pi/2$ is obtained from eq.(5) provided $0 \le \lambda_4 \le 0.4 \lambda_2$.

2.2. MODELS FOR REORIENTATIONAL MOTIONS

We shall assume that the reorientational motion of a molecule in our systems is a stochastic Markov process and neglect any inertial effects. Consequently, the Markov process need only be described in terms of the classical random Euler angles $\{ \alpha(t) \beta(t) \gamma(t) \} = \Omega (t)$. The master equation for the conditional probability $f(\Omega_0 / \Omega t)$ that the molecule has orientation Ω relative to the director at time t, given its orientation Ω_0 at t=0 is

$$\partial/\partial t \; f(\Omega_0 / \Omega t) \;\; = - \; \Gamma_\Omega \; f(\Omega_0 / \Omega t) \tag{6a}$$

$$f_0(\beta) \; = \; f_0(\Omega)/ \; 4\pi^2 \; = \; \underset{t\to\infty}{\text{Lim}} \; 1/ \; 4\pi^2 \; f(\Omega_0 / \Omega t) \tag{6b}$$

where Γ_Ω is the stochastic operator describing the orientational motion.

Two models are frequently used for the description of the rotational motion of the molecules [3-5,8]. The first, the strong collision model, postulates that the molecules change their orientation by undergoing uncorrelated jumps of arbitrary magnitude, spending on average a time τ in a given position. In the second, the rotational diffusion model, the molecules are assumed to undergo small-step, correlated, angular excursions subject to the action of an orienting potential $U(\beta)$. In view of the geometrical form of the CSL molecule, its rotational diffusion in the membrane is assumed to be cylindrically symmetric, with a rotational diffusion tensor of the form $D = \text{diag}(D_\perp, D_\perp, D_\parallel)$, where $D_\parallel$ and $D_\perp$ are the diffusion rates for rotation around the long molecular axis and rotation of that axis respectively. In this latter model, the molecule changes its macroscopic orientation as the result of many rotational jumps and the diffusion constants give the mean square angular displacement per unit time.

The strong collision model suffers from the distinct disadvantage that its application to elongated molecules is not trivial. The difficulty arises primarily in distinguishing between the rotational steps about the long molecular axis and the reorientation of this axis relative to the director. In marked contrast, the two motions are resolved within the framework of the rotational diffusion model and each is assigned its own diffusion constant, $D_\perp$ or $D_\parallel$. We shall therefore only consider the latter model further. The Markov operator is now given by

$$\underline{\Gamma}_\Omega = \underline{M} \cdot D \cdot (\underline{M} + 1/kT \ (\underline{M} \ U(\beta)) \tag{7}$$

where $\underline{M}$ is the angular momentum operator defined in the molecular frame. In view of the discussion above, Section 2.1, about the orientational distribution function, we shall choose $U(\beta)$ to take the form

$$U(\beta) = -kT \ [\ \lambda_2 \ P_2(\cos\beta) + \lambda_4 \ P_4(\cos\beta) \ ] \tag{8}$$

2.3. NUMERICAL SIMULATIONS OF ESR SPECTRA

The molecular orientations and the molecular dynamics of ESR probe molecules are reflected in the shape of the ESR absorption spectra. The general theory of the lineshapes has been described in detail elsewhere [3,5,6,8] and only the essential elements will be treated here.

The time evolution of the spin density matrix $\rho(\Omega,t)$ in the presence of an oscillating microwave field is described by the stochastic Liouville equation:

$$\partial/\partial t \ \rho(\Omega,t) = -i[\underline{H}(\Omega) + \underline{\epsilon}(t), \ \rho(\Omega,t)] - \underline{\Gamma}(\rho(\Omega,t) - \rho_0(\Omega)) \tag{9}$$

where the Euler angles Ω specify the orientation of the spin label molecule. The time independent part $\underline{H}(\Omega)$ of the one particle Hamiltonian is taken to be the sum of the electronic Zeeman and hyperfine interaction terms:

$$\underline{H}(\Omega) = \beta_e/\hbar \ \vec{H}_0 \cdot g(\Omega) \cdot \vec{S} - \gamma_e \ \vec{I} \cdot A(\Omega) \cdot \vec{S} \tag{10}$$

where $\vec{H}_0$ is the external static magnetic field and $\vec{S}$ and $\vec{I}$ are the electron and nuclear spin operators respectively. We quantize these operators along the direction of $\vec{H}_0$. The small nuclear Zeeman term is neglected in eq.(10). Moreover, we omit the nonsecular terms (that is terms proportional to S_x or S_y) from the Hamiltonian (eq.(10)) since their effect on the spectrum is negligible [9]. The principal axes of the magnetic tensors g and A are assumed to coincide and have a diagonal form in the CSL reference frame, the z-axis being parallel to the long axis of the molecule.

The microwave field $H_x(t) = 2H_1\cos\omega t$ is applied perpendicular to the external $\bar{H}_0$ field, and gives rise to the Hamiltonian $\underline{\epsilon}(t)$:

$$\underline{\epsilon}(t) = 1/2 \; \gamma_e \; H_1(S_+ \; e^{-i\omega t} + S_- \; e^{+i\omega t}) \tag{11}$$

where $S_\pm = S_x \pm iS_y$ are the spin raising and lowering operators. The operator $\underline{\Gamma}$ induces relaxation towards the equilibrium density matrix $\rho_0(\Omega)$ and is given by:

$$\underline{\Gamma} = \underline{\Gamma}_\Omega + \underline{\Gamma}_R \tag{12a}$$

where $\underline{\Gamma}_R$ is a phenomenological operator introducing inhomogeneous line broadening into the spectrum. Its operation on a spin 1/2 density matrix is defined as:

$$
\underline{\Gamma}_R
\begin{vmatrix}
\rho_{11} & \rho_{12} \\
& \\
\rho_{21} & \rho_{22}
\end{vmatrix}
=
\begin{vmatrix}
0 & \dfrac{-\rho_{12}}{T_2} \\
& \\
\dfrac{-\rho_{21}}{T_2} & 0
\end{vmatrix}
\tag{12b}
$$

This inhomogeneous broadening may be caused for example by dissolved oxygen in the sample. The equilibrium density matrix $\rho_0(\Omega)$ [10] is given by:

$$\rho_0(\Omega) = f_0(\Omega)(1 - \hbar \; \frac{H(\Omega)}{kT})/\mathrm{Tr}(1 - \hbar \; \frac{H(\Omega)}{kT}) \tag{13}$$

ESR experiments measure the microwave energy $P(\omega)$ absorbed by the sample. As the observed spectrum is inhomogeneously broadened, it can be considered to consist of a superposition of signals from a Gaussian distribution of spin packets, each absorbing a power $P_i(\omega)$ from the microwave field:

$$P(\omega) = \int_{-\infty}^{\infty} d\omega' \; \frac{1}{\sigma_G \; \sqrt{2\pi}} \; \exp \; [-\omega'^2/2\sigma_G^2] \; P_i \; (\omega + \omega') \tag{14}$$

where the Gaussian broadening σ_G is a measure of variation of the local static magnetic field. In the wings of the absorption lines, the effect of this Gaussian broadening is less pronounced than that of the Lorentzian broadening which is introduced via the relaxation operator Γ_R. Apart from this detail, T_2 and σ_G have largely similar effects and are interchangeable in this sense. A fit in absence of a Gaussian broadening results in lower values of T_2, but with essentially identical values of diffusion rates and order parameters.

The power absorbed per cycle and unit volume, $P_i(\omega)$, is given by [11]:

$$P_i(\omega) = \frac{\omega}{2\pi} \int_0^{2\pi/\omega} - M_x(t) \frac{dH_x(t)}{dt} dt \qquad (15)$$

The magnetization density $M_x(t)$ can now be evaluated from the solution $\rho(\Omega,t)$ of eq. (9):

$$M_x(t) = n\hbar\gamma_e \int_\Omega Tr[\rho(\Omega,t)S_x]f_0(\Omega)d\Omega \qquad (16)$$

where n stands for the spin label density.

The SLE can be solved by a variety of methods [12]. We have employed the method of Freed and co-workers, where the solution of $\rho(\Omega,t)$ is found by a Laplace transformation of the time variable to frequency space and the decomposition of the angular dependence in Wigner rotation matrices $D^L_{mn}(\Omega)$. The resulting eigenvalue equation can be efficiently solved by the application of the Lanczos algorithm [13,14] and requires around 100 cpu-seconds on a CYBER 855 computer. Numerically stable results were obtained with the following subset of Wigner rotation matrices: $0 \leq L \leq 10$, L even; $-4 \leq n \leq 4$, n even; and $-2 \leq m \leq 2$. The ESR absorption spectrum is now fully determined by the eigenvalue distribution in the complex { ω, T_2 } plane. It is given by the imaginary part of the complex response function $Z(\omega)$

$$P(\omega) = Im \ Z(\omega) \qquad (17a)$$

$$Z(\omega) = \sum_i | C_i |^2 / (\omega - \lambda_i) \qquad (17b)$$

where the poles are given by the eigenvalues { λ_i } and the weights C_i are the projections of the equilibrium density matrix $\rho_0(\Omega)$ on the corresponding eigenvectors. Experimentally, the first-derivative spectrum $P'(\omega)$ is observed and this is given by

$$d/d\omega \ P(\omega) = P'(\omega) = Im \sum_i | C_i |^2 / (\omega - \lambda_i)^2 \qquad (17c)$$

It is important to note that as a result of the distribution of the eigenvalues, the ESR spectrum will in general consist of numerous

Lorentzian lines. It is thus not uncommon to find quintet or higher multiplicity spectra in the slow-motion regime. However, in the motional narrowing limit all poles except three are either sufficiently far away from the real axis of the complex plane or of such small weight as to be negligible. A strictly triplet ESR spectrum is now obtained.

The lipid systems studied by us exhibited a clearly resolved triplet spectra. The question therefore arises as to whether the cumbersome slow-motion analysis of the ESR spectra is really necessary. In order to gain insight into the merits of the procedure described above it is instructive to compare the results with those obtained from the application of the motional narrowing approximation. The simplest comparison of the two methods is provided by the values for the order parameter $<P_2>$ extracted from the ESR spectra obtained from multibilayer samples at different orientations to the applied static magnetic field $\vec{H}_0$. The orientation of the sample is defined by the angle θ between its macroscopic director $\vec{n}$ and $\vec{H}_0$, fig. 4.

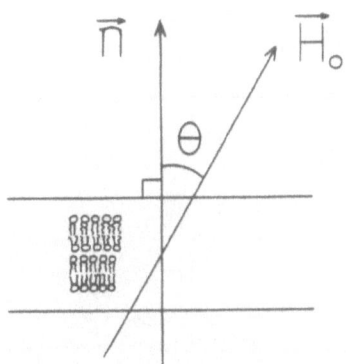

Figure 4. A schematic representation of an angle-resolved experiment. The multibilayer director $\vec{n}$ makes an angle θ with the static magnetic field $\vec{H}_0$.

To this end we simulated spectra for several values of the potential parameter λ_2 (with $\lambda_4 = 0$) over a wide range of diffusion rates. The exact result $<P_2>_e$ depends on the potential parameter λ_2 only, whereas the motional narrowing expression $<P_2>_{MN}$ is defined by [15]:

$$<P_2>_{MN} = \frac{A_{iso} - A_\|}{A_{iso} - A_{zz}} \tag{18}$$

Here A_{zz} is the z component of the hyperfine tensor A used in the simulations. $A_\|$ and $A_{iso} = 1/3(A_\| + 2A_\perp)$ are found from the line splittings $A_\|$ and $A_\perp$ of the simulated spectra at $\theta = 0°$ and $90°$

respectively. The results, fig. 5, clearly show that the motional narrowing assumption is accurate for large rotational diffusion coefficients D:

$$\lim_{D \to \infty} <P_2>_{MN} = <P_2>_e \qquad (19)$$

but seriously overestimates the molecular order for slow diffusion rates (D < 10^8 rad^{2}s^{-1}). It is important to note that this effect is more pronounced in systems with low molecular ordering. Clearly, the motional narrowing interpretation cannot be used for the lipid bilayer systems where typically at physiological temperatures <P_2> ≈ 0.4-0.7 and D ≈ 10^7-10^8 rad^{2}s^{-1}.

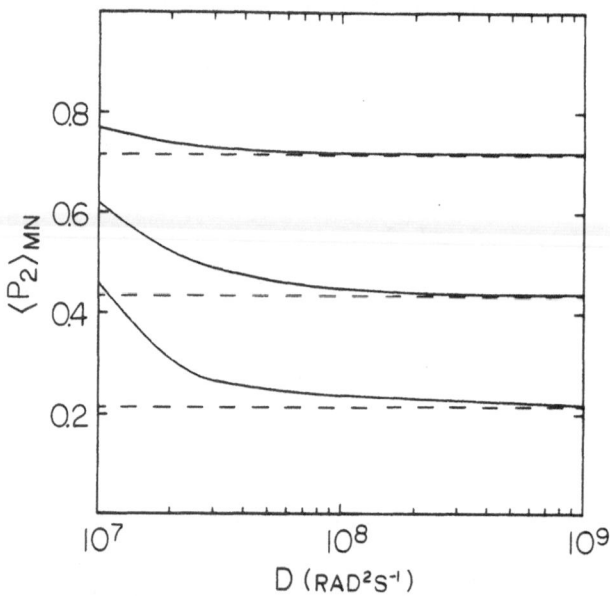

Figure 5. The dependence of the order parameter <P_2>_{MN} (———) on the diffusion constants. This parameter is calculated from the ESR spectra using eq. (18). Parameters: $D_\perp$=D and $D_\parallel$=5D. The exact order parameters <P_2>_e, eq. (19), are given by (----).

2.3.1. *Liposome and vesicle systems.* The strategy for the numerical simulations of ESR spectra described thus far can be applied only to the case of planar multibilayer samples whose director makes an arbitrary angle θ to the static magnetic field $\vec{H}_0$, fig. 4. However, in liposome and vesicle suspensions the directors to the lipid bilayers have a random distribution of orientations relative to $\vec{H}_0$, so that the samples exhibit complete orientational disorder on a macroscopic scale. In this case the observed ESR signal $\bar{P}(\omega)$ can be considered to be a superposition of the spectra of randomly distributed multibilayer samples:

$$\bar{P}(\omega) = \text{Im } \bar{Z}(\omega)$$

$$\bar{Z}(\omega) = 1/2 \int_0^\pi Z(\omega, \theta) \sin \theta \, d\theta \tag{20}$$

where $Z(\omega,\theta)$ is the response function of a multibilayer oriented at an angle θ relative to the static magnetic field.

For frequencies near resonance, where $Z(\omega,\theta)$ is a strongly peaked function of θ, a straightforward numerical evaluation of eq.(20) requires a very fine integration mesh demanding inordinate amounts of cpu time. We have therefore evaluated the integral using a subtraction method which presupposes knowledge of $Z(\omega,\theta)$ at a small number of orientations θ.

In our numerical simulations we evaluated $Z(\omega,\theta)$ at four orientations only, for $\theta = 0$, 45°, 60° and 90°. The procedure consists of fitting each of the four simulated lineshapes $Z(\omega,\theta)$, eqs. (17), by a simple triplet Lorentzian spectrum $F(\omega,\theta)$ of the form

$$F(\omega, \theta) = \sum_{j=1}^{3} R_j(\cos \theta) / [\omega - \Lambda_j(\cos \theta)] \tag{21}$$

where R_j and Λ_j are complex numbers. The background $B = Z(\omega,\theta) - F(\omega,\theta)$ can be easily integrated with a four point Gaussian integration. The integral over $F(\omega,\theta)$ can be carried out quickly on observing that while the function is strongly peaked, its residues R_j and poles Λ_j are smooth functions of $\cos \theta$, and can be interpolated as

$$p_0 + p_2 x^2 + p_4 x^4 + p_6 x^6 \; ; \; x = \cos \theta.$$

The lineshape function $F(\omega,\theta)$ can now be generated quickly at a sufficiently large number of angles. This number is strongly dependent on the linewidth of the spectrum and for our applications the generation of 30 Gaussian points proved adequate. This subtraction method affords the simulation of a complete vesicle spectrum in about 400 cpu seconds on a CYBER 855.

It must be emphasized here that our protocol is based on the observation that triplet ESR spectra are obtained for CSL molecules in liposome and vesicle systems. The procedure can be extended in an obvious way to the simulation of spectra consisting of a higher multiplet.

3. MATERIALS AND METHODS

Dimyristoyl-phosphatidylcholine (DMPC), palmitoyl-oleoyl-phospha-tidylcholine (POPC), dioleoyl-phosphatidylcholine (DOPC), and dilineoyl-phosphatidylcholine (DLPC) were purchased from Sigma Chemical Company (St. Louis, MO, USA) and used without further purification. The spin label 4',4'-dimethylspiro [5α-cholestane-3, 2'-oxazolidin]-3'-yloxy (CSL) was bought from Aldrich Chemical Company (Milwaukee, WI, USA). The purity of the lipids and the spin label was checked when necessary by high performance thin layer chromatography (HPTLC).

In all our experiments a CSL concentration of 1 mole % was used. At this concentration the spin-spin interactions between individual CSL molecules are expected [16] to produce homogeneous line broadening of ca 0.5 G. This value is comfortably exceeded by the combined effect of Lorentzian and Gaussian broadening of ca 1.5 G.

3.1. PREPARATION OF MULTIBILAYER SAMPLES

The lipid-CSL mixtures were prepared by dissolving the components in chloroform. After mixing, the chloroform was removed under vacuum. Equilibration over a saturated K_2SO_4 solution provided lipid-CSL mixtures with a water content of 24 wt %. Samples with 12 wt % were derived from this by subsequent equilibration over a saturated sodium acetate (CH_3COONa) solution. Equilibration time was longer than 16 hours. The resulting water concentration was determined gravime-trically with an estimated relative uncertainty of 10%. The hydrated lipid material was oriented between glass plates (0.2 x 4 x 8 mm) by application of shear pressure, leading to stacks of about a thousand bilayers. The macroscopic alignment was checked optically with a polarizing microscope equipped with a first order red plate. Four to six individual samples were stacked in order to improve the signal to noise ratio.

The samples were kept in the dark under a nitrogen atmosphere as much as possible. The polyunsaturated lipids DOPC and DLPC were manipulated strictly under nitrogen atmosphere to avoid oxidation of the unsaturated bonds. In spite of the fact that the method of preparation could not completely eliminate the presence of oxygen, even DLPC samples had signal losses of less than five percent over a period of four days, indicating a negligibly slow oxidation of the unsaturated bonds.

3.2. PREPARATION OF LIPOSOMES AND VESICLE SAMPLES

The lipid-CSL mixtures were prepared by dissolving the components in chloroform. After mixing, the chloroform was removed by a flow of nitrogen gas and subsequent storage under vacuum for several hours. The mixture was hydrated by the addition of a 20mM Tris buffer, pH 8.0, containing 7.5μM EDTA. The hydrated mixture, concentration 2 mg lipid/ml buffer, was homogenized with a vortex mixer for several minutes.

Liposome samples contained 20 μl of the hydrated lipid/CSL mixture in a quartz capillary. Vesicle suspensions were obtained by sonication of the vortexed lipid/CSL mixture in a bath type sonicator for 15-45 minutes. The clear suspensions were subsequently centrifuged for 1 hr at 40,000g. 20 μl of the supernatant, containing 2 mg lipid/ml, in a quartz capillary was used in the experiments.

All the preparative steps were carried out in the dark under a nitrogen atmosphere as much as possible to avoid oxidation of the lipids. The samples were used in the ESR experiments within 24 hours of their preparation.

3.3. ESR EXPERIMENTS

ESR experiments were carried out using a Varian E-9 X-band spectro-meter, equipped with a TM110 cavity. Multibilayer samples were placed in a quartz tube, above a saturated salt solution to maintain the water concentration and in a nitrogen atmosphere to prevent oxidation. The salt solution, at the bottom of the tube, was placed well away from the active region of the cavity. Samples of vesicle or liposome systems in sealed capillaries were mounted along the axis of a quartz tube. The sample temperature was regulated within 1 °C with a Varian V4540 variable temperature accessory and measured by a copper-constantan thermocouple placed above the sample, just outside the active region of the cavity.

The orientation of the sample director relative to the applied static magnetic field was varied using a home built goniometer with an accuracy of ± 1°. ESR spectra were recorded at a microwave power level of 1 to 2 mW for the multibilayer samples and 10 to 15 mW for the liposome and vesicle samples, well below saturation. A magnetic field modulation of 1.0 to 1.6 Gauss (top-top) with a frequency of 100 kHz was used to detect the first derivative of the absorption signal. The background ESR signal, arising from the quartz tube and the glass plates, was subtracted from the measurements before analysis.

4. RESULTS AND DISCUSSION

4.1. FITTING THE EXPERIMENTAL ESR SPECTRA

The usual approach for finding an agreement between the experimental data and the theoretical model is the optimization of the adjustable model parameters using least-squares techniques. The goodness-of-fit is then judged by statistical criteria such as a global minimum in the χ^2-surface. However, in our case such an approach is inappropriate in view of the long computation times involved in the simulations of the ESR spectra. The experimental spectra were fitted visually on taking into consideration the faithfulness of the reproduction of a number of the spectral features. Consequently we need to devise a strategy for the simulation of the experimental spectra. The important elements of the protocol used by us will be described below.

In general the numerical simulation of the ESR spectra requires the specification of the 4 model parameters $D_\perp$, $D\|$, λ_2 and λ_4, in addition to the components of the A- and g-tensor. Furthermore two parameters, T_2^{-1} and σ_G, are needed to describe respectively the homogeneous and inhomogeneous linewidths. It is clear that the first step must be the independent determination of the components of the magnetic interaction tensors.

The values of the components of these tensors are obtained from the simulations of the spectra of oriented systems at -25°C where the rotational motions have been effectively quenched on the ESR timescale. These rigid limit spectra can be easily simulated in terms of a superposition of signals from static molecules. The g-tensor has been found to be independent of the composition of the system, but in contrast, small but significant (ca. 5%) changes were observed in the values of the components of the A-tensor. In particular, we found it necessary to adjust the value of A_{yy} in the simulations of the actual spectra. In general we used the following tensors:

g = diag (2.0081, 2.0024, 2.0061)   and

A = diag (5.6, 34.0, 5.3).

Furthermore, we took T_2 = 2.0 • 10^{-7} sec.

4.1.1. *Multibilayer samples*. The numerical simulations of the ESR spectrum for $\theta=0°$ is the starting point of the fit protocol. This is chosen not only because the computation time of the spectrum is a factor of 10 faster than those for other orientations, but also because it is the most sensitive to changes in orientational order and dynamics in the sample. The amplitude of the centre line of the spectrum is used for scaling the simulations to the experimental lineshapes.

In the initial calculations we set $\lambda_4=0$ and choose arbitrary values for the diffusion coefficients in the slow-motional regime. The search now concentrates on the fitting of the line splittings by varying λ_2 only. When a reasonable fit has been obtained, the diffusion coefficients and σ_G are varied independently to optimize the reproduction of the line shapes. Lastly, fine-tuning of the fit is attempted by changing the value of λ_4.

The values of the parameters are now adjusted so as to improve the fit of the $\theta=90^o$ spectrum. In particular we find this spectrum to be markedly more sensitive to $D_{\parallel}$ than the $\theta=0^o$ one. Furthermore, it appears that if the lineshapes rather than the hyperfine splittings are reproduced well, than the value of A_{yy} needs to be adjusted. When both these spectra have been fitted satisfactorily, the goodness-of-fit is tested by simulations of spectra at intermediate angles, $\theta= 30^o,45^o$ and 60°. The additional features of these spectra appear to be sensitive to the value of λ_4 used in the simulation. Although all the 4 model parameters are found to affect both the hyperfine splittings and the lineshapes, we have found that they are essentially not correlated. Thus the spectral changes induced by varying one of the model parameters cannot be compensated by changing the values of the other 3 parameters.

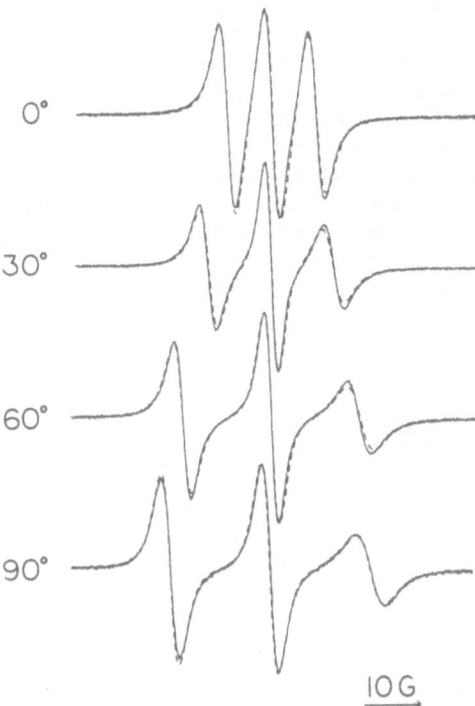

Figure 6. A comparison between the experimental (———) and simulated (- - - -) ESR spectra of CSL molecules in multibilayers of POPC (12 wt % water and T=35°C) at various orientations θ of the director relative to the static magnetic field. Fitting parameters:
$D_{\parallel}=2.0 \ 10^8$ rad^2 s^{-1}, $D_{\perp}=2.5 \ 10^7$ rad^2 s^{-1}, $\lambda_2=3.3$, $\lambda_4=0.0$ and $\sigma_G=1.2$ G.

Typical examples of the fits are shown in fig. 6. We estimate that the extracted values of the model parameters are reliable within the following bounds: λ_2:5%, λ_4:30%, $D_{//}$:30%, $D_\perp$:25% and σ_G:10%.

4.1.2. Liposome and vesicle samples. The fitting of the ESR spectra from these samples is in practice cumbersome as a result of the long computation times. Furthermore, the spectra lack the features which allow the accurate determination of λ_4 and $D_\|$ in the angle-resolved case. In particular, the hyperfine splittings showed a much smaller variation with changing order and rotational dynamics. The problem here is that the spectral details are washed out as a result of the isotropic distribution of the bilayer normals relative to the static magnetic field.

In view of these difficulties, we have carried out the simulations on fixing the ratio N of the diffusion coefficients to N= $D_\|$ / $D_\perp$ = 5 , a ratio expected from the geometrical shape of the CLS molecule [25]. In addition we have taken λ_4 = $0.4\lambda_2$, which corresponds to the broadest possible distribution function with a monotonic decrease between $\beta=0$ and $\pi/2$. In our case this choice of parameters turned out to give a satisfactory reproduction of the spectra. The fitting protocol is essentially the same as that for the multibilayer samples. However, we have found that for the systems we have studied, the hyperfine splittings are essentially dependent only on the value of order parameter $<P_2>$. This indicates that the rotational motion has significantly increased so that the motional-narrowing regime is approached. A typical example of the fits is shown in fig. 7.

We estimate the relative errors of the model parameters to be: λ_2:5%, $D_\perp$:30% and σ_G:10%.

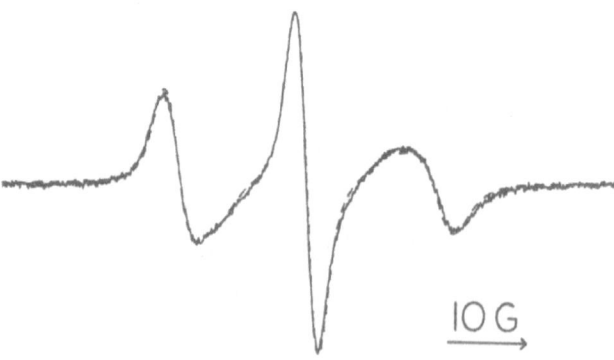

Figure 7. A comparison between the experimental (————) and simulated (----) ESR spectra of CSL molecules in vesicles of POPC at 35°C. Fitting parameters: $D_\|$=8.0 10^7 rad^2 s^{-1}, $D_\perp$=$5D_\perp$, λ_2=1.7, λ_4=$0.4\lambda_2$ and σ_G=1.2 G.

4.2. MULTIBILAYER SYSTEMS

Experiments on oriented lipid multibilayer systems of DMPC, POPC, DOPC and DLPC in the liquid-crystalline phase were carried out in the temperature range of 19-45 °C at two different water concentrations of 12 and 24 wt % . These lipids differ primarily in the number of C=C double-bonds in the fatty acid chains.

4.2.1. *Orientational order*. An increase in the water content of any of the lipid systems studied here induces a decrease in the ordering of the probe molecules, fig. 8.

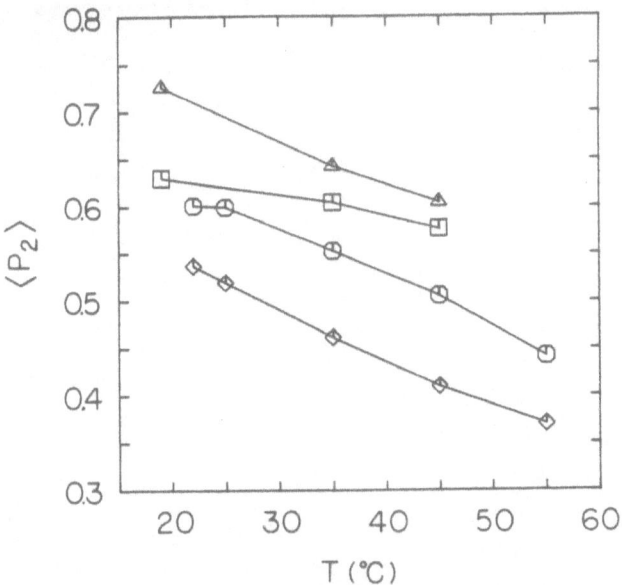

Figure 8. The temperature dependence of the order parameter $\langle P_2 \rangle$ of CSL molecules embedded in different systems of POPC: △ multibilayers, 12 wt % water; □ multibilayers, 24 wt % water; ○ liposomes; ◇ vesicles.

    The order parameters $\langle P_2 \rangle$ and $\langle P_4 \rangle$ for any lipid system exhibit a gradual decrease with increasing temperature irrespective of the hydration of the lipid headgroups. The only exception being DLPC at a water concentration of 12 wt %, where a phase separation into ordered and disordered domains is observed. We were thus unable to simulate such ESR spectra within our current model.
    An increase in the unsaturation of the hydrocarbon chains of the lipids results in a clear decrease in the order parameters $\langle P_2 \rangle$ and $\langle P_4 \rangle$. This trend is found under all our experimental conditions, in agreement with previous studies [17,18,22].

4.2.2. *Reorientational dynamics*. All the lipid systems studied here showed an increase of the diffusion parameters $D_{\parallel}$ and $D_{\perp}$ by a factor of 2 to 4 on increasing the temperature from 20 to 45 °C, fig. 9. This relatively strong temperature dependence confirms earlier results [6,19]. Moreover, the order of magnitudes of $D_{\parallel}$ and $D_{\perp}$ fall within the ESR slow-motion regime, so that an analysis of these spectra in terms of the motional-narrowing approximation is not justified. We reiterate here that reliable values of the diffusion constants can only be obtained from angle-resolved ESR experiments.

In marked contrast to the effects of temperature and hydration, we find no systematic effect of increasing unsaturation on the rotational diffusion rates. Our results thus indicate that changes in membrane molecular order and reorientational dynamics have to be considered separately and are not necessarily correlated as implied by the concept of membrane fluidity [20,21].

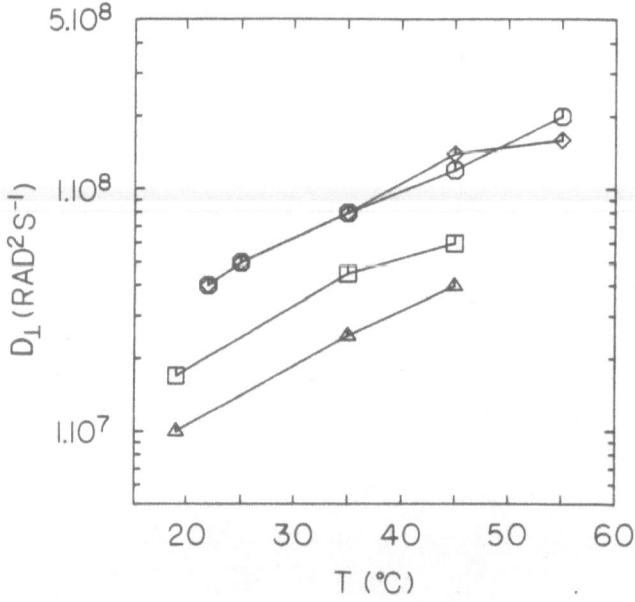

Figure 9. The temperature dependence of the rotational diffusion coefficient $D_{\perp}$ of CSL molecules embedded in different systems of POPC: Δ multibilayers, 12 wt % water; □ multibilayers, 24 wt % water; O liposomes; ◊ vesicles.

4.3. VESICLE AND LIPOSOME SYSTEMS

Liposome and vesicle systems of the lipids POPC, DOPC and DLPC were studied in the temperature range 19-55 °C.

4.3.1. *Orientational order*. The order parameter $<P_2>$ decreases gradually with increasing temperature in both the liposome and vesicle systems of all the lipids studied, fig. 8. Interestingly, the value of $<P_2>$ found for the CSL molecules in the liposomes is significantly higher than that for the CSL molecules in vesicles of the same lipid species throughout the temperature range covered. Nevertheless, increasing unsaturation of the lipid chains induces a systematic reduction in the orientational order in both liposomes and vesicle systems. It is important to note that the values of $<P_2>$ in liposomes and vesicles are lower than the values obtained from the corresponding multibilayer system. However, a simple extrapolation of the values of $<P_2>$ obtained for multibilayers at 12% and 24% water content to 35% , characteristic of the maximum hydration of liposomes [1], yielded values in excellent agreement with those found for the liposomes.
    We thus ascribe the differences in the orientational order between the planar multibilayers and liposomes to the higher hydration of the lipid headgroups in the latter systems. Furthermore, it appears that in its turn the high curvature of the bilayer in the vesicle systems also induces a reduction in the microscopic orientational order of the CSL molecules.

4.3.2. *Reorientational dynamics*. In contrast to the orientational order, similar values for the rotational diffusion coefficients are extracted from the simulations of the liposome and vesicle spectra, fig. 9. These values are a factor of 2-3 higher than those observed in the corresponding multibilayer system. Nevertheless, they can be obtained by a simple extrapolation of the multibilayer values to 35% hydration, as was done in Section 4.3.1. for $<P_2>$.
    Again, we find no clear dependence of the rates of motion on the chemical composition of the hydrocarbon chains of the lipids. It appears, therefore, that the reorientational dynamics of the CSL molecules are only affected by the temperature and the state of hydration of the lipid headgroup.

4.4. COMPARISON WITH FD TECHNIQUES

The order parameters and diffusion coefficients obtained here from ESR experiments can be compared with previous results from angle-resolved FD (AFD) experiments on oriented bilayers [22] in which 1-[4-(trime-thylammonio)phenyl]-6-phenyl-1,3,5-hexatriene (TMA-DPH) molecules were used as fluorescent probes. These molecules behave analogously to the CSL molecules as they are also expected to be anchored between the polar headgroups of the phospholipids [23]. In vesicle systems, however, the TMA-DPH molecules behave anomalously and partition between the bilayer and water phases [24]. It is also important to note that FD experiments on liposome systems are compromised by depolarization due to scattering.

The AFD experiments on multibilayer samples were analyzed with the same diffusion mechanism for reorientational motion, so that a direct comparison of both the order parameters and the diffusion coefficients can be made. As shown in Table I, the agreement between the two approaches is satisfactory for the order parameters, and within the estimated uncertainties for the diffusion rates. Note however, that only the diffusion coefficient $D_\perp$ can be determined for the TMA-DPH molecules. It indicates that, in spite of their vastly different chemical composition, both types of probe molecules undergo similar interactions with the surrounding lipid matrix. This shows that the probe molecules do not perturb the local structure to any large extent and that they indeed reflect the intrinsic behaviour of the lipid system.

TABLE I Comparison of results from ESR and AFD experiments on oriented multibilayers containing 24 wt % water. The AFD results were taken from [22] where TMA-DPH was used as a probe molecule. In the latter experiments $D_\parallel$ cannot be accessed.

lipid	experiment	T($^\circ$C)	$<P_2>$	$<P_4>$	$D_\perp(10^8 rad^2 s^{-1})$
DMPC	ESR	35	0.64	0.28	0.34
	AFD	35	0.64	0.29	0.29
POPC	ESR	19	0.63	0.27	0.17
	AFD	21	0.66	0.35	0.12
DOPC	ESR	21	0.52	0.18	0.23
	AFD	21	0.54	0.21	0.28
DLPC	ESR	21	0.50	0.20	0.20
	AFD	21	0.57	0.29	0.22

REFERENCES

(1) Silver, B.L. (1985) in The Physical Chemistry of Membranes, Solomon Press, New York.

(2) Redfield, A.G. (1965) in Advances in Magnetic Resonance vol. 1 (Waugh, J.S., ed.), pp. 1-32, Academic Press, New York.

(3) Freed, J.H., Bruno, G.V. and Polnaszek, C.F. (1971), J. Phys. Chem. 75, 3385-3399.

(4) Polnaszek, C.F., Bruno, G.V. and Freed, J.H. (1973), J. Chem. Phys. 58, 3185-3199.

(5) Freed, J.H. (1976) in Spin Labeling, Theory and Applications, (Berliner, L.J. ed.) Academic Press, New York, pp. 53-132.

(6) Lange, A., Marsh, D., Wassmer, K.H., Meier, P. and Kothe, G. (1985) Biochemistry 24, 4383-4392.

(7) Mayer, C., Müller, K., Weisz, K. and Kothe, G. (1988) Liq. Crystals 3, 797-806.

(8) Dammers, A.J. (1985) Ph.D. Thesis, Rijksuniversiteit Utrecht, Utrecht.

(9) Freed, J.H. and Fraenkel, G.K. (1963), J. Chem. Phys. 39, 326-248.

(10) Vega, A.J. and Fiat, D. (1974) J. Chem. Phys. 60, 579-583.

(11) Abragam, A. (1961) in The Principles of Nuclear Magnetism, Oxford Univ. Press, London.

(12) Schneider, D.J. and Freed, J.H. (1989) in Advances in Chemical Physics, 73, 387-453.

(13) Moro, G. and Freed, J.H. (1981) J. Chem. Phys. 74, 3757-3773.

(14) Dammers, A.J., Levine, Y.K. and Tjon, J.A. (1988) J. Chem. Phys., 89, 4505-4513.

(15) Gaffney, B.J. and McConnel, H.M. (1974), J. Magn. Reson. 16, 1-28.

(16) Sachse, J.-H., King, M.D. and Marsh, D. (1987), J. Magn. Reson. 71, 385-404.

(17) Seelig, A. and Seelig, J. (1977) Biochemistry 16, 45-50.

(18) van Ginkel, G., van Langen, H. and Levine, Y.K., Biochimie (in press).

(19) Ehrström, E. and Ehrenberg, A. (1983) Biochim. Biophys. Acta 735, 271-282.

(20) Chapman, D., Byrne, P. and Shipley, G.G. (1966) Proc. R. Soc. London, A 290, 115-142.

(21) Chapman, D. and Benga, G. (1984) in Biological Membranes V (Chapman, D. ed.) Academic Press, London, pp. 1-56.

(22) Deinum, D., van Langen, H., van Ginkel, G. and Levine, Y.K. (1988) Biochemistry 27, 852-860.

(23) Prendergast, F.G., Haugland, R.P. and Callahan, P.J. (1981) Biochemistry 20, 7333-7338.

(24) van Langen et al. (1989) Eur. Biophys. J. (in press).

(25) Rao, K.V.S., Polnaszek, C.F. and Freed, J.H. (1977) J. Phys. Chem. 81, 449-456.

ELECTROSTATIC INTERACTIONS IN SURFACTANT SOLUTION

M. DRIFFORD - P.J. DERIAN - L. BELLONI
CEA DLPC/Service de Chimie Moléculaire
CEN SACLAY
91191 GIF SUR YVETTE CEDEX
France

ABSTRACT. We present some experimental results for a surfactant solution in which the counterions contribute significantly to the scattered intensity.

The HNC approximation is applied to the primitive model of asymmetrical electrolytes that mimic "direct micellar" solutions. The structure factors $S_{ij}(q)$ are calculated for two or three components species (polyions, counterions, monomers).

Two applications are presented, the first is the investigation of a micellar solution of octyltrimethylammonium bromide (OTAB) with X-ray scattering. In this case, the high scattering length density of bromide ions contributes strongly to the total scattered intensity. The experimental curve in Small Angle X-ray Scattering presents a large intensity at high scattering vectors which is due to the contribution of bromide. We have also used the newly developed technique of Anomalous Small Angle X ray Scattering to study the bromide position correlation in micellar solutions. We observe a striking change in the shape of the peak which reflects the modification of the contribution of bromide to the scattered intensity.

To obtain a pure contribution of counterions to the scattered intensity, we have studied micellar solutions of Tetramethyl Ammonium Dodecyl Sulfate (TMADS) by Small Angle Neutron Scattering. In this case the counterion (TMA^+) contributes strongly to the neutron scattering intensity. At a particular D_2O/H_2O ratio, the micelle is perfectly matched with the solvent and one measures directly the counterion-counterion structure factor which presents a prominent peak reflecting a large accumulation of the TMA^+ ions around the micelles.

Th. Dorfmüller (ed.), Reactive and Flexible Molecules in Liquids, 309–325.
© *1989 by Kluwer Academic Publishers.*

1. INTRODUCTION

Surfactants (or amphiphilic) molecules consist of two parts, one polar (hydrophilic head) and the other one non polar (hydrophobic tail). When an amphiphile is dissolved in water, the molecules will partition between the surface and the bulk ; at increasing concentration the amphiphiles in the bulk undergo aggregation in order to reduce the contact area between the hydrophobic tails and water. The simplest possible aggregate is a globular "micelle" formed by 20-100 monomers which assemble their tails in a hydrocarbon core and expose their polar heads to the water [1,2].

The aim of the NATO conference is to discuss basic features of various "Reactive and Flexible Molecules in Liquids" and the purpose of this lecture is to show that surfactant solutions are composed of "reactive and flexible molecules in liquids". Most surfactants have a single n-alkyl chain with an ionic polar head. Thus the self-aggregation is a structure which is soft and flexible : it is a fluid like. This is because the forces that hold amphiphilic molecules together in micelles are not due to covalent bonds but arise from weaker hydrophobic hydrogen-bonding and screened electrostatic interactions. These major forces that govern the self assembly of surfactants into micelles derive from the hydrophobic interaction which induces the molecules to associate and the hydrophilic nature of polar groups which imposes the opposite requirement that they remain in contact with water.

Micellar solutions have been used extensively as media for chemical reactions. These effects have been related to the local concentration of the reactants into the micelles or at their surface. Their preferential orientation in the site of solubilization depends on the characteristics of the molecules present in the surfactant solution. Chemical reactivity studies have been also used to probe the dynamics of micellar systems and to obtain some indications as to the conformation or flexibility of the surfactant alkyl chain in micelles [3].

Thus, amphiphile chains or self aggregation of surfactants, can be considered as "flexible and reactive molecules in liquids".

Generally micellar systems contain ionizable groups on the polar heads and the solution is composed of :

- monomers species in water which are ionizable
- counter ions which are necessary to keep the electroneutrality
- micellar aggregations composed by 20-100 monomers which are stabilized by their strong mutual electrostatic interactions. The coulombic interaction is screened by mobile counter ions surrounding the macroions. As a result the polyion-polyion interaction is dominated by a double-layer repulsive interaction which, depending on ionic strength of the solution, can be long-ranged and with a magnitude of many k_BT at contact. This strong repulsive interaction between ionic micelles produces a pronounced correlation peak in the scattered intensity by small angle neutron (SANS) and X-ray (SAXS) scattering techniques.

A lot of experimentals on micellar solutions have made use of scattering techniques [4,5] and two theoretical approaches have been used :

- The first one is the most classical. In the majority of micellar solutions, the asymmetry of size and charge between macroions and counterions is so large that the experiments measure essentially the polyion-polyion correlation. Thus, for simplicity it is interesting to consider the solution as a one component model (OCM) with an effective polyion-polyion pair potential. One such potential is the so-called DLVO (Derjaguin-Landau-Verwey-Overbeek) and the effect of small ions is taken into account only through the expression of the potential [6].

- The second approach is more rigourous because all components (polyions - counterions, co ions) are considered and in this case the scattered intensity is related to all structure factors. These Equations are correctly calculated with the help of HNC (Hyper Netted Chain Equation) [7].

In the investigation of micellar systems with small aggregation number or with heavy counterions, the asymmetry is not large and the scattered intensity is related to all structure factors. Two applications are presented to show the large contribution of counterions on the scattered intensity :

- A small angle X ray scattering experiment on a micellar solution of octyltrimethylammonium bromide.

- A neutron scattering experiment on a micellar solution of tetramethyl-ammonium dodecylsulfate (TMADS).

In the present paper, we analyse the experimental data obtained on these two systems by using the wellknown primitive model and the correlation functions have been calculated with the HNC integral equation [8].

2. THEORY

A. Scattered intensity by a solution of interacting spheres

the normalized intensity scattered by a mixture of rigid spheres is given by

$$I(q) = \sum_{i,j}^{s} \sqrt{\rho_i \rho_j} \; f_i(q) f_j(q) \; S_{ij}(q) \qquad (1)$$

where ρ_i is the number concentration of species i, $f_i(q)$ its amplitude factor. The intensity scattered by a single particle i, $F_i(q) = f_i(q)^2$ is usually called the form factor.

For a spherically symmetric particle, the amplitude factor is

$$f_i(q) = \int_{(V_i)} b_i(r) \; \frac{\sin qr}{qr} \; 4\pi r^2 \; dr \qquad (2)$$

where the integral involves b_i, the electronic density in light and x-ray scattering or the scattering length density in neutron scattering. For a uniform particle, b_i is constant over the volume of the particle V_i and $f_i(q) = V_i b_i g(qr_i)$ where $r_i = \sigma_i/2$ is the radius of sphere i and

$$g(x) = 3 \; \frac{\sin x - x \cos x}{x^3} \qquad (3)$$

The spatial correlations between the different particles are introduced in the spatial structure factors $S_{ij}(q)$ which are the normalized Fourier transforms of the pair distribution functions $g_{ij}(r)$:

$$S_{ij}(q) = \delta_{ij} + (\rho_i \rho_j)^{1/2} \int_0^\infty [g_{ij}(r)-1] \times (\sin qr/qr) 4\pi r^2 dr \qquad (4)$$

The scattering vector q is related to the scattering angle θ by $q = (4\pi/\lambda)\sin \theta/2$ where λ is the wavelength of the incident radiation.

To clarify the application of Eq. (1) to aqueous solutions, the following remarks may be helpful : In theory, the sum involves the solvent and the different solutes. In practice, the solvent is treated as a continuum and Eq. (1) can represent the excess intensity with regard to the solvent if the solvent scattering length density is subtracted from the particle densities b_i. In the following, Eq. (1) represents the excess intensity and only involves the solutes ; therefore, b_i is the excess of scattering length density with regard to the solvent. This approximation is correct if the particles are much larger than the solvent particles or if the osmotic compressibility is much larger than the total compressibility of the solution.

B. HNC study in the primitive model

The functions $g_{ij}(r)$ and $S_{ij}(q)$ are deduced from the pair potentials $v_{ij}(r)$. The simplest model for the interactions in multicomponent charged systems is the so-called primitive model : the solution is assumed to be a mixture of charged hard spheres immersed in a continuous solvent of dielectric constant ϵ. Thus, the pair potential $v_{ij}(r)$ between particles of species i and j is

$$\beta v_{ij}(r) = +\infty \qquad r < \frac{\sigma_i + \sigma_j}{2}$$

$$= Z_i Z_j \frac{L_B}{r} \qquad r > \sigma_{ij} \qquad (5)$$

where $\beta = 1/k_B T$, $L_B = e^2/4\pi\epsilon_0\epsilon k_B T$ is the Bjerrum length, and Z_i is the electric charge of species i. It is worth noting that in this model the small ions are treated in the same manner as the large polyions.

The exact Ornstein-Zernike equation is a matrix relation between the Fourier transforms of the distribution functions :

$$(1 + \overline{\overline{h}})\,(1 - \overline{\overline{c}}) = \overline{\overline{1}} \qquad (6)$$

The elements $\hat{h}_{ij}(q)$ and $\hat{c}_{ij}(q)$ are the normalized Fourier transforms of the total and direct distribution functions $h_{ij}(r) = g_{ij}(r) - 1$ and $c_{ij}(r)$. The Fourier transform of the element ij is normalized by the factor $(\rho_i \rho_j)^{1/2}$.

The OZ equation is closed with the approximate hypernetted chain integral equation (HNC) :

$$g_{ij}(r) = \exp[- v_{ij}(r) + h_{ij}(r) - c_{ij}(r)] \qquad (7)$$

This equation is very accurate for charged systems and has been recently applied to highly charged solutions. Contrary to other simple approximations, this equation requires an iterative computation [8].

C. Condensation and effective system [9]

There are two possibilities for studying the solution, either with the original structural system or with a new effective system which involves larger and less charged effective polyions and less numerous counterions in order to preserve the electroneutrality condition. The physical meaning of this result is that the counterions which are close to the polyion surface feel an electrostatic attraction much larger than the thermal energy. Thus, their motion is strongly correlated to the polyion motion and a good representation is to consider them as a part of a rigid particle, the effective polyion. On the contrary, the counterions which are farther from the polyion are nearly "free".

D. Two-component "reference" system

Many micelles have a radius of about 25 Å, an aggregation number of 20-100, and have been experimentally studied in the volume fraction range of 1 % - 20 %. Therefore, we have chosen for our two-component reference case a system characterized by $\sigma_c = 5$ Å, $\sigma_p = 50$ Å, $Z_c = -1$, $Z_p = 40$, $\rho_c = 0,1$ M, and thus $\rho_p = 0.0025$ M corresponding to a volume fraction of 10 %.

The effective polyion is characterized by $\sigma_p^{eff} = 60$ Å and $Z_p^{eff} = 25.24$. The effective ionic concentration is $\rho_c^{eff} = 6.27 \times 10^{-2}$M.

We have plotted the different partial structure factors for the structural system and for the effective system, (Fig. 1a, b). The polyion-polyion structure factors are nearly identical in both approaches.

From the three structure factors it is possible now to calculate the scattering intensity with Eq.(1) provided one chooses the two scattering densities b_c and b_p.

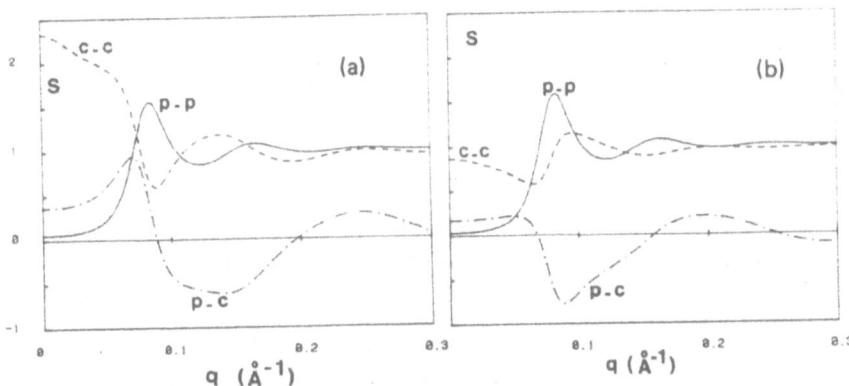

Figure 1. Partial structure factors : p = polyion ; c = counterion ρ_p = 0.0025 M (Φ = 10 %) ; T = 298 K ; ε = 78 ; σ_c = 5 Å ; Z_c = - 1. (a) Structural system : (σ_p = 50 Å ; Z_p = 40). (b) Effective system : (σ_p^{eff} = 60 Å ; Z_p^{eff} = 20.08).

3. EXPERIMENTAL RESULTS IN X-RAY SCATTERING

A. Materials

Octyltrimethylammonium bromide (OTAB) was prepared by following the reaction between octylamine and methylbromide in absolute ethanolic medium at 64°C for 3-5 days and was purified by recrystallization in diethyloxide and acetone. It is noteworthy that the recrystallizations were made in an inert atmosphere exempt of water because of the great hygroscopicity of OTAB. The purity of the final product was checked by chromatography. OTAB forms an aqueous cationic micelle having a very high critical micellar concentration (CMC) of 0.32 M in pure water. This critical micellar concentration was determined by quasielastic light scattering experiments.

The sample solution contains 1 M of OTAB and is prepared with deionized water. The sample cell is made of aluminium and has a thickness of 1 mm. The windows are in low density mylar of thickness 26 μm. Their transmission is about 0,96 for Cu $K\alpha_1$. The small angle X Ray Scattering apparatus was described in [10].

B. Results of Small Angle X-Ray Scattering (SAXS)

According to quasielastic light scattering experiments and neutron scattering (SANS) studies, the micelles of OTAB have an aggregation number of 20-30 for our sample concentration [11].

The micelle is supposed to be composed of an hydrocarbon core containing the surfactants tails $[CH_3-(CH_2)_7-]$ and of an outer head group $[-N^+-(CH_3)_3]$ region. The hydrocarbon core can be faithfully represented by a sphere with uniform density. The x-ray scattering on the micelle is largely due to the core because the head group region is heavilly wetted by water molecules and has thus a scattering length density very close to that of the solvent.

The electronic density b_p of the core is equal to the total number of electrons in the core divided by the volume of the core. The "free" OTAB monomers, which correspond to the critical micellar concentration are supposed to be spheres of uniform density. The effective steric volume of a monomer in the bulk solution was deduced from density measurement on OTAB solutions and is $v_2 = 346 \text{ Å}^3$ which gives a monomer diameter of $\sigma_m = 8.71 \text{ Å}$. The electronic density of a monomer is $b_m = 0.2835$ electrons/Å^3. The counterions Br^- have an electronic density $b_{Br} = 1.16$ electrons/Å^3 and a volume of about 31 Å^3 corresponding to a diameter of about 3.9 Å. The electronic density of water is $b_w = 0.33$ electrons/Å^3.

Our different structure factors are computed as described in Section II D and we have chosen $b_c V_c / b_p V_p = -0.04$ which corresponds to the experimental case of an x-ray scattering experiment on our micellar solution with the counterions Br^-. (The choice of the negative sign for b_p is not essential here).

The different contributions pp,pc, and cc to the scattered intensities are plotted in Figs 2 (a) and 2 (b). We note that the polyion-polyion term is not dominant although the form factor of the structural polyion at q = 0 and is 625 times larger than that of the counterion. At large q, the counterions are the only scatters. For low and intermediate q, the cross term plays an important part in the absolute value of the total intensity and shifts the position of its maximum, especially in the structural case. In Fig. 2 (b), the total intensity and the pp contribution present about the same shape for $q < 0.15$ Å^{-1} but with different amplitudes.

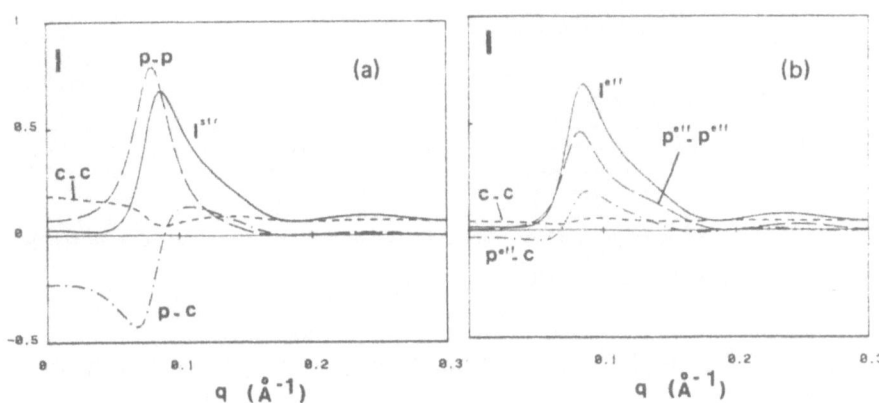

Figure 2. Contributions of the different terms in the scattered intensity for the structural (a) and effective (b) systems. ($b_c V_c / b_p V_p = - 0.04$)

The main conclusion is that for such a system it is not sufficient to consider only the polyion-polyion contribution as it is realized in the one-component approach of Hayter and Penfold [6], even if the scattering due to the condensed counterions is introduced through the form factor of the effective polyion. The contribution of the free counterions cannot be neglected even for the effective system, unless the ratio of the scattering length $V_c b_c / V_p b_p$ is very small.

To fit the experimental data, we have to specify only one parameter : the aggregation number N. The charge of the polyion which represents the micelle is exactly the structural charge, i.e., N. The best fit of the experimental scattered intensity is obtained for N = 27 (see Fig. 3). The agreement between the two curves is excellent. These values correspond closely to those determined by quasielastic light scattering experiments [11]. It is interesting to note that, since the scattered intensity is measured on an absolute scale, the comparison between the theoretical curve and the experimental one is done on an absolute scale too.

318

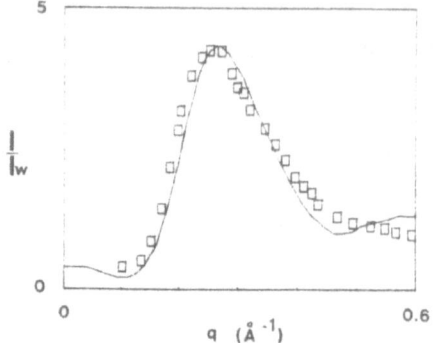

Figure 3. ☐ Experimental x-ray scattering intensity by the OTAB system. - Theoretical curve (Z = 27).

C - Results in Anomalous Small Angle X-Ray Scattering (ASAXS)

The ASAXS is a contrast variation technique shich relies on the fact that in X-ray scattering, the scattering amplitude f of an element depends on the energy of the incident photon E. Although generally very weak, that dependance beçomes considerable near the absorption edges of that element, and is maximum in a narrow range of a few tens of eV at the edge. Therefore, by varying the photon energy, it is possible to selectively probe the contribution of that atom to the scattered intensity. Changing the energy range changes the atom probed.

More precisely, f can be expressed

$$f(E) = f + f'(E) + if''(E) \qquad (8)$$

where f is the usual Thomson factor, equal to the atomic number Z in the SAXS range and f' and f" are the anomalous dispersion terms.

The ASAXS experiments were performed at LURE-DCI, Orsay, using the high resolution spectrometer D22 especially designed for ASAXS experiments [12].

Five energies were chosen, 10000, 13000, 13400, 13440 and 13470 eV, below the bromine K-edge at 13474 eV. They correspond to a variation of the tabulated values of f' from - 1.220 electron to - 7.049 and of f" from 0.866 electron to 0.503.

Experimental scattering profiles, arbitrarily normalized to the same intensity of the main peak are shown in Fig. 4 for four different energies [13]. The representation emphasizes a striking change in the shape of the peak : as the edge is approached and the contribution of bromine to the scattering decreases, the intensity of the shoulder on the high-q side relative to that of the main peak continuously decreases. Note that the profile close to the edge is qualitatively similar to what is observed in neutron scattering where the contrast of the counterion is matched to that of the solvent. Conversion to relative intensities show that the intensity of the main peak increases by more than 30 % between 10000 eV and 13470 eV. From a qualitative standpoint one can firstly conclude that the nonuniform distribution of the bromine counterions is clearly evidenced by anomalous SAXS even for such low concentration of anomalous element.

'Although it has not been possible to extract directly the partial structure factors related to the counterions but only their linear combinations, important information has been obtained by comparing with theoretical models for micellar solutions. It was found that the shell of condensed bromide ions is thinner than predicted by a purely electrostatic interaction. This can be explained by the existence of an added short-range attraction between polyion and counterion [14].

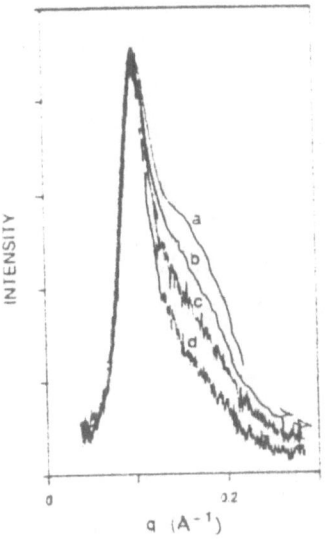

Figure 4. Scattered intensity vs. q at the energies of (a) 10,000 (b) 13,000 (c) 13,400 and (d) 13,470 eV, normalized to the same intensity of the main peak Curves (c) and (d) have been smoothed.

4. EXPERIMENTAL RESULTS IN SMALL ANGLE NEUTRON SCATTERING (SANS)

A. Materials

TMA$_h$DS$_h$ (fully protonated) and TMA$_h$DS$_d$ (fully deuterated chains) were prepared in the same way. The method consists of preparing the n-dodecylsulfuric acid from sodium dodecylsulfate (SDS) by the exchange of Na$^+$ with H$^+$ on a cationic resin (Lewatit SP 1080). The acid is quantitatively neutralized by a solution of TMAOH. Water is removed by evaporation and the crystallized TMADS is dried under vacuum at 50° C. Since all reactions are quantitative, the yield of this preparation is 100 %. The purity of TMADS was checked by atomic absorption (Perkin Elmer 2380). The final product contains less than 0.1 weight % sodium which essentially comes from the dissolution of TMAOH. The solutions are 0.1 M in TMADS. The critical micellar concentration was found to be $5.45.10^{-3}$ M by surface tension measurements and confirmed by light scattering.

B. Experiments

Neutron scattering experiments were carried out on PACE, one of the SANS instrument of the Laboratoire Léon Brillouin, CEN-Saclay (France). Three detector positions and three wavelengths (d = 1.2 m/λ = 4.67 Å, 1.8 m/5.63 Å, 3.0 m/7.96 Å) were used, giving a q-range of 0.008 to 0.3 Å^{-1}. The samples were contained in quartz cells of 1 or 2 mm path length. Measurements were done at 25° C. Scattering from the samples was corrected for detector background and sensitivity, empty cell scattering, incoherent scattering, sample transmission and cell thickness. Solvent intensity was subtracted from that of the sample. The absolute cross-section were calculated from these corrected intensities by multiplying them by the absolute scattering of pure water.

C. Results and discussion [15]

The different curves in Figure 5 correspond to different contrast, i.e. to different D$_2$O/H$_2$O compositions and provoke the following comments.

As the volume fraction of D$_2$O decreases from 100 % to 17.2 %, the experimental spectra present various heights and shapes. At 100 % in D$_2$O, the high and narrow peak contains the contribution of the different species. From 100 % to 70 %, the contribution due to the surfactant falls and then disappears. The 70 % curve represents exactly the pure contribution of the counterions since, as shown above, the signal due to the internal structure of the micelles is negligible at this contrast. Below a D_2O volume fraction of 70 %, the intensities still continue to decrease because of a partial cancellation between the different terms. Indeed, for these solvent compositions the relative scattering length densities of the micelle and of the counterions have opposite signs. For 56 % in D_2O, the cancellation is the most efficient and the scattered intensity vanishes. For lower D_2O/H_2O ratios, the intensities increase again but the shapes of the peaks are not similar. At 17.2 %, the scattered intensity is dominated by the signal due to the micelles and the shoulder visible at the right of the peak is clearly due to the polyion-polyion structure factor peak.

Theoretically, all intensities in fig. 5 can be expressed with only three partial intensities.

$$I = \Delta\rho^2_c I_{cc} + 2 \Delta\rho_c \Delta\rho_p I_{pc} + \Delta\rho^2_p I_{pp}, \qquad (9)$$

where $\Delta\rho i = \rho i - \rho s$ is the scattering length density of the species i (polyion p and counterion c) relative to the solvent s. Then, with three experimental curves corresponding to three different contrasts, it is possible to extract the three I_{ij}. We have chosen the 70, 67 and 40 % D_2O volume fraction curves.

The next stage in the treatment is to extract the three partial structure factors $S_{ij}(q)$ from the partial intensities. Now, we have to make some hypotheses on how to represent the strong accumulation of the counterions on the micellar surfaces. The simplest approach is the structural one : no counterion is explicitly attached to the micelles, all of them are in the bulk [9]. The condensation phenomenon results only *a posteriori* from the high micelle-ion correlations. Then,. each term in eq. (9) is only proportional to one partial structure factor $S_{ij}(q)$.

322

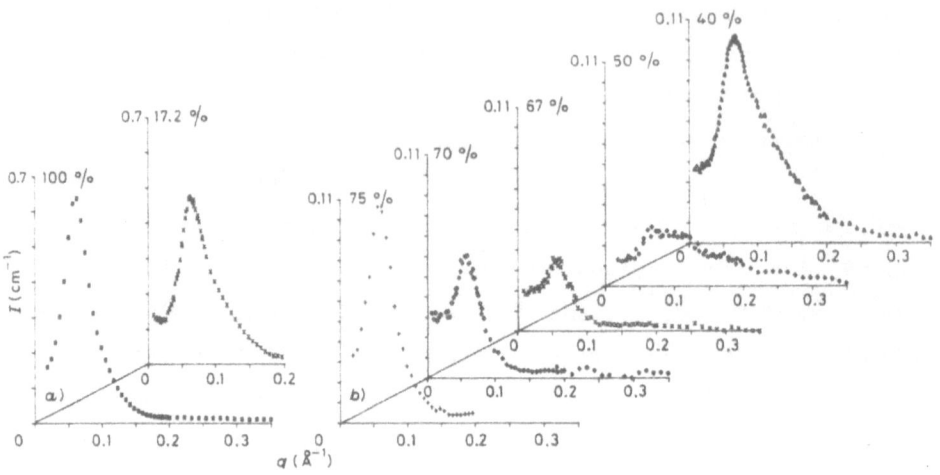

Figure 5. Absolute SANS intensities of a 0.1 M tetramethylammonium dodecylsulfate solution at different isotopic compositions for the solvent. Each curve is labelled with the D_2O volume fraction.

Over the whole experimental q-range, $f_c(q)$ is nearly constant and equal to $1.S_{cc}$, which is then simply proportional to the 70 % curve is plotted in fig. 6a on an absolute scale. This figure is perhaps the most important of this paper. Note the high values of S_{cc} ($\approx$ 11 at q = 0 and 20 for the peak) which reflect the strong correlations between the counterions which lie in the vicinity of the micellar surfaces. At high q, the measured asymptotic value of $S_{cc}(q)$ is nonzero as expected and seems to the roughly equal to the theoretical value 1.

For the two other structure factors $S_{pc}(q)$ and $S_{pp}(q)$, it is necessary to assume a value for the aggregation number N. We have chosen N = 70 which agrees with a previous work and with our theoretical investigations. Note that the precise value of N is unimportant : similar results are obtained with N = 80. N = 70 corresponds to a micellar radius of $\approx$ 19 Å.

The experimental S_{pp} presents a classical shape with a low value at $q = 0$ and a peak which results from the screened electrostatic repulsion between micelles. The cross structure factor S_{pc} presents negative values for $q \geq 0.15$ $\mathring{A}^{-1}$ in agreement with the theoretical behaviour at large q (see figures 1a and b). In the infinite wavelength limit $q = 0$, the three structure factors must verify the Stillinger-Lovett condition which expresses the electroneutrality of the solution, $S_{cc}(0) = (Z_p)^{1/2} S_{pc}(0) = Z_p S_{pp}(0)$ where Z_p is the absolute charge of the micelle. Here, in the structural approach, $Z_p = N$. The experimental values of these two ratios are, respectively, 55 ± 10 and 9 ± 2 in good agreement with the theoretical predictions.

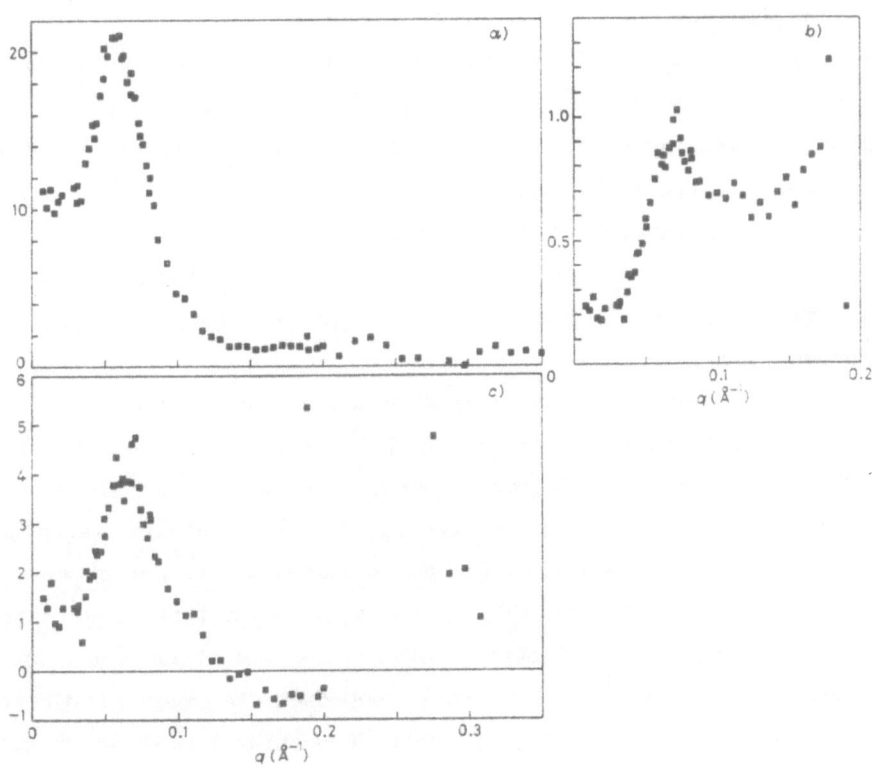

Figure 6. Counterion-counterion (a), polyion-polyion (b) and polyion-counterion (c) partial structure factors.

5. CONCLUSION

In this paper, we have investigated the contribution of the small ions to the scattered intensity in asymmetrical polyelectrolytes. Such refined study is not possible with the one-component approach and absolutely necessitates the use of the n-component primitive model. For the moment, the best relation between the structure factors and the charged hard sphere pair potentials is the HNC integral equation. As a consequence of the strong accumulation of the counterions around the polyions surface, such systems can be considered in two different ways. The structural polyion with some of the "condensed" counterions can be analyzed as a single particle : a larger and less charged effective polyion.

An application of this study has been realized by investigating a micellar solution of OTAB with X-ray scattering. The experimental spectrum presents a particular feature, a nonzero value at high q, which results from the significant contribution of the bromide ions to the intensity. The theoretical spectrum reproduces very well this special behavior. The single fit parameter is the aggregation number N = 27. Contrary to the classical procedure, the micellar charge is not an adjustable parameter but is equal to N. This allows the suppression of one extra parameter, which is somewhat new.

At this stage, we needed a neutron scattering experiments which measure only the intensity due to the small ions. We have chosen a solution of TMADS. As the solvent isotopic composition varies, the scattering spectra present various heights and shapes. At the contrast match point, the contribution due to the surfactant disappears and one observes the pure contribution due to the TMA^+ counterions. With three contrasts, we have extracted the three partial intensities and then the three partial structure factors in the structural approach. As expected, the counterion-counterion structure factor presents a prominent peak which reflects the accumulation of the TMA^+ ions around the micelles.

Acknowledgements

The authors thank M. Dubois and P. Lixon for preforming the synthesis of OTAB and TMADS - and C. Williams for performing the ASAXS experiments.

[1] Tanford C. (1973) "The hydrophobic effect : Formation of micelles and micelles and biological membranes (Willey).

[2] Degiorgio V. and Corti M. (1985) Eds Physics of Amphiphiles : Micelles, Vesicles and Microemulsions (North Holland).

[3] Feudler J.H. and Feudler E.J. (1975) "Catalysis in Micellar and Macromolecular System" (Academic Press).

[4] Cabane B. in "Surfactant Solutions. New Methods of investigation" Ed. Zana R Surfactant Sciences Series 22 57 (Deker) p. 987

[5] Chevalier Y. and Zemb Th. (1989) "Structure of micelles and microemulsions" Submitted Colloid and Interface Sci.

[6] Hayter J. and Penfold J. (1981) J. Chem. Soc. Faraday Trans. I 77, 1851.

[7] Belloni L. (1986) J. Chem. Phys. 85, 519.

[8] Belloni L. (1985) Chem. Phys. 99, 43.

[9] Dérian P.J., Belloni L., Drifford M. (1987) J. Chem. Phys. 86 (10), 5706.

[10] Zemb Th. and Charpin P. (1985) J. Phys. (Paris) 46, 249.

[11] Drifford M., Belloni L. and Dubois M. (1985) J. Colloid Interface Sci. 105, 587.

[12] Lyon O. and Williams C. (1987) Lure Activity Report.

[13] Dérian P.J. and Williams C. (1987). "Ordering and Organizing in Ionic Solutions" p. 233-240 World Scientific Publications Co and Yamada Science Foundation Osaka.

[14] Belloni L., Dérian P.J., Drifford M. To be published

[15] Dérian P.J., Belloni L., Drifford M. (1988) Europhys. Lett. 7(3), 243.

STRUCTURE AND DYNAMICS OF BRANCHED (EPOXY) POLYMERS AND KINETICS OF ITS POLYMERIZATION PROCESS

by

Benjamin Chu and Chi Wu
Department of Chemistry
State University of New York at Stony Brook
Long Island, New York 11794-3400
U. S. A.

ABSTRACT. Laser light scattering (LLS) and small-angle x-ray scattering (SAXS) studies have been made of the curing of epoxy resins from 1,4-butanediol diglycidyl ether with cis-1,2-cyclohexanedicarboxylic anhydride. The epoxy resin before its gelation threshold is soluble in methyl ethyl ketone, and scattering techniques can be used to determine the weight-average molecular weight (M_w), the fractal dimension (d_f), and the molecular weight distribution (MWD) of the branched epoxy polymer during each stage of the initial polymerization process. The MWDs obtained from LLS were compared with those determined by conventional size exclusion chromatography (SEC). From the comparison, we were able to develop a new absolute calibration procedure for SEC of specific branched polymers. By investigating the LLS envelope of the three-dimensional crosslinking process near and pass the gelation threshold and by applying the Debye-Bueche theory of light scattering for inhomogeneous solids, the structural changes of the branched epoxy polymer during the curing process could be evaluated. The change in the correlation length (a) and the mean squared average of local dielectric constant fluctuations ($\overline{\eta^2}$) could be divided into four main stages. Finally, The branching kinetics of the copolymerization reaction could be approximated by using Smoluchowski's coagulation equation.

1. INTRODUCTION

The kinetics and mechanism of copolymerization of epoxy resins with anhydrides, with or without a catalyst, have been of interest because these materials often constitute an important component in reinforced composites. However, the mechanism of the curing reaction of epoxy resins and anhydrides has been somewhat uncertain as a number of partially conflicting reaction mechanisms have been proposed in recent years[1]. With our light and x-ray scattering techniques, we find that our data, obtained from the curing of 1,4-butanediol diglycidyl ether with

327

Th. Dorfmüller (ed.), Reactive and Flexible Molecules in Liquids, 327–388.
© *1989 All Rights Reserved.*

cis-1,2-cyclohexane-dicarboxylic anhydride, in the presence of benzyl dimethyl amine as a catalyst, can best be fitted by the zeroth-order reaction. However, it is not crucial for us to know how the reaction is initiated if we focus our attention mainly to the branching kinetics, and the structure and dynamics of branched epoxy polymer products during different stages of the copolymerization reaction process.

In this article, we shall review four aspects of our studies: (i) the branching kinetics of the copolymerization reaction based on Smoluchowski's coagulation equation, (ii) the concept of fractal geometry as applied to epoxy structures, (iii) the determination of molecular weight distribution of branched epoxy polymers based on Brownian dynamics and (iv) the characterization of structural inhomogeneities based on the Debye-Bueche theory.

(i) the branching kinetics of the copolymerization reaction and the distribution of highly branched epoxy copolymers can be approximated by Smoluchowski's coagulation equation. The equation has been applied to study the structure of clusters produced by the kinetic aggregation of colloidal particles.[2-11] If we take the overall branching probability $w(i,j)$ to be proportional to the sum of active sites on the two polymers, i.e. $w(i,j) \propto (i+j)$ where i and j are the active sites on polymers i and j, respectively, the kinetic equation can be solved explicitly to obtain the weight-average molecular weight (M_w) and the molecular weight distribution (MWD) at different reaction stages.

(ii) The concept of fractal geometry[12,13] can be applied to investigations cured epoxy systems. The fractal concept has shown to be a useful approach to describe the structure of random systems, such as aggregates of colloidal silica,[14-16] branched silica condensation polymers,[17] cross-linked poly(dimethyl -siloxane),[18,19] aggregating proteins[20] and gold colloids,[21] as well as diffusion-limited[22,23] polymerization of the conducting polymer polypyrrole[24] and other growth processes, e.g., pecolation[25] and cluster-cluster aggregation.[26-28]

(iii) the molecular weight distribution (MWD) of the branched epoxy polymer formed during each reaction stage can be estimated by means of dynamic light scattering. The procedure is as follows. Estimates of the normalized characteristic line-width (Γ) distribution function, $G(\Gamma)$, can be obtained from the measured intensity-intensity time correlation function, $G^{(2)}(\tau)$, by using the Laplace inversion. $G(\Gamma)$ can then be transformed to the molecular weight distribution by incorporating information based on the static and dynamic properties of the branched epoxy polymer solution, i.e. the weight-average molecular weight (M_w), the second virial coefficient (A_2), the z-average root-mean-square radius of gyration (R_g), the diffusion second virial coefficient (k_d), the z-average translational diffusion coefficient at infinite dilution

$(\bar{D}_o^o)$, and the scaling relation $D_o^o - k_D M^{-\alpha_D}$. Knowledge gained from laser light scattering (LLS) of our broad MWD epoxy polymers is sufficient to calibrate the size exclusion chromatographic (SEC) column for specific branched epoxy polymer studies.

(iv) The laser light scattering intensity envelope in the presence of different amounts of catalyst at different temperatures (60, 70, 80 and 90°C) can be analyzed according to the Debye-Bueche theory.[29] The intensity of light scattered by an inhomogeneous medium is dependent upon the local refractive index difference in the inhomogeneous medium in terms of the mean square average of local dielectric constant fluctuations $\bar{\eta}^2$ which can be related to structural changes during the copolymerization process.

Experimentally, we used small angle x-ray scattering (SAXS) at the State University of New York (SUNY) X21A2 beamline, National Synchrotron Light Source (NSLS), Brookhaven National Laboratory (BNL) and laser light scattering (LLS) to measure the angular distribution of absolute scattered intensity ($I(\theta)$) and $G^{(2)}(\tau)$ at different reaction stages with % conversion determined by chemical analysis. From $I(\theta)$ in dilute solution, we can determine M_w, R_g, A_2 and the fractal dimension (d_f).

By using the Laplace inversion, we can transform $G^{(2)}(\tau)$ to estimate $G(\Gamma)$ which can be used to estimate the molecular weight distribution. The experimental results are then compared with the calculated values based on an analytical solution of Smoluchowski's coagulation equation, using the simplest reasonable assumption concerning the functional form of the reaction kernel in that equation. The comparison indicates that the equation provides a useful theoretical framework for the interpretation of the experimental results but requires a kernel of more general form in order to be of value in making quantitative predictions. From $I(\theta)$ measured in situ during the curing process, the correlation length (a) which characterizes the extension of the local inhomogeneous domain and the mean square average dielectric constant fluctuations ($\bar{\eta}^2$) could be used to characterize the changes in optical inhomogeneities at different copolymerization stages. The changes can be divided into four main stages as the reaction progresses.

2. EXPERIMENTAL METHODS

2.1 Materials.

1,4-butanediol diglycidyl ether (DGEB, M_w - 202.3 g/mol) and cis-1,2-cyclohexanedicarboxylic anhydride (CH, M_w - 154.2 g/mol) were purchased from Aldrich Chemical Company and used without further purification since we were able to obtain the same experimental results after both components were purified by vacuum (≈0.01 mm Hg) distillation. The catalyst (CA), benzyl dimethyl amine (M_w = 135.2 g/mol), courtesy of Gary L. Hagnauer, Polymer Research Division, Army Materials Technology Laboratory, Watertown, Mass.) was vacuum distilled before use.

2.2 Preparation of Solutions.

Known weights of CH were heated to ~ 50°C in order to melt the CH. The melted CH was then cooled to ~ 37°C, a few degrees above the melting point of CH (~ 32-34°C), and mixed well with known weights of DGEB. Then, a small amount of catalyst was added to the homogeneous liquid mixture using a Drummond digital microdispenser (± 0.01 μL). The well-mixed reaction mixture containing a molar ratio of epoxy (DGEB) : curing agent (CH) : catalyst (CA) = 1:2:0.001 was reacted at 80°C ± 0.5°C in an oil bath. Samples containing the epoxy polymer and unreacted monomers (DGEB and CH) were withdrawn from the reaction mixture during the course of the copolymerization reaction until the gel point was reached. Compositions of the reaction mixture could be analyzed chemically.[30] Portions of withdrawn samples were further dissolved in methyl ethyl ketone (MEK) for LLS measurements and in tetrahydrofuran (THF) for SEC measurements. Concentrations of the epoxy polymer ranged from ~ 1×10^{-3} g/mL for LLS experiments to ~ 1×10^{-2} g/mL for SEC experiments. Samples for LLS measurements were centrifuged at 7000 gravity and room temperatures for four hours. A middle portion of the centrifuged solution was then transferred to dust-free cylindrical light scattering cells of 10-17 mm o.d. by using a dust-free pipet.

2.3 Methods of Measurement.

A high-temperature light-scattering spectrometer was used for measurements of the angular distribution of absolute scattered intensity as well as its spectral distribution.[31] The glass jacket (4 in Figure 2 of ref 31) was modified so that the inner brass thermostat (5 in Figure 2 of ref 31) was immersed in a glass jacket containing refractive-index matching oil. This more standard arrangement permitted a reduction in the o.d. of the cylindrical light-scattering cell to between 10 and 17 mm and consequent reduction in the solution volume required to carry out light-scattering experiments. With the refractive index matching oil, we could also cover a broader scattering angular range varying from about 10° to 140°, even though the smaller accessible scattering angular range was relatively unimportant for the epoxy characterization, especially during the initial curing process. The details of laser light-scattering instrumentation have been described elsewhere.[31]

We have modified a Kratky block collimation system[32] for SAXS at SUNY X21A2 beamline, NSLS, BNL. For the epoxy polymer studies, we used a slit width of ~0.5 mm and covered a K range between ~0.07 and 3.4 nm^{-1} where the scattering wave vector $K = (4\pi/\lambda)\sin(\theta/2)$ with λ and θ being an X-ray wavelength of 0.154 nm and the scattering angle, respectively. Two ionization chambers were used to measure the sample transmission and reference intensity. The SAXS curves were corrected for detector linearity, parasitic scattering, solvent background, and sample attenuation. Desmearing was unnecessary as the incident X-ray beam had a small cross section of 0.5 x 2 mm at the sample chamber.

For SEC experiments, we used three ultrastyragel columns designated as 10^2 Å, 10^3 Å and 10^4 Å (Waters Associates) connected in series, a pump (Waters Model 590) operating at a flow rate of 1.0 mL/min, and a differential refractometer (Waters Model R401) as the detector. The chromatogram was simultaneously recorded on a strip chart recorder and a microcomputer. The sample injection volume was 50 μL with a concentration of $\approx 1 \times 10^{-2}$ g/mL, which is below the over-loading condition. All SEC experiments were performed at 45°C in order to increase the efficiency of the columns.

3. Results and Discussion.

3.1. Refractive Index Increment Measurements.

An absolute determination of the polymer molecular weight requires information on the refractive index increment $(\partial n/\partial C)_{T,P} \approx dn/dC$ as well as the Rayleigh ratio. The refractive index increment was determined according to the procedure outlined in reference 31. We did not try to isolate the epoxy polymer from varying amounts of unpolymerized DGEB and CH. Instead, we determined dn/dC for the epoxy polymers in MEK in the presence of small amounts of DGEB and CH.

We have taken into account the varying amounts of DGEB and CH by adding corresponding quantities of unreacted DGEB and CH with the solvent MEK in order to keep the ratio of MEK:DGEB:CH constant for the branched epoxy polymer solutions. Thus, for each branched epoxy polymer which we withdrew from the reaction mixture, we dissolved the polymer at different concentrations using a mixed solvent with a constant ratio of MEK:DGEB:CH. Furthermore, we took into account possible solvent preferential interactions by using two different mixed solvent ratios (MEK:DGEB:CH = 1:0.03:0.025 and 1:0.001:0.007 in volume ratios). The same results (molecular weight and radius of gyration) were obtained. Thus, we do not have to take into account the small variation in the mixed solvent composition in our analysis. dn/dC = 0.180±0.002 within the above two volume ratios of MEK:DGEB:CH at λ_o = 488 nm and 25°C and the refractive index of the corresponding solvent mixture is 1.377.

3.2. Light Scattering Intensity Measurements.

The excess absolute integrated intensity of light scattered by a dilute polymer solution has the form

$$\frac{HC}{R_{vv}(K)} = \frac{1}{M_w}\left(1 + \frac{K^2 R_g^2}{3}\right) + 2A_2C \tag{1}$$

where the optical constant $H = (4\pi^2 n^2/N_A\lambda_o^4)(dn/dC)^2$, with $\lambda(=\lambda_o/n)$, A_2, R_g^2 $(=<R_g^2>_z^{1/2})$, M_w and K being, respectively, the wavelength of light in the scattering medium, the second virial coefficient, the z-average root-mean-square radius of gyration, the weight average molecular weight and the magnitude of the momentum transfer vector. The subscripts vv denote vertically polarized incident and scattered light. From a Zimm plot as shown typically in Figure 1, we can determine M_w, R_g and A_2. The results are listed in Table I.

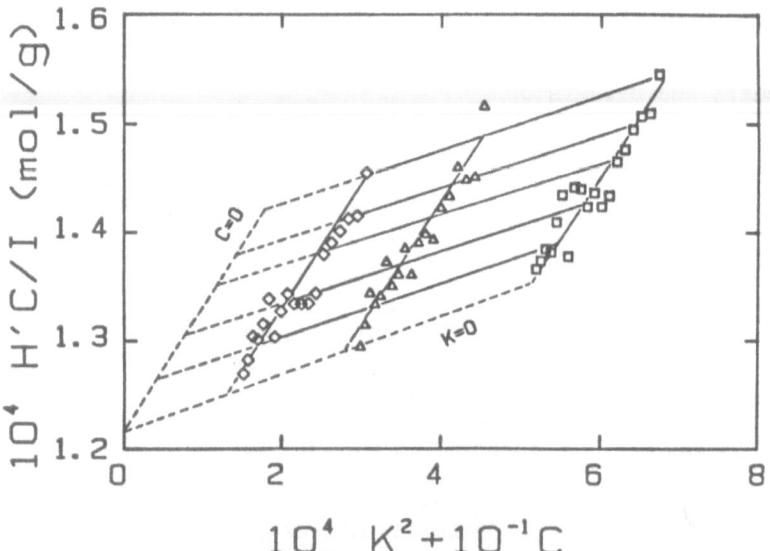

Figure 1. Typical Zimm plot of epoxy polymer in methyl ethyl ketone with traces of unreacted monomers DGEB and CH at 25°C using light scattering intensity measurements with $\lambda_o = 488$ nm sample 12 with ~45.3% CH conversion. $M_w = 3.01\times10^5$ g/mol, $A_2 = 2.88\times10^{-4}$ mL mol g^{-2}, $R_g = 25.7$ nm: (Δ) C = 1.17×10^{-4} g/mL; (□)

C = 4.27×10^{-4} g/mL; (◊) C = 7.54×10^{-4} g/mL. (Reproduction of Fig. 1 of Ref. 77)

3.3. Small Angle X-Ray Scattering Measurements.

In SAXS, the excess scattered intensity I is governed by electron density (instead of refractive index) differences between the solute and the solvent. We may write

$$\frac{H'C}{I} = \frac{1}{M_w} \left(1 + \frac{K^2 R_g^2}{3}\right) + 2A_2C \qquad (2)$$

where the optical constant H in light scattering has been replaced by an instrument constant H' which takes care of the square effect due to the electron density increment[33] and instrumentation differences between light scattering and SAXS. Without computing or measuring the electron density increment, we simply determined H' by using an epoxy polymer (sample 3) of known molecular weight (8.23×10^3 g/mol from light scattering measurements) as our SAXS calibration standard. SAXS intensity measurements could then be used to determine M_w, R_g and A_2 of other epoxy polymer samples (1,2,4-7) as shown typically in Figure 2.

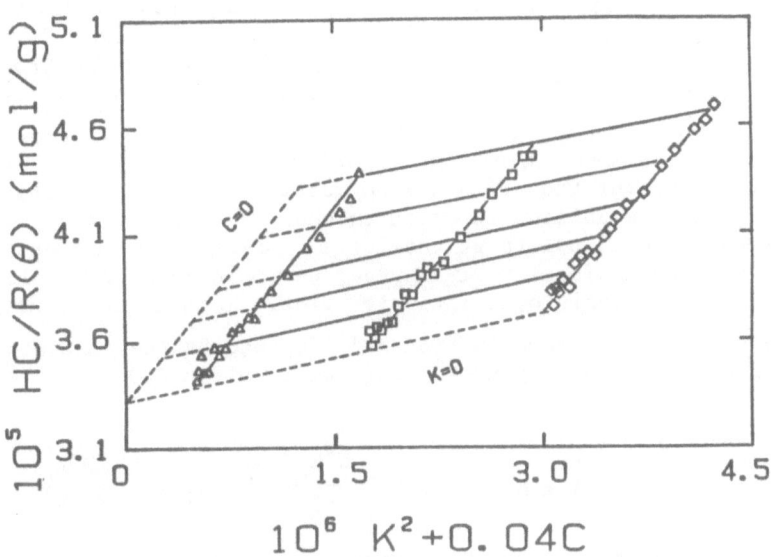

Figure 2. Typical Zimm plot of epoxy polymer in methyl ethyl ketone with traces or unreacted monomers DGEB and CH at 25°C using small angle x-ray scattering at the National Synchrotron Light

Source λ — 0.154 nm. Sample 3 with ~20% conversion. M_w — 8.23×10^3 g/mol, A_2 — 1.05×10^{-3} mL mol g^{-2}, R_g — 4.4 nm: ($\Diamond$) C — 7.89×10^{-4} g/mL; (Δ) C — 1.04×10^{-3} g/mL; ($\Box$) C — 2.51×10^{-3} g/mL. H'C/I transfers the light scattering calibration to SAXS by determining the ratio of scattering intensities in light scattering and SAXS of one epoxy polymer solution of known molecular weight. (Reproduction of Fig. 2 of Ref. 77)

The SAXS results are also listed in Table I. The agreement between light scattering and SAXS results is very good, with M_w values differing by no more than a few percent in most cases. From the initial slopes in plots of $\lim\limits_{C \to 0} HC/R_{vv}$ and $\lim\limits_{C \to 0} H'C/I$ versus K^2, we note that for sample 7, R_g — 112Å and 105Å by means of SAXS and light scattering, respectively. An R_g value of ~100Å, as determined by light scattering, has an uncertainty of about 10%, while SAXS with the intense synchrotron x-ray source yields fairly precise R_g values with uncertainties of no more than a few percent down to very small sizes (e.g. R_g (sample 1) — 29Å).
In Figure 2, the unsmoothed excess SAXS intensity has been corrected for detector nonlinearity, sample transmission, solvent background and parasitic scattering. No desmearing was required because of the small beam cross section and divergence of the incident synchrotron x-ray beam. The agreement of R_g values between light scattering and SAXS also confirms the above assumption over the K range of our measurements. It should also be noted that the signal-to-noise ratio of our SAXS measurements is better than Figure 2 suggests, as the intensity values have been greatly magnified in order to show the initial slope behavior from the first 19 data points at the lowest scattering angles. If we were to use a conventional x-ray source, we would have great difficulty in achieving such a reciprocal intensity plot. Rather, a Guinier plot of log I versus K^2, as shown in Figure 3, represents the standard method of determining R_g. According to Eq. (2), we have for C—0

$$\log(I/H'C) \approx \log M_w + \log\left(1 - \frac{K^2 R_g^2}{3}\right) \approx \log M_w - \frac{K^2 R_g^2}{3} \qquad (3)$$

The limiting slope at C—0 and $KR_g \lesssim 1$ yields the radius of gyration. The linear behavior over a much broader K-range in SAXS permits a more precise R_g determination for small R_g values ($\leq$ 100Å) by means of SAXS when compared with light scattering. In polymer characterization, light scattering and SAXS are truly complementary techniques in covering the size determination.

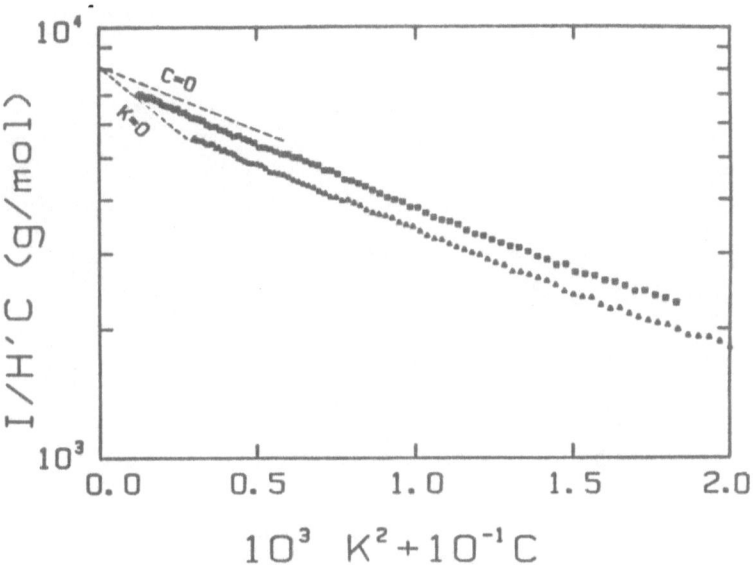

Figure 3. Guinier plots of two of the three concentrations of Fig. 2 (□) $C = 7.89 \times 10^{-4}$ g/mL; (Δ) $C = 2.51 \times 10^{-3}$ g/mL. From the initial slopes, we retrieved essentially the same values for M_w, A_2 and R_g as stated in Fig. 2. The linear behavior in a log I versus K^2 plot at $KR_g < 1$ in the SAXS measurements is clearly demonstrated. (Reproduction of Fig. 3 of Ref. 77)

For our epoxy studies, we started with monomers which eventually form very large polymer networks. Thus, during the initial stages of the epoxy polymerization process, SAXS is the proper analytical method to determine R_g of the branched epoxy polymer. Following Eqs. (1) and (3)

$$\lim_{C \to 0} \frac{M_w HC}{R_{vv}(K)} \left(= \lim_{C \to 0} \frac{M_w H'C}{I(K)} \right) \approx 1 + (R_g/3)K^2 R_g = 1 + (R_g/3)x \quad (4)$$

In plots of $\lim_{C \to 0} M_w HC/R(\theta)$ or $\lim_{C \to 0} M_w H'C/I$ versus $x(=K^2 R_g)$, as shown in Figure 4, we see that LLS measurements, as denoted by filled symbols are clearly appropriate for R_g values of a few hundred Å. It becomes increasingly more difficult at smaller R_g values because of the small $KR_g (\lesssim 1)$ ranges accessible to LLS. For sample 7 (denoted by filled and

336

hollow diamonds) we demonstrate an overlap of two independent scattering techniques on an absolute determination of R_g.

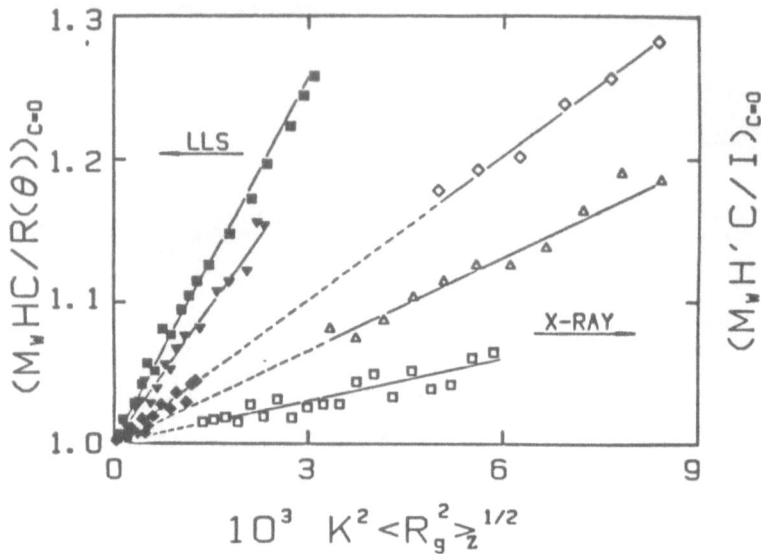

Figure 4. Plots of M_wHC/R_{vv} (for light scattering) and $M_wH'C/I$ (for SAXS) as a function of $K^2\langle R_g^2\rangle^{1/2}$. According to Eqs. 1 and 3,

$$\lim_{C\to 0} \frac{M_wHC}{R_{vv}} (= \lim_{C\to 0} \frac{M_wH'C}{I}) = 1 + \frac{R_g}{3} K^2 R_g.$$ Thus, the

slope is equal to $R_g/3$. The plots demonstrate overlapping regions of the two scattering techniques. Laser light scattering is denoted by filled symbols while SAXS is denoted by hollow symbols. (■) Sample 12, (▼) sample 10, (♦) sample 7; (◇) sample 7, (△) sample 5, (□) sample 3. Properties of various samples are listed in Table I. (Reproduction of Fig. 4 of Ref. 77)

3.4. Fractal Geometry of Branched Epoxy Polymers

The curing of epoxy resins using DGEB and CH in the presence of a small amount of catalyst CA represents a cross-linking polymerization process, as shown schematically in Figure 5. The cross linking reaction involves roughly n moles of DGEB with 2n moles of CH. The chemical reaction is known to cluster at catalytic centers and the formation of branched structures eventually link together the branched epoxy polymers to form loops or polymer networks.

The concept of fractal geometry[12,34,35] can be a useful tool to

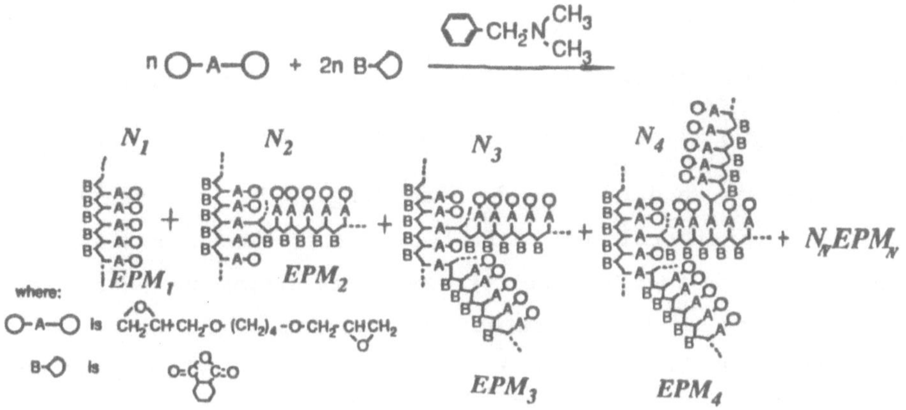

Figure 5. Crosslinking reaction of DGEB with CH in the presence of CA. (Reproduction of Fig. 1 of Ref. 79)

describe the branching structure of epoxy polymers during its curing process. The fractal dimension d_R of a molecular cluster with mass M and the radius of gyration has the relation[36] $M \sim R_g^{d_R}$ where d_R is the fractal dimension in terms of the scaling relationship between mass and radius of gyration. For polydisperse polymers, we approximate the expression to be

$$M_w \sim R_g^{d_R} \tag{5}$$

In Eq. (5), we have assumed that M_w spacings have been sufficiently far apart so that the polydispersity effect is not appreciable. It should be noted that polydispersity could act as a simple correction[37] or could affect the measured exponents d_R[38-40] as well as d_K.[41,42] The two-point density-density correlation function $C(r)$ has the form[22] $C(r) \approx C_0 r^{-\alpha}$ with the corresponding static structure factor $S(K)$ ($\sim$ the scattered intensity I), which is the Fourier transform of the pair correlation function, having a power-law relation

338

$$S(K) \sim K^{-d_K} \qquad (6)$$

The fractal dimension d_K is related to the exponent α according to $d_K = d-\alpha$ with d being the dimensionality of the embedding space or lattice containing the fractal. For a self-similar fractal $d_f = d_K = d_R$ where d_f is a general fractal dimension.[43] Measurements of the angular distribution of scattered intensity over large ranges of K with $R_g^{-1} < K < \xi^{-1}$ permit us to determine d_K according to Eq. (6) where ξ is a correlation length related to the "blob" size. Measurements of M_w and R_g of the epoxy polymer by means of Zimm plots during the polymerization process permit us to determine d_R according to Eq. (5). In this section, our main aim is to check Eqs. (5) and (6) and to find out if our epoxy polymers form self-similar fractals.

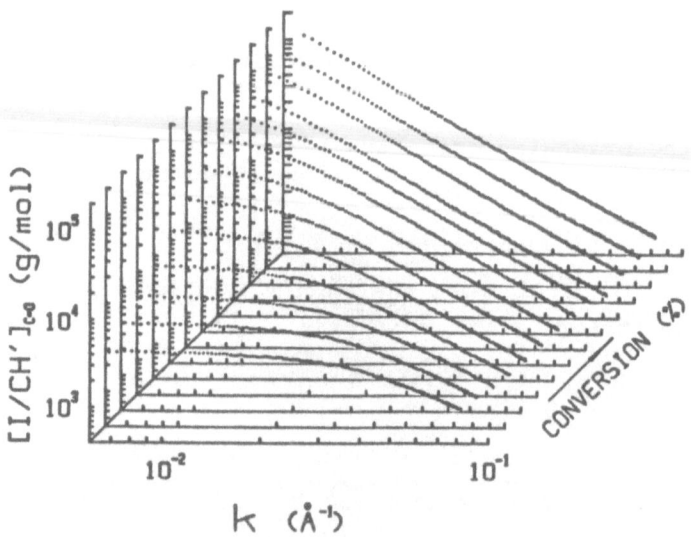

Figure 6. Structure of epoxy polymers as a function % CH conversion. $[I/CH']_{C=0}$ represents absolute SAXS intensity at infinite dilution in units of g/mol. Thus, $\lim_{K \to 0} [I/CH']_{C=0} = 1/M_w$. The scattering curves are numbered with increasing % CH conversion. Properties of the 13 samples representing epoxy polymers during different stages of the curing process are listed in Table I. (Reproduction of Fig. 6 of Ref. 77)

Figure 6 shows static structure factors $S(K)[= I/CH']$ from SAXS as a function of % CH conversion. In plotting the scattering curves, we have scaled the intensities according to Eq. (3). Thus, the y($= \lim\limits_{C \to 0} I/CH'$) axis has units of g/mol and at K = 0, denotes the M_w of the epoxy polymer as a function of % CH conversion. Such an approach is feasible up to the gelation point, beyond which the epoxy polymer can no longer be dissolved as individual macromolecules in MEK. At high % CH conversion, the epoxy polymer has reached fairly high molecular weights ($\sim 10^5$ g/mol). Thus, SAXS measures mainly the fractal geometry of the branched epoxy polymers in solution according to Eq. (6), as shown typically in Figure 7 by the hollow diamonds for sample 13 with a M_w of

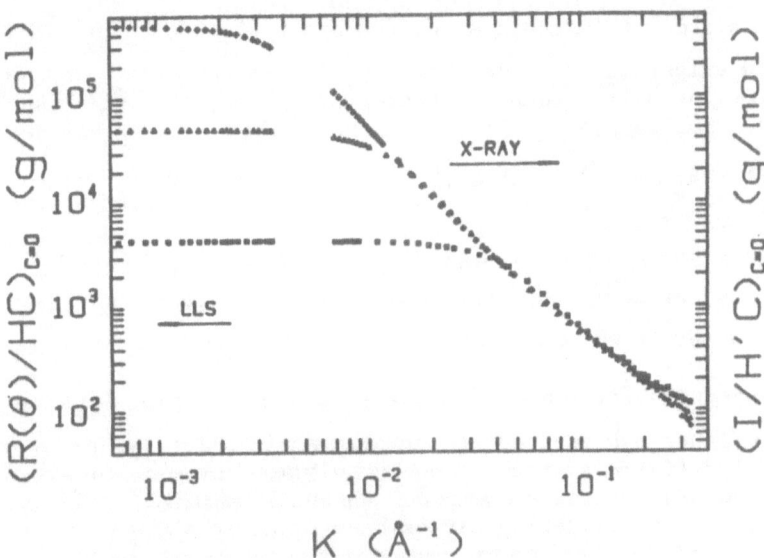

Figure 7. Log-log plots of R/HC for light scattering and I/H'C for SAXS as a function of K. (□) sample 1, (∇) sample 7, (◇) sample 13. Light scattering K range: $\sim 7 \times 10^{-4} < K < 4 \times 10^{-3}$ Å^{-1}; SAXS K range: $7 \times 10^{-3} < K < 3.5 \times 10^{-1}$ Å^{-1}. Each SAXS curve has 817 data points, of which only a fraction is plotted. The horizontal portion of the scattering curve with a small initial negative slope can be related to the radius of gyration. In the range $2/R_g < K < 1/20$ Å^{-1}, we can use the scattering curves to examine fractal geometry of the epoxy polymers. (Reproduction of Fig. 7 of Ref. 77)

5×10^5 g/mol. In Figure 7, we have also included LLS measurements (also denoted by hollow diamonds) at much smaller values of $K (< 4 \times 10^{-3}$ Å^{-1}).

It may be difficult to see from Figure 7 that the initial slope from LLS exhibiting almost horizontal behavior could easily determine the R_g value as has been demonstrated by Figure 1. As the K-range covered by SAXS is extremely broad, i.e., down to monomer dimensions, we would also expect a deviation from the fractal dimension beyond $2/R_g \lesssim K \lesssim 1/20$ $Å^{-1}$, i.e., $20K \lesssim 1$ with K expressed in $Å^{-1}$. Sample 13 represents the epoxy polymer structure just before its gelation point. At earlier times, e.g. for sample 7 with $M_w - 5 \times 10^4$ g/mol at 38.5% CH conversion, the initial slope from SAXS and that from LLS (both denoted by solid triangles) in Figure 7 could be used to determine R_g. The log-log plot of Figure 7 also demonstrates the precision with which we have to achieve in order to measure R_g of the order of 100Å. The LLS portion of the solid triangles is almost horizontal. We have also included sample 1 (denoted by hollow squares) representing only a 6.5% CH conversion in Figure 7. For sample 1, R_g(~29Å) is no longer accessible by LLS. In the SAXS region, the initial horizontal curve can be used to determine both M_w and R_g. Again, we can use Eq. 6 to determine the fractal dimension over an appropriate K range ($2/R_g \lesssim K \lesssim 1/20$ $Å^{-1}$). More importantly, we note that the scattering curves of samples 7 and 13 overlap over a broad K-range ($1.5 \times 10^{-2} < K \lesssim 3 \times 10^{-1}$ $Å^{-1}$) while the scattering curve for sample 1 with $M_w - 4 \times 10^3$ g/mol does not overlap with the other two scattering curves even in a log-log plot, indicating structural differences for the epoxy polymer between the very initial stages and before the gelation. We shall return to this point later. In Figure 7, the scattering curves have similar shapes and we want to again emphasize the complementary aspect of the two scattering techniques. For our epoxy polymerization process, we need both LLS and SAXS in order to cover the appropriate K ranges.

Now we turn our attention to the two main tasks, i.e., experimental determinations of d_K and d_R. For d_K, we can determine a value for each sample. However, for d_R, instead of doing a tedious fractionation of each polymer sample, we have chosen the approximate Eq. (5) and made an experimental determination of M_w and R_g for the epoxy polymer at 13 time intervals up to the neighborhood of the gelation threshold during the curing process.

In the d_K determination, we have considered the concentration effect as follows. Figure 8(a) shows log-log plots of I/H' versus K for sample 10 in MEK at C - ( ) 7.19×10^{-4} g/mL, (Δ) 1.14×10^{-3} g/mL, and ($\square$) 2.41×10^{-3} g/mL. As we have a total of 817 data points for each scattering curve, only a fraction of the data is plotted. It should be

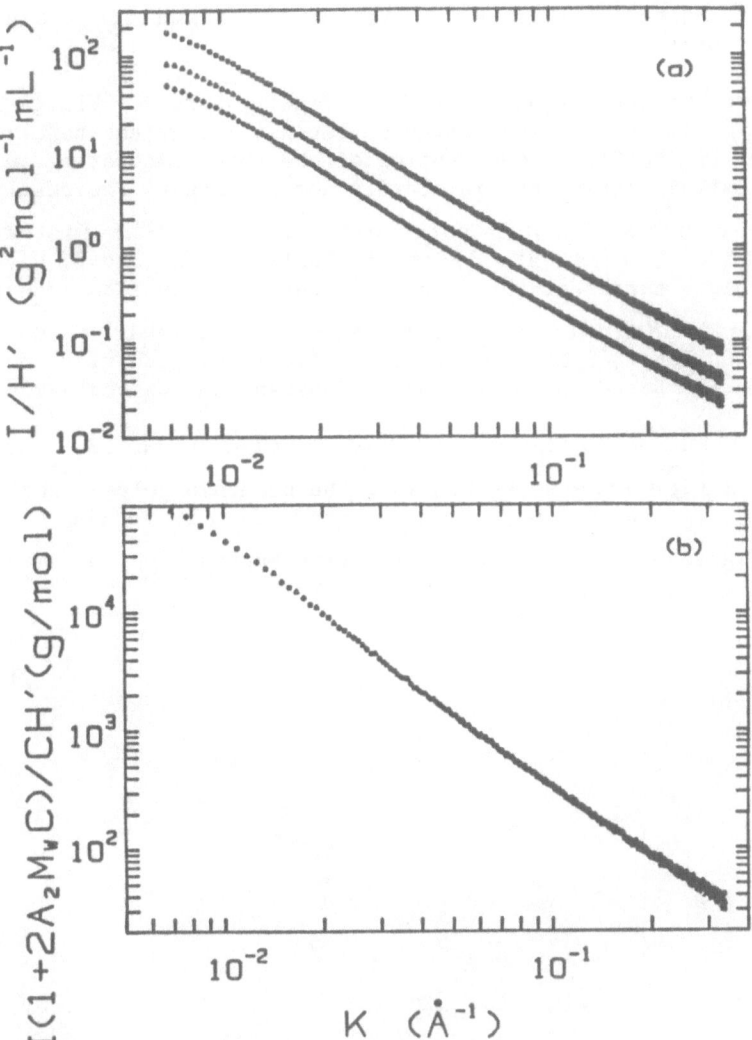

Figure 8. Concentration effect on fractal dimension of epoxy polymers.
(a) Log-log plots of I/H' versus K for sample 10 in MEK. (◊)
C = 7.19x10^{-4} g/mL; (∇) C = 1.14x10^{-3} g/mL; (□) C = 2.41x10^{-3}
g/mL. There are 817 data points on each scattering curve.
Only a fraction of data points is plotted. (b) Log-log plot
of a scaled intensity curve based on the scattering curves
from three different concentrations as shown in Fig. 8(a).

342

Same symbols as in Fig. 8(a) with $M_w = 1.42 \times 10^5$ g/mol and $A_2 = 3.63 \times 10^{-4}$ mL mol g^{-2}. Again only a small fraction of data points are plotted. (Reproduction of Fig. 8 of Ref. 77)

noted that the scattering curves are slightly curved. Figure 8(b) shows a log-log plot of scaled scattered intensities of the three scattering curves of Figure 8(a). Overlapping of the three scattering curves with zero adjustable parameters is clearly demonstrated. The results for d_K, as well as the exponent β related to the coil behavior between the entanglement points, are listed in Table II. It should be noted that our d_K values represent the fractal dimension of swollen branched epoxy polymers in solution. The variable Y in Table II represents a correlation length ξ below which the polymer coil behavior should begin to dominate. Based on the values of constant d_f, we see a cut off value of $Y^{-1} \simeq 20$Å with $d_K \simeq 2.17$ for samples 6 to 13. At earlier polymerization stages, we consider the branched polymer chains to have lower d_f values with ξ no longer related closely to the polymer mesh size. Figure 9 shows a more quantitative approach to determine the mesh

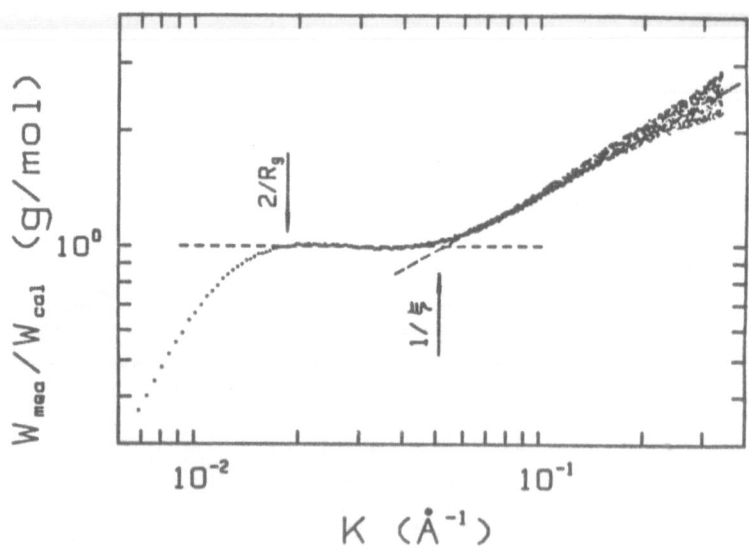

Figure 9. Plot of W_{mea}/W_{cal} versus K for sample 7. $M_w = 5.00 \times 10^4$ g/mol, $R_g = 105$ Å. $W_{mea} = (I/H'C)_{mea, C=0}$. $W_{cal} = 3.01$ K$^{-2.17}$. Slope (at K>1/ξ) = 0.49±0.06. (Reproduction of Fig. 9 of Ref. 77)

size $\xi(\approx19.2\pm3.0\text{Å})$ and to show the deviation of the measured curve from W_{cal}(g/mole) $= 3.01$ $K^{2.17}$ with K expressed in Å^{-1} for sample 7. The results are listed in Table III. As the K-range obeying the fractal dimension d_K is limited, the range of K used in the determination of d_K should always be considered with care.

Beyond the mesh size range $(K>\xi^{-1})$, we have noted an epoxy polymer coil behavior with $I(K) \sim K^{-\beta}$ and $\beta \approx 1.68\pm0.06$, as listed in Table II(b). The value $(d_f - \beta) = 2.17-1.68 = 0.49$ is the slope shown in Fig. 9 for $K>\xi^{-1}$. Daoud and Joanny[44] predicted a linear blob behavior of $\beta \sim 5/3$ in a theta solvent.

We now come to the determination of d_R. A log-log plot of $\langle R_g^2\rangle_z^{1/2}$

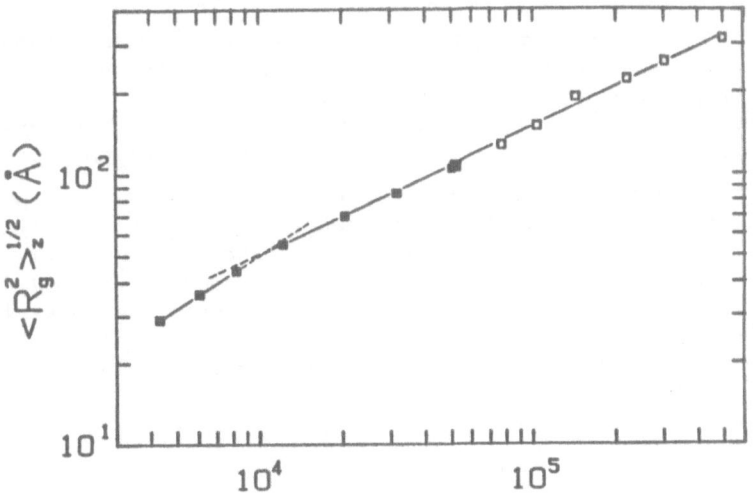

Figure 10. Log-log plot of $\langle R_g^2\rangle_z^{1/2}$ versus M_w. Solid squares denote SAXS measurements while hollow squares denote light scattering measurements. $M = k_R R_g^{d_R}$ with R_g and M expressed in units of Å and g/mol, respectively.

Fitting range	k_R	d_R
whole (13 points)	4.59±0.41	2.00±0.03
3 low points	(2.38±0.15)x10	1.55±0.02
10 higher points	3.04±0.33	2.07±0.04
9 higher points	3.25±0.76	2.06±0.05

(Reproduction of Fig. 10 of Ref. 77)

344

versus M_w, as shown in Figure 10, reveals a slight curvature in the low molecular weight region, i.e., during the initial stages of the curing process, the epoxy polymer appears to have different structures from those at later stages even before the gelation threshold. The experimental results shown in Figure 10 required a combination of SAXS (denoted by solid squares) and LLS (denoted by hollow squares) measurements. Although a least-squares fitting of all the data points using $<R_g^2>_z^{1/2} - k_R M_w^{\alpha_R}$ with $d_R - 1/\alpha_R$ shows a reasonable $d_R - 2.0$, we have approximated the slight curvature at low R_g values by breaking the curve into two straight sections near $M_w \sim 1 \times 10^4$ g/mol. Least squares fitting of the first three and four low M_w data points yields $d_R \sim 1.55$ and 1.61, respectively, while $d_R(-d_f) \sim 2.05$ for the ten higher M_w data points. Thus, the concept of self-similar fractals should not be applied uniformly to the epoxy polymer product in the very beginning of the curing process for our epoxy system.

It is interesting to note that the difference in scattering curves for low molecular weight and for high molecular weight epoxy polymers appears to occur at a fairly high molecular weight of $\sim 10^4$ g/mole. We speculate that the reason for this behavior is due to the relatively low catalyst concentration of $\sim 0.1\%$, i.e., there is only one catalyst molecule for about one thousand epoxy monomers. During the initial polymerization process, most of the epoxy polymers are relatively linear because the branching probability is fairly small.

From our light scattering intensity measurements we have also

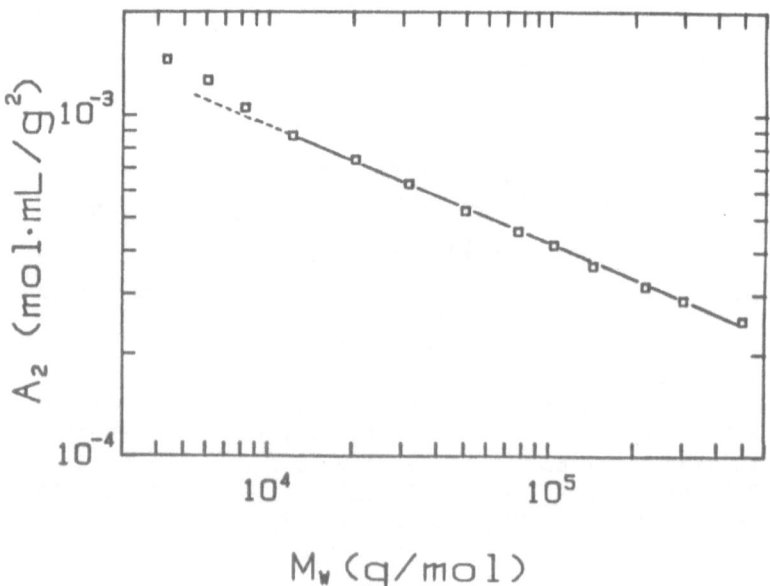

Figure 11. Log-log plots of A_2 versus M_w from light scattering intensity measurements. $A_2 = k_A M^\alpha$ with A_2 and M in units of mol mL/g^2 and g/mol, respectively.

Fitting range	k_A	α_A
whole (13 points)	$(3.02\pm0.31)\times10^{-2}$	-0.37 ± 0.01
10 higher points	$(2.19\pm0.13)\times10^{-2}$	-0.34 ± 0.01

(Reproduction of Fig. 11 of Ref. 77)

determined the second virial coefficient A_2 according to Eq. (1). Figure 11 shows a log-log plot of A_2 versus M_w. We have noted a curvature at low values of M_w, similar to the trend exhibited in Figure 10. By using

$$A_2 = k_A M_w^{\alpha_A} \tag{7}$$

we get $k_A = (3.02\pm0.31)\times10^{-2}$ and $(2.19\pm0.13)\times10^{-2}$ as well as $\alpha_A = -0.37\pm0.01$ and -0.34 ± 0.01, respectively, with all 13 data points and the 10 higher molecular weight data points. We are not aware of experimental determinations for the constants k_A and α_A in highly branched polymers. However, experimental results on star-branch polymers suggest an α_A value of -0.37 for 12-arm and 18-arm polystyrene stars,[45] in fairly good agreement with our findings.

The results of A_2 suggest that the lower M_w epoxy polymers are more swollen than the higher M_w epoxy polymers during the later stages of the curing process before the gel point. This observation agrees with our concept that during the initial polymerization stages, the epoxy polymer molecules are less branched. The smaller degree of branching in the lower M_w epoxy polymers permits easier swelling of the polymer molecules. With increasing molecular weight, the polymer molecules become more highly branched and less swollen. The concept of fractal geometry is applicable because of the agreement between $d_R(=d_f)\sim2.05$ over a substantial range of reaction time before the gel point, as shown in Table II and Fig. 10.

Beyond the gelation threshold, SAXS can be used to determine the fractal geometry of epoxy polymers as shown typically in Fig. 12. For the present epoxy system, we still require immersion of the gel particles in MEK in order to increase the electron density difference.

$d_f \sim 2.14\pm0.02$ and 2.13 ± 0.03 for $K\lesssim1/30$ Å^{-1} and $K\leq1/20$ Å^{-1}, respectively.

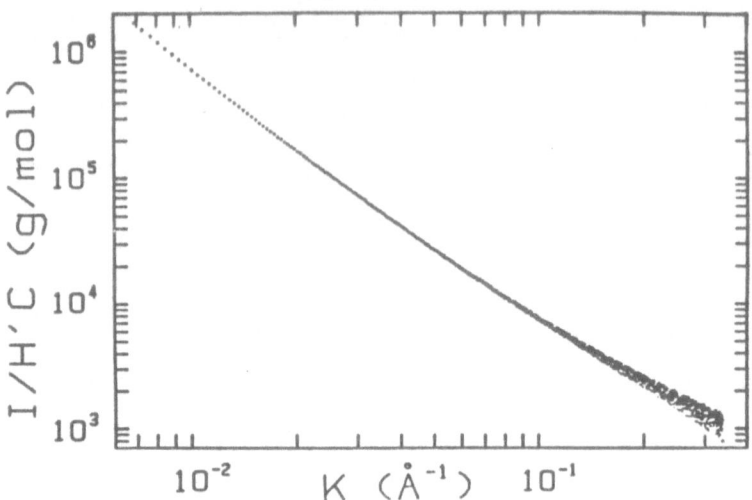

Figure 12. Log-log plots of I/H'C versus K. The epoxy gel was swollen in MEK. Least squares fitting of 817 data points to $I \propto K^{\alpha}$ yields $\alpha = 2.14\pm0.02$ and 1.69 ± 0.05 for $K\leq1/30$ Å^{-1} and $1/30\leq K\leq1/4$ Å^{-1}, respectively. (Reproduction of Fig. 13 of Ref. 77)

The linear behavior as shown in Fig. 8(b), the horizontal region as shown clearly in Fig. 9, and our knowledge on the molecular weight distribution as shown in Fig. 12, strengthen the supposition of Eq. (5), i.e., we have a K-region with a constant d_K in the polydisperse epoxy polymer and d_R can be evaluated over a broad enough M_w range even if the epoxy polymers are polydisperse.

3.5. Molecular Weight Distribution of Branched Epoxy Polymer

3.5.1. <u>Laplace Transform</u>. By following the experimental procedures for self-beating and baseline considerations, we could obtain precise measurements of the intensity-intensity time correlation function $G^{(2)}(K,\tau)$:

$$G^{(2)}(K,\tau) = A\,(1 + b|g^{(1)}(K,\tau)|^2) \tag{8}$$

where the baseline A agreed with the measured baseline $\lim\limits_{\tau\to\infty}G^{(2)}(K,\tau)$ to within about 0.1%. Figure 13 shows a typical experimental intensity-intensity photoelectron count autocorrelation function for the epoxy

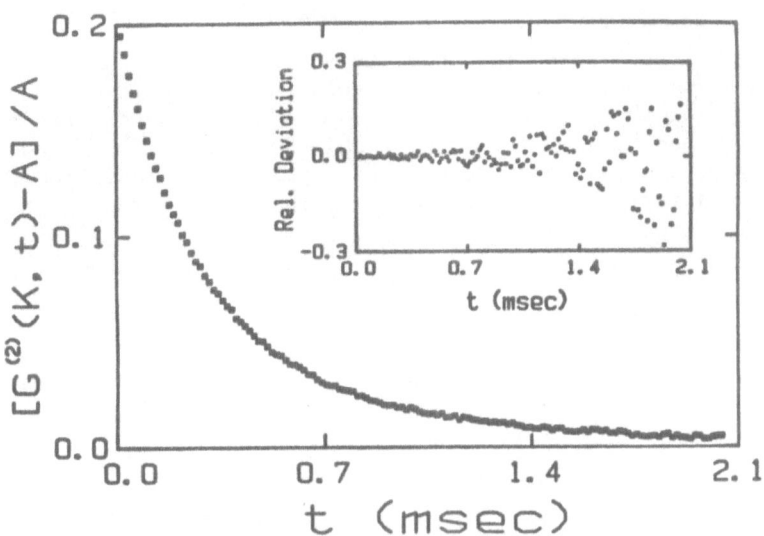

Figure 13. A typical unnormalized Intensity-intensity photo-electron count autocorrelation function. 1.11×10^{-4} g/mL of epoxy polymer sample 13 ($M_w = 4.97 \times 10^5$) in MEK measured at $\theta = 30^\circ$ and 25°C using a delay time increment $\Delta\tau = 15$ μsec. The insert is relative deviation of the measured and the computed time correlation function using the MSVD method. Relative deviation is defined as

$$1 - [b|g^{(1)}(t)|^2]_{calc.} / [b|g^{(1)}(t)|^2]_{meas.}$$

(Reproduction of Fig. 1 of Ref. 61)

polymer in MEK measured at $\theta = 30^\circ$ and 25°C using a delay time increment $\Delta\tau$ of 15 μsec and relative deviation of the measured and the computed time correlation function by using the MSVD method of Laplace inversion.

The MSVD technique[46,47,48] has been described in detail elsewhere. We outline only the essential steps which are necessary in describing our data fitting results. In the MSVD technique, we do Laplace inversion of the electric field time correlation function:

$$g^{(1)}(K,\tau) = \int_0^\infty G(K,\Gamma)e^{-\Gamma(K)\tau} \, d\Gamma \qquad (9)$$

by approximating $G(\Gamma)$ with a set of linearly or logarithmically spaced single exponentials:

$$G(\Gamma) = \sum_j P_j \, \delta(\Gamma - \Gamma_j) \qquad (10)$$

where $g^{(1)}(\tau_i) - b_i - \sum_j P_j \exp(-\Gamma_j \tau_i)$ with P_j being the weighting factors of the δ function measured at scattering vector K. The validity of the MSVD method has been tested previously.[47-50]
    We have to stress here that, in an ill-posed problem, goodness of fitting does not guarantee a correct solution to the inversion. However, the validity of the MSVD method for a unimodal characteristic line-width distribution has been tested thoroughly by using simulated data and known polymer systems under comparable experimental conditions, counting rates and statistics. Figure 14 shows typical line-width distributions

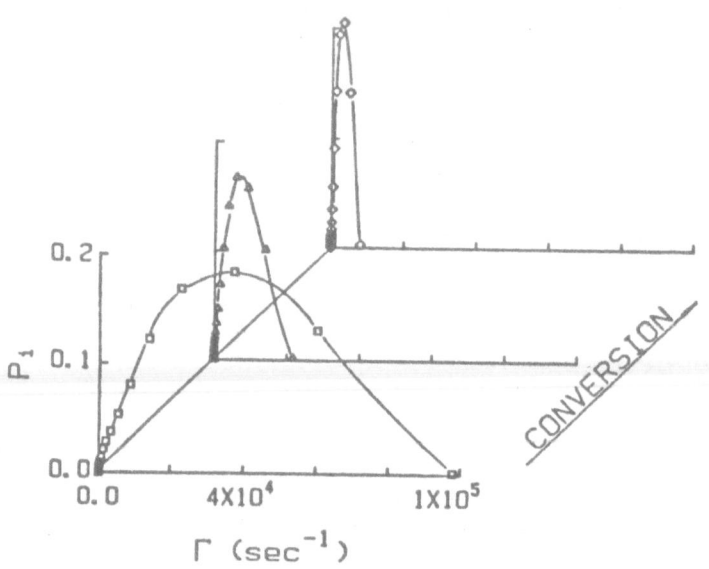

Figure 14. Plots of P_j versus Γ for epoxy polymer sample 1, sample 9 and

sample 13 in MEK at $\theta - 30°$ and $25°C$ based on the MSVD method.

Sample	Notation	M_w(g/mol)	$\bar{\Gamma}$(sec^{-1})	$\mu_2/\bar{\Gamma}^2$
1	Hollow squares	4.32×10^3	2.29×10^4	0.63
9	Hollow triangles	1.03×10^5	3.98×10^3	0.45
13	Hollow diamonds	4.97×10^5	1.79×10^3	0.40

(Reproduction of Fig. 2 of Ref. 61)

obtained by the MSVD method. In Figure 14, we clearly see that the characteristic line-width shifts to lower frequencies and the distributions become narrower with increasing % of conversion. Numerical results of the transform by the MSVD method for a set of epoxy polymers at different % conversions are listed in Table IV. It should be noted that discrete (P_js) and continuous ($G(\Gamma)$) normalized characteristic line-width distributions are not the same. In a discrete distribution in terms of P_js, we have Eq. (10). If Γ_j does not have equal spacing, which is indeed the case for our MSVD method, P_j has to be rescaled in order to convert it to a continuous distribution $G(\Gamma)$, i.e. $G(\Gamma) \neq G(\ln\Gamma)$. The rescaling is approximately done by using $G(\Gamma_j) = P_j / \Gamma_j$.

In order to transform the measured characteristic line-width distribution at finite angle and finite concentration to a molecular weight distribution, we have to know how the characteristic line-width Γ depends on the scattering angle (or K) and the polymer concentration. We experimentally determined such a relation by using

$$\bar{\Gamma} = \bar{D}_o^o K^2 (1 + fR_g^2 K^2)(1 + k_d C) \tag{11}$$

where f is a dimensionless number and depends on chain structure, polydispersity and solvent quality, and k_d is the second virial coefficient for diffusion. Figure 15 shows a typical K^2 dependence of

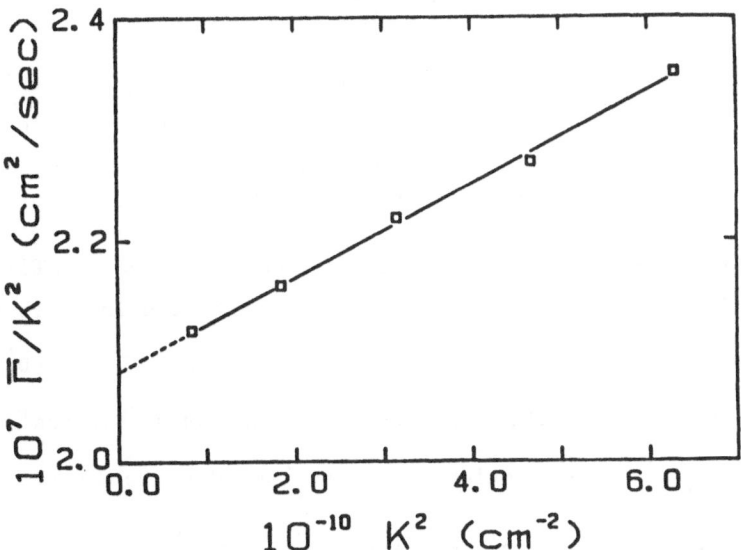

Figure 15. Plots of $\bar{\Gamma}/K^2$ versus K^2 for epoxy polymer 13 ($M_w = 4.97 \times 10^5$ and $\langle R_g^2 \rangle_z^{1/2} = 31.4$ nm) in MEK at 25°C and $\lambda_o = 488$ nm. The straight line represents
$\bar{\Gamma}/K^2 = 2.08 \times 10^{-7}(1 + 0.19 \langle R_g^2 \rangle_z K^2)$
(Reproduction of Fig. 3 of Ref. 61)

350

the z-average characteristic line-width $\bar{\Gamma}$. Figure 16 shows a typical

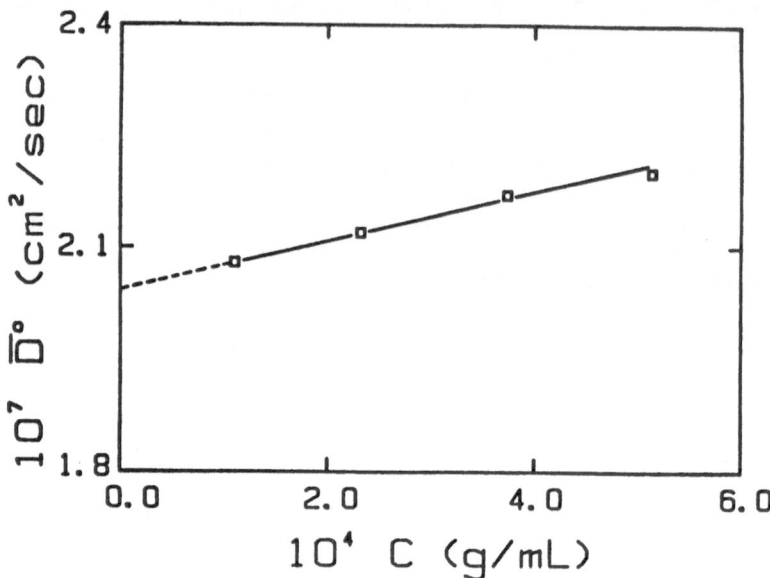

Figure 16. Plots of $\bar{D}^o$ ($= \lim_{K \to 0} \bar{\Gamma}/K^2$) versus concentrations (C) for the
same sample in Figure 15. The straight line represents
$\bar{D}^o = 2.05 \times 10^{-7}(1 + 1.41 \times 10^2 C)$
(Reproduction of Fig. 4 of Ref. 61)

concentration dependence of the translational diffusion coefficient $\bar{D}^o$
($= \lim_{K \to 0} \bar{\Gamma}/K^2$). The k_d and f values are also listed in Table IV. After

having obtained k_d and f, we can convert $G(\Gamma)$ versus Γ to $G(D_o^o)$ versus

D_o^o. Further, we can calculate the z-average translational diffusion

coefficient $\bar{D}_o^o$ at each reaction stage, as listed in Table IV. By using

the $\bar{D}_o^o$ and the Stokes-Einstein relation: $D = \dfrac{k_B T}{6\pi\eta R_h}$, we can calculate

the equivalent hydrodynamic radius R_h.

3.5.2. <u>Conversion from $G(\Gamma)$ to $F_w(M)$</u>. In previous publications,[48-50]
we converted the line-width distribution $G(\Gamma)$ to the weight average
molecular weight distribution $F_w(M)$ by using an experimentally

determined scaling relation $\bar{D}_o^o = k_D M_w^{-\alpha_D}$ where $\bar{D}_o^o$ and M_w are the z-average translational diffusion coefficient extrapolated to infinite dilution and the weight average molecular weight, respectively. A log-log plot of $\bar{D}_o^o$ versus M_w is shown on Figure 17. We find that the data

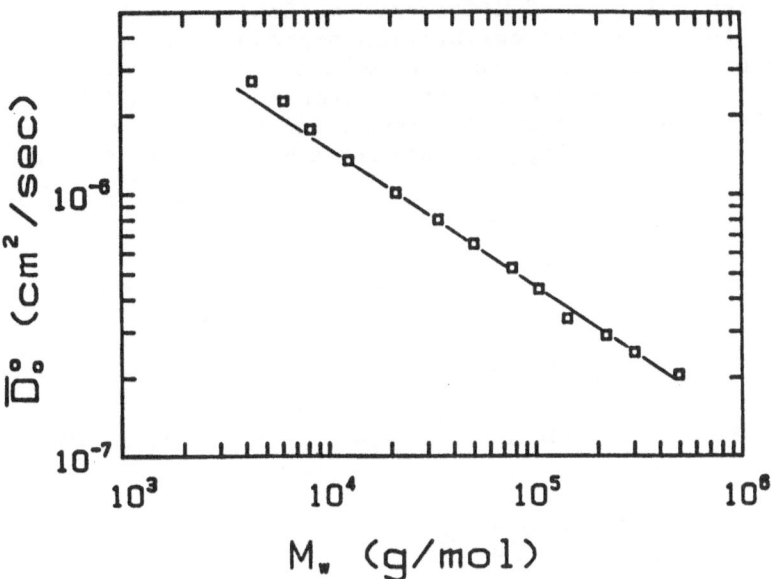

Figure 17. log-log plot of translational diffusion coefficient $\bar{D}_o^o$ versus M_w for the epoxy polymer in MEK at 25° and $\lambda_o = 488$ nm. The straight line represents

$$\bar{D}_o^o = 1.56 \times 10^{-4} M_w^{-0.508}$$

with $\bar{D}_o^o$ and M_w in units of cm^2/sec and g/mol, respectively. (Reproduction of Fig. 7 of Ref. 61)

can be represented by a straight line if we disregard the first three data points during the initial stages of the curing process. From Fig. 17, we calculated $k_D = (1.56 \pm 0.20) \times 10^{-4}$ and $\alpha_D = 0.508 \pm 0.011$ with $\bar{D}_o^o$ and M_w expressed in units of cm^2/sec and g/mol, respectively. In fact, the Γ to M conversion should use the scaling relation $D_o^o = k_D M^{-\alpha_D}$ and the condition $\Gamma = DK^2$ based on monodisperse fractions. The k_D and α_D

obtained from $\bar{D}_o^o$ and M_w are approximations to the true values since we usually do not have monodisperse polymer samples to establish the scaling relation. In other words, the narrower the molecular weight distributions of calibration samples over the same molecular weight range, the better the approximation becomes. So, the standard calibration method requires a set of narrow molecular weight distribution polymers of different molecular weights. It is very difficult to apply this calibration procedure to study the branched epoxy polymers because all epoxy polymers are very polydisperse. Thus, we need to examine the calibration problem from a slightly different viewpoint in order to achieve the conversion as precisely as possible. We applied the following procedure for a more precise determination of k_D and α_D.

For polydisperse polymer,

$$\int_o^\infty G(D_o^o)\ dD_o^o = \int_o^\infty F_z(M)\ dM = 1 \tag{12}$$

where $F_z(M)$ is the z-weighted molecular weight distribution function. By accepting the scaling relation $D_o^o = k_D M^{-\alpha_D}$, we finally obtain

$$M_w = \frac{k_D^{1/\alpha_D} \int G(D_o^o)\ dD_o^o}{\int G(D_o^o) D_o^{o\,1/\alpha_D}\ dD_o^o} \tag{13}$$

With two polymer samples of different molecular weight and distribution but obeying the same k_D and α_D, we have two $G(D_o^o)$, denoted by $G_1(D_o^o)$ and $G_2(D_o^o)$. From them, we could calculate two $(M_w)_{calc}$, denoted by $(M_{w,1})_{calc}$ and $(M_{w,2})_{calc}$. The ratio of $(M_{w,1})_{calc}$ and $(M_{w,2})_{calc}$ is

$$\frac{(M_{w,1})_{calc}}{(M_{w,2})_{calc}} = \frac{[\int G_1(D_o^o)\ dD_o^o][\int G_2(D_o^o) D_o^{o\,1/\alpha_D} dD_o^o]}{[\int G_2(D_o^o)\ dD_o^o][\int G_1(D_o^o) D_o^{o\,1/\alpha_D} dD_o^o]} \tag{14}$$

where k_D has been cancelled out. The two calculated $(M_w)_{calc}$ values have to equal the two measured M_w values. It means that we already know the value of left side of Eq. (14) experimentally. Now, we vary the value of α_D and calculate the right side of Eq. (14) until the left side equals to the right side. In this way, we are able to find the correct α_D value from two polymer samples with different and broad MWD. After we have determined α_D, we can determine the k_D value by using Eq. (13). The above procedure can be expanded easily to N samples of different molecular weight and distribution. After having determined the correct values of k_D and α_D, we can use the scaling relation $D_o^o = k_D M^{-\alpha_D}$ (not $\bar{D}_o^o = k_D M_w^{-\alpha_D}$) to convert $G(D_o^o)$ versus D_o^o to $F_w(M)$ versus M.

By using the method presented above, we obtained that $k_D = (2.14 \pm 0.32) \times 10^{-4}$ and $\alpha_D = 0.527 \pm 0.013$. The k_D and α_D obtained from average values of M_w and $\bar{D}_o^o$, as shown in Fig. 17, are different from the results obtained from M and D_o^o, which is not surprising because our epoxy polymers are quite polydisperse.

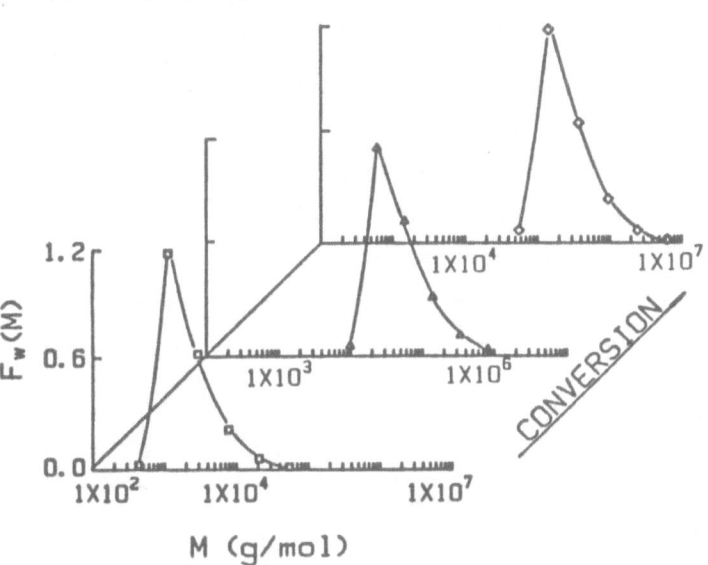

Figure 18. Typical plots of weight distributions for the same samples in Figure 14.

Sample	Notation	M_w(g/mol)	M_n(g/mol)	$M_w:M_n$
1	Hollow squares	4.32×10^3	1.30×10^3	3.32
9	Hollow triangles	1.03×10^5	3.99×10^4	2.58
13	Hollow diamonds	4.97×10^5	2.00×10^5	2.49

(Reproduction of Fig. 8 of Ref. 61)

Having computed $G(D_o^o)$ versus D_o^o, we now make use of the relation $D_{o,j}^o$ (in cm^2/sec) $= 2.14 \times 10^{-4} M_j^{-0.527}$ for each fraction of the epoxy polymer having molecular weight M_j (in g/mole). At each scattering angle, the excess Rayleigh ratio has the form: $R_{vv}(K) \propto \sum_j F_w(M_j) M_j P(M_j,K) \propto \sum_j P_j$, where $P(M,K)$ is the particle scattering factor and $F_w(M)$ is the normalized weight distribution for the epoxy polymers. The $\propto$ sign denotes that we are not concerned with the proportionality constant. At small enough scattering angles, $P_j \propto F_w(M_j) M_j P(M_j,K) \approx F_w(M_j) M_j$ as $P(M_j,K) \approx 1$. The first-order term for $P(M_j,K)$ has the form $P(M_j,K) \approx 1 - \langle R_g^2 \rangle_{z,M_j} K^2/3$. Thus, we can correct for the interference effect in the molecular weight distribution whenever it is necessary since we have the empirical scaling relation between the radius of gyration and the molecular weight, as shown in 3.4.

Figure 18 shows three typical normalized weight distributions of the epoxy polymers at three different % conversions. In Fig. 18, we have ignored the very high molecular weight tails on the distribution curves because of noise and background uncertainties. The contribution of the high molecular weight tail to the cumulative molecular weight distribution is less than 1% of the total weight fraction. From the F_w distributions, we can calculate $M_w:M_n$, which represents the polydispersity of each epoxy polymer sample. The results of $M_w:M_n$ for 13 samples at different % conversion are shown on Figure 19. It is noted

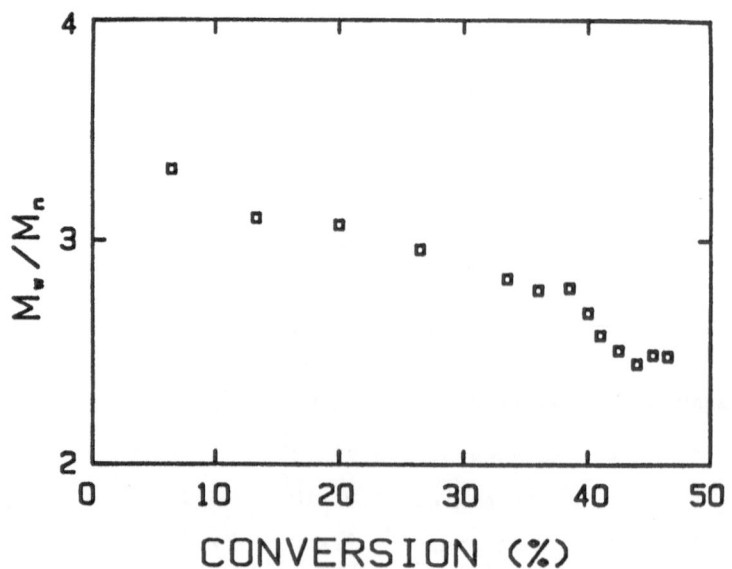

Figure 19. Plot of $M_w:M_n$ versus extent of conversion for 13 epoxy polymers at different reaction stages. $M_w:M_n$ were determined by using the MSVD method. (Reproduction of Fig. 10 of Ref. 61)

that the values of $M_w:M_n$ become smaller with increasing reaction time, suggesting that the polydispersity of the branched epoxy polymer product during the initial stages is much higher than that at the later stages near the gelation threshold. We also observed that $M_w:M_n$ approaches a constant value of ≈ 2.5 as the polymerization approaches the gelation threshold.

3.5.3. <u>SEC measurements and Comparison with LLS.</u> In order to check our determination of the molecular weight distribution by means of LLS, we used the same samples for SEC experiments. There are many calibration methods for the SEC column.[51-60] However, for branched polymers and rod-like polymers, conventional calibration methods do not apply. All reported calibration methods require at least two polymers with different molecular weights, or one sample with either two different molecular weight averages or one molecular weight plus intrinsic viscosity data, implying the use of one more instrument, such as an osmometer or a viscometer. In laser light scattering, we need only one broad MWD polymer sample and one instrument (i.e. LLS) to calibrate the SEC column. The details have been described elsewhere.[61] Figure 20 shows

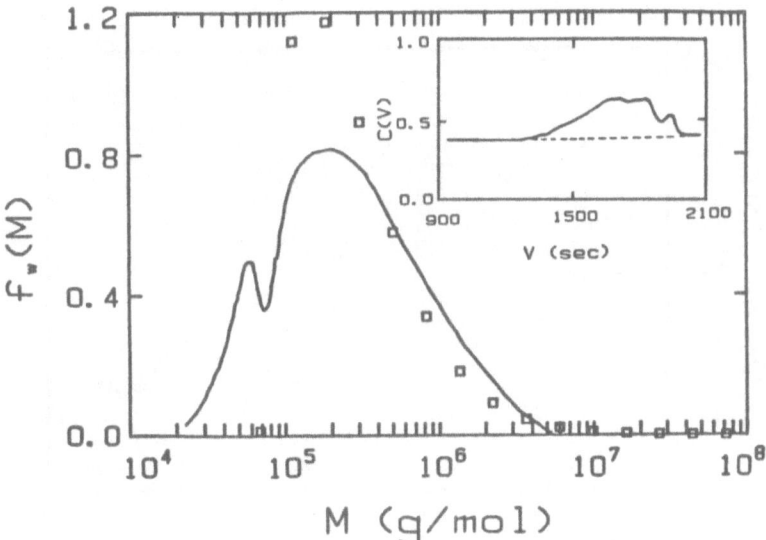

356

Figure 20. Comparison of the weight distributions $F_w(M)$ obtained for the same epoxy polymer ($M_w = 4.97 \times 10^5$ g/mole) by using the SEC experiment (continuous line) and the LLS experiment (hollow squares). (Reproduction of Fig. 11 of Ref. 61)

plots of $F_w(M)$ versus M with $F_w(M)$ from SEC denoted by the continuous line and those from LLS denoted by hollow squares. The agreement is reasonable especially if we take into account of the facts that LLS emphasizes on larger particles and SEC emphasizes on particle weight fractions. $F_w(M)$ from SEC showed a small peak around molecular weight 6×10^3, which was not observed in $F_w(M)$ from LLS. This discrepancy is not surprising because signals from LLS not only emphasized large particles but our MSVD Laplace inversion technique is designed for unimodal characteristic line-width distribution analysis. We also tried to use the CONTIN algorithm. However, with an estimated intensity ratio of only a few percent for the small to large peak ratio from SEC and a molecular weight separation of about a factor of 4 for the two peaks, it is reasonable to find that LLS failed to resolve the two peaks. The low molecular weight peak as revealed by SEC tells us additional kinetics information about the formation of the epoxy polymer network, which will be discussed in a separate article. By comparing the two MWDs from LLS and SEC, we note that $M_w:M_n$ (≈ 3.0) by SEC is larger than the value (≈ 2.5) obtained from LLS.

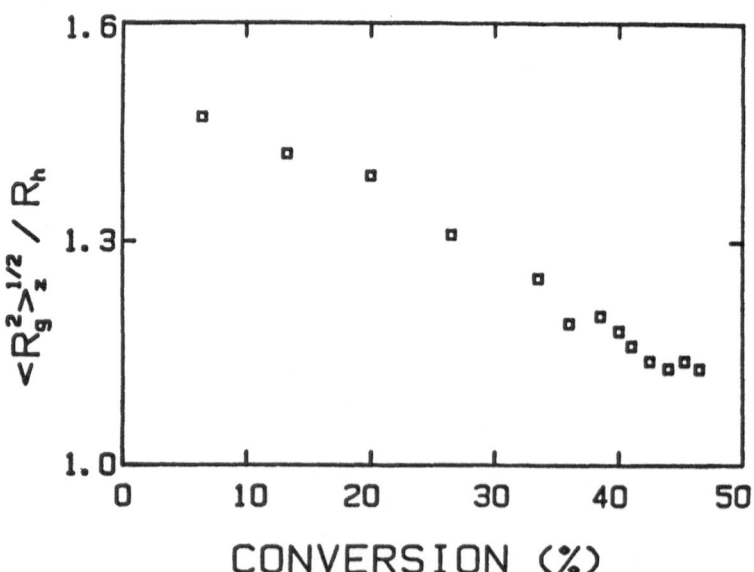

Figure 21. Plot of $\langle R_g^2 \rangle_z^{1/2}/R_h$ for 13 epoxy polymer samples at different reaction stages. (Reproduction of Fig. 12 of Ref. 61)

Figure 21 shows the ratios of the z-average root-mean-square radius of gyration and the hydrodynamic radius,[62] $\rho = R_g/R_h$, for the 13 epoxy polymer samples which we extracted during different reaction stages. ρ does not depend on the bond length and the degree of copolymerization but is a function of the branching density and polydispersity. For a linear polymer coil, $R_g/R_h \approx 1.504$. For a highly branched polymer with monodisperse primary coil chain distribution, the value of R_g/R_h decreases to 1.130.[63,64] In Table 1 of reference 63, the calculated values of ρ changed from 1.5044 to 1.1657 for polymers with the degree of branching changed from 1 (linear polymer) to 100, respectively. The value of R_g/R_h ratio obtained from our experimental results changed from 1.47 to 1.14 as the reaction approaches the gel point. The change in R_g/R_h ratio tells us that the epoxy polymer behaves close to a linear random coil during the initial reaction stages. The degree of branching increases as the polymerization reaction approaches the gel point.

The evolution of the molecular weight distribution function during the gelation transition is related to the spectacular variation of macroscopic properties of a crosslinked polymer system.[65,66] There are only small molecules inside the reaction bath at the initial reaction stage. Near the gelation threshold, a macroscopic cluster as well as precursor units and clusters of all intermediate sizes are formed. Various statistical and kinetic models have been proposed to describe the evolution of the molecule weight distribution for different gelation processes.[65-70] They all show that the molecular weight distribution of large clusters near the gelation threshold can be written in a scaling form.[71]

$$N(M, \epsilon) \approx M^{-\tau} f(M/M^*(\epsilon)) \tag{15}$$

where, $N(M, \epsilon)$ is the number of clusters with molecular weight M when the relative distance from the gel point is ϵ. $\epsilon = |p - p_c|$, where p is the extent of conversion and p_c is the extent of conversion at the gelation threshold. Eq. (15) implies that the number of clusters decreases as a power law of molecular weight with a critical exponent τ. Before the gelation threshold, a typical molecular weight $M^*(\epsilon)$ exists in the molecular weight distribution, which limits the spread of the distribution and which diverges when the gel threshold is approached: $M^* \approx \epsilon^{-1/\sigma}$ where $1/\sigma$ is a constant which is sometimes referred to as the

358

gap exponent. The crossover function f(x) describes the cutoff of the distribution for large molecules greater than M^*. Leibler, L. and Schosseler, F. experimentally demonstrated [72] a direct quantitative method of measuring the typical molecular weight (M^*) from light-scattering spectra. They showed that the molecular weight M_{max} of molecules in the distribution given the maximum scattered light intensity provides a measure of the cutoff molecular weight and can be identified with the typical molecular weight M^*. They also experimentally showed that if Eq. (15) holds for the molecular weight distribution of branched polymer molecules, the function

$$G(M/M_{max}) = M_{max}^{\tau-1} \phi(M,t) \approx (M/M_{max})^{1-\tau} f(M/M_{max}) \qquad (16)$$

is a universal function of M/M_{max} independent of the advancement of the reaction, t. Eq. (16) is valid for the high molecular weight M for which $\phi(M,t)$ is expected to be independent of the initial molecular weight distribution. τ in Eq. (16) cannot take an arbitrary value since Eq. (16) implies that the weight-average molecular weight M_w varies like $M_{max}^{3-\tau}$.

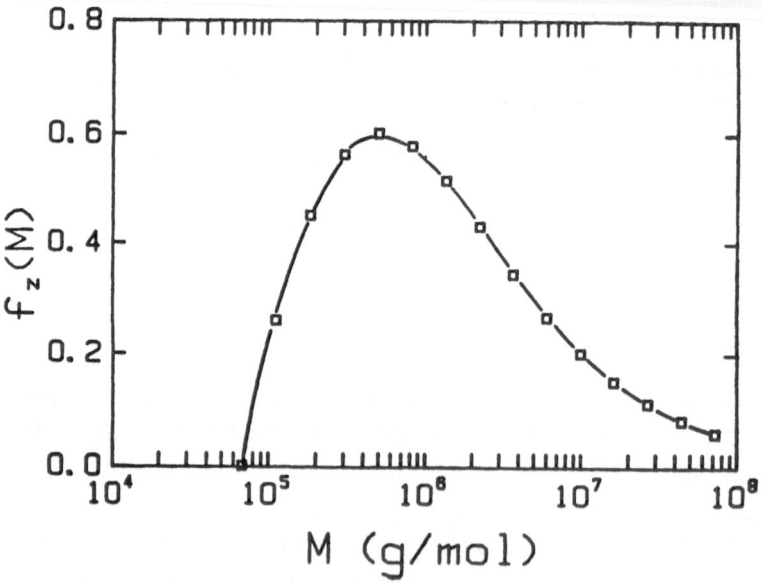

Figure 22. Plot of $f_z(M)$ versus M, based on $F_w(M)$ versus M data from LLS in Fig. 21. M_{max} is determined from the location of M at which $(f_z(M))_{max}$ occurs. (Reproduction of Fig. 13. of Ref. 61)

The scattered intensity (I) can be written as

$$I = \int_0^\infty I(M) \, dM \propto \int_0^\infty f_n(M)M^2 \, dM = \int_0^\infty f_z(M) \, dM \tag{17}$$

where $I(M)$ is the scattered intensity from the fraction of polymer with molecular weight M. Eq. (17) tells us that $(I(M))_{max}$ corresponds to $(f_z(M))_{max}$ as shown in Fig. 22. Then, the molecular weight M corresponding to $(f_z(M))_{max}$ should be M_{max}. Experimentally, we measured M_w and calculated $f_z(M)$ from the line-width measurements for different reaction stages. We obtained M_{max} from $(f_z(M))_{max}$ in the distribution. A log-log plot of M_{max} versus M_w is shown in Figure 23. The data points

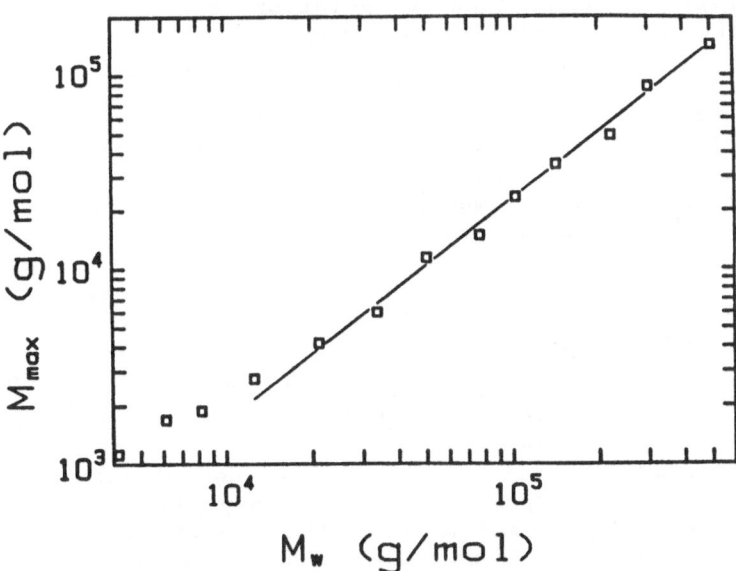

Figure 23. Log-log plot of M_{max} versus M_w. Slope $= (3 - \tau)^{-1}$ [23] $= 1.1 \pm 0.1$ and $\tau = 2.1 \pm 0.1$. (Reproduction of Fig. 14 of Ref. 61)

except the first three lower molecular weight values are essentially on a straight line, whose slope is 1.1 ± 0.1. From the slope, we calculated $\tau = 2.1 \pm 0.1$. The result further confirms the existence of a scaling law for the molecular weight distribution function in the sol phase, the main result of all theories concerned with gelation processes. Our value of the scaling constant τ is smaller than the classical mean-field prediction $\tau = 2.5$, but is closer to the value of 2.3 in reference 72.

3.6. Branching Kinetics of Epoxy Polymerization

After having obtained the molecular weight distributions at each reaction stages, we focused our attention on the branching kinetics of the epoxy polymerization because the branching kinetics could provide a theoretical background for understanding the change of the molecular weight and its distribution with the reaction time.

The copolymerization of 1,4-butanediol diglycidyl ether (DGEB) and cis-1,2-cyclohexanedicarboxylic anhydride (CH) in the presence of benzyl dimethyl amine (CA) is dominated by alternative linkages between DGEB and CH, i.e. the reactions between the epoxy resins and between the anhydrides are suppressed.[73] Highly branched epoxy copolymers are formed in the copolymerization reaction because the sum of the reaction functionalities of DGEB (two epoxy rings which can form four chemical bonds) and of CH (one anhydride ring which can form two chemical bond) is six, which is larger than the gelation criterion, i.e. a minimum total of five functionalities is needed to form a branching point. Figure 5 shows graphically how the branch points are formed in the copolymerization reaction. At the initial reaction stage, most of the polymers formed are linear. As the reaction proceeds, the degree of branching of the branched epoxy copolymer increases. Finally, a three-dimensional network is formed at the gelation threshold. We know that the triamine plays a very important role in initiating the reaction because without triamine as a catalyst, the overall reaction is very slow. But, it is not clear how the triamine starts the reaction, i.e. we do not know whether the triamine molecule first reacts with the anhydride molecule or with the epoxy resin. However, if we examine only the branching process after the very initial reaction stage, it is not important for us to know the exact initial reaction mechanism. For convenience in discussion, let us adopt Fischer's initial mechanism,[73] i.e. the triamine molecule reacts first with the anhydride molecule to form an anion (active site) on the anhydride molecule and the anion further reacts with the epoxy ring on the epoxy resin to form another anion on the epoxy resin, and so on. So the total number of active sites (i.e. the total number of CA molecules) should be constant if we assume that the triamine molecules will not transfer and the active sites will not be terminated.

By assuming that: (1) all catalyst molecules start the polymerization reaction at the same time and have the same reactivity, and they will not be terminated or transferred; (2) the reactivity is independent of polymer chain length; (3) the number of molecules undergoing the chemical reaction is counted ignoring the size of the polymer molecules; and (4) rings do not form, then we are able to present a simplified model for the branching kinetics in the copolymerization process as follows.

Let us denote as EPM_n the epoxy polymer molecule which has $(n - 1)$ branching points and n active sites, e.g., EPM_1 means a linear polymer molecule, as shown schematically in Figure 5. We further use $N_n(t)$ to denote the number of EPM_n molecules at time t (e.g., $N_1(t)$ is the number of linear polymer molecules at time t) and, $\bar{f}(EPM_n)$ to denote the average frequency with which each active site may react with EPM_n molecules at reaction time t. The average frequency $\bar{f}(EPM_n)$ can be considered to have two multiplicative parts, consisting of (1) the total number of reacted CH molecules per active site per unit time, which should be equivalent to the rate of change of CH molecules (or the rate of change of epoxy rings) divided by the total number of active sites (i.e. the initial number of CA molecules) and (2) the probability of an active site reacting with EPM_n molecules, which should be the number of EPM_n molecules divided by the total number of molecules with epoxy rings (i.e. the sum of the unreacted epoxy resin molecules and the epoxy polymer molecules). If two CH molecules react with the same single DGEB molecule instead of two DGEB molecules, it will form one branching point and consume one less DGEB molecule. So the number of unreacted epoxy resin molecules should be the difference between the initial number of the epoxy resin molecules and the number of reacted epoxy rings plus the total number of branching points. In this way, $\bar{f}(EPM_n)$ can be expressed as

$$\bar{f}(EPM_n) = \frac{-dN_{CH}/dt}{N_{CA}(0)} \cdot \frac{N_n}{N_{DGEB}(0) - pN_{CH}(0) + N_b + N_{EPM}} \quad (18)$$

where $N_{CH}(0)$, $N_{CA}(0)$ and $N_{DGEB}(0)$ are, respectively, the number of CH, CA and DGEB molecules at time t = 0. N_{CH} is the number of unreacted CH molecules at time t. The p $[= 1 - N_{CH}/N_{CH}(0)]$ is the CH extent of conversion. The minus sign "-" corresponds to the rate of change of CH molecules being positive because dN_{CH}/dt is always negative. N_{EPM} is the total number of epoxy polymer molecules. N_b is the total number of branching points on all the epoxy polymers.

$$N_b = \sum_{i=1}^{\infty} (i - 1)N_i = \sum_{i=1}^{\infty} iN_i - \sum_{i=1}^{\infty} N_i = N_{CA}(0) - N_{EPM} \quad (19)$$

By replacing N_b in Eq. (18) with Eq. (19) and p with the definition, we get

$$\bar{f}(EPM_n) = \frac{-dN_{CH}/dt \, N_n}{N_{CA}(0)N_{CH}(0)[\beta - 1 + N_{CH}/N_{CH}(0)]} \quad (20)$$

where $\beta = [N_{DGEB}(0) + N_{CA}(0)] / N_{CH}(0)$. At a fixed initial molar ratio of epoxy resin to anhydride and temperature (T), $\bar{f}(EPM_n)$ is a function of catalyst concentration and reaction time t or the extent of conversion p. Now let us consider the reaction frequency of a pair of epoxy polymer molecules (EPM_i and EPM_j). As we mentioned before, an EPM_n molecule has n active sites. So the frequency for N_i EPM_i molecules reacting with the N_j EPM_j molecules should be $iN_i\bar{f}(EPM_j)$ under the assumptions we have made here. In the same way, the frequency for N_j EPM_j molecules reacting with the N_i EPM_i molecules should be $jN_j\bar{f}(EPM_i)$. The total frequency ($F_{i,j}$) for N_i EPM_i molecules and N_j EPM_j molecules reacting with each other will be assumed to be the sum of $iN_i\bar{f}(EPM_j)$ and $jN_j\bar{f}(EPM_i)$. Using Eq. (18), we can now write

$$F_{i,j} = \frac{(-dN_{CH}/dt)(i+j)\, N_iN_j}{N_{CA}(0)N_{CH}(0)[\beta-1+N_{CH}/N_{CH}(0)]} = f_A(i+j)\, N_iN_j \quad (21)$$

where $f_A = \dfrac{-dN_{CH}/dt}{N_{CA}(0)N_{CH}(0)[\beta-1+N_{CH}/N_{CH}(0)]}$. It is obvious that $F_{i,j} = F_{j,i}$. Now we consider how N_n changes with reaction time. If an active site on EPM_n molecules reacts with an epoxy resin DGEB, the result is an increase in the length of an EPM_n molecule but N_n will not change. N_n changes only when the epoxy polymer molecules react with each other. The change in N_n consists of two parts: an increase in N_n when an EPM_i molecule reacts with an EPM_j molecule where $(i+j) = n$ and a decrease in N_n when an EPM_n molecules reacts with any other epoxy polymer molecule. Based on the above conditions, we find the reaction rate to be

$$\frac{dN_1}{dt} = -\sum_{i=1}^{\infty} F_{1,i} = -\sum_{i=1}^{\infty} f_A(1+i)\, N_1N_i \quad (22)$$

$$\frac{dN_2}{dt} = \frac{F_{1,1}}{2} - \sum_{i=1}^{\infty} F_{2,i} = f_AN_1N_1 - \sum_{i=1}^{\infty} f_A(2+i)N_2N_i \quad (23)$$

$$\frac{dN_3}{dt} = F_{1,2} - \sum_{i=1}^{\infty} F_{3,i}$$

$$= f_A(1 + 2)N_1 N_2 - \sum_{i=1}^{\infty} f_A(3 + i)N_3 N_i \qquad (24)$$

. . . .

$$\frac{dN_n}{dt} = \frac{1}{2} \sum_{i+j=n} F_{i,j} - \sum_{i=1}^{\infty} F_{n,i}$$

$$= \frac{1}{2} \sum_{i+j=n} f_A(i + j)N_i N_j - \sum_{i=1}^{\infty} f_A(n + i)N_n N_i \qquad (25)$$

Let us compare Eq. (25) with the Smoluchowski coagulation equation[74]:

$$\frac{d\nu_n(t)}{dt} = \frac{1}{2} \sum_{i+j=n} K_{i,j} \nu_i \nu_j - \nu_n \sum_{i=1}^{\infty} K_{n,i} \nu_i \qquad (26)$$

where $K_{i,j}$ is the overall collection probability and $\nu_n(t)$, with $n = 1,2,\ldots,\infty$, is the concentration of n-size "droplets". The "droplets" could be any cluster of particles undergoing the coalescence process, e.g., cloud droplets, colloidal particles and so on. It is not difficult to see that Eq. (25) has the same form as Eq. (26) if we let $f_A(i + j) = K_{i,j}$. Eq. (26) has a known explicit solution[75,76] when the overall collection probability $K_{i,j}$ is proportional to the sum of "droplet volumes" and is independent of time, i.e. $K_{i,j} = b(i + j)$ with b being a constant. The pregelation solution is

$$\nu_n(t) = \nu_\Sigma(0) \frac{(n\alpha)^{n-1}}{n!} (1 - \alpha) e^{-n\alpha} \qquad (27)$$

where $\nu_\Sigma(0) = \sum_{n=1}^{\infty} \nu_n(0)$ and $\alpha = 1 - e^{-\nu_\Sigma(0)bt}$. For our copolymerization process, however, we cannot use Eq. (27) as the solution of Eq. (25) because f_A is not a constant with respect to time. We have to rearrange Eq. (25) since f_A is a function of the reaction time t. By using the definition of f_A and p, we can transform $f_A dt$ as $d\varsigma$ by using the definitions of f_A after Eq. (21), where

$$\zeta = \frac{1}{N_{CA}(0)} \ln(\frac{1}{\beta - p}) \tag{28}$$

With Eq. (28), we rewrite Eq. (25) as

$$\frac{dN_n}{d\zeta} = \frac{1}{2} \sum_{i+j=n} (i + j)N_i N_j - \sum_{i=1}^{\infty} (n + i)N_n N_i \tag{29}$$

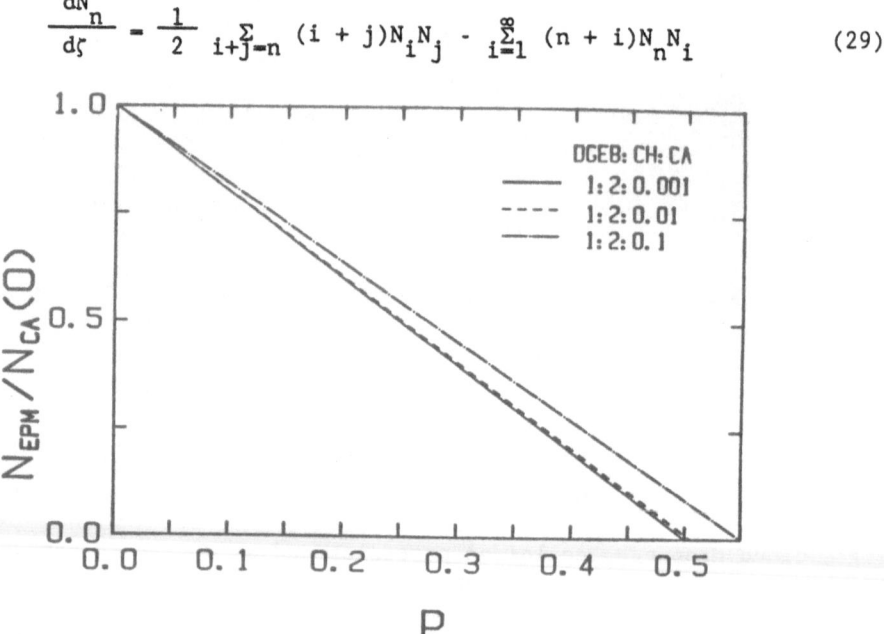

Figure 24. Plot of calculated total number of epoxy polymer molecules (EPM), N_{EPM}, normalized by initial number of catalyst molecules, $N_{CH}(0)$, versus the extent of conversion p (= 1- $N_{CH}(t)/N_{CH}(0)$) at different molar ratios of DGEB:CH:CA.

By comparing Eq. (29) with Eq. (26), we now have $K_{i,j} = (i + j)$ and can apply the solution to Eq. (25), with b and $\nu_\Sigma(0)$ replaced by 1 and $N_{CA}(0)$, respectively. Finally, we have the explicit solution to Eq.(25),

$$N_n(p) = N_{CA}(0) \frac{(n\gamma)^{n-1}}{n!} (1 - \gamma) e^{-n\gamma} \tag{30}$$

where $\gamma = 1 - e^{-N_{CA}(0)[\varsigma(p) - \varsigma(0)]}$. By using Eq. (28), we rewrite γ in the following form:

$$\gamma = \frac{pN_{CH}(0)}{N_{DGEB}(0) + N_{CA}(0)} = \frac{p}{\beta} \tag{31}$$

The total number of the epoxy polymer molecules (N_{EPM}) should be $\sum_{n=1}^{\infty} N_n(\varsigma)$, i.e.

$$N_{EPM} = \sum_{n=1}^{\infty} N_{CA}(0)\frac{(n\gamma)^{n-1}}{n!}(1 - \gamma)e^{-n\gamma} = N_{CA}(0)(1 - \gamma) \tag{32}$$

which is a linear function of the extent of conversion (p). Figure 24 shows a plot of $N_{EPM}/N_{CA}(0)$ as a function of p based on Eq. (32) for the molar ratios of DGEB:CH:CA = 1:2:0.001, 1:2:0.01 and 1:2:0.1, respectively. From Eq. (30), we can calculate N_n and the number distribution $f_N(EPM_n)$ with different extents of conversion (p) for the copolymerization reaction. Figure 25 shows a plot of $f_N(EPM_n)$ versus the number (n) of active sites on the EPM_n molecule for the same molar ratios of DGEB:CH:CA in Fig. 24 at different p values with p = 0.1, 0.2, 0.3, 0.4 and 0.45, respectively. Figure 26 graphically shows how N_n changes with p for the same molar ratios of DGEB:CH:CA in Fig. 24. From Figure 26, we note that there is a maximum number of molecules for every type of EPM_n molecule during the copolymerization process. The location

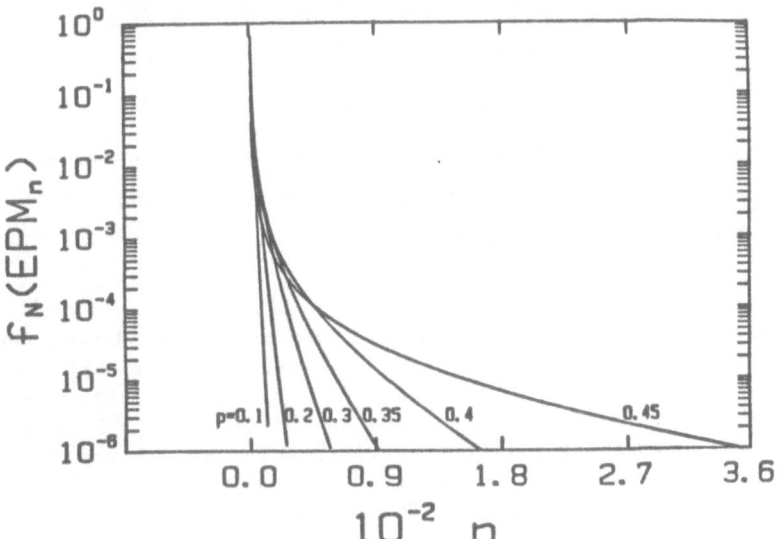

Figure 25. Plot of calculated number of epoxy polymer molecules with n
branching points (N_n) versus n at different conversion extent
(p) for the same molar ratios as in Fig. 24.

of $p_{max}(n)$ can be calculated by letting $dN_n/dp = 0$. The result is that

$$P_{max}(n) = \beta \left[1 - \left(\frac{1}{n} \right)^{1/2} \right] \tag{33}$$

When $n = 1$, $p_{max}(1) = 0$. This is appropriate because N_1 always decreases
as the reaction proceeds. When $n \to \infty$, $p_{max}(\infty) \to \beta$. This behavior can be
seen in Fig. 26.

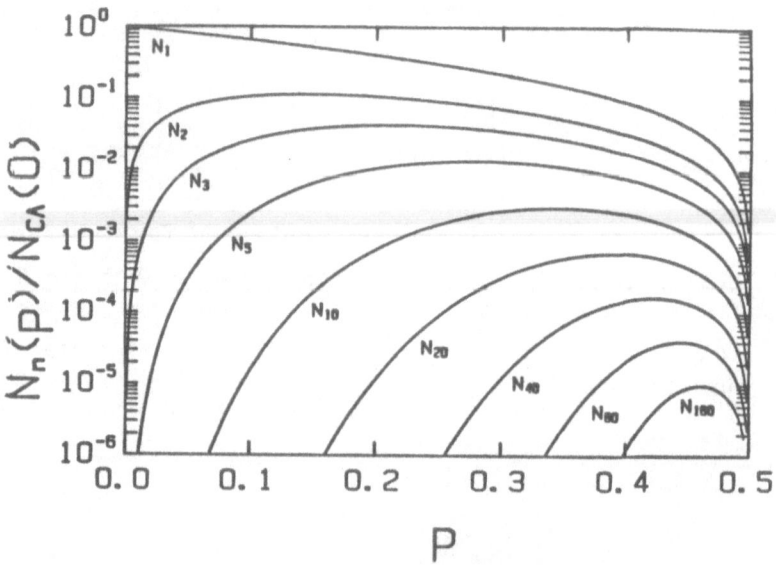

Figure 26. Plot of calculated number of epoxy polymer molecules with n
branching points (N_n) versus p for different values of n but
the same molar ratios as in Fig. 24.

By means of chemical analysis,[30] we monitored the extent of CH
conversion during the copolymerization reaction. As the reaction
approaches its gelation threshold, the chemical analysis becomes more
and more difficult and inaccurate because the reaction mixture becomes
very viscous and more gel-like. Experimentally, we defined the gelation
threshold by constantly taking out the reaction mixture from the
reaction vessel and dissolving it in MEK during the curing process until

the moment when we observed a trace amount of undissolved epoxy polymers appearing in the solution. The extent of conversion corresponding to that moment is defined as the critical conversion extent (p_c). We defined that moment as the gelation threshold. Theoretically, the gelation threshold can be defined as the point at which all epoxy polymer molecules are linked to each other to form an infinite network. This means that $N_{EPM} = 1$ at the gelation threshold. Our definition on the gelation threshold is somewhat different from what has been defined in the past. Based on the definition that the critical conversion extent (p_c) is the point at which the branching probability of the reaction is equal to the non-branching probability of the reaction. The gelation threshold should occur earlier than p_c according to our definition. Our experimental observation has indeed shown that the gelation threshold occurs before $N_{EPM} = 1$, i.e., the entire system becomes one giant polymer. By using Eq. (15) and the above definition, we can calculate that $p_c = \beta$. Both the experimentally determined p_c (expressed as hollow squares) and the calculated p_c (expressed as hollow triangles) for the molar ratio of DGEB:CH = 1:2 but different amounts of CA are plotted in

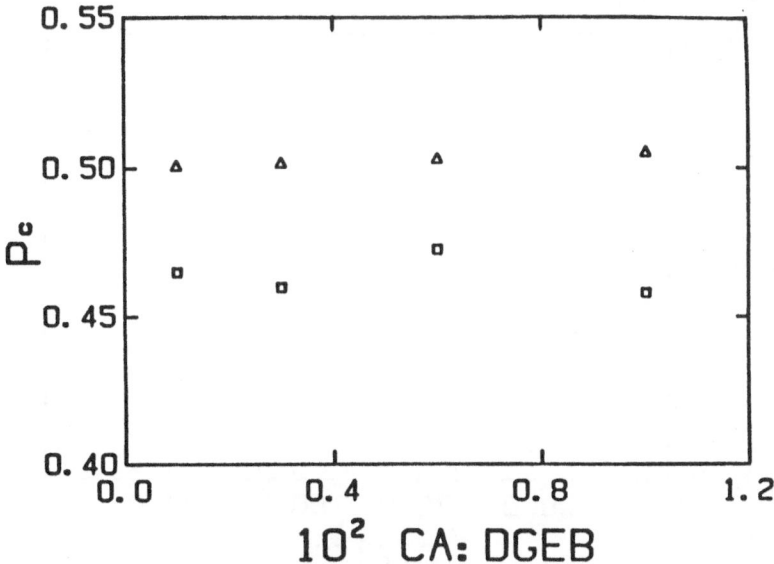

Figure 27. Plot of measured and calculated critical conversion extent (p_c) versus different molar ratios of CA:DGEB, but keeping

DGEB:CH = 1:2 and 80°C. (Reproduction of Fig. 5 of Ref. 82)

Fig. 27. The experimental results show that p_c is essentially constant, independent of the molar ratio variations in CA:DGEB from 0.001 to 0.01, as predicted by the calculated results. We also observed that the

368

experimentally determined p_c values were about 7% smaller than the calculated values. The difference existed systematically for all four different catalyst concentrations. There could be two possible reasons for this systematic deviation. (i) Our experimental criteria for p_c, i.e. the solubility test and the rate of light scattering intensity change, is not the same as the theoretical one. At the experimental p_c, the copolymerization process has not yet reached the stage of an infinite network, which is the criterion according to theory. (ii) As the reaction approaches the gelation threshold, some or all of the assumptions for the proposed branching kinetics break down, especially, the formation of rings inside highly branched epoxy polymers. The rings could produce a finite macro-size three dimensional network as evidenced by the insoluble gel particles. If we assume that f_A appearing in Eq. (21) is independent of reaction temperature, than the theoretical p_c ($= \beta$) yields a gelation threshold that is also independent of reaction temperature. Figure 28 shows a plot of the experimentally determined p_c values as a function of reaction temperature. The reaction temperature has indeed only a weak effect on the gelation threshold in terms of the extent of conversion.

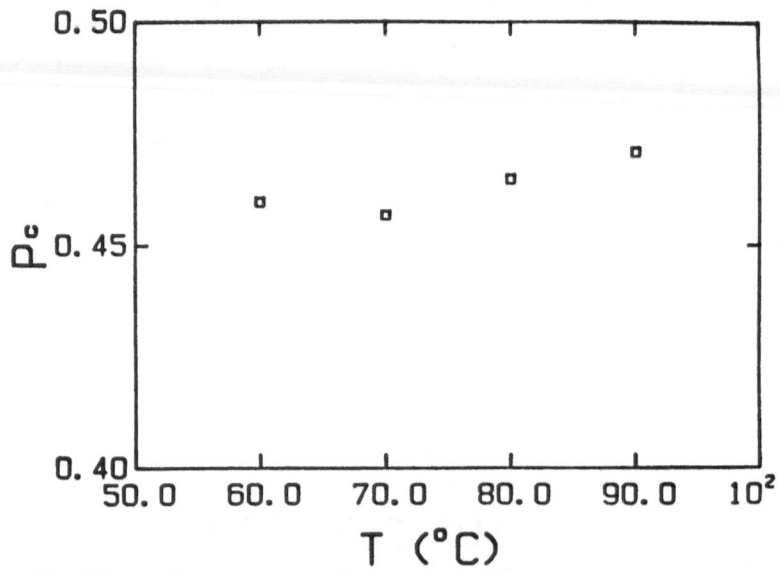

Figure 28. Plot of measured critical conversion extent (p_c) versus the reaction temperature (T) for the reaction mixture at the molar ratio of DGEB:CH:CA = 1:2:0.001. (Reproduction of Fig. 6 of Ref. 82)

By assuming that all catalyst molecules start the polymerization reaction at the same time, each active site grows at the same speed and produces the same linear length (l_o) of the subpolymer chain. The molecular weight for the subpolymer chain (l_o), denoted as $M(l_o)$, should be the total reacted weight divided by the number of active sites (i.e. $N_{CA}(0)$).

$$M(l_o) = \frac{p \, N_{CH}(0) \, (M_{DGEB} + M_{CH})}{N_{CA}(0)} + M_{CA} \qquad (34)$$

where M_{DGEB}, M_{CH} and M_{CA} are the molecular weight of DGEB, CH and CA, respectively. In this way, each EMP_n molecule has n such subpolymer chains. The molecular weight for EPM_n molecule should be n times $M(l_o)$, i.e. $M(EPM_n) = nM(l_o)$ which represents a consequence of our stringent assumptions. Together with Eq. (34), the number-average molecular weight (M_n) at different p can be written as

$$M_n = \frac{\sum_{n=1}^{\infty} M(EPM_n)N_n}{\sum_{n=1}^{\infty} N_n} = \frac{M(l_o)\sum_{n=1}^{\infty} nN_n}{\sum_{n=1}^{\infty} N_n} = \frac{M(l_o)}{(1 - \gamma)} \qquad (35)$$

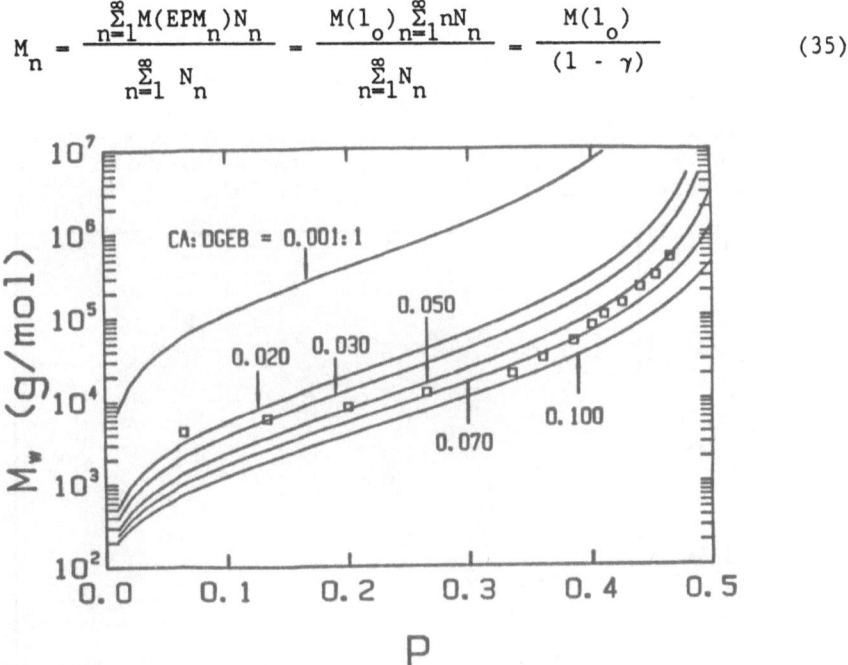

Figure 29. Plot of calculated M_w versus p for fixed molar ratio of

DGEB:CH = 1:2 but different amount of catalyst CA. Measured M_w versus p for the molar ratio of DGEB:CH:CA = 1:2:0.001 at 80°C are also plotted in the Figure. (Reproduction of Fig. 7 of Ref. 82)

and the weight-average molecular weight (M_w),

$$M_w = \frac{\sum_{n=1}^{\infty} M(EPM_n) f_w(M)}{\sum_{n=1}^{\infty} f_w(M)} = M(1_o)\sum_{n=1}^{\infty} n^2 N_n = M(1_o)\left(\frac{1}{1-\gamma}\right)^2 \quad (36)$$

Figure 29 shows how calculated M_w changes with the extent of conversion p at a fixed molar ratio of DGEB:CH = 1:2 but different amounts of catalyst CA. The measured M_w versus p for the reaction mixture at a

fixed molar ratio of DGEB:CH:CA = 1:2:0.001 at 80°C are also plotted in Figure 29. By comparison, we find that the assumptions about no termination for the active site and no catalyst transfer during the reaction process are too stringent. There is not a single calculated curve which fits the measured data over the entire measured range of p. In the actual copolymerization reaction, the catalyst molecules are transferred constantly to form new active sites.[73] This means that the reaction mixture would actually have more active sites than the number of catalyst molecules. Then the number of active sites become a function of reaction time. In Fig. 29, by varying the molar ratio of catalyst CA to DGEB in the calculation, we have simulated the catalyst transfer

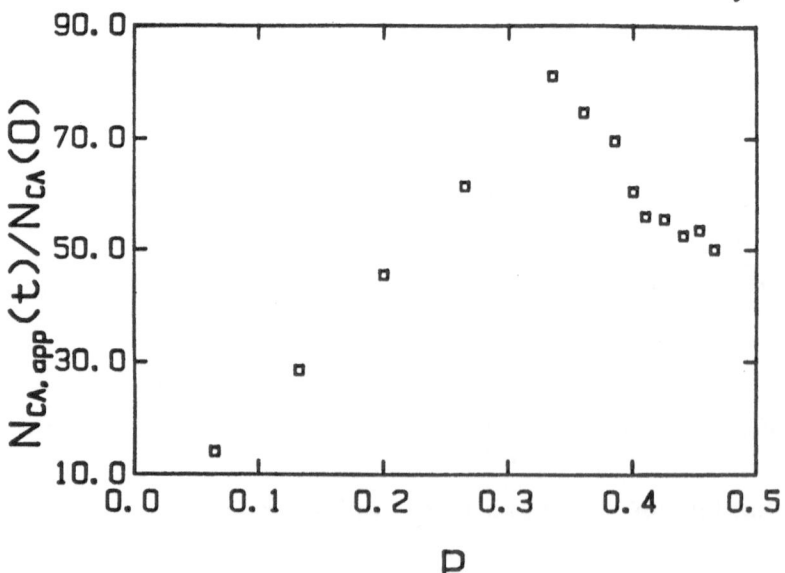

Figure 30. Plot of simulated $N_{CA,app}(0)/N_{CA}(0)$ versus p at a fixed molar ratio of DGEB:CH = 1:2 and 80°C, where the $N_{CA,app}(t)$ is the apparent total active sites in the reaction mixture at reaction time t.

during the copolymerization process for the real reaction. Thus, by comparing the calculated curves with the experimental data points, we have obtained information about how many times the catalyst molecules are transferred to form new active sites during the reaction process. The apparent number of active sites in the reaction mixture at time t is denoted as $N_{CA,app}(t)$. $N_{CA,app}(t)/N_{CA}(t=0)$, which corresponds to the average number of times each catalyst molecule has been transferred to form new active sites at reaction time t, versus the extent of conversion (p) are plotted in Figure 30. In Fig. 30, we can see that the average number of times a catalyst molecule is being transferred is a linear function of the extent of conversion. The apparent number of the active sites is about 15 times greater than $N_{CA}(t=0)$ at p = 0.07 and about 80 times greater at p = 0.35. The same phenomena was observed by Fischer.[73] We also observed that the amount of active sites decreases for p > 0.35. Why should there be a decrease in the number of active sites in the copolymerization reaction for p > 0.35? We know that the viscosity of the reaction mixture increases as the extent of conversion increases and the epoxy polymer molecules become larger and larger. Some active sites inside the larger polymer molecules will have less chance to react with other polymer molecules. The larger polymer molecules move at lower speeds than the smaller molecules implying that effectively the active sites on the larger polymer molecules have less activity than

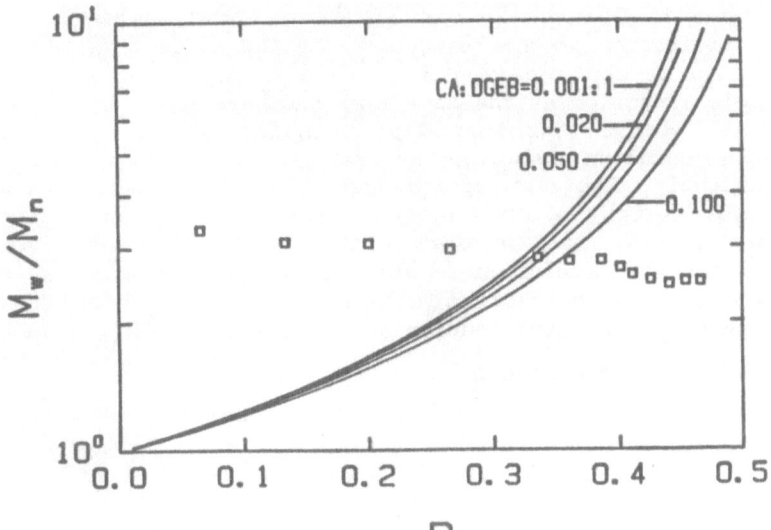

Figure 31. Plot of calculated ratio of the weight-average molecular weight to the number-average molecular weight (M_w/M_n) versus p for a fixed molar ratio of DGEB:CH = 1:2 but different amounts of catalyst CA. Measured M_w/M_n versus p for the molar ratio of DGEB:CH:CA = 1:2:0.001 at 80°C are also shown in Figure 31. (Reproduction of Fig. 9 of Ref. 82)

those on the smaller polymer molecules. Figure 31 shows that calculated ratio of M_w/M_n ($= \dfrac{1}{1-\gamma}$) versus p at a fixed molar ratio of DGEB:CH = 1:2 but different amounts of catalyst CA. The measured M_w/M_n values, denoted by hollow squares in Fig. 31, are completely different from the theoretical prediction. At low conversion extent, the calculated M_w/M_n values are much smaller than the measured values. At higher conversion extent, this relation is reversed. The difference can be explained as follows. At low conversion extent, the calculation assumes that the starting molecules are monodisperse, i.e. $M_w/M_n = 1$. For the real reaction, the epoxy polymer is already polydisperse when the branching process starts. At high conversion extent, there are not only very larger branched polymer molecules in the reaction mixture, but also very small epoxy polymer molecules, confirming a high polydispersity index. In light scattering, the presence of larger particles are emphasized, resulting in smaller measured M_w/M_n values for highly polydispersed systems with large amounts of smaller particles. The use of size-exclusion chromatography together with light scattering for analysis of branched epoxy polymer is underway.

The deviations of the calculated M_w and M_w/M_n with the measured ones as shown in Figures 29 and 31, respectively, clearly suggests a need for modification of the assumptions in the theoretical development, especially on assumption (1), i.e., all catalyst molecules start the polymerization reaction at the same time and have the same reactivity, and they will not be terminated or transferred. We have considered possible breakdowns on the transfer and termination of active sites. More importantly, from our experimental observation, we have noted that the molecular weight distribution of the epoxy polymer formed is polydisperse even in the very early reaction stages. Thus, the assumption that the catalyst molecules start the polymerization reaction at the same time must not be valid. We now modify our theory by assuming a simple Gaussian-like distribution on the number of EPM_n formed at some initial extent of conversion p_o

$$N_n(p_o) = \frac{N_{EPM}(p_o)}{\bar{n}(p_o)} \exp\left(-\frac{n}{\bar{n}(p_o)}\right) \tag{37}$$

where $N_{EPM}(p_o)$ and $\bar{n}(p_o)$ are the total number of epoxy polymer molecules at p_o and the mean active site per epoxy polymer molecule at p_o. For this particular type of initial condition, Golovin[75] and Scott[76,77,78] have shown that Eq. (29) can be solved by using Laplace transforms and a conformal tranformation, yielding an exact solution

$$N_n(p) = \left(1-\frac{p-p_o}{\beta-p_o}\right)\exp\left[-\left(\frac{p-p_o}{\beta-p_o}+1\right)\frac{n}{\bar{n}(p_o)}\right]\sum_{m=0}^{\infty}\frac{\left(\frac{p-p_o}{\beta-p_o}\right)^m[n/\bar{n}(p_o)]^{2m}}{(m+1)!\,m!} \tag{38}$$

By combining Eq. (38) with Eq. (36), we are able to calculate $f_w(M)$ at different conversion p. The calculated results are shown in Figure 32

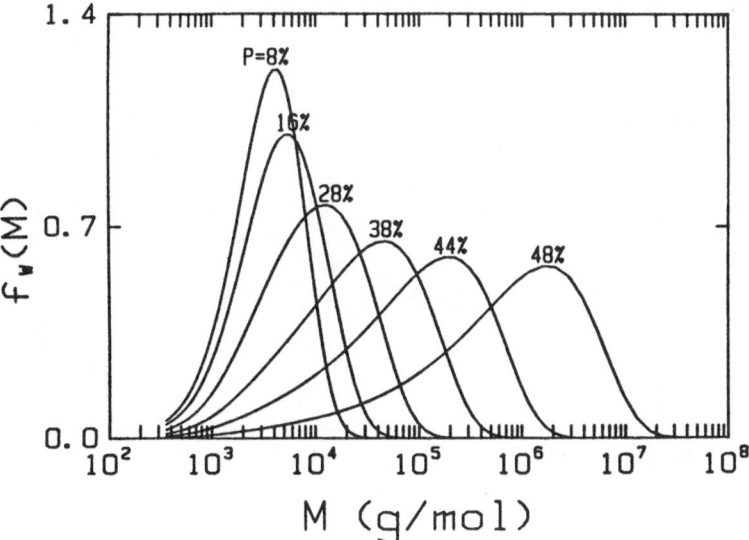

374

Figure 32. Plot of calculated weight distribution ($f_w(M)$) versus M at different extent of conversion p by using Eq. (38) where $p_0 = 0.065$, $\bar{n}(p_0) = 1$, and DGEB:CH:CA $= 1:2:0.001$. (Reproduction of Fig. 10 of Ref. 82)

where we have chosen $M_w = 4.1\times10^3$ (g/mol) at $p_0 = 0.065$, which is the lowest experimentally measured molecular weight as shown in Figure 29. In Figure 32, we see that the molecular weight distribution is spreading as the reaction approaches the gelation threshold. The long low molecular weight tails obviously give a much higher theoretical value of M_w/M_n than the measured ones because laser light scattering tends to emphasized higher molecular weight fractions. From the calculated $f_w(M)$, we can further compute out the weight-average molecular weight M_w at each reaction stage. The solid line in Figure 33 represents the calculated M_w values based on $f_w(M)$ values in Figure 31, and the hollow squares are the same experimental data as shown in Figure 29. Figure 33 shows that the agreement between the calculated curve and measured M_w values is fairly good. It suggests that the introduction of Eq. (37) has provided a realistic approach to using the Smoluchowski coagulation equation for epoxy polymerization studies.

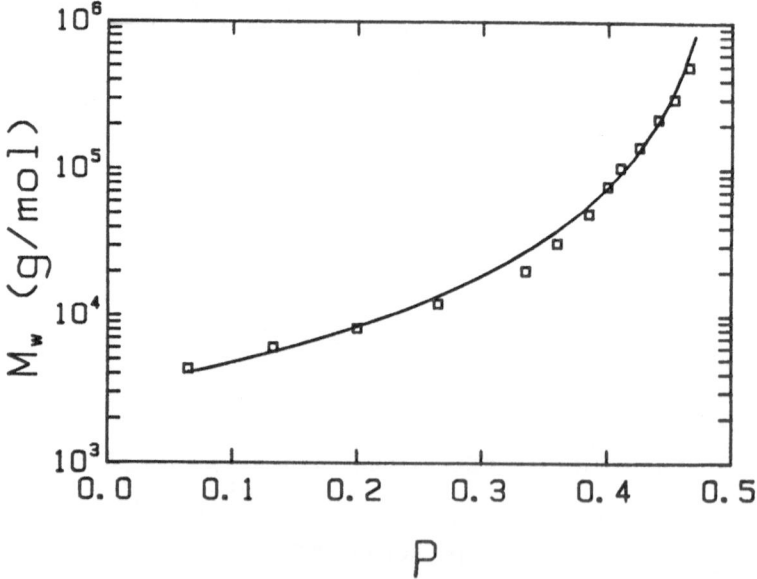

Figure 33. Plot of weight-average molecular weight M_w versus p. Solid

line represents the calculated values by using Eq. (25) where $P_o = 0.065$, $\bar{n}(p_o) = 1$, and DGEB:CH:CA $= 1:2:0.001$. Hollow squares show the real experimental data with the same ratio of DGEB:CH:CA at $80°C$. (Reproduction of Fig. 9 of Ref. 82)

3.7. Study of LLS envelope of Epoxy polymers

Debye and Bueche[29] proposed that the intensity of scattered light (i) of an inhomogeneous solid follows the expression:

$$i \propto 4\pi\bar{\eta}^2 V \int_0^\infty r^2 \gamma(r) \frac{\sin(Kr)}{Kr} \, dr \tag{39}$$

where V is the scattering volume, $\bar{\eta}^2$ is the mean square average local dielectric constant fluctuations, and $\gamma(r)$ is a correlation function defined by

$$\gamma(r) = \langle \eta_1 \eta_2 \rangle / \bar{\eta}^2 \tag{40}$$

where η_1 and η_2 are the local dielectric constant fluctuations in volume elements 1 and 2, repectively. For all pairs of volume elements separated by a scalar distance r inside the scattering volume, $\langle \eta_1 \eta_2 \rangle$ is the average value of $\eta_1 \eta_2$. If we take the correlation to be random,

$$\gamma(r) = e^{-r/a} \tag{41}$$

where a is the correlation length defining the local inhomogeneities. The Rayleigh ratio for unit scattering volume has the form

$$R_{vv}(K) = \frac{8\pi^3 \bar{\eta}^2 a^3}{\lambda_o^4 (1+K^2 a^2)^2} \tag{42}$$

We could calculate a and $\bar{\eta}^2$ from the Rayleigh ratio by plotting $(R_{vv}(K))^{-1/2}$ versus K^2.

$R_{vv}(K)_{K=0}$ at a fixed catalyst concentration (0.1%) changes with reaction time at four different temperatures. In order to find the temperature effect, we rescaled the reaction time at each temperature by multiplying it with the corresponding k_o value. The results are shown in

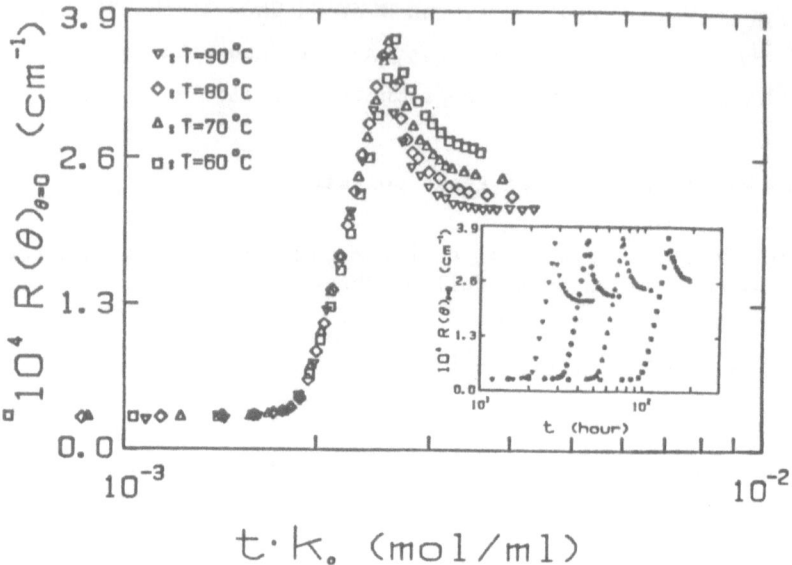

Figure 34. $R_{vv}(\theta)_{\theta=0}$ versus scaled reaction time, $k_o t$ (in units of mole/mL), at four different reaction temperatures, 60°C, 70°C, 80°C and 90°C. The molar ratio of DGEB:CH:CA = 1:2:0.001. Inset: Unscaled $R_{vv}(\theta)_{\theta=0}$ versus reaction time t (in units of hour). (Reproduction of Fig. 3 of Ref. 81)

Figure 34. We note that the temperature has little effects on $R_{vv}(K)_{K=0}$ before $R_{vv}(K)_{K=0}$ reaches its maximum value. After having passed the maximum, $R_{vv}(K)_{K=0}$ tends to behave differently at different temperatures. The higher the reaction temperature, the smaller the value of $R_{vv}(K)_{K=0}$ becomes. The phenomenon is reasonable because the system becomes more uniform and scatters less light at higher reaction temperatures. A similar phenomenon was observed when we changed the catalyst concentration, i.e., the higher the catalyst concentration, the more uniform the reaction mixture becomes and the less the scattered intensity. The optical behavior tells us that the polymerization reaction produced essentially the same degree of local dielectric constant fluctuations as long as the catalyst concentration is greater than a threshold value (say ≈ 0.3%). From the light scattering intensity studies on in situ epoxy reaction mixtures, we can make the following observations:

(1) The higher the temperature, the smaller the η^{-2} value and the larger the extension of local inhomogeneities, in terms of a.

(2) The lower the catalyst concentration, the smaller the η^{-2} values and the larger the extension of local optical inhomogeneities. This observation suggests that at lower catalyst concentrations each polymer molecule has more chance to grow bigger before it reacts with other polymer molecules.

(3) There are two different rates associated with the characteristic length increases with time. After having reached its maximum value (around 70 nm), the correlation length a gradually decreases to a constant value at around 60 nm. The changes in the η^{-2} values also show three main steps: a sharp drop, a gradual increase, and then a steady state. Each crossover point on the η^{-2} values corresponds to a change of behavior on the average characteristic length of local inhomogeneities.

By combining this observation with the knowledge we gained before, we propose that the reaction process can be divided into four main

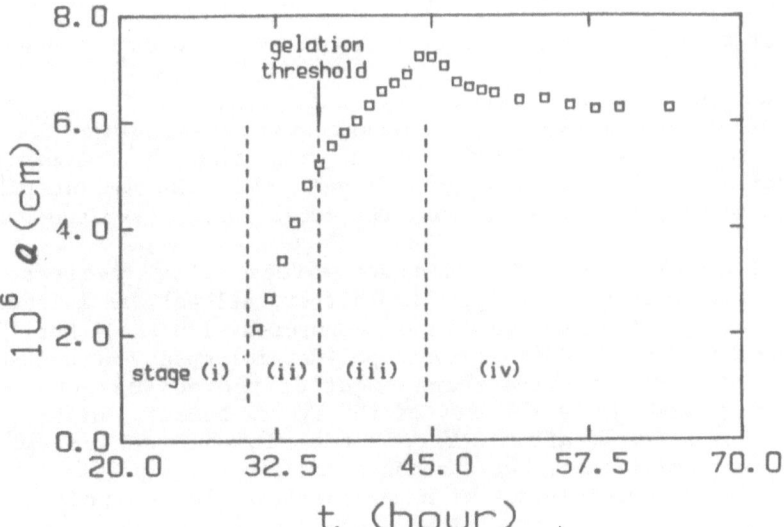

Figure 35. Correlation length (a) of local optical inhomogeneities versus reaction time t at 80°C. Catalyst concentrations is 0.1% and the molar ratio of DGEB:CH = 1:2. (Reproduction of Fig. 7 of Ref. 81)

regions. It is shown in Figure 35. (i) In the very beginning, the size of inhomogeneities is too small to be observed by laser light scattering. Most of the initial polymer product are linear or slightly branched polymer molecules. The growth of the polymer chains in this first stage is accompanied mainly by the reaction of the epoxy monomers with the curing agent. There is a gradual increase in the size of inhomogeneities through the growth of polymer molecules with some crosslinking. At the same time, the optical density difference between the polymer chains and the reaction mixture becomes larger as the polymer size becomes larger. The $\bar{\eta}^2$ value must gradually increase to its maximum value until crosslinking dominates the copolymerization reaction. In the above discussion, we have assumed that polymers are formed only at the catalytic centers and therefore, there is no increase in polymer number concentration. With increasing extent of conversion, the concentration of lower molecular weight polymers whose refractive index increment has reached a plateau value becomes higher and higher. Then the system enters the second stage. (ii) In the second stage, the polymer molecules have more chance to react with each other to form larger and more highly branched polymer molecules. It is easy to understand that the characteristic size of inhomogeneities increases very fast because the polymer chains almost double themselves in size by covalent bonding among the polymer chains. We also notice that the values of $\bar{\eta}^2$ decrease by a factor of about two. How do we interpret this observation? Let us look at the definition of $\bar{\eta}^2$. η is a local variation of the dielectric constant superimposed on the average dielectric constant ϵ. In the second stage when the polymer molecules have overlapped to form local polymer networks, the amount of polymer per unit volume is increasing, but the crosslinked polymer networks tend to shrink in size. So, the relative values of $\bar{\eta}^2$ decrease. As the polymer molecules become bigger and bigger, all polymer molecule inside the network will tend to become more connected with one other. Then, the gel state has fewer and fewer loose polymer molecules not connected with the network. (iii) In third stage, most of the polymer chains cannot move freely and the viscosity of the system becomes higher, i.e. the polymer chains are localized. Direct reaction between polymer chains becomes more difficult, but the monomers (and small polymer chains) are still moving around to react with the localized larger polymer chains. The extent of local optical inhomogeneities increases at a lower rate. At the same time, the localized polymerization reaction densifies the branched polymer chains, and thus increasing the contrast of the polymer domain to the average background of the system. However, this increase is limited, as shown by a small increase in $\bar{\eta}^2$ value during the third stage. We suggest that the crossover point for the average size of local optical inhomogeneities from one rate to another, or the lowest value of $\bar{\eta}^2$, could be considered as the gel point, which we marked out in Figure 35. (iv) In the final stage, the unreacted monomers still move inside the system to react with the crosslinked polymer chains. Experimentally we observed that the size of local optical inhomogeneities reduced to a constant value and $\bar{\eta}^2$ values remained unchanged. This observation suggests that light scattering observes different sources of local optical inhomogeneities.

4. CONCLUSIONS

Branched polymer structures and macromolecular properties of epoxy polymers in solution can be investigated using a combination of scattering techniques. By combining laser light scattering with small angle x-ray scattering, we have developed a methodology to determine the molecular weight and the molecular weight distribution of random polymer structures before its gelation threshold during the curing process. Our method differs from the usual analytical technique such as size exclusion chromatography because we do not require fractionation of random branched polymers in solution but take advantage of the fractal geometry of random structures which can be determined by SAXS. By using the LLS approach described in this article, we can study how temperature, composition and catalyst effect the epoxy polymerization process and determine the molecular weight distribution at different reaction stages without relying on the more tedious analytical technique of size exclusion chromatography. Experimentally, we have further shown that there exists a scaling relationship for the molecular weight distribution in the sol phase of the crosslinked epoxy-anhydride system. The scaling constant τ (= 2.1 $\pm$ 0.1) is very close to our previously determined fractal dimension d_f (= 2.17 $\pm$ 0.05).

The Debye-Bueche theory for inhomogeneous solids has been applied to studies of the formation of polymer networks in an epoxy polymerization reaction. As an on-line technique, laser light scattering has a unique advantage, capable of monitoring the local optical inhomogeneities during the curing process. Based on our experimental data, four main reaction stages are proposed. By combining our results with other measurements about the mechanical properties of the system, we may be able to establish a relationship between the microstructure and macroscopic properties of the cured epoxy resins. Finally, light scattering could also be used to estimate the gel point in optically clear epoxy resins.

A simplified model is proposed for the branching kinetics of the epoxy resin (DGEB) cured with the anhydride (CH) in the presence of the triamine (CA) as a catalyst. We show, for the first time, that the branching kinetics can be approximated by using the Smoluchowski coagulation equation. In a simple version of that equation used here, the overall branching probability has been taken as proportional to the sum of active sites on the two polymers. The calculated weight-average molecular weight based on the model using two different initial conditions, i.e., monodisperse polymer at p_o = 0 and polydisperse polymer at p_o - 0.065 as observed experimentally and the measured weight-average molecular weight as a function of p have been compared. The results show that the initial molecular weight distribution is not uniform. By using a Gaussian-like distribution, we can predict the nature of the epoxy polymer product from knowledge in the early stage of the polymerization process. The model also implies that the catalyst molecules do not start the reaction at the same time. Dusek et al[79] have discussed a statistical branching theory in an analysis of gelation of a similar system, diepoxide-cyclic anhydride-tert.-amine. However, their emphasis was mainly on the critical extent of conversion at the gelation threshold. The measured polydisperse index (M_w/M_n) does not agree with the calculated one because laser light scattering emphasizes larger particles. We have to use the other techniques to measured the low molecular weight tails in the distribution in order to provide a more detail test of the model. A more general form for the reaction kernel and other forms of the initial distribution should be considered. Further experimental and theoretical studies are underway.

Finally, in my lecture review of our recent epoxy works,[32,61,80-82] I should mention that Dr. Chi Wu has been the main driving force from experiments to theory and support of this project by the U.S. Army Research Office is gratefully acknowledged.

References

1. See for example, Antoon, M. K. and Koenig, J. L., J. Polym. Sci. Polym. Chem. Ed. 1981, 19, 549, and references therein.

2. D. A. Weitz and M. Y. Lin, Phys. Rev. Lett., 1986, 57, 2037.

3. R. C. Ball, D. A. Weitz, T. A. Witten and F. Leyvraz, Phys. Rev. Lett. 1987, 58, 274.

4. P. G. J. van Dongen and M. H. Ernst, J. Phys. A: Math. Gen., 1985, 18, 2779.

5. P. G. J. van Dongen and M. H. Ernst, Phys. Rev. Lett., 1985, 54, 1396.

6. M. Kolb, Phys. Rev. Lett., 1984, 53, 1653.

7. R. Botet and R. Jullien, J. Phys. A, 1984, 17, 2517.

8. F. Leyvraz, Phys. Rev. A, 1984, 29, 854.

9. M. H. Ernst, E. M. Hendriks and R. M. Ziff, J. Phys. A: Math. Gen., 1982, 15, L743.

10. M. H. Ernst, E. M. Hendriks and R. M. Ziff, Phys. Lett., 1982, 92A, 267

11. R. M. Ziff and G. Stell, J. Chem. Phys., 1980, 73, 3492.

12. Mandelbrot, B.B. "Fractals, Form, Chance and Dimension," Freeman, San Francisco, 1977; "The Fractal Geometry of Nature," Freeman, New York, 1982.

13. Stanley, H.; Reynolds, P.; Redner, S; Family, F. in Real Space Renormalization; Burkhardt, T.W.; van Leeuwen, J.M.J., Eds.; Springer: New York, 1982, Chap. 7.

14. Schaefer, D.W.; Martin, J.E.; Wiltzius, P.; Cannell, D.S. Phys. Rev. Lett. 1984, 52, 2371.

15. Martin, J.E.; Schaefer, D.W. Phys. Rev. Lett. 1984, 53, 2457.

16. Wiltzius, P. Phys. Rev. Lett. 1987, 58, 710.

17. Schaefer, D.W.; Keefer, K.D. Phys. Rev. Lett. 1984, 53, 1383.

18. Chambon, F.; Winter, H.H. Polym. Bull. (Berlin) 1985, 13, 499.

19. Muthukumar, M.; Winter, H.H. Macromolecules 1986, 19, 1284.

20. Feder, J.; Jossang, T. Phys. Rev. Lett. 1984, 53, 1403.

21. Weitz, D.A.; Huang, J.S.; Lin, M.Y.; Sung, J. Phys. Rev. Lett. 1985, 54, 416.

22. Witten, T.A.; Sander. L.M. Phys. Rev. Lett. 1981 47, 1400.

23. Meakin, P. Phys. Rev. Lett.A 1983, 27, 604, 1495.

24. Kaufman, J.H.; Baker, C.K.; Nazzal, A.I.; Flickner, M.; Melroy, O.R Phys. Rev. Lett. 1986, 56, 1932.

25. Family, F. in Random Walks and Their Applications in the Physics and Biophysical Sciences - 1982; Schlesinger, M.F., West, B.J., Eds.; AIP Conference Proceedings No. 109; Amer. Inst. Phys.: New York, 1984, p.33.

26. Meakin, P. Phys. Rev. Lett. 1983, 51, 1119.

27. Kolb, M.; Botet, R.; Jullien, J. Phys. Rev. Lett. 1983, 51, 1123.

28. Hentschel, H.G.E.; Deutch, J.M. Phys. Rev. Lett.A 1984, 29, 1609.

29. Debye, P. and Bueche, A. M., J. Appl. Phys., 1949, 20, 518

30. May, C. A. and Tanaka Y. Epoxy Resins Chemistry and Technology; Marcel Dekker: New York, 1973; p 683.

382

31. Chu, B. and Wu, C., *Macromolecules*, 1987, 20, 93.
32. Chu, B., Wu, C., Wu, D.-Q. and Phillips, J. C., *Macromolecules*, 1987, 20, 2642.
33. Kratky, O. and Oelschlaeger, H. *J. Coll. & Interface Sci.* 1969, 31, 490.
34. Sander, L.M. *Nature*, 1986, 322, 789; *Scientific American* 1986, 256, 94.
35. Meakin, P. "Fractal Aggregates and Thin Fractal Measures," to be published. See references therein.
36. Stanley, H.E. *J. Phys.* 1977, A10, L211.
37. Cotton, J.P. *J. Physique Lett.* 1980, 41, 231.
38. Daoud, M.; Family, F.; Jannink, G. *J. Phys. Lett.* 1984, 45, 199.
39. Schosseler, F.; Leibler, L. *Macromolecules* 1985, 18, 399.
40. Leibler, L.; Schosseler, F. *Phys. Rev. Lett.* 1985, 55, 1110.
41. Martin, J.E.; Ackerson, B.J. *Phys. Rev. A*, 1985, 31, 1180.
42. Bouchard, E.; Delsanti, M.; Adam, M.; Daoud, M.; Durand, D. *J. Physique*, 1986, 47, 1273.
43. Mandelbrot, B.B. in "Fractals in Physics, " Proceedings of the 6th International Symposium of ICTP, L. Pietronero and E. Tosatti, eds., North Holland, Amsterdam, 1986, p. 3,17 and 21.
44. Daoud, M.; Joanny, J.F. *J. Physique* 1981, 42, 1359.
45. Huber, K.; Burchard, W.; Fetters, L.J. *Macromolecules* 1984, 17, 541.
46. See "Proceedings of the 5th International Conference on Photon Correlation Techniques in Fluid Mechanics," Springer Series in Optical Sciences (Ed. E.O. Schulz-DuBois) Springer-Verlag, New York, 1983.
47. Chu, B., Ford, J. R. and Dhadwal, H.S., in "Methods of Enzymology", (Eds. S.P. Colowick and N.O. Kaplan) Academic Press, Inc., Orlando, Florida, Vol. 117, 1985, pp. 256-297.
48. Chu, B., Ying, Q.-C., Wu, C., Ford, J. R. and Dhadwal, H. S. *Polymer*, 1985, 26, 1408.
49. Wu, C., Buck, W. and Chu, B., *Macromolecules*, 1987, 20, 98.
50. Chu, B., Wu, C. and Buck, W., *Macromolecules*, 1989, 22 (No. 2) 831
51. Grubisic, Z., Rempp, P. and Benoit, H., *J. Polym. Sci.*, Part B, 1967, 5, 753-759.
52. Moore, J. C. and Hendrickson, J. G., *J. Polym. Sci.*, Part C, 1965, 8, 233-241.
53. Mori, S., *J. Chromatogr.*, 1978, 157, 75-84.
54. Mori, S., *J. Appl. Polym. Sci.*, 1974, 18, 2391-2397.
55. Loy, B. R., *J. Polym. Sci. Polym. Chem. Ed.*, 1976, 14, 23
56. Vrijbergen, R. R., Soeteman, A. A. and Smit, J. A. M., *J. Appl. Polym. Sci.*, 1978, 22, 1267-1276.
57. McCrackin, F. L., *J. Appl. Polym. Sci.*, 1977, 21, 191-198.
58. Mahabadi, H. K. and O'Driscoll, K. F., *J. Appl. Polym. Sci.*, 1977, 21, 1283-1287.
59. Weiss, A. R. and Cohn-Ginsberg, E. *J. Polym. Sci.*, Part B, 1969, 7, 379-381.
60. Mori, S., *Anal. Chem.* 1981, 53, 1813-1818.
61. Chi Wu, Ju Zuo, and Ben Chu, *Macromolecules*, 1989, 22 (No.2), 633

62. Burchard, W., Schmidt, M. and Stockmayer, W. H., *Macromolecules*, 1980, 13, 1265.
63. Kajiwara, K. and Burchard, W., *Polymer*, 1981, 22, 1621.
64. Gordon, M., *Proc. Roy. Soc. (London)*, 1962, 268, 240.
65. Flory, J., *J. Am. Chem. Soc.*, 63, 1941, 1096, 3083, 3091.
66. Stockmayer, W. H., *J. Chem. Phys.*, 12, 1944, 125.
67. de Gennes, P. G., *Scaling Concepts in Polymer Physics*, (Cornell Univ. Press, N.Y., 1979).
68. Stauffer, D., Coniglio, and Adam, M. *Adv. Polym. Sci.*, 1982, 44, 103.
69. Ernst, M. H., Ziff, R. M. and Hendriks, E. M., *J. Coll. Interface Sci.*, 1984, 97, 266.
70. Herrmann, H., Landau, D. P. and Stauffer, D., *Phys. Rev. Lett.*, 1983, 49, 412.
71. Stauffer, D., Phys. Rep., 1979, 54, 1.
72. Leibler, L. and Schosseler, F., *Phys. Rev. Lett.*, 1985, 55, 1110.
73. Fischer, R. F., *J. Polym. Sci.*, 1960, 44, 155.
74. Smoluchowski, M. V., *Phyzik. Z.*, 1916, 17, 585; *Z. Phys. Chem.*, 1918, 92, 129.
75. Golovin, A. M., *Bull. Akad. Sci. SSSR Geophy. Ser.*, (English Transl.), 1963, 482.
76. Scott, W. T., *J. Atmos. Sci.*, 1968, 25, 54.
77. Scott, W.T., *Analytic studies of cloud droplet coalescence. I. Tech. Rept.,* No. 9, Desert Research Institute, University of Nevada, 1965, 72.
78. Lu, B., *J. of Statistical Physics*, 1987, 49, *Nos.* 3/4, 669
79. Dusek, K., Lunak, S., Matejke, L., *Polymer Bulletin*, 1982, 7, 145.
80. Chu, B. and Wu, C., *Macromolecules*, 1988, 21, 1729.
81. Wu C., Zuo, J. and Chu, B., *Macromolecules*, 1989, 22 (No. 2), 838
82. Wu C., Chu, B. and Stell, G., submitted to *J. of Chem. Phys.* for publication.

TABLE I

Properties of Epoxy Polymer (1,4-butanediol diglycidyl ether with cis-1,2-cyclohexanedicarboxylic anhydride) during the Curing Process

Sample #	Conversion (% CH)	M_w(g/mol) X-ray	LLS	$<R_g^2>^{1/2}$(Å) X-ray	LLS	$A_2(\frac{ml\ mol}{g^2})$ X-ray	LLS
1	6.5%	4.32×10^3	4.11×10^3	29	-	-	1.45×10^{-3}
2	13.3%	6.14×10^3	6.07×10^3	36	-	-	1.26×10^{-3}
3	20.0%	8.23×10^3	8.23×10^3	44	-	1.05×10^{-3}	1.01×10^{-3}
4	26.5%	1.25×10^4	1.22×10^4	55	-	-	8.70×10^{-4}
5	33.5%	2.11×10^4	2.04×10^4	70	-	-	7.41×10^{-4}
6	36.0%	3.37×10^4	3.15×10^4	85	-	6.12×10^{-4}	6.30×10^{-4}
7	38.5%	5.60×10^4	5.00×10^4	112	105	-	5.25×10^{-4}
8	40.0%	-	7.68×10^4	-	128	-	4.57×10^{-4}
9	41.0%	-	1.03×10^5	-	150	-	4.17×10^{-4}
10	42.5%	-	1.42×10^5	-	191	-	3.63×10^{-4}
11	44.0%	-	2.19×10^5	-	222	-	3.16×10^{-4}
12	45.3%	-	3.01×10^5	-	257	-	2.881×0^{-4}
13	46.5%	-	4.97×10^5	-	314	-	2.51×10^{-4}

Table II

(a) Fractal dimension d_f of epoxy polymer solutions based on $I(K) \sim K^{d_K}$ of Eq. (6) with $(2/R_g)<K<Y$

Sample	d_f $Y^{-1}(\text{Å}) = 30$	20	15	10
13	2.17±0.01	2.16±0.01	2.14±0.02	2.10±0.03
12	2.17±0.01	2.17±0.02	2.13±0.03	2.11±0.04
11	2.18±0.02	2.17±0.02	2.14±0.02	2.12±0.03
10	2.18±0.01	2.17±0.01	2.15±0.02	2.12±0.02
9	2.17±0.02	2.17±0.02	2.14±0.03	2.10±0.03
8	2.17±0.05	2.16±0.03	2.13±0.03	2.10±0.04
7	2.18±0.05	2.17±0.04	2.14±0.04	2.11±0.05
6	2.16±0.05	2.17±0.04	2.13±0.04	2.09±0.05
5	2.15±0.07	2.16±0.05	2.13±0.04	2.10±0.05
4	-	2.12±0.06	2.10±0.05	2.05±0.06
3	-	2.05±0.08	2.00±0.06	1.94±0.06
2	-	-	2.04±0.08	1.96±0.07
1	-	-	-	1.90±0.08

Table II

(b) Coil behavior of epoxy polymers between the mesh points based on $I(K) \sim K^{-\beta}$ with $Y<K<(1/4)$ $Å^{-1}$

Sample	$Y^{-1}(Å) = 30$	β 20	15	10
13	1.74±0.06	1.67±0.05	1.64±0.06	1.59±0.09
12	1.76±0.05	1.69±0.05	1.65±0.07	1.61±0.09
11	1.75±0.06	1.67±0.05	1.63±0.07	1.60±0.09
10	1.74±0.05	1.68±0.06	1.67±0.07	1.65±0.10
9	1.73±0.06	1.67±0.05	1.65±0.07	1.62±0.10
8	1.74±0.06	1.68±0.06	1.66±0.07	1.63±0.10
7	1.72±0.07	1.67±0.06	1.65±0.07	1.62±0.09
6	1.71±0.07	1.68±0.06	1.66±0.05	1.62±0.10
5	1.70±0.06	1.68±0.06	1.65±0.06	1.61±0.10
4	1.68±0.07	1.67±0.07	1.65±0.07	1.60±0.10
3	1.68±0.06	1.67±0.07	1.64±0.06	1.58±0.11
2	1.67±0.08	1.68±0.07	1.66±0.07	1.59±0.10
1	1.65±0.08	1.67±0.07	1.66±0.07	1.59±0.11

Table III

Determination of ξ based on plots of log (W_{mea}/W_{cal}) versus log K.[a]

Sample	3[b]	4[b]	5	6	7	8	9	10	11	12	13
ξ (Å)	18.4 ±4.7	18.5 ±4.8	18.9 ±3.9	20.3 ±3.7	19.2 ±3.0	20.1 ±3.4	18.9 ±4.2	19.5 ±3.3	20.0 ±3.5	20.4 ±3.4	19.7 ±3.0

[a] ξ^{-1} is determined by the intercept between $W_{mea}/W_{cal} - 1$ and the straight line with slope ~ 0.49. [b] Estimated value

TABLE IV.

Experimental results based on the MSVD method for the epoxy polymer samples in MEK at $25^{\circ}C$ and $\lambda_{o} = 488$ nm.

Sample	conversion (CH%)	M_w(g/mol)	$\bar{D}_o^o$(cm^2/sec)	$\mu_2/\bar{\Gamma}^2$	f	k_d(mL/g)
1	6.5	4.32×10^3	2.71×10^{-6}	0.63	-	0
2	13.3	6.14×10^3	2.27×10^{-6}	0.61	-	11
3	20.0	8.23×10^3	1.77×10^{-6}	0.56	-	16
4	26.5	1.25×10^4	1.29×10^{-6}	0.53	-	22
5	33.5	2.11×10^4	1.00×10^{-6}	0.51	-	29
6	36.0	3.37×10^4	7.98×10^{-7}	0.48	-	36
7	38.5	5.00×10^4	6.45×10^{-7}	0.47	0.15	48
8	40.0	7.68×10^4	5.26×10^{-7}	0.46	0.17	63
9	41.0	1.03×10^5	4.38×10^{-7}	0.45	0.13	79
10	42.5	1.42×10^5	3.68×10^{-7}	0.43	0.18	89
11	44.0	2.19×10^5	2.97×10^{-7}	0.40	0.21	101
12	45.3	3.01×10^5	2.60×10^{-7}	0.42	0.16	132
13	46.5	4.97×10^5	2.05×10^{-7}	0.40	0.19	141

A LIGHT SCATTERING STUDY OF EPOXY RESIN POLYMERIZATION

A. BICK *and* TH. DORFMÜLLER
University of Bielefeld
Department of Chemistry
4800 Bielefeld
West Germany

ABSTRACT. The copolymerization of epoxy resins and anhydrides with tertiary amines as catalysts has been studied by Rayleigh—Brillouin scattering, using a vertical—vertical polarization geometry(I_{VV}). The obtained spectra have their origin in density fluctuations of the medium, and yield useful information on the bulk viscoelastic properties of the system in the high frequency range (hypersonic range). We thus obtain information about the velocity of sound, the phonon attenuation coefficient and the real and imaginary part of the longitudinal elastic modulus M.

1.   Introduction

We have been interested in the epoxy/anhydride/amine—system, because several conflicting reaction kinetics have been observed and different mechanisms have been proposed for this system[1]. Furthermore, this rather complex system is of considerable industrial interest.

During polymerization the viscoelastic properties undergo dramatic changes and Rayleigh—Brillouin spectroscopy has been used as an effective method to observe reacting systems of this kind. In the past the copolymerization of other epoxy—systems and the polymerization of styrene was studied by Rayleigh—Brillouin—spectroscopy[2, 3, 4]. We thus expect by monitoring the changes in viscoelastic behavior, to obtain the kinetics of the viscoelastic variables and the longitudinal relaxation time.

Additionally, differential thermal analysis (DTA) was used to determine the glass transition temperature and the reaction enthalpy of the system. This information could be related to the dynamic light scattering results. We furthermore measured the change in density and in the refractive index during the polymerization.

2.   Theoretical Background of the Rayleigh— Brillouin— Spectroscopy

Using the I_{VV} polarization geometry one generally obtains a spectrum consisting of two unshifted Lorentzians and a shifted Brillouin doublet characterized by the Brillouin shift ν_b and the Brillouin linewidth Γ.

The first of these parameters is related to the propagation velocity of hypersonic waves, the latter to the hypersonic attenuation. Figure 1 illustrates a typical Brillouin spectrum obtained for the epoxy system at 70°Celsius.

389

Th. Dorfmüller (ed.), Reactive and Flexible Molecules in Liquids, 389–398.
© *1989 by Kluwer Academic Publishers.*

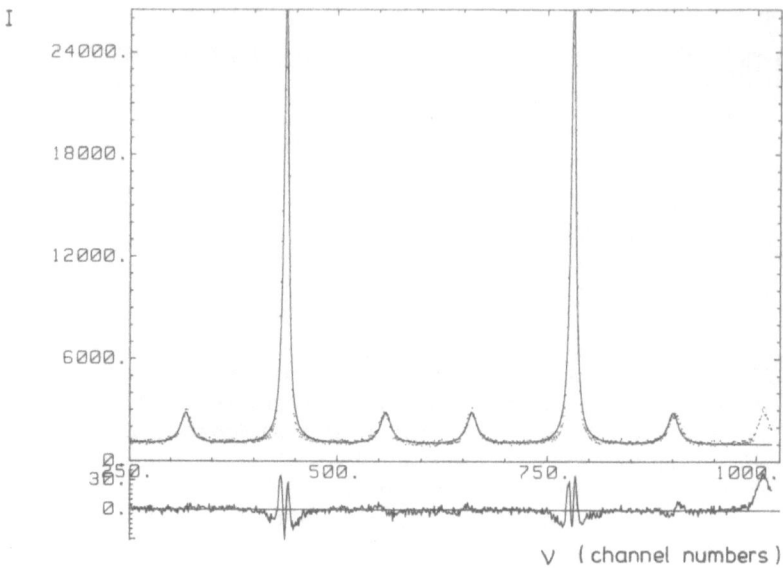

Figure 1.Rayleigh–Brillouin intensity vs. frequency

The Brillouin shift ν_b is related to the adiabatic velocity of sound (V_l) by the equation:

$$V_l = \frac{\lambda_0 \, 2\nu_b}{2n \, \sin \, (\theta/2)} \qquad (1)$$

where λ_0 is the wavelength of the incident light in vacuo, n is the refractive index and θ is the scattering angle[5,6,7].

The sound wave is damped with a factor exp $(-\Gamma t)$ in time and exp $(-\alpha r)$ in space, if the Brillouin profile is taken as a Lorentzian. In this case Γ is the half width at half height of the Lorentzian and α the phonon attenuation coefficient. The phonon attenuation coefficient is given by the relation:

$$\alpha = 2\pi \, \Gamma/V_l \qquad (2)$$

It is sometimes useful to calculate the attenuation coefficient per phonon cycle, $\lambda\alpha$:

$$\lambda\alpha = 2\pi \, \Gamma/\nu_b \qquad (3)$$

The viscoelastic parameters of matter can be calculated from the hypersonic spectroscopic parameters ν_b and Γ. The longitudinal elastic modulus is defined as $M = K + 4/3 \, G$, K representing the compressibility and G the shear modulus. M is a complex quantity, consisting of a real storage part and an imaginary loss part: $M^* = M' + iM''$. This complex quantity is used to describe the dynamic perturbation, as for example a longitudinal sound wave propagating through the medium. In a viscoelastic liquid the velocity of sound is fully characterized by the real part of the longitudinal elastic modulus M:

$$V_1(\nu_b) = \left[\frac{M'(\nu_b)}{\rho} \right]^{0.5} \tag{4}$$

The imaginary part of the modulus is related to the half width Γ by the expression:

$$\Gamma = \frac{M'' q^2 \lambda}{4 \pi c \rho} \tag{5}$$

In the case of a process with a finite set of distinct relaxation times the correct expression for the adiabatic longitudinal modulus is:

$$M^* = \gamma K_0 + M_r \int_0^\infty \frac{\rho(\tau) \, i \, 2\pi \, \nu \, \tau}{1 + i \, 2\pi \, \nu \, \tau} \, d\tau \tag{6}$$

with the relaxing part $M_r = M_\infty - \gamma K_0$

In the simplest theory which incorporates a single relaxation time, the longitudinal elastic modulus is represented as follows:

$$M = M_\infty - M_r/(1 + i\omega\tau) \tag{7}$$

The subscripts ∞ and $_0$ denote the high–frequency and the low–frequency limits respectively, $\omega = 2\pi\nu$.

Viscoelastic data is often presented in terms of the loss tangent δ, which is the imaginary part of a viscoelastic property divided by its real part. Tan δ can be approximately calculated directly from Brillouin data by the equation:

$$\tan \delta = 2\Gamma/\nu_b \tag{8}$$

This quantity differs only by a factor π from the attenuation per phonon cycle.

If, by changing the temperature or the composition, the Brillouin–frequency approaches the value of $1/\tau$ the Brillouin linewidth as a function of either of these parameters goes through a maximum (peak condition). In this case τ can be reliably estimated and the high and low frequency limits of the velocity of sound and of the longitudinal modulus can be calculated by the following formulas:

$$V_1^2 = V_0^2 + (V_\infty^2 - V_0^2)\frac{\omega^2\tau^2}{1 + \omega^2\tau^2} \tag{9a}$$

$$\alpha = \frac{(V_\infty - V_0)}{2V_1^3} \frac{\omega^2\tau}{(1+\omega^2\tau^2)} \tag{9b}$$

$$M_\infty = \rho V_\infty^2 ; \quad M_0 = \rho V_0^2 \tag{9c}$$

These relations show that if the system reaches peak conditions, the limiting values of the viscoelastic properties can be calculated rather accurately[8].

The simple equations 9a – c, resulting from the Litovitz formalism, describe the system if only one relaxation time is assumed. However, we prefer to use the calculation of the limiting values following the Rytov–formalism[9]. In this theory the phonon shift and the half width of the Brillouin peaks are represented by the following formulas:

$$2\pi\,\nu_b = \left[\frac{M_\infty\,q^2}{\rho} + 3\Gamma_b{}^2 + \frac{2\Gamma_b}{\tau}\right]^{0.5} \tag{10a}$$

$$2\pi\Gamma_b \approx \frac{M_r q^2/2\,\rho\tau}{(1/\tau^2) + (M_\infty q^2/\rho)} \tag{10b}$$

In polymers one can observe a second central feature, known as the Mountain–peak. With known relaxation time τ, the half–width of the Mountain–peak can be calculated:

$$\Gamma_\tau \equiv 1/\tau - 2\Gamma_1 \tag{11}$$

The Landau–Placzek ratio, i.e. the ratio of the integrated intensities of the Brillouin–peaks and the central peaks is very sensitive to inhomogeneities of the medium and its changes during polymerization can be used to monitor the gelation point.

The value of τ, calculated for peak– conditions, is only an averaged relaxation time, averaged over at least 2 processes and a broad molecular weight distribution of relaxing polymers[10], but it may qualitatively indicate the presence of a short–time relaxation process at the time of peak condition during polymerization.

3. Experimental

The following materials were used:

$$CH_2\!-\!CH\!-\!CH_2\!-\!O\!-\!(CH_2)_4\!-\!O\!-\!CH_2\!-\!CH\!-\!CH_2 \qquad [BGE]$$

1,4–Butanedioldiglycidylether

[CHCA]

Cis–1,2–Cyclohexanecarboxylicanhydride

$(CH_3CH_2)_3N$ [TA]

Triethylamine

The materials were purchased from Aldrich Chemical Company, [BGE] and [CHCA] were used without further purification, [TA] was distilled before use.

$3.1 \cdot 10^{-2}$ mol [BGE] and $6.2 \cdot 10^{-2}$ mol [CHCA] were heated together up to 50^0 C and then 1 ml of the mixture was filtered through a 1μ millipore filter into a dust free cuvette. Subsequently, the cuvette was centrifuged for 30 min at 4000 rpm and 2, 3, 4, 5 or 6 mol–% catalyst[TA] was added.

The cuvette was placed in a temperature controlled light scattering cell, where the following samples were measured at different temperatures:

TABLE 1. Catalyst concentrations and reaction temperatures

catalyst(mol–%)	temperature(^{0} C)
2	80, 90, 100
3	60, 70, 80, 90
4	60, 70, 80, 90
5	60, 70, 80, 90
6	60, 70, 80

A polarized monochromatic laser beam (488 nm) was scattered under an angle of 90^0 and analyzed by a planar Fabry–Perot interferometer. For each sample spectra were recorded at time intervals of approximately 50 minutes. Each spectrum was accumulated for at least 16 times. The free spectral range was 30 GHz, the finesse was between 60 and 80.

The DTA–measurements were obtained by a Mettler DTA system, refractive indices by a Bausch & Lomb Abbe–refractometer and densities were measured by dilatometry.

4.   Results and Discussion

The Brillouin shift ν_b and the half width at half height Γ_1 of the Brillouin peaks were measured. The attenuation coefficient per phonon cycle or the dynamic loss tangent can be calculated directly from the measured data. In order to calculate the sound velocity and the attenuation coefficient one has to know the scattering vector q, to calculate the longitudinal elastic moduli M$'$ and M$''$ one has to know the time dependence of the density.

The following tables and diagrams show the change of the viscoelastic properties with increasing reaction time, corresponding to an increasing degree of polymerization and increasing molecular weight. Table 2 shows the change of the velocity of sound and the change of the attenuation coefficient per phonon cycle. This is shown for a representative sample (T = 70⁰ C, 5 % catalyst):

TABLE 2. Adiabatic velocity of sound and attenuation per phonon cycle

t(min)	$V_l(m/s)$	$\alpha \cdot \lambda$
54	1400.87	0.6398
94	1419.09	0.7013
123	1434.98	0.7952
179	1516.37	0.8244
247	1766.38	0.7937
315	2135.47	0.4245
371	2322.25	0.3182
453	2378.42	0.2193
529	2390.68	0.2901
574	2393.93	0.2596

Figures 2 and 3 illustrate the data, in figure 2 is also shown a sample with higher catalyst concentration at the same temperature for comparison.

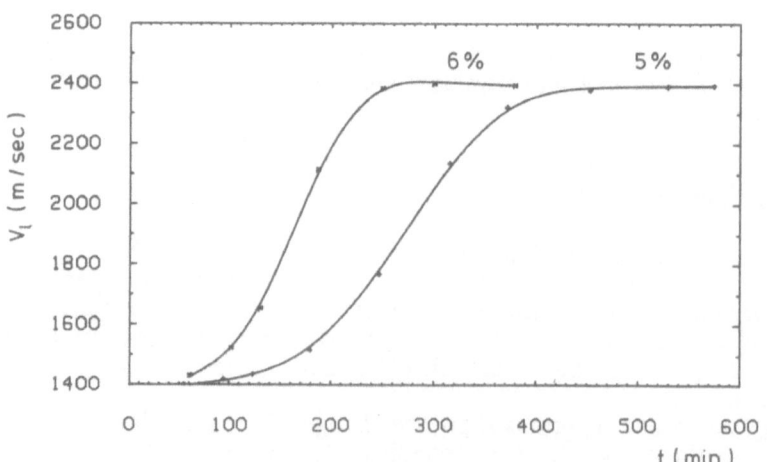

Figure 2.   Velocity of sound versus time

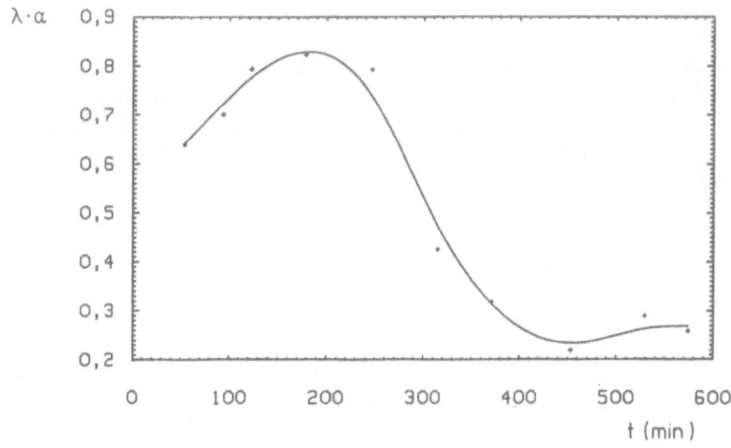

Figure 3. Attenuation per phonon cycle versus time

The velocity of sound changes by approximately a factor of 2. It starts with typical liquid data (water: 1496 m/s) and ends at typical polymer data (polystyrene: 2275 m/s at 72°C and 170 atm). The change of the sound velocity is sigmoidal, and it could be fitted to an arctan function. The same sigmoidal curves were obtained for the density change.

As expected, the change of the longitudinal storage modulus M′ likewise displays a sigmoidal progress, shown in figure 4:

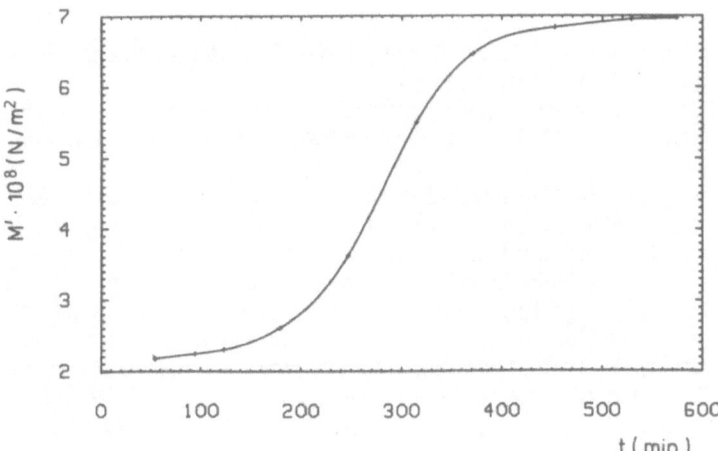

Figure 4. Longitudinal storage modolus versus time

The longitudinal loss modulus was not calculated, because the model for calculating M'' is a very crude one and we already illustrated the damping per phonon cycle in figure 3, which gives the necessary information.

The Landau–Placzek ratio (RLP) i.e. the ratio of integrated central peak intensity and the integrated intensities of the Brillouin doublet provides information on the gelation point of the sample. Its progress with increasing viscosity is illustrated in figure 5:

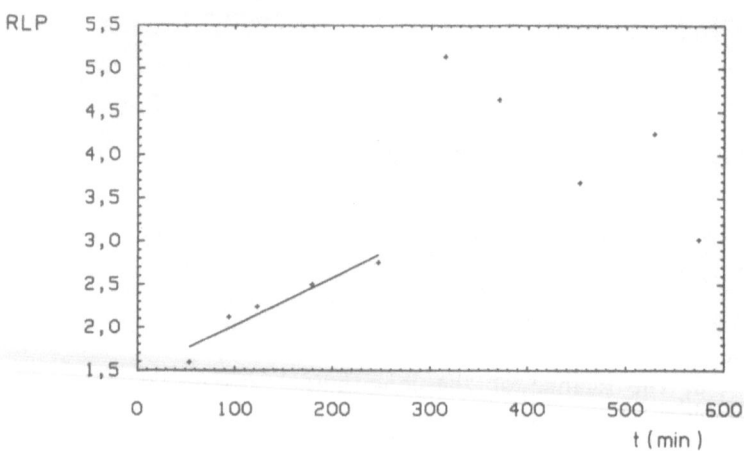

Figure 5.   Landau–Placzek ratio versus time

Following the argumentation of Kondo[3], the beginning of speckle pattern in the sample documents the start of gelation.

The glass–transition point was determined by differential scanning calorimetry for the samples with 5 % catalyst at 72 ± 2^0C and was found independent of the curing temperature.

The change in slope of the sigmoidal curves of phonon velocity were taken as half life times for a kinetic analysis of first and second order. The activation energy could be determined by calculating a rate constant by assuming both first and second order kinetics and then using an Arrhenius plot[11]. It was determined to 19.6 ± 2 kcal/mol. This is in the upper limit of activation energies for comparable systems[1]. No uniform order of the reaction could be calculated from the change of viscoelastic properties. A difficulty other authors reported before for ternary epoxy systems[1,12]. There are reasons to believe that the order of the reaction changes during polymerization. It is also possible that the system shifts from activation energy control to diffusion control as a consequence of its increasing viscosity.

The reaction enthalpy could be estimated by integrating the DTA spectrum at constant temperature. For the sample with 5% catalyst at 70^0C Δ H = 51 cal/g was calculated. This is important for the practical application of this system, because many other epoxy systems have much higher reaction enthalpies which may be a source of inhomogeneities and flaws in polymerized epoxy samples[13,14].

The Brillouin shift, the densities, the velocities of sound and the longitudinal storage moduli have a sigmoidal time dependence. At the time of peak condition we observe a maximum in production of reaction heat by DTA. Thus independent from the kind of experiment a characteristic point of the reaction was found. The temperature dependence of this point was observed. A sigmoidal time dependence of the reaction composition was also found by Tanaka and Kakiuchi by using chemical analysis[15]. Brillouin data of the binary system DGEBA/ 1.3–Diaminomethylcyclohexane by Kondo et al.[3] do not show this sigmoidal time dependence and the slow phase at the beginning of the reaction is missing. However the values for the longitudinal modulus and the sound velocity are nearly the same as ours.

The most remarkable difference between the binary and the ternary systems is the behavior of the damping coefficient. Kondo and coworkers and Stevens and coworkers[3,4] did not find a relaxation zone, they observe a monotonic decrease of $\lambda \cdot \alpha$. The ternary system shows a broad peak in the time dependence of $\lambda \cdot \alpha$. In contrast to these results we were able to find a complete relaxation zone from which we can estimate an averaged relaxation time for peak condition. At peak condition, gelation has not started and we are far away from the glass transition. We thus observed a glass–rubber–relaxation far from the glass transition. For the system at this point, the finite values for the velocity of sound and the storage modulus were calculated in the Rytov and the Litovitz–formalism as seen in the next table. We made the assumption, that the time evolution of our system simulates the temperature dependence of a chemically defined system.

TABLE 3. Data in the theories of Rytov and Litovitz. Sample: 5% catalyst, 70^0C

$t = 247$ min	$V_l = 1766\ 4$ m/s	$\alpha_{max} = 3.4 \times 10^6\ m^{-1}$	$\tau_l = 2.11 \times 10^{-11}$ s

	Rytov	Litovitz
$M_\omega\ [N/m^2]$	4.37×10^9	4.08×10^9
$V_\omega\ [m/s]$	1938.82	1874.64
$V_0\ [m/s]$	—	1651.04

We also calculated the half width of the Mountain peak, it was determined to 30 GHz.

However, the ternary epoxy resin system of low starting viscosity is of interest, because we found some effects, witch were not obtained in other epoxy systems by light scattering.

5. Literature

1)   M.K. Antoon, J.L.Koenig, J.Polym.Sci., Polym.Chem.Ed. 19, 549 (1981)

398

2)   G.C.Stevens, J.V.Champion, P.Liddell, A.Dandridge, Chem.Phys.Let. 71, 104 (1980)
3)   S.Kondo, T.Igarashi, T.Nakamura, J.Appl.Polym.Sci. 26, 2337 (1981)
4)   D.A.Jackson, G.C.Stevens, Molecular Physics 30, 911 (1975)
5)   R.Pecora(Ed.), Dynamic Light Scattering (New York: Plenum Press, 1985)
6)   D.G.Patterson, Methods of Experimental Physics 16A, 170 (1980)
7)   B.J.Berne, R.Pecora, Dynamic Light Scattering (New York: Wiley, 1976)
8)   L.Börjesson, J.R.Stevens, L.M.Torell, Polymer 28 , 667 (1987)
9)   S.M.Rytov, Sov.Phys.—JETP 31, 1163 (1970)
10)   P.J.Caroll,G.D.Patterson, J.Chem.Phys 81, 1665 (1984)
11)   A.Bick,Th.Dorfmüller, to be published
12)   C.H.Klute, W.Viehmann, J.Appl.Polym.Sci. 5, 86 (1961)
13)   L.K.Yarapov, G.S.Obanesova, Izv.Vyssh.Uchebn.Zaved, Khim.Khim.Tekhnol. 18, 1571 (1975)
14)   M.Chater, J.M.Vergnaud, D.Lalart, F.M.Dansac., Eur.Polym.J. 22, 805 (1986)
15)   Y.Tanaka, H.Kakiuchi, J.Appl.Polym.Sci. 7, 1063 (1963)

THE HYDRATED ELECTRON AND THE FEW-STEP RANDOM WALK

G. WYLLIE,
Department of Physics and Astronomy,
University of Glasgow,
Glasgow G12 8QQ,
Scotland

ABSTRACT. Two important recent articles [1,2] using a 'polymer' model
to obtain equilibrium properties of the hydrated electron also give
some hints about its kinetic properties. Random walks over short
distances give rise to non-Gaussian distributions and anomalous
diffusion. Some effects related to Lévy stable distributions and
fractal structures are discussed.

1. INTRODUCTION

The topics of this survey relate in several ways to the subjects of
this Advanced Study Institute. The hydrated electron is a highly
reactive object with a solvation structure which is not merely flexible
but mutable on the picosecond time scale. The first two papers [1,2]
that we discuss compute the structure of the electron solvation shells.
They use a variant of classical molecular dynamics in a discrete
version of the path integral method to describe the quantum behaviour
of the electron in the varying potential field of the molecules. This
amounts to representing the electron as an ideal polymer, a massive
perfectly flexible chain of identical beads with a purely harmonic
nearest-neighbour interaction.
    These papers do not address, but give some interesting hints about
the question of the random movement of the hydrated electron. In this
connection we discuss random motions leading to non-normal
distributions, and note a connection with the fractals which are
important in certain polymer structures.

2. REAL PROPERTIES OF e^-_{aq} [3]

Ionization events in an aqueous medium produce molecular ions and
electrons which after a series of secondary collisions drop to a
kinetic energy of ten or less electron volts, too low to produce
further ionization. These transfer energy to molecular motions and
thermalize in a few picoseconds in mobile states at the bottom of the
conduction band, 1.2 eV below the energy of the free electron in
vacuum.

Th. Dorfmüller (ed.), Reactive and Flexible Molecules in Liquids, 399–405.
© 1989 by Kluwer Academic Publishers.

Unless trapped by other reactions at this stage, each then digs its own local potential pit by rearrangement of surrounding water molecules and settles in about four picoseconds in the relatively immobile fully solvated state.

The hydrated electron has a strong and very broad optical absorption peaking at 1.73 eV with half maximum absorption at 1.4 eV and 2.25 eV. There has been a good deal of argument about the extent to which this width is to be assigned to fluctuating structure in the initial state or to a broad distribution of immediately accessible traps in the upper band. The absorption is quite sensitive to pressure, the maximum shifting to 2 eV at 6.3 kbar.

The half-life of the hydrated electron is 230 μs in pure water at room temperature, mostly owing to the reaction $e^-_{aq} + H_3O^+ \rightarrow H_2O + \frac{1}{2}H_2$ with rate constant 2×10^{10} $M^{-1}s^{-1}$. It is stable in special media such as alkaline glasses. ESR measurements indicate that in such media the first hydration shell of the electron contains six H_2O molecules bond orientated so that one OH bond on each H_2O is directed radially inwards to secure maximum immersion of the proton in the electron charge cloud.

The heat of hydration of the electron is about 1.7 eV, the standard entropy of hydration, at $-2k_B$, is small and negative, and the increase in volume is only about 0.2 of the water molecular volume – surprisingly much less than the corresponding increase of 68 ml/mole for the ammoniated electron. The diffusivity is larger than the self-diffusivity of water molecules –

$$D(e^-_{aq}) \simeq 5 \times 10^{-9} m^2 s^{-1} \simeq 2DS(H_2O) \simeq D(OH^-) \simeq \frac{1}{2}D(H_3O^+)$$

so diffusion is mostly by charge transfer and not by block molecular movement.

The ratio of the rate constant, 14×10^{10} $M^{-1}s^{-1}$, for the neutralization of hydroxonium by hydroxyl ions to that quoted above for neutralization by the electron is one of several features of the kinetics awaiting more complete molecular description. A limited amount of evidence on molecular movement in the hydration shell comes from the recent paper [2] of Schnitker and Rossky (but see Appendix A).

3. COMPUTED PROPERTIES OF $(H_2O)_n^-$ [1]

The following energies were calculated for a stable internal electron state in a cluster of 128 water molecules and one electron at 300K.

Electron-water mean interaction	$\langle V \rangle$ =	-5.9 eV
Electron excess kinetic (localisation) energy	K =	2.35 eV
Electron vertical binding energy	$\langle V \rangle + K$ =	-3.5 eV
Cluster excess energy (water interaction)	E_c =	2.6 eV
Adiabatic binding energy	$K + \langle V \rangle + E_c$ =	-1.0 eV

For clusters of 11 and 19 water molecules the computation obtained stable electron binding, in agreement with experiment and contrary to previous simpler calculations, but, interestingly, for surface rather than internal electron states.

For the large cluster, the radius of gyration of the electron density is 2 Å, nearly the radial distance of the innermost H atoms from the mass centre of the 'electron' chain. For charged (c) clusters of 128 and 64 molecules, and a neutral (n) cluster of 64 molecules, the numbers of molecules in successive radial shells are –

Shell	1	2	3
Radial distance	$0 - 7a_0$	$7 - 10a_0$	$10 - 15a_0$ ($a_0 = 0.53$Å)
n = 128 (c)	6	16	54
n = 64 (c)	6	15	32
n = 64 (n)	8	13	35

illustrating nicely the small excess volume of the hydrated electron. The pentagonal rings commonly seen in computed water structures are frequent in and outside shell 3.

Barnett et al. give a rather brief discussion of the orientation distribution. It can be inferred from the histograms in their paper that in shell 1 bond orientation is nearly complete at n = 32 and n = 64, but falls to 50%, with about 10% dipole orientation , at n = 128. There is no clear evidence of bond orientation in the outer shells. In shell 2, dipole orientation is about 60% at n = 32 and 128, with a curious substantial drop at n = 64, and in shell 3 dipole orientation is 30% at n = 64 and 128. These results suggest strong competition between a compatible hydrogen bonding of molecules and the incompatible preferred orientations, bonding in shell 1 and dipole in shells 2 and 3, during the build up of a cluster. Such a balance of strong interactions is concordant with the appearance of large structural fluctuations in thermal equilibrium.

Motion of such a hydrated electron through the solvent then requires quite extensive co-ordinated rotation and translation of molecules in the solvation shells.

4. COMPUTED PROPERTIES OF $[(H_2O)_{300}^-]_\infty$

Schnitker and Rossky [2] simulated an electron in bulk water, using periodic boundary conditions with a truncated octahedral cell containing 300 molecules. Results are generally compatible with those of Barnett et al. but they found a large slow fluctuation during a run which simulated 60 ps of H_2O time. In successive thirds of this run the electron excitation energy had values 2.4, 2.25 and 2.0 eV, with 5 to 6 molecules in a shell of radius 4 Å in the first period, and only 2 to 4 in the same shell in the third period.

It must be emphasized that the method used is designed to give equilibrium results quickly and reliably and is not to be relied on for kinetics. Although the calculated response time of the electron chain was only a few picoseconds, the chain contains most of the mass of the system and artificially slow motions are not obviously impossible. Moreover, the simulation was allowed to run under its own dynamics for only 0.5 ps at a time before all velocities were replaced by random choice from a Boltzmann distribution. This destruction of velocity correlations means that all low frequency motions are heavily overdamped.

Nevertheless, these elegant and important results strongly suggest that the real hydrated electron is liable to large fluctuations of structure on the time scale 10^{-12} to 10^{-11} s. The considerable changes in the electron-water interaction on the one hand and in the sum of excess water-water interaction and electron localisation energy on the other then compensate in a nearly constant total energy.

The pattern of diffusion of the hydrated electron is not established by these calculations. If we suppose that a recognizably new solvation site is most likely to be centred in the second solvation shell, at a distance of, say, 4 Å from the previous centre, a jump rate of 2×10^{11} s^{-1} would be needed to give the observed diffusivity. This would be compatible with an 'amoeboid' movement of the solvation structure, but electron tunnelling over somewhat larger distances can not be excluded.

These uncertainties, compounded with the fact that the hydrated electron is often produced not far from its point of final reaction, raise the question whether the transport of the electron is legitimately described by the usual diffusion equation. Since the characteristic solution of the diffusion equation has the form of the normal distribution, we may equally ask whether a random walk of short total range may give rise to something other than a normal distribution. In general, it will. In fact, the sum of a few steps will only be normally distributed if the individual steps have such a distribution. In other cases, normal distribution arises in the limit of many steps provided the central limit theorem is applicable.

5. THE FEW-STEP RANDOM WALK

If we suppose that every step of a random walk has the same probability distribution, it is well known that the distance travelled in N steps will be normally distributed if each step has a normal distribution, whatever the value of N. The central limit theorem assures us that, for a wide range of other density functions for the individual steps, the N-step distance tends to a normal distribution as N becomes indefinitely large.

Domb and Offenbacher [4] have developed corrections to the normal approximation, for finite N, for a one-dimensional random walk with probability density function $p(x)$ for the length x of individual steps. The moment generating function for the single steps is

$$M(t) \equiv \int_{-\infty}^{\infty} p(x) \exp(xt) dx,$$

and the cumulant generating function is

$$K(t) \equiv \ln M(t).$$

Defining $r = X/N$ where X is the total displacement after N steps, the probability density $P_N(r)$ is then given by

$$2\pi P_N(r) = \int \exp\{N[K(t) - rt]\} dt$$

integrated round a contour for t equivalent to a circle round the origin for $z = \exp(t)$.

Then a point t_0 which gives a maximum real part for $[K(t)-rt]$ gives a sharp maximum of the integrand for moderately large N, and for each such point

$$(2\pi N|K_2|)^{1/2}\exp\{N[rt_0-K(t_0)]\}P_N(r) \sim$$

$$\sim 1 + N^{-1}[K_4/(8K_2^2) - 5K_3^2/(24K_2^3)] + N^{-2}[-K_6/(48K_2^3) +$$

$$+ 35K_4^2/(384K_2^4) + 7K_3K_5/(96K_2^4) - 35K_3^2K_4/(192K_2^5) +$$

$$+ 385K_3^4/(1152K_2^6)] + O(N^{-3})$$

where $K_n = d^nK(t)/dt^n$ at $t = t_0$. The left hand side of the above equation embodies the normal approximation to P_N. If the K_n can be found, even the N^{-1} correction gives a notable improvement for modest values of N.

If $k_n = d^nK(t)/dt^n$ at $t = 0$, $r = k_1 + k_2t_0 + k_3t_0^2/2! - - -$, which may be invertible. The first two terms in fact give the normal approximation. Note that if several saddle points t_0 exist, the contribution of each must be included in P_N.

This all depends on the existence of the K_n, that is, of the moments of the one-step distribution. Distributions with ill-behaved moments are not unfamiliar. They are best known as distributions in time.

To take an example from reaction processes, if a particle diffuses in one dimension, starting from the origin, the mean time taken for it to reach for the first time a boundary at a or one at -b, $(a,b > 0)$, is $ab/2D$. This evidently goes to infinity as, say, b goes to infinity.

The probability density of the time t to first passage at a, starting at the origin, with the whole left half line available for escape, is

$$(4\pi Dt^3)^{-1/2}a \exp(-a^2/4Dt)$$

which has no finite moments at all. But it does have a characteristic function $M(ik) = \exp(-Wk^{1/2})$ and belongs to the interesting class of Lévy stable distributions. These have characteristic functions $\exp(-Wk^\alpha)$, where $0<\alpha\leq2$ – with certain optional complications [5] (see appendix B) – and have the very important property that the sum of several random variables, each of which has the same Lévy distribution, has a Lévy distribution with the same value of α. Particular members of the class are, the normal distribution with $\alpha = 2$, Holtsmark with $\alpha = 3/2$, Cauchy with $\alpha = 1$ and passage time with $\alpha = 1/2$. The Holtsmark distribution relates to the gravitational field within stellar clusters, and the Cauchy is familiar as the Lorentz line shape function.

The divergent moments of the Lévy distributions correspond to high probabilities for large values of the variable. Their stability on addition makes them convenient for sums of few elements. They are just what we require for the description of random walks on fractals, which correspond, over a suitable range of scale, to the physically important critical percolating structures and to the structures of a number of polymers.

Shlesinger and Klafter [6] give a particularly lucid account of the one-dimensional random walk, the Weierstrass flight, for which

$$p(x) \propto \Sigma_{j=0}^{\infty} n^{-j} [\delta(x-b^j)+\delta(x+b^j)]$$

$$M(ik) \propto \Sigma \, n^{-j} \cos(b^j k)$$

and $\quad \langle x^2 \rangle \propto \Sigma \, (b^2/n)^j$, and diverges for $b^2 > n$.

Then $M(ik)$ is proportional to $\exp(-k^\alpha)$ with $\alpha = \ln(n)/\ln(b)$, and the track of the flight is of dimension α. This situation, in which a step of length b^j is available with probability proportional to n^{-j}, is typical of fractal structures which are at least statistically self-similar on change of scale. In condensed matter, the molecular structure of course sets a lower limit to the scale at which self-similarity can be relevant, but there is frequently a range, from this limit up to a macroscopic correlation length, over which a fractal description is useful. As has been described in other talks to this school, branching polymers may show such behaviour. More relevant for diffusion kinetics in general is the fact that critical percolating clusters are generally fractals.

Percolating clusters as a rule are richly supplied with dead ends which act as reservoirs for diffusing material but offer no through path. Removal of such decorations leaves the backbone of the cluster. One simple construction for a network which simulates a critical percolating backbone is to take a given segment and erect an equilateral triangle on its middle third: one then continues the construction on each of the five equal segments thus generated, and so indefinitely. Since each step down in length scale by a factor three multiplies the number of elements by five, the fractal dimension of this net is $\ln5/\ln3 =1.465$. Different choices for the steps allowed from different levels of vertex on the net lead to different diffusion kinetics. The treatment is complicated by the fact that there are two different types of vertex at each level of the scale.

Let me finally raise the speculation that in a number of kinetic systems the interesting action involves transport on tracks, defined by the pattern of thermodynamic fluctuations, which are just percolating under the given conditions. It does appear that critically percolating tracks typically occupy about 15% of the volume of the system, so these are pretty large fluctuations. If, on the other hand, we are concerned with transport on a permanent structure which has its fractal character built in, the only relevant fluctuations are those involved in individual steps. The fractal structure of such tracks will support diffusive modes with a relaxation spectrum $G(\tau) \propto \tau^{-1-\delta}$, where δ, the spectral dimension, is generally less than the fractal dimension and is frequently close to 4/3. 'Anomalous' kinetic behaviour will then cease to appear unusual.

APPENDIX A [7]

The Hamiltonian function used in the computations of references [1] and [2] is

$$H = \Sigma_{i>j}U(R_i,R_j) + P^{-1}\Sigma_{i,\alpha}V_{i\alpha}(R_i,r_\alpha) +$$

$$+ 2\pi^2 Pm_{\bullet}\Sigma_\alpha(r_\alpha-r_{\alpha+1})^2/(\beta^2 h^2) + \tfrac{1}{2}\Sigma_{\delta,i}m_\delta v_{\delta,i}{}^2 + \tfrac{1}{2}m'_\bullet\Sigma_\alpha v_\alpha{}^2$$

where $\beta^{-1} = k_B T$, $m_\bullet$ is the true electron mass, i,j denote molecules, δ denotes nuclei within molecules, α $(1\leqslant\alpha\leqslant P)$ lists successive beads on the closed chain representing the electron, and $V_{i\alpha}$ is a suitable approximation for the electron-molecule interaction.

For the calculation of equilibrium space distributions, the values of the masses m_δ and $m'_\bullet$ may be chosen arbitrarily. If the m_δ are set at the true nuclear masses, a convenient choice of $m'_\bullet$ will be such as to make the oscillation frequencies of the electron chain fall in the frequency band of the molecular motions, so as to ensure good energy transfer.

The mass of each bead is proportional to P, so the mass of the 'electron' chain goes as P^2. In fact, for P = 1,000 at 300K the total chain mass is 26,000 a.m.u. against a water mass, in the cell used by Schnitker and Rossky, of 5,400 a.m.u. In view of the additional computational destruction of velocity correlations, kinetic interpretation of these results must be very cautious.

APPENDIX B

Feller [5] shows that the logarithm of the most general stable characteristic function is of the form $a\psi(k) + ibk$, where $a>0$, b is real, and
$$\psi = -|k|^\alpha(1 \mp i\delta\tan\tfrac{1}{2}\pi\alpha) \qquad \alpha \neq 1$$
$$\psi = -|k|(1 \pm i\delta\ln|k|) \qquad \alpha = 1$$
the upper sign applying when k > 0. $0 < \alpha \leqslant 2$, $-1 \leqslant \delta \leqslant 1$.

REFERENCES

1) Barnett,R.N.,Landman,U.Cleveland,C.L.& Jortner,J. (1988)
   J.Chem.Phys.,**88**,pp.4429-4447
2) Schnitker,J.& Rossky,P.J. (1986) J.Chem.Phys.,**86**,pp.3471-3485
3) Webster,B.C. (1979) Ann.Rep.Prog.Chem.,**76C**,pp.287-313
4) Domb,C.& Offenbacher,E.L. (1978) Am.J.Phys.,**46**,pp.49-56
5) Feller,W. (1966) Introduction to probability theory and its
   applications (Wiley, New York) v.II,p.542
6) Shlesinger,M.F.& Klafter,J.(1986) in 'On growth and form',
   eds. Stanley,H.E.& Ostrowsky,N.;Martinus Nijhoff,Dordrecht,1986;
   N.A.T.O. A.S.I.,Cargèse 1985;pp.279-283
7) Parrinello,M.& Rahman,A. (1984) J.Chem.Phys.,**80**,pp.860-867

Femtosecond Laser Spectroscopy and Dynamics

in Dense Media

by

Kathy Carpenter and Geraldine Kenney–Wallace

Lash Miller Laboratories

University of Toronto

Toronto M5S 1A1

Femtosecond laser spectroscopy and nonlinear optical or photophysical measurements up to 100 picoseconds have become a powerful tool to explore dynamics, photophysics, photochemistry and photobiology of molecules in dense media. A review of contempory details of the experimental design concepts in ultrafast pump–probe laser spectroscopy links the observables with the underlying theory, where the curreṇ limits of time resolution approach a few optical cycles. Selected studies are then outlined to illustrate some of the scientific questions currently being investigated in our research group and several other laser laboratories, whose focus is on the time evolution of dynamical events in dense media.

Th. Dorfmüller (ed.), Reactive and Flexible Molecules in Liquids, 407–438.

1. INTRODUCTION

Following almost two decades of intense research activity on lasers and quantum electronics to generate increasingly shorter laser pulses and the emerging apllications of ultrashort laser pulses, time resolved spectroscopy in the femtosecond(fs)/picosecond(ps) regime has become a powerful technique for studying dynamical molecular processes in solids and liquids that occur on the 10^{-14}s to 10^{-10}s time scale. New insights gained from such studies [1—6], in conjuction with the wealth of information available from frequency domain light scattering and NMR experiments, I.R. studies of vibrational relaxation processes, and molecular dynamic computer simulations [7—10], has enabled scientists from diverse disciplines to begin to better understand the effect that internal (intermolecular forces) and external (E field) stimuli have on nuclear motions and electronic processes in all states of matter. Some of the frequency and time domain experimental techniques that are commonly employed to study these ultrafast dynamical processes are depolarization spectroscopy, nonlinear optical Kerr responses, four wave mixing processes (4WM) to study phase conjugation and coherence phenomena, fluorescence depolarization spectroscopy to look at excited state dynamics, coherent antistokes Raman scattering, analagous coherent scattering processes, and I.R. studies of vibrational relaxation in the liquid and solid state.

The majority of time—domain experiments in ultrafast laser spectroscopy are conceptualized within the framework of pump—probe spectroscopy. We briefly outline the methodology of this category of experiments. In pump—probe spectroscopy, the initial laser pulse resonantly pumps the atom or molecule into an excited state, following which the excited system is probed by a second, much attenuated laser pulse. By splitting off a fractional intensity from the initial pump pulse, and diverting the newly created probe pulse through a variable optical delay line prior to its arrival at the excited target system, we obtain internal synchronization of the two pulses and an ability to map out the full relaxation of the excited molecular system. The latter may undergo dynamical change for up to 10^{-8} seconds or longer after a fs exciting pulse. Since the atoms or molecules are initially prepared by the pump pulse in a nonequilibrium state, whose polarization, absorption, or radiative properties are different from the ground state, the system has been "labelled" and the probe pulse monitors the labelled molecules at successive delay times after the initial event. The "label" created varies: it can be absorption or emission intensity, Raman, polarization dependence, or phase coherence of the initial atom or molecule, or of the ensemble. Alternately, nonresonant interactions such as field—induced alignment effects in optical Kerr studies can be similarly investigated in the pump—probe format.

The delay time, τ, is obtained by varying the length of the optical delay line. The precision of the temporal measurement of the relaxation event under study is thus entirely controlled by the precision with which we can determine τ. This in turn depends upon the accuracy of distances to ±1 micron across several meters of optical delay lines, which contains numerous optical elements such as filters, prisms and polarizers. Each element introduces group velocity dispersion for a given laser wavelength and thus further optical delays of as much as several picocoseconds in a single 2 mm filter. The challenge of pump—probe spectroscopy is to design experiments with as much precision of measurement as high resolution laser spectroscopy and as much stable pulse stability as x—ray or neutron scattering.

2.GENERATION, COMPRESSION, AND AMPLIFICATION OF ULTRASHORT PULSES

The actual technical aspects of generating ultrashort laser pulses needed for the experiments reflects the strong interplay between ultrafast phenomena and the techniques based upon them. Therefore, in this section the generation of ultrashort pulses and their diagnostics will be outlined because it is vital the experimentalists be aware of both the power and the limitations of the measurements underway and that theoreticians also be aware of the links between the observables and the models that are subsequently constructed. The two most versatile sources of ultrafast dye laser pulses are presently (1) the actively modelocked continuous wave (CWML) argon ion Nd:YAG synchronously pumped dye lasers. The 1970's had witnessed substantial pioneering work in picosecond spectroscopy carried out upon solid state, mode—locked Nd:glass, Nd:YAG and ruby lasers. The gradual transformation to CWML systems in the 1980's has slowly revealed the merits of using different single—shot to high—repetition rate approaches, as the emphasis moved from generating the actual reproducible laser pulse to enhancing the signal:noise (S:N) ratio in the experiment itself. Concomitantly, KHz and MHz phase modulation techniques gained renewed interest. It is impossible in a brief space to adequately cover the broad range of lasers and techniques currently in use. As a result, we have chosen to describe as representative of ultrafast laser spectroscopy the lasers and synchronous pumping techniques developed and ususually built in our laboratory. Additional discussions of measurment, theory and experiment are referenced elsewhere [1—6,18].

Figure 1 illustrates a system designed and built for tunable pump—probe femtosecond and picosecond laser spectroscopy, based on Nd:YAG master oscillator [11,12]. The principles of the scheme, including pulse amplification and optical pulse compression (OPC), are generic and apply to argon ion or krypton CWML lasers as well.

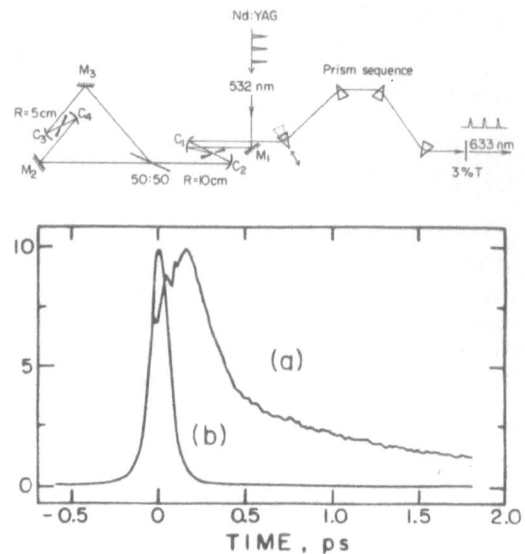

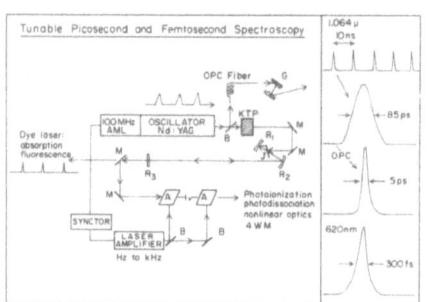

Figure 1. Femtosecond laser system for pump—probe spec—troscopy = see text for de—tails.

Figure 2a. ARR Anti—resonant ring dye laser configuration as described in text [13].

Figure 2b. CS_2. Femtosecond Kerr response from CS_2 with 70 fs pulses (trace a) with super—imposed $G_0^{(2)}$ signal from KDP crystal (trace b) [14].

The train of typically 85 ps pulses at 1.064 μ emerging from a CW acousto—optically mode—locked (AML) Nd:YAG has a ns interpulse spacing determined by the mode—locking frequency. The latter usually ranges from 75 MHz to 100 MHz for AML and this translates to 8 ns to 12 ns time delay between pulses, thus imposing temporal boundary conditions on the lifetimes of phenomena being investigated. Average powers P of $\leq$ 20 watts ensure peak—powers of kW per pulse. Therefore, second harmonic generation (SHG, See Figure 1), obtained through passage of the polarized I.R. pulse through a 5 mm path length of a nonlinear crystal (KTP), is quite efficient, providing TEM_{00} mode quality is good and phase matching in KTP is optimized. Such lasers produce $P_{2\omega} \leq$ 2W of green at $\lambda = 532$ nm before KTP damage thresholds are met. These green pulses can be utilized to synchronously pump a dye laser cavity in a folded, linear configuration (R_1, R_2, R_3) as Figure 1 shows and generate 200—300 fs tunable laser pulses. The pulse train (with identical interpulse spacing) emerges from R_3 with 0.5 nJ per pulse or kW peak power [12]. When an intiresonant ring dye laser is synchronously pumped as illustrated in Figure 2, it generates 60 fs pulses [11, 13] with an average power of $\leq$ 120 mW before amplification.

In both types of dye lasers, saturable absorbers such as DODCI have been introduced either into the jet (J) with the dye gain medium, or as a second jet in the cavity, with the purpose of further reducing the pulse width of the produced visible pulses at wavelengths $590 \leq \lambda \leq 635$ nm for Rhodamine 6G. While most ultrafast fs spectroscopy relies on Rh 6G or other visible dyes to generate visible laser pulses, a broader range of dyes is now available for IR and UV picosecond work in particular. Picosecond laser photophysics and photochemistry can employ the full UV to IR spectrum by judicious selection of dyes and use of nonlinear sum–difference frequency mixing schemes [2—5].

If amplified pulses are required, then these fs (or ps) pulses can be used to seed an optical amplication (A) scheme comprising several stages of optical gain, up to 10^6. The laser dye in A is pumped independently by a second and synchronous (SYNCTOR) laser, such as ns Nd:YAG at 10Hz, or a ns Copper Vapour laser amplifier at up to KHz repetition rate. The stored energy in A is transferred to the fs laser pulse as it propagates along the amplifier chain but timing is absolutely crucial to avoid loss of stored energy as spontaneous emission from A.

Alternatively, the fs pulse is directed into a regenerative amplifier seeded by the residual IR pulses from the Nd:YAG, after generating the green pulses via SHG. Comprising a Nd:YAG rod in an optical cavity, whose switch out capability s controlled by a pockels cell driven at rates up to KHz, the REGEN technique generates high power trains of IR and thus green pulses to pump the A gain cells.

These various schemes for amplication are discussed elsewhere in detail [6] and are pragmatically evolving as the barriers (in terms of material thermal response, gain, acoustic problems, and electronic jitter) are understood. At these very fast timescales and high repetition rates, where stationary–state conditions clearly do not always apply in the circuits or the gain medium, much research and development work is still needed. Even the laser cavity operating parameters for fs pulses are not yet fully understood in a quantitiative way for synchronous pumping.

Optical pulse compression (OPC) via pulse passage through a single mode polarization preserving optical fiber, can be used to compress in one step 0.03A, 5 ps, Nd:YAG IR pulses to 6A, 5 ps and the 300 fs, 26 cm—1 visible dye laser pulses to 1000 cm—1 and $\geq$ 8 fs, respectively. The spectral components of these pulses, broadened by self–phase modulation (SPM) and group velocity dispersion (GVD), are reconstructed in a linear dispersive line such as a grating pair, G [15]. OPC has been used extensively and forms an interesting subject for study in its own right [16]. However, for experiments of chemical interest, it is prudent to achieve a wise trade–off between spectral band–width and pulse duration. Hence the window 10^{-13} to 10^{-12} s has particular appeal for photoselective experimentation [17].

As previosly mentioned the anti–resonant (ARR) ring dye laser shown in Figure 2 [11, 13] is an excellent method for producing ultrashort and very stable fs laser pulses. In this dye configuration the back reflector of a normal linear cavity has been replaced by a thin dielectric or pellicle beam splitter (50:50) that directs pulses into a ring configuration, in which a saturable absorber jet is centrally located. The resulting fs, counter propagating pulse replicas collide at the same point in space and time in the (typically) 25 micron thick jet of etheylene glycol containing DODCI. The subsequent nonlinear saturable absorption discriminates against the transmission of the low intensity wings of the fs laser pulse in favor of the peak intensity. After many transit times around the gain cavity in the ring, a stable train of 60 fs pulses of 0.25 nJ per pulse emerges. Focussed down, these pulses can generate 150 MW·cm^{-2} and readily drive nonlinear processes, thus avoiding the further need for fs pulse amplification. The pulse characteristics are further refined by dispersion compensating elements in the optical gain cavity, and 4 prisms play that role in an inverted sequence as shown. Finally the pulse duration can be varied by translating one prism across the beam waist of the cavity, allowing typically a range of 60—300 fs pulse duration to be generated in a controlled manner.

3. PULSE DIAGNOSTICS AND TIMING

Of primary importance in the application of ultrafast spectroscopy to the study of chemical systems, is the ability to first of all characterise the laser pulses themselves, and to determine the timescale for $\tau = 0$ reference. The various single pulse and CW pulse real–time and sampling photodiode and streak camera techniques employed are discussed in the literature cited [1—6,11—18] and we ill only briefly outline the most frequently used pulse autocorrelation technique which measures the properties of ensembles of pulses and is therefore better suited to high repetition rate laser systems.

Photodiodes are currently limited to 40 ps rise time, and streak camera determinations are capable of resolving $\leq$ 1 ps. However, autocorrelation techniques based on the second order autocorrelation $G_o^{(2)}$ of the laser pulses are the most widely used [2,3,12]. They rely on the intrinsically instanteneous response of the electronic nonlinear polarization $P_{2\omega}^{(2)}$ generated in a non centrosymmetric crystal, such as potassium dihydrogen phosphate, to produce the second harmonic radiation at 2ω when two separate beams of intensity I(t) and amplitude E_o at frequency ω overlap in space and time in the crystal.

$$P_{2\omega}^{(2)} = \frac{1}{2}\chi_{ijk}^{(2)} E_o^2 (1 - \cos 2\omega t) \tag{1}$$

$$I_{2\omega}(t) = \int_{t=0}^{\infty} I_{\omega}(t) I_{\omega}'(t + \tau)dt \tag{2}$$

$$I_{2\omega}(t) = G_o^{(2)}(\tau) + r(t) \tag{3}$$

The resultant intensity of the signal $I_{2\omega}(t)$ is the convolution of the 2 pulses $I_{\omega}(t)$ and $I_{\omega}'(t + \tau)$ and the integral is proportional to the background free second order autocorrelation $G_o^{(2)}$. The additional term r(t) is the rapidly varying, phase interference factor which is included for completeness but actually <r(t)> is optically averaged to zero. Finally the output from the PMT is recorded in real time on an XY oscilloscope, and the profile of $G_o^{(2)}$ is captured from which we may deconvolve to obtain the original function I(t) and thus determine the pulse width. Three very important points must be made here. First, a second–order autocorrelation function is symmetric about τ. Thus we cannot obtain information on the skewedness of the pulse in this manner. $G_o^{(3)}$ is the necessary measurement for that information and requires a 3 photon correlation. Second, we must assume a pulse shape in order to deduce the pulse width in time, based on transform–limited bandwidth considerations. This choice could lead to a factor of 2 uncertainty in the real value on going from (for example) gaussian where $\Delta\nu\Delta\tau$ = 0.4413 to exponential where $\Delta\nu\Delta\tau$ = 0.1103 . Third these are ensemble measurements, not single pulse measurements, and the measurement in reality carries information not only on the pulse envelope $G_p(\tau)$ but also on pulse substructure $G_n(\tau)$, which is the Fourier transform of the spectrum of the laser output. The influence of the latter as coherence spikes on pulse determinations and spectroscopic data;

$$G^{(2)}(\tau) = G_p^{(2)}(\tau)\,[1 + G_o^{(2)}\,G_n^{(2)}(\tau)] \tag{4}$$

should be recognized as a continuing and central point of discussion in ultrafast pulse diagnostics.

The autocorrelation $G_o^{(2)}$ from the laser pulses emerging from the ARR design of Figure 2a is 100 fs (fwhm) and, assuming $sech^2$ shape, the pulse width is 65 fs. The trace in Figure 2b is the transmission through an optical Kerr cell containing CS_2 at 298K (see below) and the $G^{(2)}$ trace is superimposed to illustrate the contrast between steep laser pulse and relatively slower grow in of the signal, and subsequent rapid decay of the induced anisotropy [14].

While pulse diagnostics play a crucial role in interpreting the time–resolved responses, since the pulse duration has to be deconvolved from the signal to obtain the molecular response function, the need for good disgnostics must be coupled with a need for precision of measurement. The following five points are an intrinsic part of the design of any ultrafast laser–induced experiment. The five do not represent an exhaustive list but a priority list, and are: (i) group velocity v_g = $(\partial k/\partial\omega)^{-1}$,where k is the wave vector $(2\pi/\lambda)$ (ii) GVD or group velocity dispersion, which leads to different frequencies travelling at different velocities, $(\partial v_g^{-1}/\partial\lambda) = (\partial^2 k/\partial\omega^2)$, and often of pulse distortion, (iii) frequency–time bandwidth relations $\Delta\tau\Delta\nu$, (iv) transit times across the sample for fs ultrashort pulses, and (v) time delays experienced in common focussing optics for two colour experimental pump–probe configuration.

In this chapter, we first review some of the laser techniques currently used in pump–probe spectroscopy and then examine several applications to molecular dynamics in liquids and solutions, to electron localization and solvation, to biomolecular systems, and to semiconductors. Finally we conclude with a brief section on the latest results from some of our nonlinear laser spectroscopy to show how applying this new perspective on dynamical problems to a wide range of molecules reveals the microscopic details of the dynamical structures in liquids and disordered media. This chapter is not intended to be a comprehensive review of all the work which has been recently reported but rather a selected survey and a stimulus to the reader.

4.APPLICATIONS OF ULTRAFAST LASER SPECTROSCOPY

Now that some of the essential elements of ultrafast laser techniques have been described, a number of important relaxation phenomena in solids and liquids that are currently being investigated, using a variety of ultrafast spectroscopic techniques will be reviewed. In particular reorientation of dye molecules in liquids, solvation dynamics of electrons and solute molecules in liquids, charge transfer events in photosynthetic systems, electronic transport events in semiconductors and optical Kerr responses in simple liquids will be discussed in the remainder of this chapter. We begin with the essence of chemical reactions. How does a molecule move and how do those motions influence the outcome of a chemical event?

4a.SOLUTION DYNAMICS

An understanding of molecular motions associated with chemical reactions in solution is required for a complete microscopic description of solution phase reaction dynamics. A major contemporary challenge is to link the macroscopic and empirical observations of rate constants or scattering cross–sections of the ensemble to the detailed microscopic dynamics characterizing state–selected activity in a given atom or molecule. From a thermodynamic point of view, macroscopic properties of the solvent are thought to influence the free energy surface of the reaction. For example, if a reaction proceeds via a transition state which is less polar than the reactants, in a polar solvent the reactants will be stabilized relative to the transition state and the observed rate constant for chemical reaction will decrease. If solvent fluctuations are comparable or faster than reaction rates, then the details of molecular interactions and structure of the surrounding solvent can no longer be ignored. In this microscopic picture one must include coupling of individual solvent motions to the reactive potential energy surface in order to describe completely the dynamics of the chemically reacting system.

Theoretical approaches in the past have generally modelled the solvent as a structureless fluid with a frequency dependent dielectric constant $\epsilon(\omega)$ [19,20,21,22]. In this continuum treatment $\epsilon(\omega)$ is usually expressed in the Debye form;

$$\epsilon(\omega) = \epsilon_\infty + \frac{\epsilon_0 - \epsilon_\infty}{1 + i\omega\tau_D} \tag{5}$$

where τ_D is the Debye relaxation time and ϵ_0 and ϵ_∞ are the zero and infinite dielectric constants respectively. In many polar aprotic solvents such as DMSO and acetonitrile, the dielectric response appears to be well described by a single Debye time. In alcohols, on the other hand, the response is generally described as a sum of two or three dispersions.

In addition to τ_D, there exists a second relaxation time τ_1, which is commonly used as a guage of dynamical solvent effects. This longitudinal relaxation time is derived from the macroscopic dielectric relaxation time(s) τ_D, via;

$$\tau_1 = (\epsilon_\infty/\epsilon_0)/\tau_D \tag{6}$$

Thus in polar solvents where $\epsilon_0 >> \epsilon_\infty$, τ_1 is predicted to be much shorter than τ_D. The question arises in studying picosecond dynamics of rotating or reacting molecules, to what extent does the probe molecule behave as though it is embedded in a continuum, and to what extent a more molecular dynamic approach, with highly oscillatory fluctuations in the local structure, dominates events.

4b.TIME DEPENDENT STOKES–SHIFT

Upon optical excitation, the fluorescence spectrum of a polar probe red shifts in time as the solvent re–equilibrates to the new excited charge distribution. Time dependent Stokes–shift thus provides a direct measure of the dynamics of solvent relaxation around probe molecules which have been perturbed from equilibrium solvation. The choice of probe and solvent are key points in determining the validity of relating time dependent measurements to solvation dynamics. Ideally for transient solvation studies, the primary solute–solvent interactions should be dipolar interactions rather than specific interactions like hydrogen bonding: also, the probe molecules should have a dipole moment that has a very different magnitude and/or direction in the ground state as compared to the excited state so that upon optical excitation the solute–solvent system is taken out of a state of equilibrium and time dependent spectral shifts will monitor solvent relaxation. Problems involved with using dye molecules for transient solvation studies include identifying the precise attachment of solvent molecules to the dye chromophores and an accurate determination of the magnitude and orientation of the transition dipole.

Polar aprotic solvents are good choices for solvation studies for two reasons. First, problems of multiple dielectric relaxation mechanisms and specific hydrogen bonding are avoided in these liquids, and secondly, solvents of this type are close to those used in basic theoretical models of dynamic and static solvation [19,20,21,23,24]. Time dependent studies therefore offer an important opportunity to evaluate these theoretical models.

As a matter of book—keeping, the following Stokes—shift correlation function has been introduced [19,20,25];

$$C(t) = \frac{\nu(t) - \nu(\infty)}{\nu(0) - \nu(\infty)}$$

(7)

where $\nu(t)$, $\nu(0)$, and $\nu(\infty)$ are the emission maxima at times t, zero and infinity respectively. According to dielectric continuum theories, this solvent correlation function is predicted to relax on a time scale approximately equal to the solvent longitudinal relaxation time τ_l [19,20]. If the response of the solvent decays as a sum of exponentials, then the spectral shift correlation function is also thought to decay as a sum of exponentials. However, this is not always found experimentally and the importance of τ_l as a guage of solvent dynamics has been discussed by several workers. Calef and Wolynes [24,26] for example suggest that at larger distances from a solute the solvent should approximate a continuum and relaxation is expected to be on the τ_l time scale. Near the solute charge, however, the solvent dipoles are expected to rotate on a time scale similar to τ_D. Van der Zwan and Hynes [27] point out that solvent relaxation may also occur by a phenomena known as polarization diffusion. This describes the component of the polarization response arising from translational motion of the solvent dipoles rather than the rotational motion and is therefore not detected in bulk dielectric measurements. As a result, polarization diffusion could lead to substantial acceleration of solvent relaxation in certain solvents which would not be expected based on continuum theory. The two examples just cited above offer a demonstration of the failure of long range continuum models to predict transient solvation dynamics in many solute—solvent systems since molecular interactions and structures in the inner solvation shells are ignored in these models [11,12,14,17].

In order to demonstrate the degree to which continuum models adequately describe solvation dynamics, some specific experimental examples will be discussed. In agreement with continuum theories, Castner et al. [28] show that for the probe dye molecule LDS—750 in both aprotic and alcohol solvents, solvation occurs on a time scale roughly given by the longitudinal relaxation time. Safarzadeh—Amiri [29] examined the probe trans—4—dimethyl—amino—4'—cyanostilbene in n—butanol between 250 and 300 K. They find solvation times approximately equal to τ_l over this temperature range. Yeh et al. [30] also find C(t) decays in accordance with continuum theory for the probe 4—amino pthalimide in several alcohols near room temperature.

In contrast, experiments conducted by several workers show that i) the average time constant observed is in general greater than τ_l, ii) the relaxation dynamics are determined by C(t) are nonexponential and iii) as the dielectric constant of the solvent increases, τ_l and the observed solvation time become more dissimilar. For example, in the case of LDS–750 solvation dynamics in nitrobenzene and butanol are found to be nonexponential [28]. This the authors attribute to the influence of secondary dielectric dispersion. However, for the same probe, the relaxation time in methanol is significantly shorter than that calculated from continuum theory [28]. It is believed that polarization–diffusion is responsible for the faster relaxation in this solvent. Stokes correlation functions for the probe 1–amino–napthalene in several alcohols [31] are found to be similar to those of Coumarin 153. This is an indication that solvent relaxation does not seem to depend on the details of the solute. Solvation times of Coumarin 311 in these same solvents were found to be generally longer than τ_l [31] which further emphasises the inadequacy of continuum theory in describing solvent relaxation Finally, the picture proposed by Calef and Wolynes has been applied to experimental studies of Coumarin 153 in 1–propanol over a wide temperature range [28], and 4–aminopthalimide in alcohols [30]. Solution dynamics were correctly predicted using their model in both cases.

From the experiments just outlined, it is apparent that deviations from continuum theory are found for many solute–solvent systems. It can thus be concluded that general aspects of the solvent molecularity cannot be neglected when discussing solvation dynamics.

4c. REORIENTATION DIFFUSION

Experimentally, insight into specific solute–solvent interactions have been gained through time resolved studies of reorientational diffusion of dye molecules in solution. In general, dye molecules make good probes for reorientation studies because of their typically large absorption cross sections at wavelengths convenient for laser excitation. However, they should be as rigid as possible in order to avoid complication in the interpretation of results due to molecular rearrangements [32]. For example, dye molecules Resorufin and Cresyl Violet are fairly rigid thus rendering them good probes for dynamical studies. Fluorescein and BBOT on the other hand, are capable of internal rotations and are therefore poor candidates for reorientational experiments.

The most frequently applied model used to describe rotational motion in solution is based on Debye–Stokes–Einstein (DSE) theory [33] which assumes a large probe molecule diffusing in a structureless continuum. In this hydrodynamic picture, the reorientational time constant τ_{rot} is predicted to vary linearly with the macroscopic shear viscosity η. The validity of applying such an approach to the rotation of small molecules in solution is questionable since in this particular situation the rotational reorientation time is expected to be influenced by solute–solvent and solvent–solvent interactions. Therefore approaches have been made which incorporate molecular properties into the hydrodynamic formalism: i.e. theoretical concepts of the DSE model have been generalized to include changing boundary conditions from stick to slip [34], selective solute–solvent size [35], free spaces in the solvent structure [36], and deviations from spherical shape [34]. Inclusion of these variable properties into hydrodynamic theory has resulted in the derivation of a modified DSE equation which may be expressed as [37];

$$\tau_{rot} = \frac{V\eta(fC)}{K_B T} \tag{8}$$

where T is the temperature, K_B is Boltzman's constant, and V is the volume swept out by the solute as it reorients in solution. If the reorientation behaviour is anisotropic, then the shape of the volume may be modelled as either an oblate or a prolate ellipsoid. The shape factor f which takes into account deviations from spherical shape for a solute has been tabulated by Perrin [38] for a series of ellipsoids. Unfortunately, there is no accepted convention in use for the concept of molecular volume which in turn leads to significant discrepancies between the results cited in the literature. For example, three laboratory studies of CV^+ in ethanol all yielded different results for τ_{rot} from the same raw data [40,41,42].

This example illustrates the importance of choosing a model which will accurately describe the molecular volume of the solvent for a particular solvent–solute system.

The factor C in equation (8) is used to take into account hydrodynamic boundary conditions at the surface of the solute. In general, it is believed that non–polar molecules give reorientation times in agreement with slip boundary conditions (Figure 3a.) [43]; i.e. tangential velocity of solvent molecules relative to probe is zero, while reorientation of macromolecules in solution are adequately described by stick boundary conditions (Figure 3b.) [44,45]. Stick boundary conditions have also been used to explain the motion of charged dye molecules in alcohols.

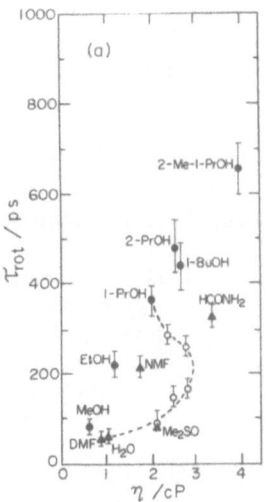

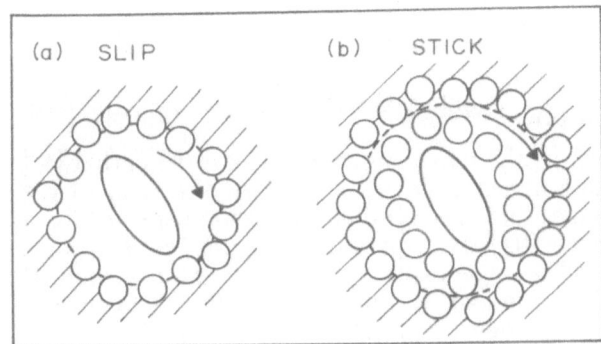

Figure 3a. Boundary conditions for hydrodynamic theory. a) slip: free rotation of probe within solvent continuum. b) stick: adjacent sol—vent rotates with probe.

Figure 4. Hydrodynamic plot of τ_{rot} vs. solvent shear viscosity η for Res—orufin in pure and mixed solvents.

Reorientational diffusion times of molecules in the excited state are measured using a technique called fluorescence depolarization, while ground state reorientation times are calculated from data obtained by transient absorption studies. In both cases a polarized pump beam (ω_1) is used to deplete selectively the ground electronic state via a $S_o \longrightarrow S_n$ absorption, thus creating an orientational anisotropy in both the ground and excited state populations. As the ground state molecules reorient to their normal isotropic distributions, the resulting decay in the anisotropy is monitored by the transient absorption of a secondary probe pulse polarized either $\perp$ or Π with respect to the pump ω_1. Spontaneous emission collected at angles $\perp$ and Π relative to the excited state transition dipole is the analagous experimental signal needed to determine the decay of the excited state anisotropy. The polarization anisotropy is expressed by the difference in emission for the excited state;

$$R(t) = [I_\Pi(t) - I_\perp(t)]/[I_\Pi(t) + 2I_\perp(t)] \tag{9}$$

and transmission for the ground state;

$$R(t) = T_\Pi - T_\perp \tag{10}$$

for the Π and $\perp$ cases.

In addition to its dependence on purely spectroscopic data, the polarization anisotropy R(t) for both states is directly related to a dipole correlation function P_2 which contains all necessary reorientational information;

$$R(t) \propto <P_2 \, [\mu(0).\mu(t)]>e^{-t}/\tau_{G.S.R} \tag{11}$$

The ground state recovery time $\tau_{G.S.R.}$ is obtained independently from the data

polarized at 54.7^O, otherwise known as the magic angle. Chuang et Eisenthal [46] have related the experimentally determined R(t) to the relative directions of the pumped and probed transition dipoles and the rotational diffusion constant for a general ellipsoid. In their picture, R(t) is expected to decay as a double exponential for oblate behaviour and a single exponential for prolate behaviour if the transition dipole is oriented along the long axis of the molecule. The converse is expected if the transition dipole is along the short axis. Blanchard et al. [47] examined the rotational diffusion behaviour of Cresyl Violet (CV) in isoviscous solvents. They found that R(t) decayed as a single exponential in both ethylene glycol at 26^OC and 1–dodecanol at 37° C but a two component decay is observed in 1–dodecanol at 26° C. They conclude that the shape of the volume swept out by the CV probe changes as a function of temperature in 1–dodecanol. This is a clear indication that the coupling of transient solvent structure to solute dynamics cannot be neglected.

When consideration of microscopic properties of solution becomes necessary, hydrodynamic models actually have limited predictive power. Therefore, alternative microscopic theories have been proposed which take into account collisional effects and hydrogen bonding interactions. For example, Hynes et al. [48] argue that the dominant effect responsible for reorientational motion is collisional relaxation with the first solvation shell. In an attempt to understand the extent to which microscopic properties influence a rotating molecule, the dynamics of rotational motion has been studied as a function of several solution properties which include solute shape [36,49,50], solute charge [36,41,42,50,51] solvent [36,52,53], and temperature [54].

The charge on the probe molecule is expected to play a large role in its reorientational dynamics. For example, Templeton and Kenney–Wallace [54a] found that in Me_2SO, there exists a 5–fold difference in the reorientation time of

the probe R^- as compared to that of CV^+. This is expected since solvation of the cation by the S=O moiety has been known to be much stronger than solvation of the anion by adjacent methyl groups [55,56]. Spears et al. [57] measured τ_{rot} for

Resorufin in several polar aprotic and H–bonding solvents. From their data it is apparent that in changing solvent from dipolar aprotic molecules to hydrogen bonding molecules, τ_{rot} increases largely. This the authors attribute to negatively

charged oxygens at each end of Resorufin coordinating to the H–bonding molecules so that the reorientation data actually result from rotation of a "new molecule" with an average solvent coordination. Rotation of this solvated species will be much less affected by the molecular charge since long range charge–solvent forces are adequately shielded by the primary solvation shell. These issues have been considered in depth by Kirk Gladstone, (PhD Toronto 1987). Such ideas are further supported by other studies [42,58,59] which reveal that anions without hindrance to solvent hydrogen bonding show "stick" behaviour in polar aprotic solvents while in H–bonding solvents "superstick" behaviour is observed.

Several picosecond studies have provided evidence that molecules in the excited state reorient much slower than their ground state counterparts [60,62]. Since the decay of the anisotropy is dependent on both transition polarization and rotational diffusion constants, they all assume that changes in the diffusion constant upon excitation are responsible for observed differences in reorientation times and that the transition polarization essentially remains the same at all wavelengths. Blanchard et al. however, looked at the anisotropy decay of Cresyl Violet in methanol at 5 different wavelengths, three ground state and two in the excited state [63]. Experimentally a different decay of the anisotropy is observed at each wavelength. In particular they find that the R(t) decay time is different for each of 2 non–overlapped ground state measurements. This can only be due to a wavelength dependence of the transition polarization since the diffusion constant must remain the same in a given state. It is therefore concluded that CV in methanol reorients identically in its ground and excited states. When comparing ground state τ_{rot} values with those in the excited state for the rigid planar molecule Resorufin in various solvents, we find that upon electronic excitation little change occurs in the solute–solvent interactions responsible for rotational relaxation.

Recently Fleming et al. [28] examined rotational dynamics of Coumarin 153 in a number of polar solvents. Based on hydrodynamic theory, one would anticipate the rotational anisotropy of this molecule to decay approximately as a single exponential. However, in n–propanol the deconvoluted anisotropy decay is fit to a double exponential form. The slow component correlates well with solvent viscosity and is thus ascribed to normal diffusional rotation. The faster component is thought to arise from rapid rotation of the transition dipole of Coumarin 153 in response to changes in solvent environment.

If the solute molecule has a non zero dipole moment, it can be viewed as a dipole in a cavity of radius r_o embedded in a dielectric continuum. The presence of this dipole causes a polarization to be induced in the surrounding medium. If the response of the medium is not instantaneous, the induced polarization lags behind the movement of the dipole and the probe experiences a net retarding torque. This type of solvent–solute interaction is known as dielectric friction and has been the subject of several theoretical investigations [64—66]. Mathematically this dielectric friction is handled through inclusion of an additive correction to the usual DSE equation;

$$\tau_{rot} = \frac{V\eta(fC)}{K_B T} + \frac{\xi_{DF}}{K_B T} \tag{12}$$

where the dielectric friction constant is given by;

$$\xi_{DF} = \frac{6\mu^2(\epsilon_0 - 1)\tau}{r_0^3(2\epsilon_0 + 1)^2} \tag{13}$$

In addition to longitudinal and transverse dielectric properties of the solvent [64,67], translational motion of the probe and solvent dipoles [64] and the rotational motion of the probe are known to contribute to relaxation of the medium. Therefore, the torque is expected to be dampened on these time scales. For large probe ions, translational motion of the probe and solvent can be neglected; the former because the probe rotates and translates slowly relative to solvent motion and the latter because movement of solvent molecules over distances comparable to the dimensions of the probe are slow.

If hydrodynamic theory is viewed as one extreme of transient solvation dynamics and specific molecular short range interactions represents the other, then dielectric friction models can be viewed as medium range in length and time scale.

In a series of papers on the relative impact of continuum versus molecular dynamic arguments to explain rotational data, Templeton, and Kenney–Wallace measured the reorientation time of Resorufin, thionine and Cresyl Violet in amides, alcohols, and water alcohol binary systems [40]. While good agreement with hydrodynamic theories is seen for the pure alcohol systems a curvilinear dependence of τ_{rot} as a function of η is observed in the binary propanol/water system (Figure 4.). When dielectric friction is applied to this system, the curvilinear dependence is predicted at least qualitatively. In another of their studies the reorientational dynamics of the same dye molecules in various Me_2SO/H_2O mixtures were examined. These systems also show a curvilinear dependence of τ_{rot} on η but dielectric friction alone is not able to explain the data [68]. This is a strong indication that molecular details of the solvent cannot be neglected when discussing solvation dynamics. Philips et al. [69,70] measured the reorientation time of Rhodamine 6G as a function of viscosity. The viscosity was varied either by changing the solvent in a series of alcohols or by changing the pressure of a single solvent, ethanol. In the solvent experiments τ_{rot}/η increases with η while in the high pressure experiment τ_{rot}/η is independent of η. By applying dielectric friction, the differences are explained. As an additional test they looked at the reorientational dynamics of R6G in ethanol as a function of added salt. With added salt the value of τ_{rot}/η decreases as a function of η; the implication being that dielectric friction is reduced in electrolyte solutions. Most recently Paone and Kenney–Wallace [70 a] have shown that the dynamics of resorufin in electrolytes must be viewed through competitive solvation arguments, because dielectric friction alone does not fully explain the trends in data.

As a final word on the subject of dielectric friction, associated liquids are known to exhibit hydrogen bonding between solvent and probe molecules. Therefore, in studying rotational diffusion, one must distinguish between dielectric and hydrogen bonding contributions.

5.ULTRAFAST ELECTRON SOLVATION AND TRAPPING

Transient solvation of electrons in polar media has been examined by many research groups and several review articles have been published on the subject [71—73]. On the basis of experiments in alcohols and water [74—78] it has been suggested that solvation occurs by the following sequence of events. At subpicosecond times the injected electron scatters through the conduction band of the fluid at high mobility rate. The electron in this extended state then relaxes to the bottom of the conduction band, which has a characteristic energy V_0. The

quasifree electron is now thermalized and has access to a tail of localized states immediately below the conduction band. Once in this quasilocalized state, the electron is assumed to be trapped and, if the residence time of the trapped state overlaps molecular reorientation times, full solvation follows by configurational relaxation of solvent molecules in the immediate field of the excess electronic charge. Based on a continuum model for the solvent, this relaxation is predicted to occur on a time scale equal to τ_l. However, substantial picosecond absorption

experimental data in the 1970's and modelling of electron–dipole interactions led to the conclusion that it is the short range dynamical structure of the fluid which provides both localization sites for the initial trapping and which governs the subsequent solvation processes in response to the electron perturbation [14,74]. These are the concepts which underlie the cluster model of electron solvation [75—78] and link the solvation dynamics to the intrinsic microscopic motions of the molecules in the host liquid. Recently the fs librational motions of molecules have been linked to the initial trapping step of excess electrons [78].

Since the localized electron states have high absorption cross sections, picosecond transient absorption spectroscopy provides a good tool for measuring the dynamics of electron solvation. In this type of experiment, the pump pulse is responsible for injection of the electron into the media via either direct 2 photon ionization of the neat liquid or photoionization of a solute molecule. A second probed pulse monitors the state of the electron as a function of time after injection.

At short times, probably $<< 10-^{12}$s, an absorption band in the I.R. is observed which is assigned to the trapped, quasilocalized state. This band then decays and an analagous one grows in the visible. This second band is characteristically broad ($\simeq$ 1eV fwhm), structureless and assymetric around the energy maximum and is typical of the electron in a configurationally relaxed, equilibrium state. The

lifetime of electrons ranges from $10-^9$ s to $10-^3$ s in some liquids; the enormous reactivity of this species has been well documented and for e^-_{aq} it appears that its

424

kinetics are more studied than those of the H atom.

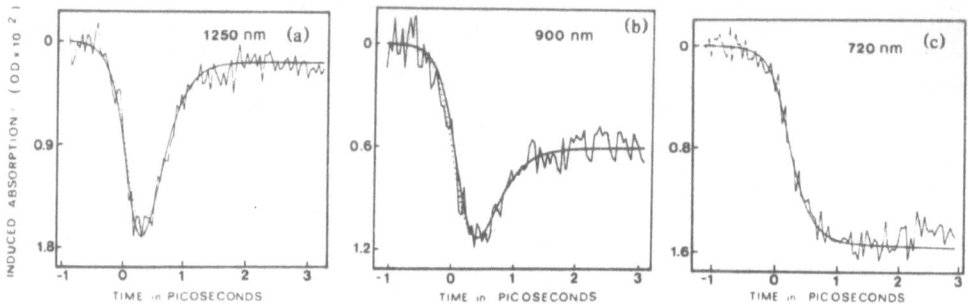

Figure 5. Typical kinetics of induced absorption vs time in liquid water, taken at different wavelengths. The instantaneous responses (dotted lines) have been recorded by exciting in the same conditions liquid n–heptane. The smooth lines represent the computed best fits assuming aa appearance time $T_1 = 110$ fs for the ir species and its relaxation $T_2 = 240$ fs towards the solvated state. The kinetics at 1.25 μm (a) and 0.72 μm (c) reflects mostly the evolution of the ir and solvated species respectively, while the intermediate case at 0.9 μm (b) corresponds to a 2/3 proportion of these species.

Recently Migus et al. [79] used femtosecond absorption spectroscopy to examine the localization and solvation of electrons in pure water. From their characteristic absorption spectra (Figure 5) it is apparent that the electron induced by 2 photon ionization of the solvent thermalizes in 110 fs within the conduction band and from there relaxes with a time constant $T_2 = 240$ (their notation) fs towards the fully solvated state. This second time constant T_2 is in excellent agreement with the characteristic longitudinal relaxation time τ_1 of water which at first glance suggests that a continuum description is all that is needed to adequately describe the solution dynamics of this system. However, continuum theory also predicts a continuous shift in the absorption spectra, which was not observed. In fact, the authors' finding of a stepwise process is a good indication that the structural relaxation of the surrounding molecules, which favours the nonradiative electronic transition to the ground e_s^- state, is limited to an extremely small number of motions. Their observations are in agreement with the data and the two step cluster model of electron solvation built upon the ps alcohol:alkanes studies by Kenney–Wallace and Jonah during the 1970's. More recently, molecular dynamics simulations conductivity, and fs molecular dynamics results on electron localization, solvation and mobility issues have been reviewed by Kestner, Schmidt, and Kenney–Wallace as part of the NATO ASI on electrical properties of the liquid state.

6.BIOMOLECULES AND ULTRAFAST PROCESSES

Photosynthesis in green plants and in most bacterial systems begins with the uptake of light energy by a number of pigment molecules specifically bound to polypeptides within an antenna complex [80]. The pigments become excited and transfer their excitation to so called special pairs [81,82]. Following this initial excitation the newly created charge migrates along a chain of donor and acceptor pigment molecules. Once this sequence of events is complete the potential energy remaining in the stabilized molecules may be used in metabolic reactions. The pigment molecules involved in all the aforementioned photosynthetic processes are contained in what are known as reaction centers (RC) which have been isolated for

several photosynthetic bacteria. A detailed understanding of the extremely fast e^- transfer reactions occuring in the reaction centers immediately following absorption is essential for the formation of a complete picture of the mechanisms involved in natural photosynthetic processes. In recent years, ultrafast spectroscopy techniques, combined with a knowledge of the RC structure have provided the tools needed to substantially characterize photosynthetic events occuring in a variety of reaction centers [83—89]. In particular the reaction centers of purple bacteria have been studied extensively [89—92]. Structurally they are known to include 4 bacteriochlorophyll molecules, two of them being accessory monomers (BC) and the other two forming the special pair, P, 2 bacteriopheophytin (BPh) molecules, 2 quinone molecules, and a nonheme Fe atom all attached to a membrane with protein α–helices. Recently, an x–ray crystallographic study of the specific purple bacteria RC, Rhodopseudomonas R–virides [93] revealed the pigments to be arranged in 2 chains with an appropriate 2–fold rotational symmetry. The 2 bacteriochlorophyll making up the special pair are 3 A apart and the so called accessory monomers are 11 A from these. The bacteriopheophytin are located 10 A from the accessory monomers and 17 A from P. The charge transfer process in the reaction center was studied by Zinth et al. [89] using ultrafast femtosecond optical pump probe absorption experiments. Four transient species were seen during the first 10 ps after excitation. These are assigned as follows: 1) excitation of the special pair followed by a fast charge separation, to form the P^{+-} state, occurs within the time frame (150 fs) of the laser pump pulse 2) within the next picosecond the electron rapidly passes to the neighboring BC monomer to form a transient P^+BC^- state. Subsequent passage of the e^- to the pheophytin results in a new state P^+BPh^- being established within 5 PS. Schematically the proposed sequence is as follows:

$$P \xrightarrow{h\nu} P^* \xrightarrow{0.1ps} P^{+-} \xrightarrow{0.1ps} P^+BC^- \xrightarrow{5ps} P^+BPh^-$$

in another experiment by Zinth et al. [90] the dynamical absorption changes in

R–virides reaction centers were compared to those of the purple bacteria, R–sphaeroides. Identical behaviour was found for the 2 types of RC's which forced the author to conclude that pigment arrangements and interactions between the pigments in one system are very similar to those of the other. Meech et al. [91] used subpicosecond photon echo spectroscopy to further resolve the dynamical events occurring in the initial excitation of the RC. A rapid decay of 25 fs in the signal was observed for both R–virides and Rhodobacter (Rb) sphaeroides which was independent of excitation wavelength. It was suggested that the mixed state formed on this fast time scale contains major contributions from the P–Frenkel dimer and intradimer charge transfer state. A similar ultrafast decay was obtained for R–virides using hole burning experiments thus further confirming the existence of such primary events [94, 95].

Photosynthetic activities in halobacteria are based on the membrane protein bacteriorhodopsin (BR) which acts as a light driven protein pump converting light energy into an electrochemical gradient across the cell membrane [96]. This gradient is subsequently used to synthesize adenosine triphosphate (ATP) and to drive vital metabolic processes necessary for normal functioning of the bacteria. Structurally BR contains only one chromophore, a retinal molecule which is linked via a protonated Schiff base to Lysine 216 in the primary structure of the protein. Upon prolonged illumination the retinal molecule adopts an all trans configuration, while in the absence of light a thermodynamic equilibrium consisting of 50% of the all trans retinal and 50% of the 13–cis, 15–cis retinal structure is established [97, 98]. the light and dark adapted forms are displayed in figure 6. In most halobacterial strains BR forms trimers which are arranged in a 2 dimensional hexagonal lattice within the cell membrane.

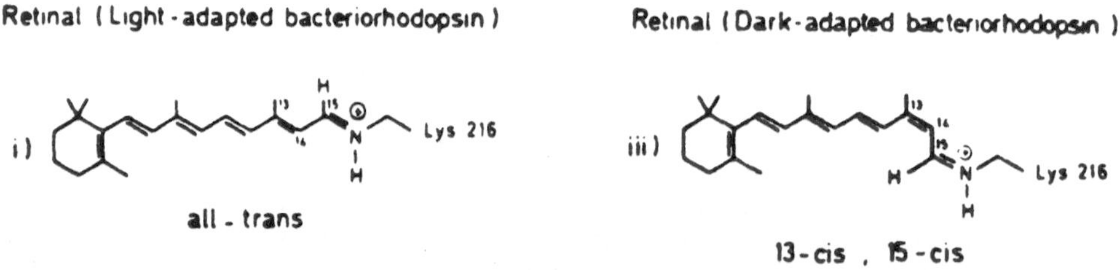

Retinal (Light - adapted bacteriorhodopsin)   Retinal (Dark - adapted bacteriorhodopsin)

Figure 6. Structures of retinal in (i) light–adapted and in (iii) dark–adapted bacteriorhodopsin before light absorption.

The first step in the photochemistry of BR is the absorption of a quanta of light which is then followed by the promotion of a BR monomer into the first excited electronic state (S_1). From there a proton pumping cycle occurs through a series of reactions involving the retinal molecule. The time needed for this series of events to occur is on the order of a few picoseconds, which implies a need for ultrafast techniques to monitor the formation and disappearance of intermediates involved in the photochemical cycle. It is now believed that the primary photophysical event in the light cycle involves a trans to cis isomerization about the C_{13}—C_{14} bond [99—102]. The isomerization is accompanied by separation of the protonated Schiff base from a negatively charged counterion. In an experiment carried out by Polland et al. [103], the photochemistry of BR was compared to that of a retinal analog modified to prohibit transformation to a 13–cis configuration. The analog exhibited a 12–fold enhancement in its fluorescence quantum yield as compared to the normal BR monomer. This experimental example clearly supports the idea of isomerization being the initiating event in the light adapted BR photochemical cycle. Transient absorption studies [104—106] of the light adapted form reveal time scale for isomerization to be within 100 fs at which point the minimum of the S_1 potential surface is reached. Radiationless

coupling to a ground state photoproduct J follows with a time constant of 750 fs [103, 104, 106, 107, 108] (figure 7). J has been determined to decay to a relatively long–lived bathochromic intermediate K within 3—5 ps [103, 107, 108]. From this state regeneratin of the all trans retinal is completed within a few milliseconds via a pathway involving intermediates denoted as L, M, and O which have been characterized by their absorption spectra [109, 110].

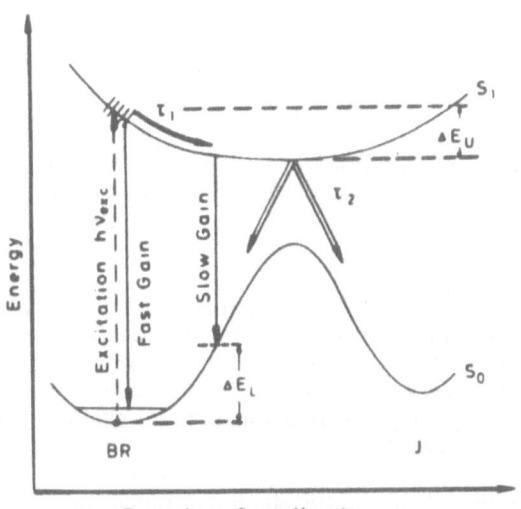

Figure 7. Model of the ground state (S_0) and excited state

(S_1) energy surfaces as a function of the reaction coordinate.

The first motion leads the system with the time constant of 200 fs to the bottom of the S_1 potential surface. The reaction proceeds

to the ground state photoproduct J or the initial state of BR with the time constant of 500 fs.

Petrich et al. [105] performed identical experiments on the dark adapted form of BR and found similar time constants for the formation of J and K counterparts. They conclude that similarity of time constants for both the light adapted and dark adapted intermediates is due to equivalent separations between the protonated Schiff base of the retinal and the counterion rather than to the attainment of equivalent structures as proposed by Trissl and Gartner [111].

As a final remark, Kolling et al. [107] compared the transient absorption spectra of a deuterated BR monomer with that of a normal protonated monomer. No isotope effect was observed for the $S_1{\to}J$ and $J{\to}K$ steps which suggests that

proton transfer across the membrane is not a primary process in the photochemical cycle of BR.

7. NONLINEAR LASER SPECTROSCOPY: A NEW PERSPECTIVE ON DYNAMICAL EVENTS

When a laser pulse is incident on a polarizable liquid, the polarization (P) induced in the medium may be expressed as a series in ascending powers of the applied field, i.e.

$$P\ (t) = \chi^{(1)}_{ij} E(\omega,t) + \chi^{(2)}_{ijk} E(\omega,t)\ E(\omega,t)$$
$$+ \chi^{(3)}_{ijkl} E(\omega,t)\ E(\omega,t)\ E(\omega,t) \ ... \tag{15}$$

The coefficients $\chi^{(2)}$, $\chi^{(3)}$ etc. are characteristic properties of the system of interest and depend intrinsically on the dynamical behaviour of a molecule under the influence of the electric field. Since these coefficients provide a weighting factor for responses higher order in the field strength, they are known as nonlinear susceptibilities.

Spectroscopic studies involving nonlinear properties [1—6,11—18] have been applied to aid in the understanding of the operation of several devices such as optical switches, potential components for optical computers, media for image reconstruction and 3–D image propagation, and unstable devices whose behaviour often has a mathematical morphological relationship to chaotic systems and oscillating chemical reaction systems.

In recent years, semi–insulating materials have functioned as important components in a wide range of useful devices including high speed electronics and optoelectronics, high frequency oscillators, Gunn devices, and high speed semiconductor lasers. As a result considerable effort has been devoted to understanding the dynamical events that occur in these solid state materials once subjected to an external field. Since transient species associated with such materials typically have lifetimes on the order of a few ps, ultrafast techniques are needed to experimentally determine their physical and temporal properties. It is beyond the scope of this chapter to elaborate on the diverse range of time dependent studies which are on going in this field of research. Therefore the reader should consult references [3,4,5,6] for further information.

7a.OPTICAL KERR EFFECT

Up to this point the review has focussed on experimental techniques which monitor changes in the nuclear coordinates of molecules in liquids through an averaged correlation function of the bulk <u>linear</u> susceptibility (related to the linear molecular polarizability) or dipole moments;

$$< [\chi_{ij}(t) \cdot \chi_{kl}(t = 0)] > \qquad (16)$$
$$< [\mu(t) \cdot \mu(t = 0)] > \qquad (16')$$

With the ability to generate ultrashort laser pulses, the applicability of higher order responses to the understanding of molecular dynamics has gained importance over the past few years.

The optical Kerr effect is an example of a dynamical response which occurs in liquids through the third order susceptibility. The majority of experiments are usually performed upon transparent media, far from resonances (eg lossless media) thus only the real part of the complex susceptibility will be utilised. In this optical Kerr effect, a temporary transient birefringence is induced in the liquid by the passage of an intense ultrashort laser pulse. The physical mechanism governing the birefringence is initiated through the induction of a temporary dipole moment as the polarized laser pulse $E(\omega,t)$ interacts with the molecular nonlinear polarizability. The induced dipole, U_{ind}, in turn interacts with the field giving rise to a torque. The net result is that a significant population of the ground state molecules preferentially align their axes of maximum polarization with the polarization of the laser field i.e. for a brief moment, the liquid acquires the character of a uniaxial crystal. This birefringence can be probed by a second polarized laser pulse (in a pump probe configuration). The dynamical observable of interest is the phase shift $\partial\phi(t)$ experienced by the probe pulse as it passes through a preferentially aligned zone of path length l;

$$\partial\phi(t) = \frac{2\pi l}{\lambda} \partial n (t) \qquad (17)$$

If the liquid under investigation possesses a number of dynamically distinct responses to the applied field and if the responses all occur on separable time scales, then the effective change in birefringence may be expressed as a linear superposition of dynamically independent contributions i.e.,

$$\Delta n(t) = \Sigma_i \Delta n_i(t) \qquad (18)$$

Assuming an arbitrary shaped laser pulse, this change in refractive index $\Delta n(t)$, is monitored in time through a convolution of the pump pulse I_p with the Kerr response function;

$$\Delta n(t) = I_p (t') r_i(t - t')dt'; I_p(t) = < E(\omega,t)^2 > \qquad (19)$$

This leads to an expression for the transmission (T) of the probe pulse that is a function of the overlap of pump and probe responses which occur at an identical sample position in space and time;

$$T(\tau) = \Sigma_i \int_{-\infty}^{+\infty} G_0^{(2)}(\tau)\, r_i(t - t')dt' \tag{20}$$

The $G_0^{(2)}$ autocorrelation function that appears in the above summation is usally determined as described in [13].Following deconvolution from the transmission signal (typically shown in (b) of Figure 10) the residual Kerr profile contains valuable and novel dynamical information intrinsic to the atomic and molecular liquid or dense material under study. Schematically the O.K.E. experiment (figure 10) is set up to operate as follows: a KHz modulated train of polarized pump pulses, from a synchronously pumped dye laser, generates an anisotropy in the sample material. The probe pulses, polarized at $45°$ relative to the pump, sample the rise and decay of the induced birefringence [11,12,13] in the sample medium. This is accomplished by observing the rotation of the plane of polarization of the transmitted probe pulses as a function of time. The resultant signal is detected through crossed polarizers (P_1, P_2) into a fast photomultiplier (PMT) which in turn is coupled to a phase sensitive detection system. The use of optical heterodyne techniques [3,13] allows a linear measurement to be made, despite the fact that Δn scales quadratically with the applied field E. The S:N is enhanced by KHz modulation

In Figures 8 and 9 we show recent fs Kerr results from this laboratory illustrating the experimental precision with which electronic and nuclear motions can now be separated and examined on the 10^{-14} s to 10^{-11} s timescale [11—14]. We are motivated by the fact that prevailing diffusion and hydrodynamic—based theories cannot predict these ultrafast molecular events at the earliest times [118], but that a transition between the microscopic realm of high frequency, oscillatory dynamical intermolecular interactions and the macroscopic realm of slowly varying, space and time averaged interactions must occur in real systems in this time window. Thus we have set out a comprehensive program of exploring molecular liquids of varying structure, symmetry and equilibrium packing properties. The highlights of results on CS_2, $CHCl_3$ and CCl_4 are now presented [118—120] below, which are consistent with other independent observations on CS_2 or comparable organic systems [112,121,113,114]. The reader is referred to the journal literature for this rapidly developing area of fs spectroscopy to molecular dynamics and to reactive and flexible molecular systems of photochemical and photobiological interest.

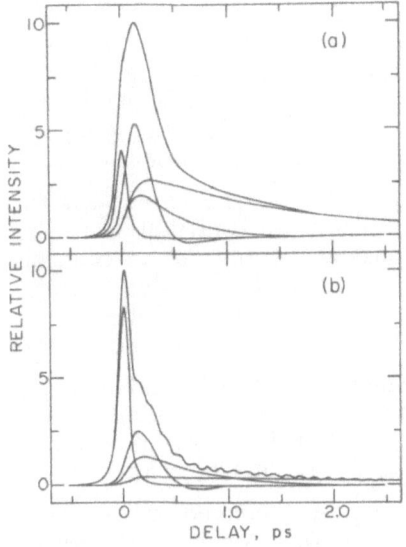

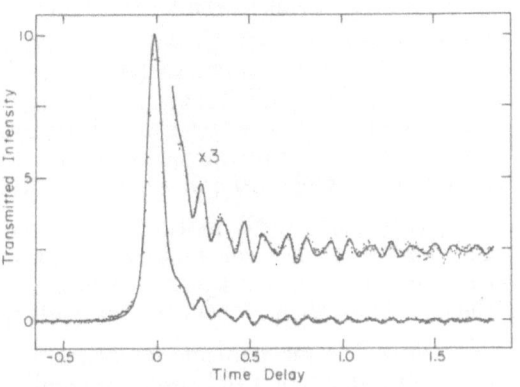

Figure 8.Femtosecond Kerr responses from a CS_2 and b) CHCl$_3$, showing traces com— prising several dynamically distinct electronic and nu— clear motions.

Figure 9. Femtosecond Kerr response from CCl$_4$ at 295 K with Raman induced vibrational oscillation superimposed on the nonlinear polarization.

In the case of CS_2 (shown in Figure 2a and 8a) it is quite clear that it is possible to observe the separate contributions of fs laser–induced electronic responses from nuclear responses in the fs time domain. The temporal assymmetry in the rise and fall of the overall Kerr response in Figure 2a reflects a notable ~ 70 fs delay between the peak response of the molecular system and that of the purely electronic (trace (b)), superimposed as an in situ measurement of $G_0^{(2)}$ with a 100 μ KDP wafer [14]. This first illustrated the inertial effect, and has now been seen in all systems to date. The ultrafast leading edge in all systems represents the electronic nonlinearity, instantaneous in the applied fs laser field and in CHCl$_3$ and CCl$_4$ similar responses are recorded. The tall inset peak in Figure 8a and 8b, overlapping the leading edge and then falling fast under the curve represents a scalar contribution of $\tau = 100$ fs $= G_0^{(2)}$ to the overall response. The second inset curve, showing the 70 fs lag, corresponds to the first nuclear and indeed dominant motion on this fs time scale, namely the ultrafast librational motion. In CS_2, this motion is centered at a frequency of ~ 38 cm^{-1} with τ_2 ~ 135 fs. The signal

corresponds to an ensemble of molecules driven coherently by the fs laser pulse and coherently librating in a distribution of intermolecular potentials representing the local degree of structural disorder in the fluid. The degree of inhomogeneous broadening is estimated to be ~ 40 cm^{-1} and in recent dilution studies [120] the influence of the changing nearest neighbour interactions was clearly seen. In CHCl$_3$

the librational motion shown in Figure 8b is characterized at 31.2 cm^{-1} with the inhomogeneity estimated to be 39.3 cm^{-1} (fwhm). The third contribution to the overall signals in all liquids arises from translational anisotropy due to collision induced effects falling under the "interaction—induced" umbrella discussed in detail elsewhere by Madden in particular [1,114]. These time constants in all liquids we have studied range from 350 fs $< \tau_3 <$ 700 fs. While clearly in, in a real liquid, the

molecules execute all motions and experience sequences of elastic and inelastic collisions (etc) in a continually fluctuating environment, these separations are for the present time useful mental constraints with which to build a new microscopic theory. The linkages between τ_2 and τ_3, in terms of the loss of coherence via

dephasing, the mechanisms for dephasing, and the degree of inhomogeneity of the intermolecular potential wells in which coherent libration occurs, are part of ongoing research questions [120]. This is also the period of transition from the microscopic to the macroscopic response of the system. Finally, the molecules undergo rotationally diffusive motions and these showing the inertial effect rise slowly and relax, extending into picoseconds. For CS$_2$ τ_4 = 1.61 ps, for CHCl$_3$ τ_4

= 2.0 ps. However, in CCl$_4$ and CHCl$_3$ the relaxations into ps are accompanied by

regular oscillatory "beats" riding on the nonlinear signals.

These Tetrahertz (THz)resonances are Raman—induced phenomena. Here reported are the first <u>direct</u> observations of the temporal evolution of coherently—excited, <u>intramolecular</u> normal mode vibrations by Nelson et al. using a transient grating technique on CH$_2$Br$_2$ [115] and in CCl$_4$ and CHCl$_3$ via Kerr by

McMorrow et al. [13,119]. More recently such "beats" have been observed superimposed on complex temporal profiles in coheren antistokes Raman spectroscopy (CARS 116) and earlier were reported on excited electronic state relaxations of Nile Blue molecules in solution, in transmission correlation techniques [116].

These normal modes in CHCl$_3$ at 7.78 THz are identified as ν_6 = 262 cm^{-1}

(6.54 THz); in CCl$_4$ both ν_2 = 218 cm^{-1} and 314 cm^{-1} (9.95 THz) are observed.

These intramolecular responses decay rapidly through both homogeneous and inhomogeneous dephasing processes and are the consequence of Raman excitations using the spectral bandwidth of the fs laser pulse. Such observations open up an exciting and novel opportunity to probe shifts in the frequency skeletal motions during probe molecule or ion solvation, clustering in liquids, the formation and dynamics of of clusters in jets and ultrafast ps and fs measurements [122—125] on intramolecular energy redistribution, as well as following pathways of reactive, photochemical, and photofragmentation pathways.

In concluding this brief expose to nonlinear spectroscopy and ultrafast phenomena, we have merely attempted to highlight novel directions as they are emerging. Femtosecond and picosecond laser spectroscopy have gone far beyond the technological challenges of the past decade and now offer an opportunity to integrate questions of fundamental significance in many different fields into a oherent and interdisciplinary set of studies in the 10^{-14} s to 10^{-10} s time domain.

8.REFERENCES

1. Applications of Picosecond Spectroscopy to Chemistry, ed. K.B.Eisenthal (D.Reidel Publishing Company, Boston, 1984).
2.
3. M.D.Levenson, Introduction to Nonlinear Laser Spectroscopy (Academic Press, N.Y.1982).
4. J.P.Hansen and I.MacDonald Optical Phase Conjugation, ed. R.A.Fisher, (Academic Press, N.Y. 1983).
5. Ultrafast Phenomena, (Springer–Verlag); a research series published biannually since 1978. These books comprise the invited papers and proceedings of the International Ultrafast Phenomena meetings with the conference chairman as editors.
6. Ultrafast Phenomena, Special Issue of International Quantum Electronics, IEEE JQE 24 (1988) ed. A.Johnson and references therein to the various papers on pulse generation, amplification, nonlinear optical measurements and dynamics.
7. B.J.Berne and R.Pecora, Dynamic Light Scattering, (Wiley N.Y.1976).
8. Molecular Liquids = Dynamics and Interactions ed.A.J.Barnes, W.J.Orville–Thomas and J.Yarwood (Dreidel Publishing Company, 1984).
9. B.Berne in Advanced Treatise on Physical Chemistry, vol VIII B ed. D.Henderson, Academic Press 1971; R.G.Gordon, Adv.Magnetic Res. $\underline{3}$, 1(1968).
10. J.D.Simon, Acc.Chem. Res. 21(3) $\underline{128-134}$ (1988).
11. G.A.Kenney–Wallace, Ultrafast Laser Spectroscopy and Applications in Lasers in Chemistry, ed. K.Evans (Marcel Dekker, N.Y. 1988).
12. E.Quitevis, P.M.Kroger, C.Kalpouzos, G.A.Kenney–Wallace, Can.J.Chem.$\underline{61}$, 975 (1983).
13. D.McMorrow, W.T.Lotshaw and G.A.Kenney–Wallace IEEE J.Quant.Elec. $\underline{24}$, 443(1988).
14. G.A.Kenney–Wallace, M.Golombok, and T.Dickson, J.Chem.Soc.Faraday $\underline{83}$, 1825 (1987).
15. J.A.Valdmanis, R.L.Fork, and J.P.Gordon Optic. Lett.$\underline{10}$, 131 (1985) and references therein.
16. E.B.Treacy IEEE J. Quant. Elec. QE–5, 454, 1959; B.Nikolaus and D.Grischkowsky , Applied Phys. Lett. $\underline{43}$, 228 (1983) and references therein
17. G.A.Kenney–Wallace, E.Quitevis, and E.Templeton, Reviews of Chem. Intermediates $\underline{6}$, 197 (1985).
18. A.E.Siegman, Appl.Phys.Lett.$\underline{30}$, 21 (1977); R.Trebino, C.E.Barber, and A.E.Siegman, IEEE J. Quant. Elec. QE–22 1413(1986).
19. B.Bagchi, D.W.Oxtoby, and G.R.Fleming, Chem. Phys. $\underline{86}$, 257(1984).
20. G.Van der Zwan, and J.T.Hynes, J. Phys. Chem. $\underline{89}$, 4181(1985).
21. H. Sumi, and R.A.Marcus, J. Chem. Phys. $\underline{84}$, 4272(1986).
22. P.G.Wolynes, J. Chem. Phys. $\underline{86}$, 5133(1987).
23. S.R.Flom, and R.F.Barbara, J. Phys. Chem. $\underline{89}$, 4489(1985).
24. D.F.Calef, and P.G.Wolynes, J. Chem. Phys. $\underline{78}$, 4145(1983).
25. Y.T.Mazurenko, and N.G.Bakhshiev, Opt. Spectrosc., $\underline{28}$, 490(1970).
26. D.F.Calef, and P.G.Wolynes, J. Phys. Chem. $\underline{87}$, 3387(1983).
27. G.Van der Zwan, and J.T.Hynes, Chem. Phys. Lett. $\underline{101}$, 367(1983).

28. E.W.Castner Jr., M.Maroncelli, and G.R.Fleming, J. Chem. Phys. 86(3), 1090(1987).
29. A.Safarzadeh–Amiri, Chem. Phys. Lett., 125(3), 272(1986).
30. S.W.Yeh, L.A.Philips, S.P.Webb, L.F.Buhse, and J.H.Clark, in Ultrafast Phenomena IV, ed. D.H.Auston and K.B.Eisenthal (Springer–Verlag), Berlin, 359(1984).
31. M.Maroncelli, and G.R.Fleming, J. Chem. Phys. 86(11), 6221(1987).
32. V.Nagarajan, A.M.Brearley, T.J.Kang, and P.F.Barbara, J. Chem. Phys. 86, 3183(1983).
33. A.Einstein, Ann. Phys. 19, 289(1906).
34. C.M.Hu, and R.Zwanzig, J. Chem. Phys. 60, 4354(1974).
35. A.Gierer, and K.Z.Wertz, Naturforsch A 8, 532(1953).
36. A.Von Jena, and H.E.Lessing, Chem. Phys. Lett. 78, 187(1981).
37. L.A.Philips, S.P.Webb, S.W.Yeh, and and J.H.Clark, J. Phys. Chem. 89, 17(1985).
38. F.J.Perrin, Phys. Radium 5, 497(1934).
39. T.Tao, Biopolymers 8, 609(1969).
40. E.F.Gudgin Templeton, and G.A.Kenney–Wallace, The Journal of Phys. Chem.90, 5441(1986).
41. D.Waldeck, A.J.Cross Jr., D.McDonald, and G.R.Fleming, J. Chem. Phys. 74, 3381(1981).
42. A.Von Jena, and H.E.Lessing, Chem. Phys. 40(3), 245(1979).
43. D.Kivelson, and P.A.Madden, Annu. Rev. Phys. Chem. 31, 523(1980).
44. B.Kowert, and D.Kivelson, J. Chem. Phys. 64, 5206(1976).
45. K.B.Eisenthal, Acc. Chem. Res. 8, 62(1975).
46. T.J.Chuang, and K.B.Eisenthal, J.Chem.Phys.57, 5094(1972).
47. G.J.Blanchard, J. Chem. Phys. 87(12), 6802(1987).
48. J.T.Hynes, R.Kapral, and M.Weinberg, J. Chem. Phys. 69, 2725(1978).
49. M.J. Sanders, and M.J.Wirth, Chem. Phys. Lett. 101, 361(1983).
50. G.R.Fleming, A.E.W.Knight, J.M.Morris, and R.J.Robinson, Chem. Phys. Lett.51, 399(1978).
51. D.H.Waldeck, W.T.Lotshaw, D.B.McDonald, and G.R.Fleming, Chem. Phys. Lett. 88, 297(1982).
52. R.S.Moog, M.D.Ediger, S.G.Boxer, and M.D.Fayer, J. Phys. Chem. 86, 4694(1982).
53. K.G.Spears, and L.E.Cramer, Chem. Phys. 30, 1(1978).
54. D.H.Waldeck, and G.R.Fleming, J. Phys. Chem. 85, 2614(1981).
54a. E.F.Gudgin, and G.A.Kenney–Wallace AIP Conf. Proc. 1986, 146 (Adv. Laser Sci.–1), 600–3.
55. Y.K.Lau, P.P.S.Saluja, and P.Kebarle, J. Am. Chem. Soc. 102, 7429(1980).
56. R.R.Hentz, and G.A.Kenney–Wallace, J.Phys.Chem.78, 514(1974).
57. K.G.Spears, and K.M.Steinmetz, J.Phys.Chem.89, 3623(1985).
58. D.P.Millar, R.Shah, and A.H.Zewail, Chem.Phys.Lett.66, 435(1979).
59. D.Waldeck, A.J.Cross Jr., D.B.McDonald, and G.R.Fleming, J.Chem.Phys.74, 3381(1891).
60. D.Reiser, and A.Laubereau, Ber Bunsenges, Phys. Chem. 86, 1106(1982).
61. D.Reiser, and A.Laubereau, Opt. Commun. 42, 329(1982).
62. D.Reiser, and A.Laubereau, Chem. Phys. Lett. 92, 297(1982).
63. G.J.Blanchard, and M.J.Wirth, J. Chem. Phys. 82(1), (1985).
64. P.Madden, and D.Kivelson, J. Phys. Chem. 86, 4244(1982).

436

65. T.W.Nee, and R.Zwanzig, J. Chem. Phys. 52, 6353(1970).
66. J.B.Hubbard, and P.G.Wolynes, J. Chem. Phys. 69, 998(1979).
67. P.A.Madden, and D.Kivelson, Adv. Chem. Phys. 56, 467(1984).
68. E.F.Gudgin Templeton, and G.A. Kenney–Wallace, J. Phys. Chem. 90,
69. L.A.Philips, S.P.Webb, S.W.Yeh, and J.H.Clark, J. Phys. Chem. 89, 17(1985).
70. L.A.Philips, S.P.Webb, and J.H.Clark, J. Chem. Phys. 83(11), 5810(1985).
70a. G.A.Kenney–Wallace, S.Paone and C.Kalpouzos, Farad. Discuss. Chem. Soc. 85, 1 (1988)
71. E.J.Hart, M.Anbar, The Hydrated Electron, Wiley, N.Y. (1970).
72. J.Jortner, and N.R.Kestner eds., Electrons in Fluids, (Springer–Verlag), N.Y. (1973); L.Kevan, and B.C.Webster eds. Electron – Solvent and Anion – Solvent Interactions, Elsevier, N.Y.(1976).
73. Liquid State and its Electrical Properties, ed. L.Luessen and L.Christophoron NATO ASI in press, 1988.
74. G.A.Kenney–Wallace, Adv. Chem. Phys. 57, 535(1981).
75. G.A.Kenney–Wallace, and C.D.Jonah, J. Phys. Chem. 86, 2572(1982).
76. G.A.Kenney–Wallace, Can. J. Chem. 55, 2009 (1977).; G.A. Kenney–Wallace, and C.D.Jonah, Chem. Phys. Lett. 39, 596(1976).
77. G.A.Kenney–Wallace, G.E.Hall, L.A.Hunt, and K.Sarantidis, J. Phys. Chem. 84, 1145(1980).
78. G.A.Kenney–Wallace, C.Kalpouzos and W.T.Lotshaw, Int. J. Radiat. Phys. Chem. 32, 573 (1988).
79. A.Migus, Y.Gauduel, J.L.Martin, and A.Antonetti, Phys. Rev. Lett. 58(15), 1559(1987).
80. J.P.Thornber, R.J.Cogdell, B.K.Pierson, and R.E.B.Seftor, J.Cellular Biochemistry 23, 159(1983).
81. R.M.Pearlstein, Photochem.Photobiol.35, 835(1982).
82. A.Yu Borisov, Energy – Migration Mechanisms in Antenna Chlorophylls in The Photosynthetic Bacteria, eds. R.K.Clayton, and W.R.Sistrom, N.Y., London 323(1978).
83. J.L.Martin, J.Breton, A.J.Hoff, A.Migus, and A.Antonetti, Proc. Natl. Acad. Sci. USA 83, 957(1986).
84. J.Breton, J.L.Martin, A.Migus, A.Antonetti, and A.Orzag, in Ultrafast Phenomena V, eds. G.R.Fleming, A.E.Siegman, (Springer–Verlag), Berlin 393(1986).
85. S.V.Chekalin, Y.A.Matveets, A.Y.Shkuropatov, V.A.Shuralov, and A. P. Yartsev, Febs. Lett. 216, 245(1987).
86. N.W.T.Woodbury, M.Becker, D.Middendorf, and W.W.Parson, Biochemistry 24, 7516(1985).
87. J.Breton, J.L.Martin, A.Migus, A.Antonetti, and A.Otszag, Proc. Natl. Acad. Sci. USA 83, 5121(1986).
88. J.Breton, J.L.Martin, J.Petrich, A.Migus, and A.Antonetti, FEBS Lett. 20, 37(1986).
89. W. Zinth, M. C. Nuss, M. A. Franz, W. Kaiser, and H. Michel, in Antennas and Reaction Centers of Photosynthetic Bacteria, ed. M. E. Michel–Beyerle (Springer–Verlag), Berlin, 286(1985).
90. W. Zinth, J. Dobler, and W. Kaiser in Ultrafast Phenomena V (Springer –Verlag) 379(1986).

91. S. R. Meech, A. J. Hoff, and D. A. Wiersma in <u>Ultrafast Phenomena V</u> (Springer–Verlag) 384(1986).
92. S.V.Chekalin, Yu.A.Matveets, and A.P.Yartsev, Rev. Phys. Appl. 22, 1761(1987).
93. J.Deisenhofer, O.Epp, K.Miki, R.Huber, and H.Michel, J. Mol. Biol. 180, 385(1984).
94. S.R.Meech, A.J.Hoff, and D.A.Wiersma, Chem. Phys. Lett. 121, 287(1985).
95. S.R.Meech, A.J.Hoff and D.A.Wiersma, Proc. Natl. Acad. Sci. USA 83(24) 9464(1986).
96. J.M.Fang, J.D.Carriker, V.Balogh–Nair, and K.Nakanishi, J. Am. Chem. Soc. 105, 5162(1983).
97. B.Aton, A.G.Doukas, R.H.Callender, B.Becher, and T.G.Ebrey, Biochemistry 16, 29(1977).
98. C.L.Hsieh, M.A.El–Sayed, M.Nicol, M.Nagumo and J.H.Lee, Photochem. Photobiol.38, 83(1983).
99. M.S.Braiman and R.A.Mathies, Proc. Natl. Acad. Sci. USA 79, 403(1982).
100. S.O.Smith, J.Lugtenburg, and R.A.Mathies, J. Membrane Biol.85, 95(1985).
101. S.O.Smith, I.Hornung, R.Van der Steen, M.S.Braiman, J.Lugtenburg, J.A.Pardoen, and R.A.Mathies, Proc. Natl. Acad. Sci. USA 83 967(1986).
102. K.Gerwert and F.Siebert, EMBO J.5, 805(1986).
103. H.J.Polland, M.A.Franz, W.Zinth, W.Kaier, E.Koelling and D.Oesterhelt, Biophys. J. 49, 651(1986).
104. R.A.Mathies, C.H.Brito Cruz, W.T.Polland, and C.V.Shank, Reports 777(1988).
105. J.W.Petrich, J.Breton, J.L.Martin, and A.Antonetti, Chem. Phys. Lett. 137(4), 369(1987).
106. J.Dobler, W.Zinth, W.Kaiser, and D.Oesterhelt, Chem. Phys. Lett. 144(2), 215(1988).
107. M.C.Nuss, W.Zinth, W.Kaiser, E.Koelling, and D.Oesterhelt, Chem. Phys. Lett. 117, 1(1985).
108. A.V.Sharkov, A.V.Pakulev, S.V.Chekalin, and Y.A.Matveets, Biochim. Biophys. Acta. 808, 94(1985).
109. R.H.Lozier, R.A.Bogomolni, and W.Stoeckenius, biophys. J. 15, 955(1975).
110. M.C.Kung, D.Devault, B.Hess, and D.Oesterhelt, Biophys. J. 15, 907(1975).
111. H.W.Trissl and W.Gartner, Biochemistry, to be published.
112. S.Ruhman, L.R.Williams, and K.A.Nelson, J. Phys. Chem.91, 2237 (1987).
113. S.Rhuman, B.Kohler, A.G.Joly and K.A.Nelson, Chem. Phys. Lett. 141, 16(1987).
114. P.A.Madden, and D.Tildesley, Mol. Phys. 49, 193 (1983) and P.A.Madden in ref. 1 (p 431–474).
115. S.Rhuman, A.G.Jolly and K.A.Nelson, J. Chem. Phys. 86, 6563 (1987).
116. M.J.Rosker, F.W.Wise and C.L.Tang, Phys. Rev. Lett. 57, 321(1986); J. Chem. Phys. 86, 2827 (1987).
117. R.Leonhardt, W.Holzapfel, W.Zinth and W.Kaiser, Chem. Phys. Lett. 133, 373 (1987).
118. G.A.Kenney–Wallace in ref. 1, (p 139–162).
119. D.McMorrow, W.T.Lotshaw, and G.A.Kenney–Wallace, Chem. Phys. Lett. 145, 309 (1988).

438

120. C.Kalpouzos, D.McMorrow, W.T.Lotshaw and G.A.Kenney—Wallace, Chem. Phys. Lett. 150, 138 (1988); ibid, J. Phys. Chem. 91, 2082 (1987).
121. W.T.Lotshaw, D.McMorrow, C.Kalpouzos, and G.A.Kenney—Wallace, Chem. Phys. Lett. 136, 323 (1987).
122. D.R.Demmer, J.W.Hager, G.W.Leach, and S.C.Wallace, Chem. Phys. Lett., 136, 329 (1987).
123. G.A.Bickel, D.R.Demmer, G.W.Leach, and S.C.Wallace, Chem. Phys. Lett. 145, 423 (1988).
124. P.M.Felker and A.H.Zewail, J. Chem. Phys. 82, 3003 (1985).
125. P.M.Felker and A.H.Zewail J.Chem.Phys. 82, 2994 (1985); P.M.Felker and A.H.Zewail J. Chem. Phys. 82, 2961 (1985).

ELECTROPHORETIC LIGHT SCATTERING REVISITED: THE ROLE OF ELECTRO-OSMOSIS

H. VERSMOLD and T. PALBERG
Lehrstuhl für Physikalische Chemie II
Rheinisch-Westfälische Technische Hochschule Aachen
Templergraben 59
D-5100 Aachen
F.R. Germany

ABSTRACT

For the understanding of the reactivity of macromolecules in solution it is important to know the charge of the molecules. Information on the charge can be obtained from electrophoresis experiments. The present paper is concerned with electrophoretic light scattering in the presence of electro-osmosis. Experimental spectra obtained by using a rectangular scattering cell can fully be accounted for by theoretical expressions also derived in the paper. The dependence of electrophoretic - electro-osmotic light scattering spectra on (a) the electro-osmotic velocity, (b) the electric field strength, (c) the particle diffusion coefficient, and (d) the electrophoretic velocity is discussed. Quantitative agreement with results obtained by the method of microelectrophoresis is obtained.

Since the well-established method of microelectrophoresis is time consuming, tedious, and cannot be applied to particles with a diameter less than about 300 nm, the extended light scattering method described in this paper may become a quick and reliable means to determine electrokinetic parameters of macromolecular solutions.

Th. Dorfmüller (ed.), Reactive and Flexible Molecules in Liquids, 439–453.
© 1989 by Kluwer Academic Publishers.

1. INTRODUCTION

One important aspect of reactive and flexible polyelectrolyte molecules in liquid phases is their interaction via long ranged electrostatic forces. If such molecules are sufficiently charged, the electrostatic repulsion prevents the molecules from comming sufficiently close together and a bond cannot be formed. For an understanding of the structure, stability and reactivity of such systems it is therefore prerequisite to know the particle's charge under various physical and chemical conditions.

Electrophoretic light scattering (ELS) is presently considered as a rather well established method for the determination of the electrophoretic mobility and total charge of macromolecules in suspension [1-7]. Theoretical and experimental aspects of this method have been reviewed recently [3,4].

In order to induce electrophoretic movement in a cell, an electric field E has to be applied. This field, however, does not only induce an electrophoretic migration, but also causes electro-osmotic flow due to double layers close to the cell walls. In a closed cell a stationary flow field of the solvent results, in which the electrophoretic movement of the particles takes place [2]. At present the electro-osmotic flow in a cell is usually considered an unwanted complication. Various techniques have been applied to eliminate or minimize the perturbations resulting from electro-osmosis, the most satisfactory of which appears to be focusing at the stationary layer [2,8]. For cells of small diameter, however, strong focusing is required, which may cause complications in the heterodyne scattering experiment due to a large uncertainty of the scattering vector Q. Other precautions like coating of the cell walls [2,9] or choosing a particular cell geometry or electrode arrangement [4] have been considered as well. Although suited to suppress electro-osmosis, these latter methods have their own drawbacks: For example, coating of the cell walls with a polyelectrolyte often results in a condensation of the polyelectrolyte also on the surface of the particles under investigation with a drastic change of the surface charge of the particles. Uzgiris [4] used a scattering cell with a very narrow gap between the electrodes for which electro-osmosis becomes negligible. We observed severe discrepancies between quantitative electrophoretic mobilities as obtained in such a cell and those determined by the well established method of microelectrophoresis. The main reason for this discrepancy appears to be electrode polarisation which leads to a poorly defined electric field in the narrow gap between the electrodes.

In view of the difficulties just mentioned we started a systematic investigation of electrophoretic light scattering in the presence of electro-osmosis. The aim of this investigation is to show that the two electrokinetic effects, electrophoresis and electro-osmosis, can be treated in a symmetrical manner and can be measured simultaneously in an ELS experiment.

2. THEORETICAL

The theory of dynamic light scattering (DLS) is now well established and documented [1,10]. Here, we recall merely few basic facts in order to make the application of DLS to electrophoretic and electro-osmotic light scattering more transparent.

For simplicity the colloidal particles are taken as monodisperse and spherical. Further, we assume that the diameter σ of the particles is small compared with the wavelength λ of the incident light. The complex amplitude of the electric field from N scattering particles with scattering amplitudes f is

$$E(\vec{Q},t) = \sum_{i=1}^{N} f \cdot \exp \left\{ i \, \vec{Q} \cdot \vec{r}_i(t) \right\} , \tag{1}$$

where $\vec{r}_i(t)$ is the position vector of particle i at time t, and $\vec{Q}$ is the scattering vector, $|Q| = (4\pi n/\lambda) \sin(\theta/2)$, with θ the scattering angle and n the index of refraction of the suspension.

The heterodyne dynamic light scattering experiment [1] provides an experimental measure of $g^{(1)}(\vec{Q},t)$, the normalized autocorrelation function of the electric field amplitude E

$$g^{(1)}(\vec{Q},t) = \frac{<E(\vec{Q},\tau) \cdot E^*(\vec{Q},t+\tau)>}{<E|(\vec{Q},\tau)|^2>} = F(\vec{Q},t)/S(\vec{Q}) . \tag{2}$$

$F(\vec{Q},t)$, the dynamic structure factor, is defined as

$$F(\vec{Q},t) = \frac{1}{N} <\sum_{i,j}^{N} \text{ex} \left\{ i \, \vec{Q} \cdot (\vec{r}_i(0) - r_j(t)) \right\}> , \tag{3}$$

and the structure factor, $S(\vec{Q})$, is related to $F(\vec{Q},t)$ as follows

$$S(\vec{Q}) = F(\vec{Q},0) . \tag{4}$$

For a disordered colloidal suspension different particles i and j in Eq.(3) are uncorrelated and $F(\vec{Q},t)$ simplifies to [1]

$$F(\vec{Q},t) = \frac{1}{N} <\sum_{i=1}^{N} \exp \left\{ i \, \vec{Q} \cdot (\vec{r}_i(0) - \vec{r}_i(t)) \right\}> . \tag{5}$$

In contrast to the usual self-diffusion treatment, the correlation functions from different particles i will not be identical due to the electro-osmotic velocity profile in the scattering volume. With the assumption that the electro-kinetic effects and the particle diffusion are independent, we can write

$$\vec{r}_i(t) = \vec{r}_i(0) + \Delta\vec{r}_{d,i}(t) + \Delta\vec{r}_{ek,i}(t) \quad , \tag{6}$$

where $\Delta\vec{r}_{d,i}(t)$ and $\Delta\vec{r}_{ek,i}(t)$ are the diffusional and the electro-kinetic displacement of particle i in the time t. $\Delta\vec{r}_{ek,i}(t)$ is related to the electro-kinetic particle velocity $\vec{v}_{ek,i} = \vec{v}_{p,i}$ as

$$\Delta\vec{r}_{ek,i}(t) = \vec{v}_{p,i} \cdot t \quad . \tag{7}$$

With Eqs. (6) and (7) the dynamic structure factor becomes [1]

$$F(\vec{Q},t) = 1/N \sum_{i=1}^{N} <\exp\{i\ \vec{Q}\cdot\vec{r}_{d,i}(t)\}\ \exp\{i\ \vec{Q}\cdot\vec{v}_{p,i}\cdot t\}>$$

$$= 1/N \exp\{-Q^2 Dt\} \sum_{i=1}^{N} \exp\{i\ \vec{Q}\cdot\vec{v}_{p,i}\cdot t\} \quad . \tag{8}$$

Since we are dealing with a large number of scattering particles, we introduce a continuous velocity distribution $P(v_p)$ and obtain

$$F(\vec{Q},t) = 1/N \exp\{-Q^2 Dt\} \int dv_p\ P(v_p)\ \exp\{i\ \vec{Q}\ \vec{v}_p\ t\} \quad . \tag{9}$$

The heterodyne correlation function is then proportional to Re $(F(Q,t))$ with the corresponding power spectrum [1]

$$I(\omega) \sim \int dv_p\ P(v_p) \left\{ \frac{Q^2 D}{(\omega - \vec{Q}\cdot\vec{v}_p)^2 + (Q^2 D)^2} + \frac{Q^2 D}{(\omega + \vec{Q}\cdot\vec{v}_p)^2 + (Q^2 D)^2} \right\} . \tag{10}$$

In order to proceed the particle velocity distribution $P(v_p)$ must be calculated. We refer to a rectangular scattering cell of height 2 h, length 2 l, and depth 2 d in the z-, x-, and y- direction. The electrodes are placed at x = $\pm$ l and the scattering volume extends from y = + d to - d. The length 2 l is taken as much larger than the depth 2 d and the height 2 h.

Dropping the index i, we can write for the particle velocity in the x-direction

$$v_p = v_1(y) + v_E \quad , \tag{11}$$

with v_E the electrophoretic drift velocity and $v_1(y)$ the electro-osmotic velocity, which depends on the position y of th scattering particle. The velocity profiles $v_1(y)$ and v_p are drawn schematically in Fig. 1.

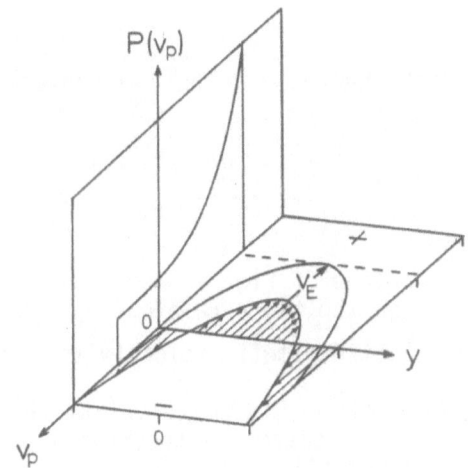

Figure 1.   The velocity profiles $v_1(y)$ and $v_p(y)$ (horizontal plane) and the corresponding particle velocity distribution $P(v_p)$ (vertical plane)(Ref.14)

The number of particles with a given velocity v_p will be proportional to dy/dv_p , i.e. the particle velocity distribution $P(v_p)$ is given by $P(v_p) \sim dy/dv_p$. For rectangular scattering cells with a hight to depth ratio larger than 20 the elaborate hydrodynamic calculations of Komagata [11] lead to the velocity profile in the x-y-plane

$$v_1(y) = v_{eo} \left(1 - 3 \cdot \left(\frac{1-y^2/d^2}{2-384/\pi^5 k} \right)\right) . \tag{12}$$

Here, v_{eo} is the electro-osmotic velocity in the vicinity of the cell walls, $\pm d$ is the extension of the cell in the y-direction and $k = h/d$ with $\pm \bar{h}$ the extension of the cell in the z-direction.

In order to proceed we rewrite Eq. (12) as follows

$$v_1(y) = \frac{v_{eo}}{2} \left(\frac{3 \, y^2}{d'^2} - e \right) \tag{13}$$

with

$$d'^2 = N \, d^2/2 \quad , \tag{14}$$

$$N = 2 - \frac{384}{\pi^5 k} \tag{15}$$

$$e = 6/N - 2 \ . \tag{16}$$

Next, the particle velocity distribution can be calculated with the result

$$P(v_p) = (6 \ v_{eo} \ 2 \ (\frac{1+e/2}{3}))^{-1/2} \ (v_p - v_E + \frac{v_{eo} \cdot e}{2})^{-1/2} \ . \tag{17}$$

$P(v_p)$ is sketched in the vertical plane of Fig. 1. As Eq.(17) indicates, the velocity distribution depends on the two parameters v_E, the electro-phoretic velocity, and v_{eo}, the electro-osmotic velocity in the vicinity of the cell wall.

Inserting Eq.(17) into Eq.(10) allows us to calculate the combined electrophoretic-electro-osmotic light scattering spectrum $I(\omega)$. Due to the two terms in braces in Eq.(10) the spectra are composed of two sub-spectra, which extend symmetrically to positive and negative frequencies.

Next, theoretical spectra for several characteristic situations will be considered. The new aspect of our treatment is the additional influence of the electro-osmosis on light scattering spactra. In Fig. 2 spectra are shown which were calculated with constant diffusion coefficient $D = 1.02 \cdot 10^{-8}$ cm^2 s^{-1} and constant electrophoretic velocity $v_E = -50$ μm s^{-1} and the following electro-osmotic velocities v_{eo}: (a) 12.5 μm s^{-1}, (b) 25 μm s^{-1}, (c) 50 μm s^{-1}, (d) 100 1m s^{-1}. It is obvious from Fig. 2 that the pronounced peak of the spectra moves to higher frequency with increasing v_{eo}.

In Fig. 3 the influence of the electric field strength on the spectral band shape is reproduced. These spectra where calculated with $D = 1.02 \cdot 10^{-8}$ cm^2 s^{-1}, $\mu_E = -6.25$ μm s^{-1} / V cm^{-1}, $\mu_{eo} = 3.125$ μm s^{-1}/ V cm^{-1}, and electric field strength (a) 4 V cm^{-1}, (b) 8 V cm^{-1}, (c) 12 V cm^{-1}, (d) 16 V cm^{-1}, (e) 20 V cm^{-1}.

Fig. 4 shows the influence of the particle diffusion coefficient on the electrokinetic spectra. In this case the electrophoretic velocity $v_E = -100$ μm s^{-1} and the electro-osmotic velocity $v_{eo} = 50$ μm s^{-1} were kept constant. The calculations were carried out with the following diffusion coefficients D (a) $0.6 \cdot 10^{-8}$ cm^2 s^{-1}, (b) $1.2 \cdot 10^{-8}$ cm^2 s^{-1}, (c) $2.5 \cdot 10^{-8}$ cm^2 s^{-1}, (d) $5 \cdot 10^{-8}$ cm^2 s^{-1}, and (e) $10 \cdot 10^{-8}$ cm^2 s^{-1}.

The most striking variation of the spectra occurs for a constant diffusion coefficient D and constant v_{eo} but variable electrophoretic

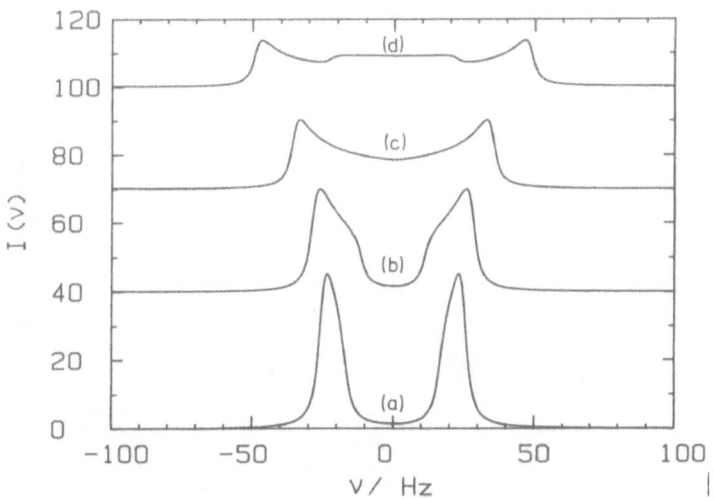

Figure 2. Influence of the electro-osmotic velocity: Computed electro-kinetic spectra with constant electrophoretic velocity $v_E = -50$ µm s^{-1} and constant diffusion coefficient $D = 1.02 \cdot 10^{-8}$ cm^2 s^{-1} and the following electro-osmotic velocities v_{eo}: (a) 12.5 µm s^{-1}, (b) 25 µm s^{-1}, (c) 50 µm s^{-1}, (d) 100 µm s^{-1}.

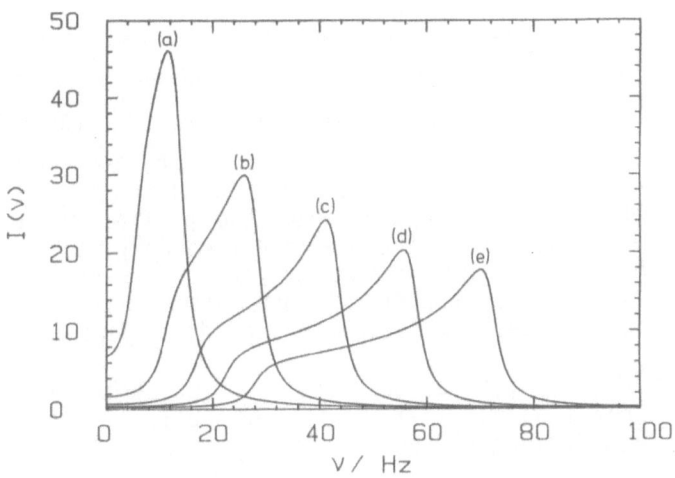

Figure 3. Influence of the electric field strength: Computed electro-kinetic spectra with $D = 1.0 \cdot 10^{-8}$ cm^2 s^{-1}, $\mu_E = -6.25$ µm s^{-1} / V cm^{-1}, $\mu_{eo} = 3.125$ µm s^{-1} / V cm^{-1} and the following electric field strength: (a) 4 V cm^{-1}, (b) 8 V cm^{-1}, (c) 12 V cm^{-1}, (d) 16 V cm^{-1}, (e) 20 V cm^{-1} (Ref.14)

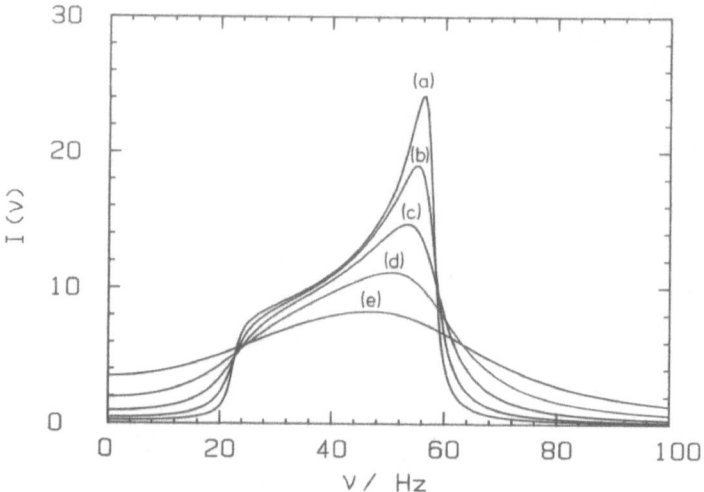

Figure 4.   Influence of the particle diffusion coefficient: Computed spectra with constant v_E = 100 µm s^{-1}, v_{eo} = 50 µm s^{-1} and the following diffusion coefficients d: (a) $0.6 \cdot 10^{-8}$ cm^2 s^{-1}, (b) $1.2 \cdot 10^{-8}$ cm^2 s^{-1}, (c) $2.5 \cdot 10^{-8}$ cm^2 s^{-1}, (d) $5 \cdot 10^{-8}$ cm^2 s^{-1}, and (e) $10 \cdot 10^{-8}$ cm^2 s^{-1} (Ref.14)

velocity v_E. Typical spectra are shown in Fig. 5, which were calculated with D = $1.02 \cdot 10^{-8}$ cm^2 s^{-1}, v_{eo} = 50 µm s^{-1} and the following v_E values (a) -25 µm s^{-1}. (b) -12.5 µm s^{-1}, (c) -6.25 µm s^{-1}, (d) 0 µm s^{-1}, (e) 6.25 µm s^{-1}, (f) 12.5 µm s^{-1}, (g) 25 µm s^{-1}, (h) 100 µm s^{-1}. Although the spectra are displayed in the positive and negative frequency range in order to make their development from the partial spectra (Eq.(10)) apparent, we recall that only the positive frequency range is experimentally accessible. It is interesting to note the close resemblence of some of the spectra with powder pattern solid state nuclear magnetic resonance spectra. A detailed discussion of the spectra will be given below.

Having developed a general description of electrokinetic light scattering (ELS) from a rectangular scattering cell, we turn next to the experimental verification of such spectra.

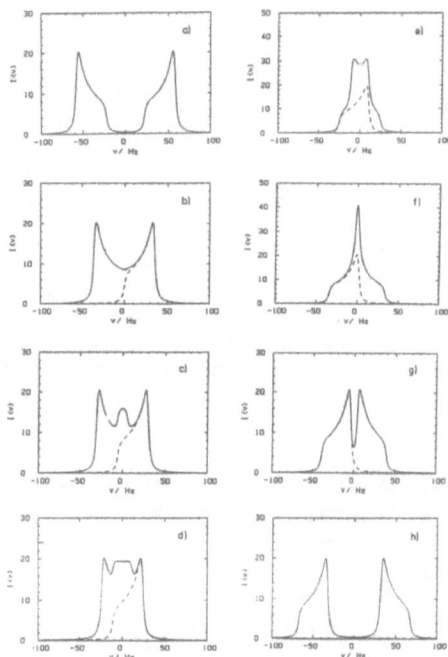

Figure 5.   Influence of the electrophoretic velocity: Computed spectra with constant $D = 1.02 \cdot 10^8$ cm^2 s^{-1} and $v_{eo} = 50$ μm s^{-1} and the following v_E: (a -25 μm s^{-1}, (b) -12.5 μm s^{-1}, (c) -6.25 μm s^{-1}, (d) 0 μm s^{-1}, (e) 6.25 μm s^{-1}, (f) 12.5 μm s^{-1}, (g) 25 μm s^{-1}, (h) 100 μm s^{-1} (Ref.14)

3. EXPERIMENTAL

Fig. 6 shows the schematic experimental light scattering arrangement. The incident beam, supplied by a He-Ne laser, Spectra Physics Mod. 124 B, passes a beam splitter B. This main beam is then reflected by mirror M_1 parallel to the optical axis of lens L_1. The reference beam, emerging from the beam splitter B passes a variable attenuator A and is then mirrored by M_2 and M_3 also parallel to the optical axis of L_1. The parallel main and reference beams after passing L_1 meet in the focus of this lens. The light scattering cell is placed such that the origin of the coordinate system introduced previously and the focus of lens L_1 coincide. The reference beam is directed towards the detector, which consists of a Malvern Instruments Photomultiplier Assembly PCS 5. The scattered light from particles in the main beam and the field of the refer-

448

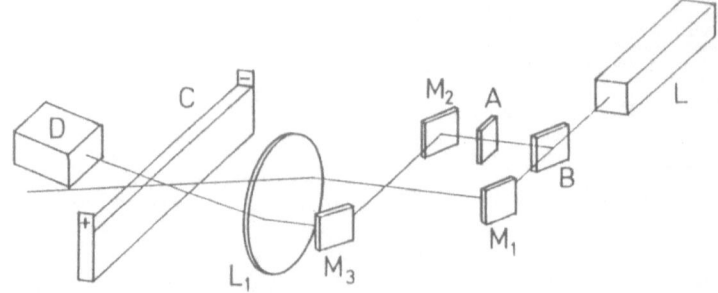

Figure 6.   Schematic experimental light scattering set up: A = attenuator, B = beam splitter, C = light scattering cell, D = detector, L = laser, L_1 = lens, and M_1, M_2, M_3 = mirrors (Ref.14)

ence beam mix at the photocathode of the photomultiplier tube to produce a beat signal. This beat signal is fed into an Ono Sokki Mod. CF-300 Fast Fourier Transform (FFT) analyser which allows direct determination of light scattering spectra. The FFT analyser is interfaced to a HP Mod. 310 desk computer. With this computer spectrum simulations and the comparison of theoretical and experimental spectra can be carried out.

The rectangular electrophoretic cells for the light scattering investigations were identical with those commonly used in microelectrophoresis experiments [2]. Typical dimensions are length 2l = 40 mm, height 2h = 10 mm, and depth 2d = 1 mm; the distance of the platinated platinum electrodes is about 70 mm. The experiments were carried out with square wave electric fields up to 50 V/cm in the frequency range 0.1-5 Hz. The cells were kept at room temperature 21°C. Convection due to joule heating was carefully avoided. The scattering angle was θ = 12.2° in all experiments.

For suspensions with particle diameters larger than about 300 nm observation through an ultramicroscope is possible. The present light scattering electrophoresis experiments were checked against results obtained with a Rank Brother Mark II microelectrophoresis apparatus. The Mark II microelectrophoresis apparatus also turned out as very helpful for measuring the geometrical dimensions of the cells. The effective field E_{eff} in the cells was determined, as usual, by conductivity experiments [2]. As mentioned above, severe discrepancies occured between electrophoretic mobilities μ_E as obtained by the light scattering method with a Uzgiris type cell [4] and those obtained by microelectrophoresis in a conventional flat cell.

The experiments reported here were performed with standard Dow polystyrene latex material of nominal diameter 481 nm, purchased from Serva Chemicals, Heidelberg. The particle diameter was checked by photon correlation spectroscopy with a spectrometer descibed previously [12]. The hydrodynamic diameter of the particles σ_H = 485 $\pm$ 2 nm was slightly larger than the nominal diameter 481 nm reported by the manufacturer. The particle number concentration of the suspension was adjusted to $n \simeq 2.0 \cdot 10^{18}/m^3$.

4. RESULTS AND DISCUSSION

In this section we consider a quantitative comparison of electrokinetic light scattering with microelectrophoresis measurements. One important difference between the microelectrophoresis and our light scattering set up, even if identical cells and electrodes are used, lies in the different imaging of the moving particles in the cell. The ultramicroscope of the microelectrophoresis allows to observe particles in a narrow xz-plane. The velocity in the cell can then be determined by scanning through the y-direction. Typical velocity profiles, obtained for a polystyrene latex suspension with particle diameter 481 nm and applied fields (a) 4 V cm^{-1}, (b) 8 V cm^{-1}, (c) 12 V cm^{-1}, (d) 16 V cm^{-1}, (e) 20 V cm^{-1} are shown in Fig. 7. By contrast, in our heterodyne light

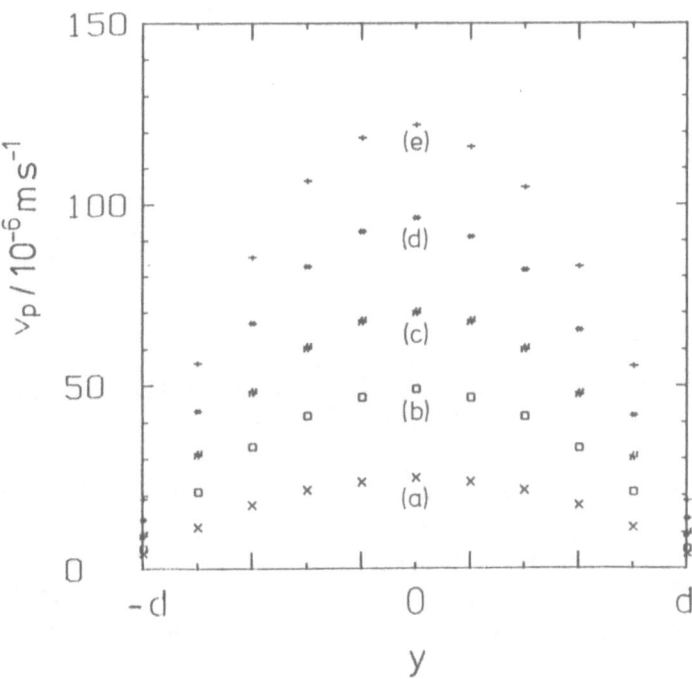

Figure 7.   Particle velocity profiles as determined by the microelectrophoresis method for particles with σ = 481 nm and applied fields: (a) 4 V cm^{-1}, (b) 8 V cm^{-1}, (c) 12 V cm^{-1}, (d) 16 V cm^{-1}, (e) 20 V cm^{-1} (Ref. 14)

450

scattering set up no focusing to a particular plane in the cell takes place. All particles positioned in the main laser beam contribute equally to the scattered field at the detector. From the full scans of the velocity profiles shown in Fig. 7 the velocity distributions $P(v_p)$ at the given electric fields can be calculated. The only missing parameter for a calculation of the electrokinetic light scattering spectra remains the diffusion coefficient D of the particles.

In Fig. 8 the dotted curves represent experimental light scattering spectra of the same sample for which microelectrophoresis data are

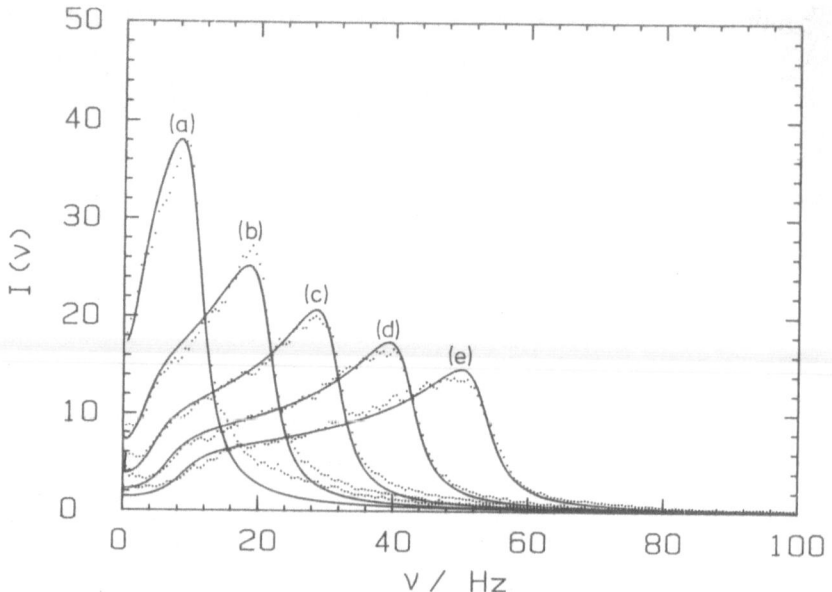

Figure 8.   Comparison of experimental (············) with theoretical (————) electrokinetic light scattering spectra. Details are given in the text (Ref.14)

shown in Fig. 7. The applied fields were again (a) 4 V cm^{-1}, (b) 8 V cm^{-1}, (c) 12 V cm^{-1}, (d) 16 V cm^{-1}, and (e) 20 V cm^{-1}. The solidly drawn lines in Fig. 8 are theoretical spectra which were obtained by optimizing the parameters v_E, v_{eo}, and D. First, we note that a convincing agreement between the experimental and the theoretical spectra is achieved. In Fig. 9 the electrophoretic velocity $-v_E$ (upper data points) and the electro-osmotic velocity v_{eo} (lower data points) as obtained by microelectrophoresis (o) and by light scattering experiments (x) are shown. Again a satisfactory agreement between the results of the two experimental methods has been achieved.

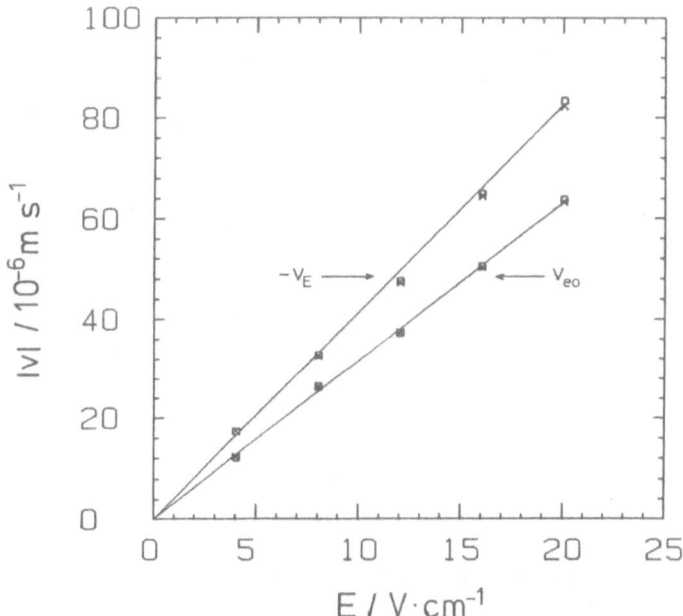

Figure 9.   Electrophoretic velocity -v_E (upper data points) and electro-osmotic velocity v_{eo} (lower data points) as obtained by
light scattering (x) and by microelectrophoresis (o) as a
function of the applied field E (Ref.14)

In order to obtain an optimal representation of the experimental
spectra by the theoretical spectra in Fig. 8, the diffusion coefficient
D also was taken as an adjustable parameter. Clearly, we would expect D
to be independent of the electrical field E. In Fig. 10 we show the
fitted diffusion coefficients D as a function of the applied field,
which are definitely not independent of E. It is interesting to note,
however, that with vanishing field E the diffusion coefficient extrap-
olates to D = 1.09 $\cdot$ 10^{-8} cm^2 s^{-1}, a value close to D = 1.02 $\cdot$ 10^{-8} cm^2
s^{-1}, the diffusion coefficient of particles with the hydrodynamic diame-
ter 481 nm. We are thus forced to conclude that an additional broadening
mechanism, different from simple diffusion is operative as the electric
field is turned on. Although details of this field dependent broadening
mechanism are presently unknown, one simple possible explanation could
be size and/or charge polydispersity, which would lead to a distribution
of the electrophoretic mobility of the particles and thus to a field
dependent broadening of the spectra. Systematic investigations of this
phenomenon are in preparation.

452

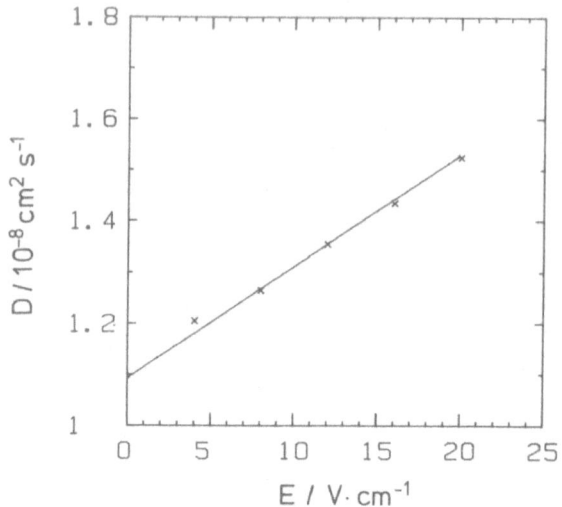

Figure 10. Variation of the fitted diffusion coefficient D with the applied electric field E (Ref.14)

5. CONCLUSION

In this paper electrophoretic light scattering in the presence of electro-osmosis is investigated both theoretically and experimentally. For
a rectangular scattering cell we observed a wide range of inhomogeneously broadened light scattering spectra which can be fully accounted for
by the theoretical expressions given above. For a selected macromolecular system it is shown that the electrokinetic light scattering results are in quantitative agreement with results abtained by the method
of microelectrophoresis.
    Since the well-established method of microelectrophoresis is time
consuming, tedious and cannot be applied to particles with diameters
less than about 300 nm, the extended light scattering technique described in this paper may become a quick and reliable method for the
characterisation and determination electrokinetic parameters of complex
macromolecules in solution.

ACKNOWLEDGMENTS

Financial support of this work by the Deutsche Forschungsgemeinschaft
and by the Fonds der Chemischen Industrie is greatefully acknowledged.

REFERENCES

1. B.J. Berne and R. Pecora, "Dynamic Light Scattering", J. Wiley & Sons, New York, 1976

2. E.J. Hunter, "Zeta Potential in Colloid Science", Academic Press, London, 1981

3. B.R. Ware, Adv. Colloid Interface Sci., **4**, 1 (1974)

4. E.E. Uzgiris, Adv. Colloid Interface Sci., **14**, 75 (1981)

5. B.R. Ware and W.H. Flygare, Chem. Phys. Lett., **12**, 81 (1971)

6. D.B. Sattelle, W.I. Lee, and B.R. Ware, Editors, "Biomedical Application of Laser Light Scattering", Elsevier, Amsterdam, 1982

7. W.L.K. Schwoyer, Editor, "Polyelectrolytes for Water and Wastewater Treatment", CRC press, 1981

8. E. Hantz, A. Cao, P. Depraetere, and E. Tallandier, J. Phys. Chem., **89**, 5832 (1985)

9. J.W. Klein, B.R. Ware, G. Barclay and H.R. Petty, Chem. Phys. Liqids, **43**, 13 (1987)

10. H.Z. Cummins and E.R. Pike, Editors, "Photon Correlation and Light Beating Spectroscopy", Plenum Press, New York, 1974

11. S. Komagata, Researches Electrotech. Lab. Tokyo Comm. No. 348 (1933); see also B.W. Currie, Phil. Mag., **12**, 429 (1931)

12. W. Härtl and H. Versmold, Z. Phys. Chem. N.F., **153**, 1 (1987)

13. A. Homola and R.O. James, J. Colloid Interface Sci., **59**, 123 (1977)

14. T. Palberg and H. Versmold, submitted

I. SUBJECT INDEX

The asterisc indicates references to a whole article

II. INDEX OF SUBSTANCES

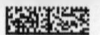